FUNDAMENTALS OF MATRIX ANALYSIS WITH APPLICATIONS

FUNDAMENTALS OF MATRIX ANALYSIS WITH APPLICATIONS

EDWARD BARRY SAFF

Department of Mathematics
Center for Constructive Approximation
Vanderbilt University
Nashville, TN, USA

ARTHUR DAVID SNIDER

Department of Electrical Engineering
University of South Florida
Tampa, FL, USA

Published by John Wiley & Sons, Inc., Hoboken, New Jersey
Published simultaneously in Canada

For general information on our other products and services or for technical support, please contact our Customer Care Department within the United States at (800) 762-2974, outside the United States at (317) 572-3993 or fax (317) 572-4002.

Wiley also publishes its books in a variety of electronic formats. Some content that appears in print may not be available in electronic formats. For more information about Wiley products, visit our web site at www.wiley.com.

Library of Congress Cataloging-in-Publication Data:

Saff, E. B., 1944–
Fundamentals of matrix analysis with applications / Edward Barry Saff, Center for Constructive Approximation, Vanderbilt University, Nashville, Tennessee, Arthur David Snider, Department of Electrical Engineering, University of South Florida, Tampa, Florida.
    pages   cm
  Includes bibliographical references and index.
  ISBN 978-1-118-95365-5 (cloth)
1. Matrices.   2. Algebras, Linear.   3. Orthogonalization methods.   4. Eigenvalues.
I. Snider, Arthur David, 1940–   II. Title.
  QA188.S194 2015
  512.9'434–dc23
                                                                    2015016670

Printed in the United States of America

10 9 8 7 6 5 4 3 2

CONTENTS

PREFACE   **ix**

PART I   INTRODUCTION: THREE EXAMPLES   **1**

1   Systems of Linear Algebraic Equations   **5**

   1.1 Linear Algebraic Equations, 5
   1.2 Matrix Representation of Linear Systems and the Gauss-Jordan
       Algorithm, 17
   1.3 The Complete Gauss Elimination Algorithm, 27
   1.4 Echelon Form and Rank, 38
   1.5 Computational Considerations, 46
   1.6 Summary, 55

2   Matrix Algebra   **58**

   2.1 Matrix Multiplication, 58
   2.2 Some Physical Applications of Matrix Operators, 69
   2.3 The Inverse and the Transpose, 76
   2.4 Determinants, 86
   2.5 Three Important Determinant Rules, 100
   2.6 Summary, 111
   Group Projects for Part I
   A. LU Factorization, 116
   B. Two-Point Boundary Value Problem, 118
   C. Electrostatic Voltage, 119

D. Kirchhoff's Laws, 120

E. Global Positioning Systems, 122

F. Fixed-Point Methods, 123

PART II   INTRODUCTION: THE STRUCTURE OF GENERAL
SOLUTIONS TO LINEAR ALGEBRAIC EQUATIONS   **129**

3 Vector Spaces   **133**

3.1 General Spaces, Subspaces, and Spans, 133

3.2 Linear Dependence, 142

3.3 Bases, Dimension, and Rank, 151

3.4 Summary, 164

4 Orthogonality   **165**

4.1 Orthogonal Vectors and the Gram–Schmidt Algorithm, 165

4.2 Orthogonal Matrices, 174

4.3 Least Squares, 180

4.4 Function Spaces, 190

4.5 Summary, 197

Group Projects for Part II

A. Rotations and Reflections, 201

B. Householder Reflectors, 201

C. Infinite Dimensional Matrices, 202

PART III   INTRODUCTION: REFLECT ON THIS   **205**

5 Eigenvectors and Eigenvalues   **209**

5.1 Eigenvector Basics, 209

5.2 Calculating Eigenvalues and Eigenvectors, 217

5.3 Symmetric and Hermitian Matrices, 225

5.4 Summary, 232

6 Similarity   **233**

6.1 Similarity Transformations and Diagonalizability, 233

6.2 Principle Axes and Normal Modes, 244

6.3 Schur Decomposition and Its Implications, 257

6.4 The Singular Value Decomposition, 264

6.5 The Power Method and the QR Algorithm, 282

6.6 Summary, 290

7   Linear Systems of Differential Equations   **293**

7.1 First-Order Linear Systems, 293
7.2 The Matrix Exponential Function, 306
7.3 The Jordan Normal Form, 316
7.4 Matrix Exponentiation via Generalized Eigenvectors, 333
7.5 Summary, 339
Group Projects for Part III
A.   Positive Definite Matrices, 342
B.   Hessenberg Form, 343
C.   Discrete Fourier Transform, 344
D.   Construction of the SVD, 346
E.   Total Least Squares, 348
F.   Fibonacci Numbers, 350

ANSWERS TO ODD NUMBERED EXERCISES   **351**
INDEX   **393**

PREFACE

Our goal in writing this book is to describe matrix theory from a geometric, physical point of view. The beauty of matrices is that they can express so many things in a compact, suggestive vernacular. The drudgery of matrices lies in the meticulous computation of the entries. We think matrices are beautiful.

So we try to describe each matrix operation pictorially, and squeeze as much information out of this picture as we can before we turn it over to the computer for number crunching.

Of course we want to be the computer's master, not its vassal; we want to know what the computer is doing. So we have interspersed our narrative with glimpses of the computational issues that lurk behind the symbology.

Part I. The initial hurdle that a matrix textbook author has to face is the exposition of Gauss elimination. Some readers will be seeing this for the first time, and it is of prime importance to spell out all the details of the algorithm. But students who have acquired familiarity with the basics of solving systems of equations in high school need to be stimulated occasionally to keep them awake during this tedious (in their eyes) review. In Part I, we try to pique the interests of the latter by inserting tidbits of information that would not have occurred to them, such as operation counts and computer timing, pivoting, complex coefficients, parametrized solution descriptions of underdetermined systems, and the logical pitfalls that can arise when one fails to adhere strictly to Gauss's instructions.

The introduction of matrix formulations is heralded both as a notational shorthand and as a quantifier of physical operations such as rotations, projections, reflections, and Gauss's row reductions. Inverses are studied first in this operator context before addressing them computationally. The determinant is cast in its proper light as an important concept in theory, but a cumbersome practical tool.

Readers are guided to explore projects involving LU factorizations, the matrix aspects of finite difference modeling, Kirchhoff's circuit laws, GPS systems, and fixed point methods.

Part II. We show how the vector space concepts supply an orderly organizational structure for the capabilities acquired in Part I. The many facets of orthogonality are stressed. To maintain computational perspective, a bit of attention is directed to the numerical issues of rank fragility and error control through norm preservation. Projects include rotational kinematics, Householder implementation of QR factorizations, and the infinite dimensional matrices arising in Haar wavelet formulations.

Part III. We devote a lot of print to physical visualizations of eigenvectors—for mirror reflections, rotations, row reductions, circulant matrices—before turning to the tedious issue of their calculation via the characteristic polynomial. Similarity transformations are viewed as alternative interpretations of a matrix operator; the associated theorems address its basis-free descriptors. Diagonalization is heralded as a holy grail, facilitating scads of algebraic manipulations such as inversion, root extraction, and power series evaluation. A physical experiment illustrating the stability/instability of principal axis rotations is employed to stimulate insight into quadratic forms.

Schur decomposition, though ponderous, provides a valuable instrument for understanding the orthogonal diagonalizability of normal matrices, as well as the Cayley–Hamilton theorem.

Thanks to invaluable input from our colleague Michael Lachance, Part III also provides a transparent exposition of the properties and applications of the singular value decomposition, including rank reduction and the pseudoinverse.

The practical futility of eigenvector calculation through the characteristic polynomial is outlined in a section devoted to a bird's-eye perspective of the QR algorithm. The role of luck in its implementation, as well as in the occurrence of defective matrices, is addressed.

Finally, we describe the role of matrices in the solution of linear systems of differential equations with constant coefficients, via the matrix exponential. It can be mastered before the reader has taken a course in differential equations, thanks to the analogy with the simple equation of radioactive decay. We delineate the properties of the matrix exponential and briefly survey the issues involved in its computation.

The interesting question here (in theory, at least) is the exponential of a defective matrix. Although we direct readers elsewhere for a rigorous proof of the Jordan decomposition theorem, we work out the format of the resulting exponential. Many authors ignore, mislead, or confuse their readers in the calculation of the generalized eigenvector Jordan chains of a defective matrix, and we describe a straightforward and foolproof procedure for this task. The alternative calculation of the matrix exponential, based on the primary decomposition theorem and forgoing the Jordan chains, is also presented.

Group projects for Part III address positive definite matrices, Hessenberg forms, the discrete Fourier transform, and advanced aspects of the singular value decomposition. Each part includes summaries, review problems, and technical writing exercises.

EDWARD BARRY SAFF
Vanderbilt University
edward.b.saff@vanderbilt.edu

ARTHUR DAVID SNIDER
University of South Florida
snider@usf.edu

PART I

INTRODUCTION: THREE EXAMPLES

Antarctic explorers face a problem that the rest of us wish we had. They need to consume lots of calories to keep their bodies warm. To ensure sufficient caloric intake during an upcoming 10-week expedition, a dietician wants her team to consume 2300 ounces of milk chocolate and 1100 ounces of almonds. Her outfitter can supply her with chocolate almond bars, each containing 1 ounce of milk chocolate and 0.4 ounces of almonds, for $1.50 apiece, and he can supply bags of chocolate-covered almonds, each containing 2.75 ounces of chocolate and 2 ounces of almonds, for $3.75 each. (For convenience, assume that she can purchase either item in fractional quantities.) How many chocolate bars and covered almonds should she buy to meet the dietary requirements? How much does it cost?

If the dietician orders x_1 chocolate bars and x_2 covered almonds, she has $1x_1 + 2.75x_2$ ounces of chocolate, and she requires 2300 ounces, so

$$1x_1 + 2.75x_2 = 2300. \tag{1}$$

Similarly, the almond requirement is

$$0.4x_1 + 2x_2 = 1100. \tag{2}$$

You're familiar with several methods of solving simultaneous equations like (1) and (2): graphing them, substituting one into another, possibly even using determinants. You can calculate the solution to be $x_1 = 1750$ bars of chocolate and $x_2 = 200$ bags of almonds at a cost of $\$1.50x_1 + \$3.75x_2 = \$3375.00$.

Fundamentals of Matrix Analysis with Applications,
First Edition. Edward Barry Saff and Arthur David Snider.
© 2016 John Wiley & Sons, Inc. Published 2016 by John Wiley & Sons, Inc.

But did you see that for $238.63 *less*, she can *meet or exceed* the caloric requirements by purchasing 836.36... bags of almonds and no chocolate bars? We'll explore this in Problem 20, Exercises 1.3. Simultaneous equations, and the *linear algebra* they spawn, contain a richness that will occupy us for the entire book.

Another surprising illustration of the variety of phenomena that can occur arises in the study of differential equations.

Two differentiations of the function $\cos t$ merely result in a change of sign; in other words, $x(t) = \cos t$ solves the second-order differential equation $x'' = -x$. Another solution is $\sin t$, and it is easily verified that every combination of the form

$$x(t) = c_1 \cos t + c_2 \sin t,$$

where c_1 and c_2 are arbitrary constants, is a solution. Find values of the constants (if possible) so that $x(t)$ meets the following specifications:

$$x(0) = x(\pi/2) = 4; \tag{3}$$
$$x(0) = x(\pi) = 4; \tag{4}$$
$$x(0) = 4; \ x(\pi) = -4. \tag{5}$$

In most differential equations textbooks, it is shown that solutions to $x'' = -x$ can be visualized as vibratory motions of a mass connected to a spring, as depicted in Figure I.1. So we can interpret our task as asking if the solutions can be timed so that they pass through specified positions at specified times. This is an example of a *boundary value problem* for the differential equation. We shall show that the three specifications lead to entirely different results.

Evaluation of the trigonometric functions in the expression $x(t) = c_1 \cos t + c_2 \sin t$ reveals that for the conditions (3) we require

$$c_1 \cdot (1) + c_2 \cdot (0) = 4$$
$$c_1 \cdot (0) + c_2 \cdot (1) = 4 \tag{3'}$$

with the obvious solution $c_1 = 4$, $c_2 = 4$. The combination $x(t) = 4\cos t + 4\sin t$ meets the specifications, and in fact, it is the only such combination to do so.

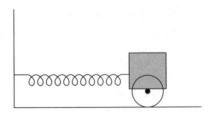

Fig. I.1   Mass-spring oscillator.

Conditions (4) require that

$$c_1 \cdot (1) + c_2 \cdot (0) = 4,$$
$$c_1 \cdot (-1) + c_2 \cdot (0) = 4, \tag{4'}$$

demanding that c_1 be equal both to 4 and to -4. The specifications are incompatible, so no solution $x(t)$ can satisfy (4).

The requirements of system (5) are

$$c_1 \cdot (1) + c_2 \cdot (0) = 4,$$
$$c_1 \cdot (-1) + c_2 \cdot (0) = -4. \tag{5'}$$

Both equations demand that c_1 equals 4, but no restrictions at all are placed on c_2. Thus there are infinite number of solutions of the form

$$x(t) = 4\cos t + c_2 \sin t.$$

Requirements (1, 2) and (3′, 4′, 5′) are examples of *systems of linear algebraic equations*, and although these particular cases are quite trivial to analyze, they demonstrate the varieties of solution categories that are possible. We can gain some perspective of the complexity of this topic by looking at another application governed by a linear system, namely, Computerized Axial Tomography (CAT).

The goal of a "CAT" scan is to employ radiation transmission measurements to construct a map of flesh density in a cross section of the human body. Figure I.2 shows the final resuls of a scan through a patient's midsection; experts can detect the presence of cancer tumors by noting unusual variations in the density.

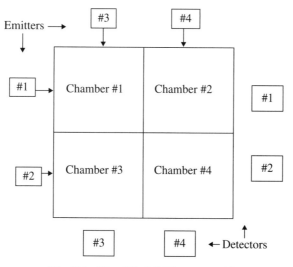

Fig. I.2   Simplified CAT scan model.

A simplified version of the technology is illustrated in Figure I.2. The stomach is modeled very crudely as an assemblage of four chambers, each with its own density. (An effective three-dimensional model for detection of tiny tumors would require millions of chambers.) A fixed dose of radiation is applied at each of the four indicated emitter locations in turn, and the amounts of radiation measured at the four detectors are recorded. We want to deduce, from this data, the flesh densities of the four subsections.

Now each chamber transmits a fraction r_i of the radiation that strikes it. Thus if a unit dose of radiation is discharged by emitter #1, a fraction r_1 of it is transmitted through chamber #1 to chamber #2, and a fraction r_2 of that is subsequently transmitted to detector #1. From biochemistry we can determine the flesh densities if we can find the transmission coefficients r_i.

So if, say, detector #1 measures a radiation intensity of 50%, and detectors #2, #3, and #4 measure intensities of 60%, 70%, and 55% respectively, then the following equations hold:

$$r_1 r_2 = 0.50; \quad r_3 r_4 = 0.60; \quad r_1 r_3 = 0.70; \quad r_2 r_4 = 0.55.$$

By taking logarithms of both sides of these equations and setting $x_i = \ln r_i$, we find

$$
\begin{aligned}
x_1 + x_2 \qquad\qquad &= \ln 0.50 \\
x_3 + x_4 &= \ln 0.60 \\
x_1 \qquad + x_3 \qquad &= \ln 0.70 \\
x_2 + \qquad x_4 &= \ln 0.55,
\end{aligned}
\qquad (6)
$$

which is a system of linear algebraic equations similar to (1, 2) and $(3', 4', 5')$. But the efficient solution of (6) is much more daunting—awesome, in fact, when one considers that a realistic model comprises over 10^6 equations, and that it may possess no solutions, an infinity of solutions, or one unique solution.

In the first few chapters of this book, we will see how the basic tool of linear algebra—namely, the matrix–can be used to provide an efficient and systematic algorithm for analyzing and solving such systems. Indeed, matrices are employed in virtually every academic discipline to formulate and analyze questions of a quantitative nature. Furthermore, in Chapter Seven, we will study how linear algebra also facilitates the description of the underlying structure of the solutions of linear differential equations.

1

SYSTEMS OF LINEAR ALGEBRAIC EQUATIONS

1.1 LINEAR ALGEBRAIC EQUATIONS

Systems of equations such as (1)–(2), (3′), (4′), (5′), and (6) given in the "Introduction to Part I" are instances of a mathematical structure that pervades practically every application of mathematics.

Systems of Linear Equations

Definition 1. An equation that can be put in the form

$$a_1x_1 + a_2x_2 + \cdots + a_nx_n = b,$$

where the coefficients a_i and the right-hand side b are constants, is called a **linear algebraic equation** in the variables $x_1, x_2, \ldots, x_n$. If m linear equations, each having n variables, are all to be satisfied by the same set of values for the variables, then the m equations constitute a **system of linear algebraic equations** (or simply a **linear system**):

$$a_{11}x_1 + a_{12}x_2 + \cdots + a_{1n}x_n = b_1$$
$$a_{21}x_1 + a_{22}x_2 + \cdots + a_{2n}x_n = b_2$$
$$\vdots$$
$$a_{m1}x_1 + a_{m2}x_2 + \cdots + a_{mn}x_n = b_m \qquad (1)$$

Fundamentals of Matrix Analysis with Applications,
First Edition. Edward Barry Saff and Arthur David Snider.
© 2016 John Wiley & Sons, Inc. Published 2016 by John Wiley & Sons, Inc.

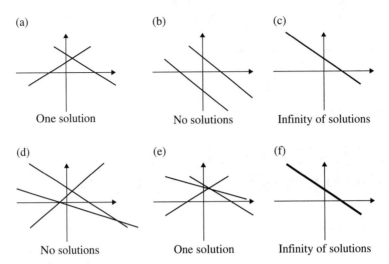

Fig. 1.1   Typical graphs of linear systems with two unknowns.

We assume that the reader has had some experience solving linear systems of two equations in two unknowns, such as (1, 2) in the preceding section, although the lack of solutions to (4′) and the multiplicity of solutions to (5′) may have come as a surprise. The graph of the solutions to a *single* equation in two unknowns is, of course, a straight line (hence the designation "linear") and the graphs in Figure 1.1(a–c) illustrate why a system of *two* such equations typically has one solution, but special alignments can result in zero or an infinite number of solutions. If a third equation is added to the system (Figure 1.1(d–f)), typically there will be no solutions, although accidental alignments can result in one, or an infinity of, solutions.

The graph of the solutions to a "linear" algebraic equation in three unknowns is not literally a line, but a plane, and Figure 1.2 reinforces our expectation that systems with three unknowns can again possess zero, one, or an infinity of solutions.

Although the graphs provide much insight into the nature of the solutions of linear systems, they are obviously inadequate as tools for finding these solutions when more than three unknowns are present. We need an analytic technique for solving simultaneous linear equations that is efficient, accurate, foolproof, and easy to automate. After all, the task of solving linear systems is performed billions of times each day in computers around the world. It arises in an enormous number of applications. The allocation of medical resources such as hospital beds, antibiotics, health personnel, etc. after a disaster involves dozens of variables. Displaying and/or saving an image on a computer screen may require thousands of equations converting the red/green/blue intensities to the wavelet coefficients for the JPEG 2000 image compression format; similar considerations hold for face recognition and other image processing schemes. The algorithm employed in CAT scans, we have noted, can involve millions of unknowns. In fact, the task of solving large systems of linear equations is intimately

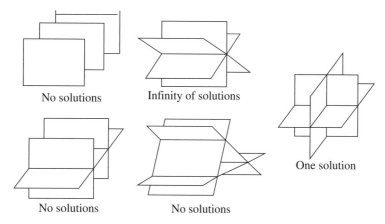

No solutions   Infinity of solutions

No solutions   No solutions

One solution

Fig. 1.2   Graph of three equations in three unknown.

entwined with the history and evolution of the electronic computer. So bear in mind that although you will be solving small linear systems by hand as you learn the basics, in practice you will usually be entrusting your analyses to software. In such applications it is absolutely essential that the computer's speed be coupled with a precise understanding of what it is doing, and that is the goal of our exposition in the next few sections.

We are going to focus on one algorithmic procedure for solving linear systems. It is known as *Gauss elimination* (in tribute to the great German mathematician Carl Friedrich Gauss (1777–1855), although his other works dwarf this effort). It[1] is used almost universally for solving generic linear systems. (To be sure, faster algorithms exist for systems with special structures.)

The Gauss elimination algorithm, "by the book," is inflexible. Just the same, anytime you're solving a system by hand and you see a shortcut, by all means take it; most of the time you will save time and work. However, there will be occasions when you get nonsense (see Problems 15–18 for examples), and then you must back up and follow Gauss's rules rigidly.

The basic idea of Gauss elimination is to add multiples of earlier equations to subsequent equations, choosing the multiples so as to reduce the number of unknowns. (Equivalently, one can say we are *subtracting* multiples of earlier equations from subsequent equations. We'll employ whichever interpretation seems appropriate.) The resulting system of equations can then be solved in reverse order. We demonstrate with a system of three equations in three unknowns before giving the general formulation.

---

[1]Or an equivalent formulation called "LU decomposition," described in Group Project A, Part I.

Example 1.   Solve the system

$$x_1 + 2x_2 + 2x_3 = 6$$
$$2x_1 + x_2 + x_3 = 6$$
$$x_1 + x_2 + 3x_3 = 6. \tag{2}$$

Solution. We eliminate x_1 in the second equation by adding (-2) times the first equation (or subtracting 2 times the first equation):

$$x_1 + 2x_2 + 2x_3 = 6$$
$$-3x_2 - 3x_3 = -6$$
$$x_1 + x_2 + 3x_3 = 6.$$

Similarly x_1 is eliminated from the third equation by adding (-1) times the first;

$$x_1 + 2x_2 + 2x_3 = 6$$
$$-3x_2 - 3x_3 = -6$$
$$-x_2 + x_3 = 0. \tag{3}$$

Next x_2 is eliminated from the third equation by adding $(-1/3)$ times the second:

$$x_1 + 2x_2 + 2x_3 = 6$$
$$-3x_2 - 3x_3 = -6$$
$$2x_3 = 2. \tag{4}$$

The third equation only has one unknown and its solution is immediate:

$$x_3 = 1.$$

But now, the second equation has, effectively, only one unknown since we can substitute 1 for x_3. Hence, its solution is

$$x_2 = -\frac{1}{3}[-6 + (3 \times 1)] = 1. \tag{5}$$

And substitution for x_3 and x_2 in the first equation yields

$$x_1 = 6 - 2 \times (1) - 2 \times (1) = 2. \qquad\blacksquare$$

To maintain focus on the methodology of Gauss elimination, we usually contrive examples like the above, with integer or small-fraction constants. But sometimes the transparency of simplified examples can obscure the underlying algorithm. For example, if you examine the system (3), you may see that adding (2/3) times the second equation to first would have enabled us to conclude $x_1 = 2$ immediately. Obviously we can't rely on such serendipity in general. So we include the following unwieldy example to focus your attention on the regimented steps of Gauss elimination. Don't bother to follow the details of the arithmetic; just note the general procedure.

Example 2. Solve the linear system

$$0.202131x_1 + 0.732543x_2 + 0.141527x_3 + 0.359867x_4 = 0.171112$$
$$0.333333x_1 - 0.112987x_2 + 0.412989x_3 + 0.838838x_4 = 0.747474$$
$$-0.486542x_1 + 0.500000x_2 + 0.989989x_3 - 0.246801x_4 = 0.101001$$
$$0.101101x_1 + 0.321111x_2 - 0.444444x_3 + 0.245542x_4 = 0.888888$$

Solution. To eliminate x_1 from the second equation, we add $(-0.333333/0.202131)$ times the first. Similarly we add $(0.486542/0.202131)$ times the first equation to the third, and add $(-0.101101/0.202131)$ times the first to the fourth. These three operations produce (to 6 digits)

$$0.202131x_1 + 0.732543x_2 + 0.141527x_3 + 0.359867x_4 = 0.171112$$
$$-1.32101x_2 + 0.179598x_3 + 0.245384x_4 = 0.465294$$
$$2.26317x_2 + 1.33063x_3 + 0.619368x_4 = 0.512853$$
$$-0.045289x_2 - 0.515232x_3 + 0.065545x_4 = 0.803302$$

(Roundoff effects will be discussed in Section 1.5.)

Next, we eliminate x_2 from the third and fourth equations by adding, in turn, the multiples $(2.26317/1.32102)$ and $(-0.045289/1.32102)$ of the second equation. And so on. Continuing with the forward part of the algorithm we arrive at the system

$$0.202131x_1 + 0.732543x_2 + 0.141527x_3 + 0.359867x_4 = 0.171112$$
$$-1.32101x_2 + 0.179598x_3 + 0.245384x_4 = 0.465294$$
$$1.63832x_3 + 1.03176x_4 = 1.30999$$
$$0.388032x_4 = 1.20425 \qquad (6)$$

And performing "back substitution," that is solving for the variables in reverse order from the bottom up, we obtain the solution. It turns out to be

$$x_4 = 3.10348$$
$$x_3 = -1.17003$$
$$x_2 = 0.065189$$
$$x_1 = -4.09581 \qquad \blacksquare$$

We now give a brief exposition of the steps of the Gauss elimination process, as demonstrated in these examples. It may have occurred to you that the algorithm, as described so far, has a flaw; it can be foiled by the untimely appearance of zeros at certain critical stages. But we'll ignore this possibility for the moment. After you've become more familiar with the procedure, we shall patch it up and streamline it (Section 1.3). Refer to Example 2 as you read the description below.

Gauss Elimination Algorithm (without anomalies) (n equations in n unknowns, no "zero denominators")

1. Eliminate x_1 from the second, third, ..., nth equations by adding appropriate multiples of the first equation.
2. Eliminate x_2 from the (new) third, ..., nth equations by adding the appropriate multiples of the (new) second equation.
3. Continue in this manner: eliminate x_i from the subsequent equations by adding the appropriate multiples of the ith equation.

    When x_{n-1} has been eliminated from the nth equation, the forward part of the algorithm is finished. The solution is completed by performing back substitution:

4. Solve the nth equation for x_n.
5. Substitute the value of x_n into the $(n-1)$st equation and solve for x_{n-1}.
6. Continue in this manner until x_1 is determined.

Problems 15–18 at the end of this section demonstrate some of the common traps that people can fall into if they *don't* follow the Gauss procedure rigorously. So how can we be sure that we haven't introduced new "solutions" or lost valid solutions with this algorithm? To prove this, let us define two systems of simultaneous linear equations to be equivalent if, and only if, they have identical solution sets. Then we claim that the following three basic operations, which are at the core of the Gauss elimination algorithm, are guaranteed not to alter the solutions.

Basic Operations

Theorem 1. If any system of linear algebraic equations is modified by one of the following operations, the resulting system is equivalent to the original (in other words, these operations leave the solution set intact):

(i) adding a constant multiple of one equation to another and replacing the latter with the result;
(ii) multiplying an equation by a nonzero constant and replacing the original equation with the result;
(iii) reordering the equations.

We leave the formal proof of this elementary theorem to the reader (Problem 14) and content ourselves here with some pertinent observations.

- We didn't reorder equations in the examples of this section; it *will* become necessary for the more general systems discussed in Section 1.3.
- Multiplying an equation by a nonzero constant was employed in the back substitution steps where we solve the single-unknown equations. It can't compromise us, because we can always recover the original equation by dividing by that constant (that is, multiplying by its reciprocal).
- Similarly, we can always restore any equation altered by operation (*i*); we just add the multiple of the other equation back in.

Notice that not only will we eventually have to make adjustments for untimely zeros, but also we will need the flexibility to handle cases when the number of equations does not match the number of unknowns, such as described in Problem 19.

Exercises 1.1

1. Solve the following linear systems by back substitution (reordering of equations may be helpful).

(a)
$$
\begin{aligned}
2x_1 + x_2 \quad\quad - x_4 &= -2 \\
3x_2 + 2x_3 + x_4 &= 2 \\
x_3 + 2x_4 &= -1 \\
4x_4 &= 4
\end{aligned}
$$

(b)
$$
\begin{aligned}
5x_1 \quad\quad\quad &= 15 \\
-2x_1 - 3x_2 \quad &= -12 \\
x_1 + x_2 + x_3 &= 5
\end{aligned}
$$

(c)
$$
\begin{aligned}
- x_3 + 2x_4 &= 1 \\
4x_1 + 2x_2 \quad\quad + x_4 &= -3 \\
- 2x_4 &= -2 \\
x_2 + 2x_3 + 3x_4 &= 5
\end{aligned}
$$

(d)
$$
\begin{aligned}
2x_1 + x_2 \quad\quad &= 3 \\
3x_1 + 2x_2 + x_3 &= 6 \\
4x_1 \quad\quad\quad &= 4
\end{aligned}
$$

2. For each of the following systems, graph the linear equations and determine the solution(s), if any.

(a)
$$
\begin{aligned}
x + 2y &= 1 \\
x - y &= 0
\end{aligned}
$$

(b)
$$
\begin{aligned}
x + 2y &= 1 \\
x - 4y &= 0 \\
x - y &= 0
\end{aligned}
$$

(c)
$$
\begin{aligned}
x + 2y &= 1 \\
x + 3y &= 1 \\
x + 4y &= 1
\end{aligned}
$$

(d)
$$
\begin{aligned}
x + 2y &= 1 \\
-2x - 4y &= -2 \\
\tfrac{1}{2}x + y &= \tfrac{1}{2}
\end{aligned}
$$

In Problems 3–6, use Gauss elimination to solve the given linear system.

3. $3x_1 + 8x_2 + 3x_3 = 7$
$2x_1 - 3x_2 + x_3 = -10$
$x_1 + 3x_2 + x_3 = 3$

4. $2x_1 - 4x_2 + 3x_3 = 0$
$2x_1 + 5x_2 - 2x_3 = 0$
$5x_1 + 4x_2 - 6x_3 = 0$

5. $x_1 + x_2 + x_3 + x_4 = 1$
$x_1 \qquad\quad + x_4 = 0$
$2x_1 + 2x_2 - x_3 + x_4 = 6$
$x_1 + 2x_2 - x_3 + x_4 = 0$

6. $x_1 - x_2 - x_3 - x_4 - x_5 = -1$
$-x_1 - x_2 - x_3 - x_4 + x_5 = -1$
$x_1 - x_2 + x_3 - x_4 + x_5 = 1$
$x_1 - x_2 + x_3 - x_4 - x_5 = -1$
$x_1 + x_2 + x_3 + x_4 + x_5 = 3$

7. Solve the system

$$0.123x_1 + 0.456x_2 = 0.789$$
$$0.987x_1 + 0.654x_2 = 0.321$$

by Gauss elimination, using a calculator and retaining 3 significant digits at each stage. What is the total number of multiplications, divisions, and (signed) additions required? (*Hint:* Try to organize your calculations so that you minimize your effort. Don't bother to compute numbers which you know will turn out to be zero.)

8. Show that if $ae - bd \neq 0$, then solving the system

$$ax + by = c$$
$$dx + ey = f$$

for x and y in terms of the other symbols yields

$$x = \frac{ce - bf}{ae - bd}, \quad y = \frac{af - cd}{ae - bd}.$$

9. Use the formula derived in Problem 8 to solve the system in Problem 7. What is the number of divisions, multiplications, and additions that you performed?

10. Show that if $D := afk + bgi + cej - cfi - bek - agj$ is not zero, then solving the system

$$ax + by + cz = d$$
$$ex + fy + gz = h$$
$$ix + jy + kz = l$$

for x, y, z in terms of the other symbols yields

$$x = \frac{dfk + bgl + chj - dgj - bhk - cfl}{D},$$
$$y = \frac{ahk + dgi + cel - chi - dek - agl}{D},$$
$$z = \frac{afl + bhi + dej - dfi - ahj - bel}{D}.$$

(*Hint*: Eliminate x from the last two equations and then apply the formula derived in Problem 8 to solve for y and z.)

11. Solve the system

$$0.123x_1 + 0.456x_2 + 0.789x_3 = 0.111$$
$$0.987x_1 + 0.654x_2 + 0.321x_3 = 0.444$$
$$0.333x_1 - 0.555x_2 - 0.777x_3 = 0.888$$

by Gauss elimination, using a calculator and retaining three significant digits. What is the total number of multiplications, divisions, and (signed) additions required? (See Problem 7.)

12. Apply the formula derived in Problem 10 to solve Problem 11. What is the number of multiplications, divisions, and (signed) additions required?

13. Use a calculator to solve the following systems of equations by back substitution. What is the total number of multiplications, divisions, and (signed) additions required? (See Problem 7.)

(a)

$$1.23x_1 + 7.29x_2 - 3.21x_3 = -4.22$$
$$2.73x_2 + 1.34x_3 = 1.11$$
$$1.42x_3 = 5.16$$

(b)

$$0.500x_1 + 0.333x_2 + 0.250x_3 + 0.200x_4 = 1$$
$$+0.111x_2 + 0.222x_3 + 0.333x_4 = -1$$
$$0.999x_3 + 0.888x_4 = 1$$
$$+0.250x_4 = -1$$

14. Give a formal proof of Theorem 1. (*Hint*: You must show that if the x_i's satisfy a particular linear system, then they also satisfy the new linear system that results by performing one of the "sanctioned" operations on the old system. Then show that if they satisfy the new system, they also satisfy the old.)

15. Starting with the linear system (2), derive the following system by subtracting the second equation from the first, the third from the second, and the first from the third:

$$-x_1 + x_2 + x_3 = 0$$
$$x_1 \quad\quad - 2x_3 = 0$$
$$-x_2 + x_3 = 0.$$

Is this procedure consistent with the rules for Gauss elimination? Explain. Note that the new system admits more solutions than (2); for example, it sanctions $x_1 = x_2 = x_3 = 0$.

16. (a) Why does squaring an equation change its solution set? Specifically, consider the equation $4x = 8$.

    (b) Why does Gauss elimination sanction multiplying equations by *nonzero* constants only?

17. For the nonlinear system of equations

$$x^2 - y^2 + 2y = 1$$
$$x + 2y = 5$$

consider the following "derivation":

$$x^2 = y^2 - 2y + 1 = (y - 1)^2$$
$$x = y - 1$$
$$(y - 1) + 2y = 5$$
$$y = 2$$
$$x = 2 - 1 = 1$$

This solution does satisfy the original system, but so does $x = -3, y = 4$. Why was one solution lost? (Graph the original equations.)

18. For the nonlinear system of equations

$$\sqrt{x} + 1 = y$$
$$x - y^2 = -1 + 2y$$

consider the following "derivation":

$$\sqrt{x} = y - 1$$
$$x = y^2 - 2y + 1$$
$$(y^2 - 2y + 1) - y^2 = -2y + 1 = -1 + 2y$$
$$y = 1/2$$
$$x = (1/2)^2 - 2(1/2) + 1 = 1/4.$$

Now observe that this solution does not solve the original equation:

$$\sqrt{x} + 1 = \sqrt{1/4} + 1 = 3/2 \neq 1/2 = y$$

Where is the logical error? (Graph the original equations.)

19. Consider the problem of finding the equation of a straight line passing through two given points; in other words, one must determine the values of a, b, and c so that $ax + by = c$ is satisfied by two specific pairs $x = x_1, y = y_1$ and $x = x_2, y = y_2$. Show that this problem can be formulated as a system of two simultaneous linear equations in the three unknowns a, b, and c. This system will always have an infinite number of solutions; why?

20. The nonlinear equation $y = ax^2 + bx + c$ has a parabola as its graph. Consider the problem of passing a parabola through the three points $x_0 = 0, y_0 = 1$; $x_1 = 1$, $y_1 = 2$; and $x_2 = 2, y_2 = 1$. Write down a system of three simultaneous linear equations for the unknown coefficients a, b, and c. Then solve this system.

21. The generalization of Problem 20 is known as the **Lagrange Interpolation Problem**. The coefficients $a_0, a_1, \ldots, a_n$ are to be chosen so that the graph of the nth degree polynomial

$$y = a_0 + a_1 x + a_2 x^2 + \cdots + a_n x^n$$

passes through the $n + 1$ points $(x_0, y_0), (x_1, y_1), \ldots, (x_n, y_n)$. Express this as a system of $n + 1$ simultaneous linear equations for the unknowns a_i.

While the examples presented so far have had real numbers as the coefficients in the equations, Gauss elimination can be carried out in any number system that supports the basic operations of addition, subtraction, multiplication, and division. In Problems 22 and 23, apply the algorithm to solve the given system of linear equations over the complex numbers.

22.
$$(1 + i)x_1 + (1 - i)x_2 = i$$
$$2x_1 + \quad ix_2 = 1$$

23.
$$(1 + i)x_1 + \quad 2x_2 \qquad\qquad = 2i$$
$$x_1 + \quad ix_2 + \quad (2 - i)x_3 \quad = 1 + i$$
$$(-1 + 2i)x_2 + \quad (-2 - 3i)x_3 \quad = 0.$$

24. Complex equations can be expressed with real numbers by separating the real and imaginary parts.
(a) Show that the system in Problem 22 can be expressed (with $x = u + iv$) as

$$u_1 - v_1 + u_2 + v_2 = 0$$
$$u_1 + v_1 - u_2 + v_2 = 1$$
$$2u_1 \qquad\qquad - v_2 = 1$$
$$2v_1 + u_2 \qquad = 0$$

(b) Show that the system in Problem 23 can be expressed as

$$
\begin{aligned}
u_1 - v_1 + 2u_2 &= 0 \\
u_1 + v_1 \quad + 2v_2 &= 2 \\
u_1 \quad - v_2 + 2u_3 + v_3 &= 1 \\
v_1 + u_2 \quad - u_3 + 2v_3 &= 1 \\
- u_2 - 2v_2 - 2u_3 + 3v_3 &= 0 \\
2u_2 - v_2 - 3u_3 - 2v_3 &= 0.
\end{aligned}
$$

25. What inhibits the application of the Gauss elimination algorithm to the algebraic system composed of just the integers? Prove that $6x - 4y = 9$ has no solutions among the integers.

26. Prove that if a linear system of n equations in n unknowns has only rational numbers (i.e., ratios of integers) as coefficients and as right-hand sides and if Gauss elimination proceeds "normally" (no unwanted zeros), then the solutions are rational numbers also.

27. (For students with experience in modular arithmetic) Use Gauss elimination to solve the following system in the integers modulo 7.

$$
\begin{aligned}
2x + 4y &= 3 \\
3x + 2y &= 3.
\end{aligned}
$$

Does the system have a solution in the integers mod 6?

28. Show that a general system of n equations in n unknowns *having the structure of the systems in Problem* 13 can be solved using back substitution with n divisions, $n(n - 1)/2$ multiplications, and $n(n - 1)/2$ additions. Then show that the total number of divisions, multiplications, and additions that are required to solve n equations in n unknowns using Gauss elimination, assuming no zero denominators occur, is $n(n + 1)/2$ divisions, $(n - 1)n(2n + 5)/6$ multiplications, and $(n - 1)n(2n + 5)/6$ additions.

29. Use the formulas derived in Problem 28 to estimate the computer times required to solve representative linear systems arising in the following disciplines. Compare the performance of a typical PC, which performs 5×10^9 operations (signed additions, multiplications, or divisions) per second, with that of the Tianhe-2 computer developed by the National University of Defense Technology in Changsha city in central China, 3.38×10^{16} operations per second (Forbes, November 17, 2014).

Thermal stress (10,000 unknowns);

American Sign Language (100,000 unknowns);

Chemical plant modeling (300,000 unknowns);

Mechanics of composite materials (1,000,000) unknowns;

Electromagnetic modeling (100,000,000 unknowns);

Computational fluid dynamics (1,000,000,000 unknowns).

(Fortunately most of these systems have special structures that accommodate algorithms faster than Gauss elimination.)

30. Show that an exchange of rows i and j in a matrix can be achieved by using only the other two elementary row operations, as follows. Subtract row i from row j, add the resulting row j to row i, subtract the resulting row i from row j, and multiply row j by (-1).

1.2 MATRIX REPRESENTATION OF LINEAR SYSTEMS AND THE GAUSS-JORDAN ALGORITHM

The system of linear equations that we solved in Example 1 of the preceding section,

$$\begin{aligned} x_1 + 2x_2 + 2x_3 &= 6 \\ 2x_1 + x_2 + x_3 &= 6 \\ x_1 + x_2 + 3x_3 &= 6, \end{aligned} \tag{1}$$

can be represented in an obvious way by a 3-by-4 (some authors write 3×4) rectangular array, or **matrix**[2]:

$$\begin{bmatrix} 1 & 2 & 2 & \vdots & 6 \\ 2 & 1 & 1 & \vdots & 6 \\ 1 & 1 & 3 & \vdots & 6 \end{bmatrix}. \tag{2}$$

Note the use of the dots to indicate the equal signs. The coefficients of the variables appear in a submatrix called, appropriately, the **coefficient matrix**:

$$\begin{bmatrix} 1 & 2 & 2 \\ 2 & 1 & 1 \\ 1 & 1 & 3 \end{bmatrix} \tag{3}$$

---

[2]The term "matrix" was coined in 1850 by the English mathematician James Joseph Sylvester, the founder of the *American Journal of Mathematics*.

In general, a system of m simultaneous linear equations in n unknowns,

$$\begin{aligned}
a_{11}x_1 + a_{12}x_2 + \cdots + a_{1n}x_n &= b_1 \\
a_{21}x_1 + a_{22}x_2 + \cdots + a_{2n}x_n &= b_2 \\
&\vdots \\
a_{m1}x_1 + a_{m2}x_2 + \cdots + a_{mn}x_n &= b_m,
\end{aligned} \tag{4}$$

will be represented by an m-by-$(n + 1)$ **augmented coefficient matrix** of the form

$$\mathbf{A} = \begin{bmatrix}
a_{11} & a_{12} & \cdots & a_{1n} & \vdots & b_1 \\
a_{21} & a_{22} & \cdots & a_{2n} & \vdots & b_2 \\
\vdots & & \ddots & \vdots & \vdots & \vdots \\
a_{m1} & a_{m2} & \cdots & a_{mn} & \vdots & b_m
\end{bmatrix}, \tag{5}$$

which we write in compact notation as $[\mathbf{A}|\mathbf{b}]$. $\mathbf{A}$ is the m-by-n coefficient matrix and $\mathbf{b}$ is the m-by-1 column matrix of the b_i's. (We denote matrices by boldface capitals.) The ordering of the subscripts demonstrates how the entries in a matrix are addressed; a_{ij} denotes the element in the ith row and jth column,[3] and is the coefficient in the ith equation of the jth variable (except for $j = n + 1$, when a_{ij} is b_i).

The three basic operations of Gauss elimination become operations on the rows of the augmented coefficient matrix. Let's see how the solution procedure of the example looks in matrix form.

Example 1.   Carry out the solution of systems (1) and (2), reproduced below.

$$\begin{aligned}
x_1 + 2x_2 + 2x_3 &= 6 \\
2x_1 + x_2 + x_3 &= 6 \\
x_1 + x_2 + 3x_3 &= 6
\end{aligned}
\qquad
\begin{bmatrix}
1 & 2 & 2 & \vdots & 6 \\
2 & 1 & 1 & \vdots & 6 \\
1 & 1 & 3 & \vdots & 6
\end{bmatrix} \tag{6}$$

Solution. Notice that the elimination of x_1 is accomplished, in the matrix form, by adding the appropriate multiple of the first *row* to the subsequent ones. This gives

$$\begin{aligned}
x_1 + 2x_2 + 2x_3 &= 6 \\
-3x_2 - 3x_3 &= -6 \\
- x_2 + x_3 &= 0
\end{aligned}
\qquad
\begin{bmatrix}
1 & 2 & 2 & \vdots & 6 \\
0 & -3 & -3 & \vdots & -6 \\
0 & -1 & 1 & \vdots & 0
\end{bmatrix} \tag{7}$$

---

[3]Sometimes, for clarity, we use a comma: $a_{i,j}$.

The forward part is completed with the elimination of x_2:

$$\begin{aligned} x_1 + 2x_2 + 2x_3 &= 6 \\ -3x_2 - 3x_3 &= -6 \\ 2x_3 &= 2 \end{aligned} \qquad \left[\begin{array}{ccccc} 1 & 2 & 2 & \vdots & 6 \\ 0 & -3 & -3 & \vdots & -6 \\ 0 & 0 & 2 & \vdots & 2 \end{array} \right] \qquad (8)$$

Back substitution proceeds the same in both cases and we obtain

$$x_3 = 1, x_2 = 1, x_1 = 2. \qquad (9)$$

■

The following terminology is helpful in describing the matrix manipulations pertinent to Gauss elimination.

Diagonal Matrices

Definition 2. The **diagonal** of an m-by-n matrix $\mathbf{A}$ is the set of elements $\{a_{ii} : i = 1, 2, \ldots, \min\{m, n\}\}$. A **diagonal matrix** is a matrix with all entries equal to zero except those on the diagonal.

Some examples of diagonal matrices are

$$\begin{bmatrix} 2 & 0 & 0 & 0 & 0 \\ 0 & 1 & 0 & 0 & 0 \\ 0 & 0 & 2 & 0 & 0 \\ 0 & 0 & 0 & 0 & 0 \\ 0 & 0 & 0 & 0 & -1 \end{bmatrix}, \quad \begin{bmatrix} 1 & 0 & 0 & 0 & 0 \\ 0 & 2 & 0 & 0 & 0 \\ 0 & 0 & 3 & 0 & 0 \end{bmatrix}, \quad \begin{bmatrix} 4 & 0 & 0 \\ 0 & 4 & 0 \\ 0 & 0 & 4 \\ 0 & 0 & 0 \\ 0 & 0 & 0 \end{bmatrix}.$$

Trapezoidal Matrices

Definition 3. An **upper trapezoidal matrix A** has all the entries below the diagonal equal to zero, that is, $a_{ij} = 0$ if $i > j$. If $\mathbf{A}$ is square and upper trapezoidal, it is called **upper triangular**.

Examples of upper trapezoidal matrices are

$$\begin{bmatrix} 1 & 1 & 1 & 1 & 1 \\ 0 & 1 & 1 & 1 & 1 \\ 0 & 0 & 1 & 1 & 1 \end{bmatrix}, \quad \begin{bmatrix} 1 & 2 & 3 \\ 0 & 4 & 5 \\ 0 & 0 & 6 \\ 0 & 0 & 0 \\ 0 & 0 & 0 \end{bmatrix}, \quad \begin{bmatrix} 1 & 2 & 3 & 4 & 5 \\ 0 & 0 & 3 & 4 & 5 \\ 0 & 0 & 3 & 4 & 5 \\ 0 & 0 & 0 & 4 & 5 \\ 0 & 0 & 0 & 0 & 5 \end{bmatrix},$$

where the last matrix is upper triangular.

Lower trapezoidal and lower triangular matrices are defined similarly; some examples are as follows:

$$\begin{bmatrix} 1 & 0 & 0 & 0 & 0 \\ 1 & 1 & 0 & 0 & 0 \\ 1 & 1 & 1 & 0 & 0 \end{bmatrix}, \quad \begin{bmatrix} 1 & 0 & 0 \\ 2 & 3 & 0 \\ 4 & 5 & 6 \\ 8 & 6 & 9 \\ 9 & 6 & 8 \end{bmatrix}, \quad \begin{bmatrix} 1 & 0 & 0 & 0 & 0 \\ 1 & 2 & 0 & 0 & 0 \\ 1 & 2 & 3 & 0 & 0 \\ 1 & 2 & 3 & 4 & 0 \\ 1 & 2 & 3 & 4 & 5 \end{bmatrix}.$$

Important special cases of upper trapezoidal matrices are the **row matrices** or **row vectors**,

$$[1\ 2], \quad [1\ 2\ 3], \quad [1\ 0\ 0\ 0\ 0\ 1],$$

and the **column matrices** or **column vectors** are special instances of lower trapezoidal matrices,

$$\begin{bmatrix} 1 \\ 2 \end{bmatrix}, \quad \begin{bmatrix} 1 \\ 2 \\ 3 \end{bmatrix}, \quad \begin{bmatrix} 1 \\ 0 \\ 0 \\ 0 \\ 0 \\ 1 \end{bmatrix}.$$

Elementary Row Operations

Definition 4. The following operations on a matrix are known as **elementary row operations**:

(i) adding (or subtracting) a multiple of one row to another and replacing the latter with the result;

(ii) multiplying a row by a nonzero constant and replacing the original row by the result;

(iii) switching two rows.

These are, of course, the matrix analogs of the operations listed in Theorem 1 and they are guaranteed not to alter the solution set. With the new terminology, we can summarize our deliberations by saying *the forward part of the Gauss elimination algorithm uses elementary row operations to reduce the augmented coefficient matrix to upper trapezoidal form.* This is achieved by adding appropriate multiples of the ith row to the subsequent rows, so as to form zeros in the ith column below the diagonal (for $i = 1$ to $n - 1$).

In the matrix below, the entries marked with # signs are targeted for "annihilation" by adding multiples of the second row.

$$\begin{bmatrix} X & X & X & X & \vdots & X \\ 0 & X & X & X & \vdots & X \\ 0 & \# & X & X & \vdots & X \\ 0 & \# & X & X & \vdots & X \end{bmatrix}.$$

In order to help the reader see which of the various elementary row operations is being employed in a specific computation, we introduce a shorthand:

(i) to add (-3) times row #2 to row #4 and replace the latter with the result, write

$$(-3)\rho_2 + \rho_4 \rightarrow \rho_4 \ (\text{"row 4"});$$

(ii) to multiply row #3 by 7 and replace the original row by the result, write

$$7\rho_3 \rightarrow \rho_3;$$

(iii) to switch rows #2 and #4, write

$$\rho_2 \leftrightarrow \rho_4.$$

We employ this shorthand in the next example; it motivates a very useful extension of the matrix representation.

Example 2. Solve the following linear systems, which have identical coefficient matrices.

$$\begin{bmatrix} 2 & 1 & 4 & \vdots & -2 \\ 3 & 3 & 0 & \vdots & 6 \\ -1 & 4 & 2 & \vdots & -5 \end{bmatrix}; \quad \begin{bmatrix} 2 & 1 & 4 & \vdots & 7 \\ 3 & 3 & 0 & \vdots & 6 \\ -1 & 4 & 2 & \vdots & 5 \end{bmatrix}. \tag{10}$$

Solution. Observe that the Gauss elimination algorithm calls for the same sequence of elementary row operations for both systems. First we add $(-3/2)$ times the first row to the second, and add $1/2$ times the first to the third, *in both cases*. This gives

$$
\begin{array}{l}
(-3/2)\rho_1 + \rho_2 \to \rho_2 \\
(1/2)\rho_1 + \rho_3 \to \rho_3
\end{array}
\quad
\begin{bmatrix}
2 & 1 & 4 & \vdots & -2 \\
0 & 3/2 & -6 & \vdots & 9 \\
0 & 9/2 & 4 & \vdots & -6
\end{bmatrix}
\;;\;
\begin{bmatrix}
2 & 1 & 4 & \vdots & 7 \\
0 & 3/2 & -6 & \vdots & -9/2 \\
0 & 9/2 & 4 & \vdots & 17/2
\end{bmatrix} .
$$

Next we add (-3) times the second row to the third row—again, *for both systems*—yielding

$$
(-3)\rho_2 + \rho_3 \to \rho_3
\quad
\begin{bmatrix}
2 & 1 & 4 & \vdots & -2 \\
0 & 3/2 & -6 & \vdots & 9 \\
0 & 0 & 22 & \vdots & -33
\end{bmatrix}
\;;\;
\begin{bmatrix}
2 & 1 & 4 & \vdots & 7 \\
0 & 3/2 & -6 & \vdots & -9/2 \\
0 & 0 & 22 & \vdots & 22
\end{bmatrix} .
$$

Back substitution then proceeds normally and we obtain

$$
\begin{array}{ll}
x_3 = -33/22 = -3/2 & x_3 = 1 \\
x_2 = (2/3)(9-9) = 0 \quad ; & x_2 = (2/3)(-9/2+6) = 1 \\
x_1 = (1/2)(-2+6-0) = 2 & x_1 = (1/2)(7-4-1) = 1.
\end{array}
\tag{11}
$$

■

The point that we wish to make here is that the forward part of the Gauss algorithm only "looks at" the *coefficient* matrix (submatrix) in selecting its multipliers for the eliminations. The data on the right are passively swept along by the process. So the efficient way to handle such situations is to augment the (common) coefficient matrix with *two* columns, one for each right-hand side. Thus, the calculations are more efficiently expressed as follows:

$$
\begin{bmatrix}
2 & 1 & 4 & \vdots & -2 & 7 \\
3 & 3 & 0 & \vdots & 6 & 6 \\
-1 & 4 & 2 & \vdots & -5 & 5
\end{bmatrix}
\quad
\begin{array}{l}
(-3/2)\rho_1 + \rho_2 \to \rho_2 \\
(1/2)\rho_1 + \rho_3 \to \rho_3
\end{array}
\quad
\begin{bmatrix}
2 & 1 & 4 & \vdots & -2 & 7 \\
0 & 3/2 & -6 & \vdots & 9 & -9/2 \\
0 & 9/2 & 4 & \vdots & -6 & 17/2
\end{bmatrix}
$$

$$
(-3)\rho_2 + \rho_3 \to \rho_3
\quad
\begin{bmatrix}
2 & 1 & 4 & \vdots & -2 & 7 \\
0 & 3/2 & -6 & \vdots & 9 & -9/2 \\
0 & 0 & 22 & \vdots & -33 & 22
\end{bmatrix}
\tag{12}
$$

and the final result (11) is obtained by back substitution for each right-hand side.

There is a minor modification to Gauss elimination that renders the back-substitution step unnecessary; it is known as the **Gauss-Jordan algorithm**. Before we eliminate the

"3" and the "-1" in the first column of system (10), we multiply the first row by (1/2):

$$(-1/2)\rho_1 \rightarrow \rho_1 \begin{bmatrix} 1 & 1/2 & 2 & \vdots & -1 & 7/2 \\ 3 & 3 & 0 & \vdots & 6 & 6 \\ -1 & 4 & 2 & \vdots & -5 & 5 \end{bmatrix}.$$

The replacement of the leading "2" by a "1" makes the elimination of the "3" and "-1" a little easier for hand computation. We add the multiples (-3) and (1) times the first row to the second and third rows:

$$\begin{matrix} (-3)\rho_1 + \rho_2 \rightarrow \rho_2 \\ \rho_1 + \rho_3 \rightarrow \rho_3 \end{matrix} \begin{bmatrix} 1 & 1/2 & 2 & \vdots & -1 & 7/2 \\ 0 & 3/2 & -6 & \vdots & 9 & -9/2 \\ 0 & 9/2 & 4 & \vdots & -6 & 17/2 \end{bmatrix}.$$

Similarly, we multiply the second row by (2/3)

$$(2/3)\rho_2 \rightarrow \rho_2 \begin{bmatrix} 1 & 1/2 & 2 & \vdots & -1 & 7/2 \\ 0 & 1 & -4 & \vdots & 6 & -3 \\ 0 & 9/2 & 4 & \vdots & -6 & 17/2 \end{bmatrix}$$

before eliminating the "9/2" by adding ($-9/2$) times the second row to the third row.

$$(-9/2)\rho_2 + \rho_3 \rightarrow \rho_3 \begin{bmatrix} 1 & 1/2 & 2 & \vdots & -1 & 7/2 \\ 0 & 1 & -4 & \vdots & 6 & -3 \\ 0 & 0 & 22 & \vdots & -33 & 22 \end{bmatrix}.$$

But at this stage, we also eliminate the "1/2" in the first row by adding ($-1/2$) times the second row:

$$(-1/2)\rho_2 + \rho_1 \rightarrow \rho_1 \begin{bmatrix} 1 & 0 & 4 & \vdots & -4 & 5 \\ 0 & 1 & -4 & \vdots & 6 & -3 \\ 0 & 0 & 22 & \vdots & -33 & 22 \end{bmatrix}.$$

In terms of equations, we have eliminated x_2 from all equations below *and above* the second.

Finally we multiply the third row by (1/22) and add the multiples (-4), (4) of the (new) third row to the first and second (eliminating x_3 from these equations):

$$(1/22)\rho_3 \rightarrow \rho_3$$

$$
\begin{bmatrix}
1 & 0 & 4 & \vdots & -4 & 5 \\
0 & 1 & -4 & \vdots & 6 & -3 \\
0 & 0 & 1 & \vdots & -3/2 & 1
\end{bmatrix}
\begin{matrix}
(-4)\rho_3 + \rho_1 \rightarrow \rho_1 \\
4\rho_3 + \rho_2 \rightarrow \rho_2
\end{matrix}
\begin{bmatrix}
1 & 0 & 0 & \vdots & 2 & 1 \\
0 & 1 & 0 & \vdots & 0 & 1 \\
0 & 0 & 1 & \vdots & -3/2 & 1
\end{bmatrix}.
$$

Now the solutions (11) are displayed explicitly.

The rules for the Gauss-Jordan algorithm are easily obtained by modifying those of Gauss elimination. Expressed in terms of equations, they are

---

Gauss-Jordan Algorithm (without anomalies) (n equations in n unknowns, no zero denominators)

1. Multiply the first equation by the reciprocal of the coefficient of x_1.
2. Eliminate x_1 from all equations except the first by adding (or subtracting) appropriate multiples of the (new) first equation.
3. Multiply the (new) second equation by the reciprocal of the coefficient of x_2.
4. Eliminate x_2 from all equations except the second by adding appropriate multiples of the second equation.
5. Continue in this manner: multiply the kth equation by the reciprocal of the coefficient of x_k and eliminate x_k from all other equations by adding appropriate multiples of the kth equation.

The resulting system (or augmented matrix) will then display the solution explicitly.

The Gauss-Jordan algorithm[4] is suitable for hand calculation with small systems, but the apparent simplicity of its final format is deceiving; its complete execution actually requires more calculations than does Gauss elimination. It has value as a theoretical construct, however, and we shall have occasion to refer to it again in subsequent discussions. Like Gauss elimination, it can be foiled by inconvenient zeros, necessitating the remedial measures that are described in the following section.

---

[4]Wilhelm Jordan (Yor' - dan) was a 19th-century German geodist. The algorithm is often attributed to his contemporary, the French mathematician Camille Jordan (Jor - dan'), who conceived the *Jordan form* described in Section 7.3.

Exercises 1.2

In Problems 1–4, solve the systems of equations represented by the following augmented coefficient matrices.

1. $\begin{bmatrix} 2 & 1 & \vdots & 8 \\ 1 & -3 & \vdots & -3 \end{bmatrix}$ **2.** $\begin{bmatrix} 1 & 1 & 1 & \vdots & 0 \\ 1 & 0 & 1 & \vdots & 3 \\ 0 & 1 & 1 & \vdots & 1 \end{bmatrix}$

3. $\begin{bmatrix} 4 & 2 & 2 & \vdots & 8 & 0 \\ 1 & -1 & 1 & \vdots & 4 & 2 \\ 3 & 2 & 1 & \vdots & 2 & 0 \end{bmatrix}$ **4.** $\begin{bmatrix} 1 & -1 & 0 & 1 & \vdots & 4 & 0 \\ 1 & 1 & 1 & -2 & \vdots & -4 & 2 \\ 0 & 2 & 1 & 3 & \vdots & 4 & 2 \\ 2 & 1 & -1 & 2 & \vdots & 5 & 3 \end{bmatrix}$

5. In most computer languages, the elements of an *m*-by-*n* matrix are stored in consecutive locations by columns, that is, the addressing sequence is as shown:

$$\begin{bmatrix} \#1 & \#6 & \#11 \\ \#2 & \#7 & \#12 \\ \#3 & \#8 & \#13 \\ \#4 & \#9 & \#14 \\ \#5 & \#10 & \#15 \end{bmatrix}$$

(a) In a 7-by-8 matrix what are the addresses of $a_{1,7}$, $a_{7,1}$, $a_{5,5}$, and $a_{7,8}$?

(b) In a 5-by-12 matrix what are the row and column numbers of the data at addresses #4, #20, and #50?

(c) Given the matrix dimensions *m* and *n*, what formula does the computer use to calculate the address of $a_{i,j}$?

(d) Given *m* and *n*, how does the computer calculate the row and column for the datum at address #*p* ?

6. Analogous to the elementary row operations are the elementary column operations:

(i) adding a multiple of one column to another and replacing the latter with the result;

(ii) multiplying a column by a nonzero constant and replacing the original column with the result;

(iii) switching two columns.

These operations are of little value in equation solving, but they have interesting interpretations.

(a) Show that (i) corresponds to replacing some x_i by a combination $x_i + cx_j$.

(b) Show that (ii) corresponds to "rescaling" one of the unknowns (e.g., replacing x_3 by x_3/c).

(c) Show that (iii) corresponds to renumbering the unknowns.

7. Solve the following systems using the Gauss-Jordan algorithm:

(a) $\begin{bmatrix} 3 & 8 & 3 & : & 7 \\ 2 & -3 & 1 & : & -10 \\ 1 & 3 & 1 & : & 3 \end{bmatrix}$ (b) $\begin{bmatrix} 1 & 2 & 1 & : & 1 & 0 & 0 \\ 1 & 3 & 2 & : & 0 & 1 & 0 \\ 1 & 0 & 1 & : & 0 & 0 & 1 \end{bmatrix}$

8. A classic problem in numerical analysis is least squares approximation by polynomials. Here, a given function is to be approximated by a polynomial of degree, say, 2:

$$f(x) \approx a_0 + a_1 x + a_2 x^2,$$

and the coefficients a_i are to be selected so as to minimize the *mean square error*

$$\varepsilon = \int_0^1 [f(x) - (a_0 + a_1 x + a_2 x^2)]^2 dx.$$

(a) Since ε is a function of the coefficients a_0, a_1, and a_2, calculus directs us to the simultaneous solution of the *variational equations*

$$\frac{\partial \varepsilon}{\partial a_0} = 0, \quad \frac{\partial \varepsilon}{\partial a_1} = 0, \quad \frac{\partial \varepsilon}{\partial a_2} = 0 \tag{13}$$

for the values of a_0, a_1, and a_2 minimizing ε. Work out these equations and show that they are *linear* in the unknowns a_0, a_1, and a_2.

(b) Assemble the augmented coefficient matrix for the system and evaluate the integrals in the coefficient submatrix. Show that the latter is the 3-by-3 matrix

$$\begin{bmatrix} 1 & 1/2 & 1/3 \\ 1/2 & 1/3 & 1/4 \\ 1/3 & 1/4 & 1/5 \end{bmatrix}.$$

(c) Derive the coefficient submatrix for the unknown polynomial coefficients in the least square approximation by polynomials of degree 3.

9. A minor improvement in the Gauss-Jordan algorithm, as stated in the text, goes as follows: instead of eliminating the unknown x_j from all equations other than the jth equation, eliminate it only in the equations following the jth, reducing the system to upper triangular form; then working from the bottom up, eliminate $x_n, x_{n-1}, \ldots, x_2$ from the earlier equations. Why is this more efficient? (*Hint*: If you don't see this immediately, try it on a system of four equations in four unknowns.)

1.3   THE COMPLETE GAUSS ELIMINATION ALGORITHM

In this section, we will modify the basic Gauss algorithm to accommodate linear systems that have no solutions or an infinite number of solutions, systems for which the number of equations and unknowns don't match, and systems that appear to call for division by zero. We begin by considering a system in which there are more equations than unknowns.

Example 1.   Solve the system of 4 equations in 3 unknowns represented by

$$\begin{bmatrix} 1 & 0 & 1 & \vdots & 0 \\ 1 & 1 & 1 & \vdots & 1 \\ 0 & 1 & 1 & \vdots & 1 \\ 3 & 3 & 6 & \vdots & 3 \end{bmatrix}. \tag{1}$$

Solution.   We proceed to eliminate variables according to the forward part of the Gauss algorithm. Elimination of x_1 and x_2 produces

$$\begin{matrix} (-1)\rho_1 + \rho_2 \to \rho_2 \\ \\ (-3)\rho_1 + \rho_4 \to \rho_4 \end{matrix} \quad \begin{bmatrix} 1 & 0 & 1 & \vdots & 0 \\ 0 & 1 & 0 & \vdots & 1 \\ 0 & 1 & 1 & \vdots & 1 \\ 0 & 3 & 3 & \vdots & 3 \end{bmatrix} \quad \begin{matrix} (-1)\rho_2 + \rho_3 \to \rho_3 \\ (-3)\rho_2 + \rho_4 \to \rho_4 \end{matrix} \quad \begin{bmatrix} 1 & 0 & 1 & \vdots & 0 \\ 0 & 1 & 0 & \vdots & 1 \\ 0 & 0 & 1 & \vdots & 0 \\ 0 & 0 & 3 & \vdots & 0 \end{bmatrix}.$$

At this point notice that the third and fourth equations say the same thing; namely, that $x_3 = 0$. Thus, the fourth equation contains no new information. This becomes even more obvious if we proceed in strict accordance with Gauss' procedure, which calls for the subtraction of three times the third row from the fourth. Elimination of x_3 results in

$$\begin{matrix} \\ \\ (-3)\rho_3 + \rho_4 \to \rho_4 \end{matrix} \quad \begin{bmatrix} 1 & 0 & 1 & \vdots & 0 \\ 0 & 1 & 0 & \vdots & 1 \\ 0 & 0 & 1 & \vdots & 0 \\ 0 & 0 & 0 & \vdots & 0 \end{bmatrix}. \tag{2}$$

The solution of this system is easily obtained by back substitution:

$$\begin{aligned} x_3 &= 0 \\ x_2 &= 1 \\ x_1 &= 0. \end{aligned} \tag{3}$$

This must also be the solution for the original system (1) since we have proceeded in accordance with the rules in Theorem 1 (Section 1.1).

The fact that the fourth equation ultimately became $0 = 0$ (see (2)) indicates that in the original system (1), the fourth equation was redundant. After all, if the fourth row was reduced to zero by subtracting off multiples of previous rows, then it must have been expressible as a sum of multiples of those previous rows. Consequently, the fourth equation could have been deduced from the others; it carried no new information. In fact, for the system (1), the fourth row could have been derived by adding the first row to the third, and multiplying by 3.[5]

By generalizing this idea, we uncover a valuable feature of the Gauss algorithm: *when a row of zeros appears in the augmented coefficient matrix during the forward part of Gauss elimination, the original equation corresponding to that row is redundant* (derivable from the others).

A redundant system of equations in two unknowns was depicted in the solution graphs in Figure 1.1e. Any two of the straight lines specifies the solution; the third line conveys no new information.

Another situation is illustrated in the next example, in which the linear system differs from (1) only by the inclusion of a fifth equation.

Example 2. Perform Gauss elimination on the system

$$
\begin{bmatrix}
1 & 0 & 1 & \vdots & 0 \\
1 & 1 & 1 & \vdots & 1 \\
0 & 1 & 1 & \vdots & 1 \\
3 & 3 & 6 & \vdots & 3 \\
1 & 2 & 2 & \vdots & 1
\end{bmatrix} . \tag{4}
$$

Solution. Eliminating $x_1, x_2,$ and x_3 in turn produces

$$
\begin{array}{c}
(-1)\rho_1 + \rho_2 \to \rho_2 \\
\\
(-3)\rho_1 + \rho_4 \to \rho_4 \\
(-1)\rho_1 + \rho_5 \to \rho_5
\end{array}
\begin{bmatrix}
1 & 0 & 1 & \vdots & 0 \\
0 & 1 & 0 & \vdots & 1 \\
0 & 1 & 1 & \vdots & 1 \\
0 & 3 & 3 & \vdots & 3 \\
0 & 2 & 1 & \vdots & 1
\end{bmatrix}
\begin{array}{c}
\\
(-1)\rho_2 + \rho_3 \to \rho_3 \\
(-3)\rho_2 + \rho_4 \to \rho_4 \\
(-2)\rho_2 + \rho_5 \to \rho_5
\end{array}
\begin{bmatrix}
1 & 0 & 1 & \vdots & 0 \\
0 & 1 & 0 & \vdots & 1 \\
0 & 0 & 1 & \vdots & 0 \\
0 & 0 & 3 & \vdots & 0 \\
0 & 0 & 1 & \vdots & -1
\end{bmatrix}
$$

[5]Problem 16 spells out how one can derive the expression of the redundant equation in terms of the others by keeping track of the elementary row operations that led to its "annihilation."

$$\begin{array}{c} \\ \\ \\ (-3)\rho_3 + \rho_4 \rightarrow \rho_4 \\ (-1)\rho_3 + \rho_5 \rightarrow \rho_5 \end{array} \begin{bmatrix} 1 & 0 & 1 & \vdots & 0 \\ 0 & 1 & 0 & \vdots & 1 \\ 0 & 0 & 1 & \vdots & 0 \\ 0 & 0 & 0 & \vdots & 0 \\ 0 & 0 & 0 & \vdots & -1 \end{bmatrix}.$$

As before we see that the original fourth equation has been exposed as redundant, but more interesting is the final equation, $0 = -1$! Of course, no set of values of x_1, x_2 and x_3 will make 0 equal -1. Consequently (4) has *no solution*, and hence by Theorem 1 (Section 1.1), the *original* system has no solution.

The fact that the *left*-hand side of the fifth equation in (4) was reduced to zero by subtracting multiples of other equations indicates, as before, that its original left-hand side was expressible as a sum of multiples of the other left-hand sides. However, the corresponding combination of the other *right*-hand sides evidently didn't match up. So the fifth equation is incompatible with the others, and we say the system is **inconsistent**. In fact, if we add the original equations #2 and #3 in (4), we deduce that the combination $x_1 + 2x_2 + 2x_3$ equals 2, whereas the fifth equation says it equals 1. Gauss elimination exposed the inconsistency. *If, at any stage, the Gauss algorithm produces a row that states $0 = c$ for a nonzero constant c, then the linear system has no solution* (is inconsistent).

An inconsistent system of three equations in two unknowns was depicted graphically in Figure 1.1d. The fact that the third line "misses the mark" demonstrates its incompatibility.

Next we turn to the problem of unwanted zeros in the Gauss algorithm. It arises right at the start in the following example.

Example 3. Solve the system

$$\begin{bmatrix} 0 & 2 & 1 & 1 & 0 & \vdots & 0 \\ 2 & 4 & 4 & 2 & 2 & \vdots & 0 \\ 3 & 6 & 6 & 0 & 0 & \vdots & 0 \\ 0 & -2 & -1 & -2 & 2 & \vdots & 1 \\ 0 & 2 & 1 & 2 & 4 & \vdots & 1 \end{bmatrix}. \tag{5}$$

Solution. As we have formulated it, the Gauss algorithm calls for multiplying the first row by $(-2/0)$, as a preliminary to eliminating x_1 from the second row. Naturally this cannot be done; in fact there is no way row #1 can be used to eliminate x_1 from row #2, because of the 0 in the $(1, 1)$ position. Of course the cure is obvious; we'll use *another* row to eliminate x_1. We can achieve this while maintaining the "spirit" of the Gauss algorithm by switching the first row with either the second or third (obviously not the fourth or fifth!) and proceeding as usual.

Switching the first and third rows results in

$$\rho_1 \leftrightarrow \rho_3 \quad \begin{bmatrix} 3 & 6 & 6 & 0 & 0 & \vdots & 0 \\ 2 & 4 & 4 & 2 & 2 & \vdots & 0 \\ 0 & 2 & 1 & 1 & 0 & \vdots & 0 \\ 0 & -2 & -1 & -2 & 2 & \vdots & 1 \\ 0 & 2 & 1 & 2 & 4 & \vdots & 1 \end{bmatrix} . \tag{6}$$

Elimination of x_1 then proceeds normally as follows:

$$(-2/3)\rho_1 + \rho_2 \to \rho_2 \quad \begin{bmatrix} 3 & 6 & 6 & 0 & 0 & \vdots & 0 \\ 0 & 0 & 0 & 2 & 2 & \vdots & 0 \\ 0 & 2 & 1 & 1 & 0 & \vdots & 0 \\ 0 & -2 & -1 & -2 & 2 & \vdots & 1 \\ 0 & 2 & 1 & 2 & 4 & \vdots & 1 \end{bmatrix} . \tag{7}$$

The tactic we have just performed is known as **pivoting**; by using the *third* row to eliminate x_1, we have designated row #3 as the **pivot row** and the coefficient of x_1 in this row (namely, 3) is the **pivot element**. In all our previous examples, the pivot elements occurred along the diagonal of the coefficient matrix, so our pivoting strategy did not call for switching any rows. *In general when one uses row #i to eliminate the coefficient of x_j from other rows, the ith row is called a pivot row and a_{ij} is a pivot element.* Commonly one says "we pivot on the element a_{ij}."

Returning to (7) in the example, we see that the elimination of x_2 presents the same dilemma as did x_1, and we resolve it by pivoting on the 2 in the third row. That is, we switch the second and third rows and then eliminate x_2 as usual:

$\rho_2 \leftrightarrow \rho_3$

$$\begin{bmatrix} 3 & 6 & 6 & 0 & 0 & \vdots & 0 \\ 0 & 2 & 1 & 1 & 0 & \vdots & 0 \\ 0 & 0 & 0 & 2 & 2 & \vdots & 0 \\ 0 & -2 & -1 & -2 & 2 & \vdots & 1 \\ 0 & 2 & 1 & 2 & 4 & \vdots & 1 \end{bmatrix} \quad \begin{matrix} \rho_2 + \rho_4 \to \rho_4 \\ (-1)\rho_2 + \rho_5 \to \rho_5 \end{matrix} \quad \begin{bmatrix} 3 & 6 & 6 & 0 & 0 & \vdots & 0 \\ 0 & 2 & 1 & 1 & 0 & \vdots & 0 \\ 0 & 0 & 0 & 2 & 2 & \vdots & 0 \\ 0 & 0 & 0 & -1 & 2 & \vdots & 1 \\ 0 & 0 & 0 & 1 & 4 & \vdots & 1 \end{bmatrix}$$

Now we encounter a totally new situation. On the one hand, there is no row that we can switch with row #3 to avoid a zero division because the coefficient for x_3 is zero in all the subsequent rows. On the other hand, there is no *need* to eliminate x_3, for the very same reason! Since there's nothing we can do about x_3, we skip it and go on to x_4. The role of x_3 will become clear later when we perform back substitution.

We use the third row (not the fourth) as the pivot row to eliminate x_4 from subsequent rows; then we eliminate x_5.

$$
\begin{array}{c}
(1/2)\rho_3 + \rho_4 \to \rho_4 \\
(-1/2)\rho_3 + \rho_5 \to \rho_5
\end{array}
\qquad
\left[
\begin{array}{ccccc:c}
3 & 6 & 6 & 0 & 0 & 0 \\
0 & 2 & 1 & 1 & 0 & 0 \\
0 & 0 & 0 & 2 & 2 & 0 \\
0 & 0 & 0 & 0 & 3 & 1 \\
0 & 0 & 0 & 0 & 3 & 1
\end{array}
\right]
$$

$$
(-1)\rho_4 + \rho_5 \to \rho_5
\qquad
\left[
\begin{array}{ccccc:c}
3 & 6 & 6 & 0 & 0 & 0 \\
0 & 2 & 1 & 1 & 0 & 0 \\
0 & 0 & 0 & 2 & 2 & 0 \\
0 & 0 & 0 & 0 & 3 & 1 \\
0 & 0 & 0 & 0 & 0 & 0
\end{array}
\right]. \tag{8}
$$

Gauss elimination has exposed the fifth equation as redundant.

Back substitution produces

$$
x_5 = 1/3
$$
$$
x_4 = -1/3.
$$

Now there is no equation for x_3. The second equation

$$
2x_2 + x_3 = -x_4 = 1/3 \tag{9}
$$

tells us the values of the *combination* $(2x_2 + x_3)$ but says nothing about x_2 or x_3 separately. Evidently x_3 can take on any value, as long as x_2 is adjusted to make the combination $(2x_2 + x_3)$ equal to $1/3$. So we enforce this by writing

$$
x_3 \text{ arbitrary}
$$

$$
x_2 = \frac{1}{2}\left(\frac{1}{3} - x_3\right) = \frac{1}{6} - \frac{1}{2}x_3. \tag{10}
$$

Back substitution of these expressions into the first equation gives

$$x_1 = -2x_2 - 2x_3 = -2\left(\frac{1}{6} - \frac{1}{2}x_3\right) - 2x_3 = -\frac{1}{3} - x_3. \tag{11}$$

Equations (10) and (11), and hence the original equations (5), have an infinite number of solutions—one for each value of x_3. Since we are free to choose x_3, it is customary to call x_3 a *free variable* and display the solutions as

$$x_1 = -\frac{1}{3} - x_3$$

$$x_2 = \frac{1}{6} - \frac{1}{2}x_3$$

$$x_3 \text{ free}$$

$$x_4 = -\frac{1}{3}$$

$$x_5 = \frac{1}{3}.$$

It will prove convenient for later purposes to display these solutions by introducing a "dummy variable" t for x_3 and writing

$$x_1 = -\frac{1}{3} - t, \quad x_2 = \frac{1}{6} - \frac{1}{2}t, \quad x_3 = t, \quad x_4 = -\frac{1}{3}, \quad x_5 = \frac{1}{3}. \tag{12}$$

We call (12) a *parametrization* of the solution set; all the solutions of the original system (5) are generated by letting the parameter t range through the real numbers. ∎

Notice that there is some flexibility in interpreting the format (8); in particular, its second equation (9). We could equally well have taken x_2 as the free variable and then adjusted x_3 to make the combination $(2x_2 + x_3)$ equal 1/3. Then (10) and (11) would be replaced by

$$x_3 = \frac{1}{3} - 2x_2 \tag{13}$$

$$x_1 = -2x_2 - 2\left(\frac{1}{3} - 2x_2\right) = -\frac{2}{3} + 2x_2, \tag{14}$$

and the tabulation would take the following parametric form (we use s instead of t to distinguish the parametrizations):

$$x_1 = -\frac{2}{3} + 2s, \; x_2 = s, \; x_3 = \frac{1}{3} - 2s, \; x_4 = -\frac{1}{3}, \; x_5 = \frac{1}{3}. \tag{15}$$

The equations (15) must, of course, describe the same solution set as (12). In Problem 17, the reader is requested to construct yet another parametrization with x_1 as the free variable. Since the values of x_4 and x_5 are fixed, however, they can never serve as free variables. (Problem 18 addresses the question of how one can determine, in general, whether or not two different parameterizations describe the same solution set.)

We summarize the considerations of this section, so far, as follows: *when a variable is eliminated "prematurely" in the forward part of the Gauss procedure, it becomes a free variable in the solution set. Unless the system is inconsistent, the solution set will be infinite.*

The final case to consider in this section arises when the number of equations is fewer than the number of unknowns. The following example demonstrates that this introduces no new complications.

Example 4. Solve the following the linear system of three equations in six unknowns:

$$\begin{bmatrix} 1 & 2 & 1 & 1 & -1 & 0 & \vdots & 3 \\ 1 & 2 & 0 & -1 & 1 & 0 & \vdots & 2 \\ 0 & 0 & 1 & 2 & -2 & 1 & \vdots & 2 \end{bmatrix}.$$

Solution. Elimination of x_1 and x_3 (skipping x_2) produces

$$(-1)\rho_1 + \rho_2 \to \rho_2 \quad \begin{bmatrix} 1 & 2 & 1 & 1 & -1 & 0 & \vdots & 3 \\ 0 & 0 & -1 & -2 & 2 & 0 & \vdots & -1 \\ 0 & 0 & 1 & 2 & -2 & 1 & \vdots & 2 \end{bmatrix}$$

$$\rho_2 + \rho_3 \to \rho_3 \quad \begin{bmatrix} 1 & 2 & 1 & 1 & -1 & 0 & \vdots & 3 \\ 0 & 0 & -1 & -2 & 2 & 0 & \vdots & -1 \\ 0 & 0 & 0 & 0 & 0 & 1 & \vdots & 1 \end{bmatrix}.$$

Back substitution tells us that

$$
\begin{aligned}
x_6 &= 1 \\
x_5 &\text{ is free} \\
x_4 &\text{ is also free (!)} \\
x_3 &= 1 - 2x_4 + 2x_5 \\
x_2 &\text{ is free (!!)} \\
x_1 &= 3 - 2x_2 - (1 - 2x_4 + 2x_5) - x_4 + x_5 \\
    &= 2 - 2x_2 + x_4 - x_5.
\end{aligned}
\tag{16}
$$

Thus, we have a three-*parameter solution set* which we can write as

$$x_1 = 2 - 2t_1 + t_2 - t_3, \quad x_2 = t_1, \quad x_3 = 1 - 2t_2 + 2t_3, \quad x_4 = t_2, \quad x_5 = t_3, \quad x_6 = 1. \quad \blacksquare$$

We have now covered the basic techniques needed for solving linear algebraic systems. After you have practiced with the problems in this section, you should feel confident that, given enough time (and perhaps some computational assistance), you will be able to determine the solution set for any system of simultaneous linear equations. Although we shall, on occasion, have a few more things to say about this subject, from here on we presume the right to call on your "Gauss Elimination Equation Solver" whenever the analysis of any problem has been reduced to a linear system of equations.

Exercises 1.3

In Problems 1–11, solve the given system:

1. $\begin{bmatrix} 1 & 1 & 1 & 1 & \vdots & 1 \\ 2 & 2 & -1 & 1 & \vdots & 0 \\ 1 & 0 & 0 & 1 & \vdots & 0 \\ 1 & 2 & -1 & 1 & \vdots & 0 \end{bmatrix}$
2. $\begin{bmatrix} 1 & 1 & 1 & 1 & \vdots & 1 \\ 2 & 2 & -1 & 1 & \vdots & 0 \\ -1 & -1 & 0 & 1 & \vdots & 0 \\ 0 & 0 & 1 & -1 & \vdots & 0 \end{bmatrix}$

3. $\begin{bmatrix} 1 & 1 & 1 & 1 & \vdots & 1 \\ 2 & 2 & -1 & 1 & \vdots & 0 \\ -1 & -1 & 0 & 1 & \vdots & 0 \\ 0 & 0 & 3 & -4 & \vdots & 11 \end{bmatrix}$
4. $\begin{bmatrix} 3 & -1 & -2 & \vdots & 0 \\ 2 & -1 & -1 & \vdots & 0 \\ 1 & 2 & -3 & \vdots & 0 \end{bmatrix}$

5. $\begin{bmatrix} 1 & 3 & 1 & \vdots & 2 \\ 3 & 4 & -1 & \vdots & 1 \\ 1 & -2 & -3 & \vdots & 1 \end{bmatrix}$
6. $\begin{bmatrix} 1 & -1 & 1 & 1 & \vdots & 0 \\ 2 & 1 & -1 & 1 & \vdots & 0 \\ 4 & -1 & 1 & 3 & \vdots & 0 \end{bmatrix}$

7. $\begin{bmatrix} 1 & -1 & -2 & 3 & \vdots & 1 & 0 & 1 \\ 1 & 2 & 1 & -2 & \vdots & 1 & 0 & -1 \\ 2 & 1 & -1 & 1 & \vdots & 2 & 0 & 1 \end{bmatrix}$
8. $\begin{bmatrix} 2 & 4 & 6 & 3 & \vdots & 1 & 0 & 1 \\ -4 & -8 & 8 & 4 & \vdots & 3 & 0 & 1 \\ 2 & 4 & 26 & 13 & \vdots & 6 & 0 & 1 \end{bmatrix}$

9. $\begin{bmatrix} 1 & -1 & 2 & 0 & 0 & \vdots & 1 \\ 2 & -2 & 4 & 1 & 0 & \vdots & 5 \\ 3 & -3 & 6 & -1 & 1 & \vdots & -2 \end{bmatrix}$
10. $\begin{bmatrix} 3 & 12 & 0 & 10 & \vdots & 16 \\ -5 & -20 & 1 & -17 & \vdots & -26 \\ 1 & 4 & 0 & 3 & \vdots & 3 \end{bmatrix}$

11. $\begin{bmatrix} 0 & 2 & 1 & 1 & 0 & \vdots & 0 & 1 & 0 \\ 2 & 4 & 4 & 2 & 2 & \vdots & 0 & 1 & 0 \\ 3 & 6 & 6 & 0 & 0 & \vdots & 0 & 1 & 0 \\ 0 & -2 & -1 & -2 & 2 & \vdots & 1 & 0 & 0 \\ 0 & 2 & 1 & 2 & 4 & \vdots & 1 & 0 & 0 \end{bmatrix}$ (recall Example 3)

12. *Prove*: if matrix **A** is changed into matrix **B** by a sequence of elementary row operations, then there is another sequence of elementary row operations that changes **B** back into **A**. How is the second sequence related to the first?

In Problems 13–15, determine what values of α (if any) will make the following systems have no solution, one solution, an infinity of solutions.

13. $\begin{bmatrix} 1 & 1 & 1 & \vdots & 1 \\ -2 & 7 & 4 & \vdots & \alpha \\ 0 & 3 & 2 & \vdots & 2 \end{bmatrix}$ **14.** $\begin{bmatrix} 2 & 0 & \alpha & \vdots & 2 \\ 1 & 1 & 1 & \vdots & 1 \\ 4 & -2 & 7 & \vdots & 4 \end{bmatrix}$

15. $\begin{bmatrix} 2 & 1 & 2 & \vdots & 1 \\ 2 & 2 & \alpha & \vdots & 1 \\ 4 & 2 & 4 & \vdots & 1 \end{bmatrix}$

16. To see how an equation, exposed as redundant by Gauss elimination, is implied by the other equations, follow these steps:

(a) *Prove*: if the forward part of the Gauss elimination algorithm can be performed without pivoting, then every row of the final augmented coefficient matrix equals the corresponding row in the original matrix, plus a sum of multiples of the preceding rows in the original matrix. How is this statement modified if pivoting is performed?

(b) Show how, by keeping track of the multipliers and pivot exchanges employed in the elimination process, one can explicitly express the final rows in terms of the original rows. (*Hint*: It may be helpful for you to trace the elimination steps for a particular example such as Example 3 or 4.)

(c) Apply your method to Example 1 to derive the statement, made in the text, that the original fourth row equals three times the sum of the first and third rows.

17. Express the solution set for (5) with x_1 as the parameter. (*Hint*: It may be easiest to start from (8).)

18. To prove that two different parametrizations, such as (12) and (15), describe the same solution set, one must show that (i) for any value of s in (15), a value can be found for t in (12) so that the two solutions are the same and (ii) conversely, for any t in (12) an s can be found for (15) that makes the solutions match. One way to do this is to consider the system formed by equating the corresponding parametric expressions for each variable.

$$-1/3 - t = -2/3 + 2s$$
$$1/6 - t/2 = s$$
$$-1/3 = -1/3$$
$$1/3 = 1/3$$

Now the parametrizations are equivalent if this system is consistent (i.e., has solutions) both when t is considered as the unknown (and s known but arbitrary) and when s is considered as the unknown. Thus, the equivalence can be established by two applications of the forward part of the Gauss algorithm.

(a) Carry this out to prove that (12) and (15) are equivalent parametrizations.

(b) Show that the following parametrization (see Problem 17) is equivalent to (12):

$$x_1 = r$$
$$x_2 = 1/3 + r/2$$
$$x_3 = -r - 1/3$$
$$x_4 = -1/3$$
$$x_5 = 1/3.$$

(c) Show that the following parametrization is not equivalent to (12).

$$x_1 = -2/3 + u$$
$$x_2 = u$$
$$x_3 = 1/2 - 2u$$
$$x_4 = -1/3$$
$$x_5 = 1/3.$$

Exhibit a particular solution to (5) which cannot be described by the above parametrization.

(d) Show that the following parametrization is equivalent to (16).

$$x_1 = s_1$$
$$x_2 = 5/4 - s_1/2 - s_2/4$$
$$x_3 = s_2$$
$$x_4 = s_3$$
$$x_5 = -1/2 + s_2/2 + s_3$$
$$x_6 = 1.$$

19. Without performing any computations, decide whether the following systems have no solution, one solution, or infinitely many solutions.

(a)
$$\begin{bmatrix} 2 & 0 & 1 & 1 & \vdots & \pi \\ 0 & 3/2 & 5 & 22 & \vdots & 0 \\ 0 & 0 & 22 & -3 & \vdots & 8 \end{bmatrix}$$
(b)
$$\begin{bmatrix} 1 & -9 & 6 & \vdots & 13 \\ 0 & 1 & 9 & \vdots & 9 \\ 0 & 0 & 1 & \vdots & -7 \\ 0 & 0 & 0 & \vdots & 0 \end{bmatrix}$$

(c)
$$\begin{bmatrix} 1 & 8 & 94 & \vdots & 0 \\ 0 & 1 & 32 & \vdots & 12 \\ 0 & 0 & 1 & \vdots & 0 \\ 0 & 0 & 0 & \vdots & 0 \\ 0 & 0 & 0 & \vdots & -6 \end{bmatrix}$$
(d)
$$\begin{bmatrix} 3 & 9 & -6 & 0 & 0 & \vdots & 0 \\ 0 & 2 & 1 & 7 & -3 & \vdots & 0 \\ 0 & 0 & 0 & 2 & 8 & \vdots & 6 \\ 0 & 0 & 0 & 0 & 3 & \vdots & 4 \\ 0 & 0 & 0 & 0 & 0 & \vdots & 0 \end{bmatrix}$$

20. Let's return to the Antarctic expedition dietary problem in the "Introduction to Part I". We formulated equations (1) and (2) that said the chocolate/almond purchases should exactly match the caloric requirements. But in fact the stated requirements are *minimum* consumption levels; there is no harm in exceeding them. Thus, equations (1) and (2) are more honestly expressed as inequalities:

$$1x_1 + 2.75x_2 \geq 2300 \text{ (at least 2300 ounces of chocolate)}$$
$$0.4x_1 + 2x_2 \geq 1100 \text{ (at least 1100 ounces of almonds)}.$$

Now if we introduce the variable x_3 which measures the "slack" in the first inequality, i.e., the extent to which its left member exceeds its right, then we can replace the inequality by an equation

$$1x_1 + 2.75x_2 - x_3 = 2300$$

with the understanding that the slack variable x_3 is nonnegative. Similarly, the second inequality is rewritten using the nonnegative slack variable x_4:

$$0.4x_1 + 2x_2 - x_4 = 1100, \ x_4 \geq 0.$$

(a) Show that the solutions to these equations can be expressed parametrically as

$$x_1 = 1750 + \frac{20}{9}x_3 - \frac{55}{18}x_4, \quad x_2 = 200 - \frac{4}{9}x_3 + \frac{10}{9}x_4.$$

(b) Show that the cost of the order is

$$1.50x_1 + 3.75x_2 = 3375 + \frac{15}{9}x_3 - \frac{5}{12}x_4.$$

(c) Keeping in mind that the slack variables x_3 and x_4 must be nonnegative, argue that the cost can be decreased by choosing $x_3 = 0$ and making x_4 as large as possible. Since the purchases x_1 and x_2 cannot be negative, the parametric equation for x_1 limits x_4. Show that the minimum cost is \$3136.36 (rounded to cents), in accordance with the comment in the "Introduction to Part I".

1.4   ECHELON FORM AND RANK

To get an overview on how Gauss elimination works in general, it is instructive to consider the structure of the augmented coefficient matrix after the forward phase has been completed, i.e., just prior to back substitution. The forms of the matrices at this stage, for the examples that we have studied in this chapter, are summarized below. Only the pivot elements and the zeros corresponding to "eliminated" elements are written out, since the other entries have no bearing on the structure. (In other words, the X's may, or may not, be zero.)

$$
(a) \quad
\begin{bmatrix}
0.202131 & X & X & X & \vdots & X \\
0 & -1.32101 & X & X & \vdots & X \\
0 & 0 & 1.63823 & X & \vdots & X \\
0 & 0 & 0 & 0.388032 & \vdots & X
\end{bmatrix}
$$

$$
(b) \quad
\begin{bmatrix}
2 & X & X & X & \vdots & X \\
0 & 3/2 & X & X & \vdots & X \\
0 & 0 & 22 & X & \vdots & X
\end{bmatrix}
\qquad
(c) \quad
\begin{bmatrix}
1 & X & X & \vdots & X \\
0 & 1 & X & \vdots & X \\
0 & 0 & 1 & \vdots & X \\
0 & 0 & 0 & \vdots & 0
\end{bmatrix}
$$

$$
(d) \quad
\begin{bmatrix}
1 & X & X & \vdots & X \\
0 & 1 & X & \vdots & X \\
0 & 0 & 1 & \vdots & 0 \\
0 & 0 & 0 & \vdots & 0 \\
0 & 0 & 0 & \vdots & -1
\end{bmatrix}
\qquad
(e) \quad
\begin{bmatrix}
3 & X & X & X & X & \vdots & X \\
0 & 2 & 1 & X & X & \vdots & X \\
0 & 0 & 0 & 2 & X & \vdots & X \\
0 & 0 & 0 & 0 & 3 & \vdots & X \\
0 & 0 & 0 & 0 & 0 & \vdots & 0
\end{bmatrix}
$$

$$
(f) \quad
\begin{bmatrix}
1 & X & X & X & X & X & \vdots & X \\
0 & 0 & -1 & X & X & X & \vdots & X \\
0 & 0 & 0 & 0 & 0 & 1 & \vdots & X
\end{bmatrix}
$$

The first matrix displays the form when there are no "complications"—same number of equations as unknowns, no zero pivots. The coefficient submatrix is upper triangular and the augmented matrix is upper trapezoidal. However, the subsequent examples reveal that the arrangement will be more complicated than this, in general. Ignoring the right-hand sides, we see that the zero entries of the coefficient matrix exhibit a staircase, or *echelon*, structure, with the steps occurring at the pivot elements.

Row-Echelon Form

> **Definition 5.** Let **A** be an m-by-n matrix with the following property: starting from the second row, the first nonzero entry in each row lies to the right of the first nonzero entry in the preceding row. Then **A** is said to be a **row-echelon matrix**, or a matrix in row-echelon form.[6]

Note that all the *coefficient* matrices that appear in (a) through (f) above are in row-echelon form and so are the augmented matrices except for (d). Indeed, one more row switch will bring (d) into conformity:

$$\text{(d')} \quad \begin{bmatrix} 1 & X & X & \vdots & X \\ 0 & 1 & X & \vdots & X \\ 0 & 0 & 1 & \vdots & 0 \\ 0 & 0 & 0 & \vdots & -1 \\ 0 & 0 & 0 & \vdots & 0 \end{bmatrix}$$

Using the concept of row-echelon form, we summarize our equation-solving procedure as follows.

Synopsis of the Gauss Elimination Algorithm

> In the forward part of the algorithm, elementary row operations are performed on the augmented coefficient matrix to reduce the coefficient submatrix to row-echelon form. Within the (resulting) coefficient submatrix, the first nonzero entry in each row is a pivot element and its corresponding variable is labeled a *pivot variable*; all other variables are *free variables*. The rows of the augmented matrix that state $0 = 0$ are ignored. If any row states $0 = c$ and c is nonzero, the system is inconsistent. Otherwise one proceeds with back substitution, solving the

---

[6]The corresponding form for the Gauss-Jordan algorithm is often called the **row-reduced echelon form**. It differs from row-echelon form in that the entries *above*, as well as below, the first nonzero entry in each row are zero, and this nonzero entry is 1.

equations in succession from the last to the first and expressing the pivot variables in terms of the free variables.

For future reference, we record some useful facts that arise as "spinoffs" of the Gauss process. Theorem 2 summarizes our deliberations to this point.

Reduction to Row-Echelon Form

Theorem 2. Any matrix can be reduced to row-echelon form[7] by a sequence of elementary row operations.

Rank of Row-Echelon Matrices

Definition 6. If a matrix is in row-echelon form, we say that its *rank* equals the number of nonzero rows.[8]

Note that rank of a row-echelon matrix cannot exceed the number of its columns.

Suppose that a linear system of equations with coefficient matrix $\mathbf{A}$ is represented by the augmented matrix $[\mathbf{A}|\mathbf{b}]$ and let $[\mathbf{A}^{ech}|\mathbf{b}']$ denote the resulting row-echelon matrix after appropriate row operations have been performed on the original (augmented) matrix. From the ranks of $[\mathbf{A}^{ech}|\mathbf{b}']$ and $\mathbf{A}^{ech}$, we can glean some important information about the numbers of solutions for the system. For example, the inconsistency of the system (d') can be attributed to the fact that after reduction to row-echelon form the rank of the augmented matrix exceeds the rank of the coefficient matrix.

Furthermore, generalizing from the matrices in (a) and (c), we observe that if a system is consistent, it has a *unique* solution if the number of columns of $\mathbf{A}^{ech}$ (i.e., the number of unknowns) equals its rank; in this case we say the coefficient matrix $\mathbf{A}^{ech}$ has *full rank*. (See, for example, the matrices (a) and (c).) But the matrices in (b), (e), and (f) demonstrate that if the system is consistent, it has an infinite number of solutions whenever the number of columns of $\mathbf{A}^{ech}$ exceeds its rank. To summarize:

Characterization of Number of Solutions of a Linear System by Rank

Theorem 3. For a system of linear equations represented by the augmented matrix $[\mathbf{A}|\mathbf{b}]$, let $[\mathbf{A}^{ech}|\mathbf{b}']$ be an echelon form resulting from applying elementary row operations to $[\mathbf{A}|\mathbf{b}]$.

---

[7]Or row-reduced echelon form.
[8]In Section 3.3, we will generalize this definition to all matrices.

- If the rank of $\left[\mathbf{A}^{\text{ech}}|\mathbf{b}'\right]$ exceeds the rank of $\mathbf{A}^{\text{ech}}$, the linear system is inconsistent (has no solutions).
- If the ranks of $\left[\mathbf{A}^{\text{ech}}|\mathbf{b}'\right]$ and $\mathbf{A}^{\text{ech}}$ are equal, then the linear system is consistent, having

  (i) exactly one solution when the (common) rank equals the number of unknowns (which is the same as the number of columns of $\mathbf{A}^{\text{ech}}$);

  (ii) infinitely many solutions if the (common) rank is less than the number of unknowns.

Example 1. A linear system of equations in x_1, x_2, x_3, x_4, and x_5 is represented by the augmented coefficient matrix

$$\begin{bmatrix} 3 & 9 & -6 & 0 & 0 & \vdots & 1 \\ 0 & 2 & 7 & 8 & 3 & \vdots & 5 \\ 0 & 0 & 0 & 2 & 6 & \vdots & 9 \\ 0 & 0 & 0 & 0 & 0 & \vdots & 0 \end{bmatrix}. \tag{1}$$

Verify that

$$\# \text{ columns } \mathbf{A}^{\text{ech}} = \text{rank} \left[\mathbf{A}^{\text{ech}}|\mathbf{b}'\right] + \# \text{ free parameters} \tag{2}$$

(the last term is the number of free parameters in the solution set).

Solution. The number of nonzero rows in (1) is clearly 3; so 3 is the rank. Back substitution would dictate that x_5 is free, x_4 is not, x_3 is free, and x_2 and x_1 are not; the number of free parameters is 2. Therefore, the right-hand member of (2) is $3 + 2 = 5$, which is the number of columns of $\mathbf{A}^{\text{ech}}$.   ∎

In fact, equation (2) holds for *all* consistent systems in row-echelon form. (See Problem 16.) Section 3.3 will establish an important generalization of this equation.

A useful spinoff of Theorem 3 occurs when the system is homogeneous, in accordance with the following:

Linear Homogeneous System

Definition 7. A system of m equations in n unknowns that can be put in the form

$$
\begin{aligned}
a_{11}x_1 + a_{12}x_2 + \cdots + a_{1n}x_n &= 0 \\
a_{21}x_1 + a_{22}x_2 + \cdots + a_{2n}x_n &= 0 \\
&\vdots \\
a_{m1}x_1 + a_{m2}x_2 + \cdots + a_{mn}x_n &= 0,
\end{aligned}
\tag{3}
$$

where the right-hand sides are all zero, is called a **linear homogeneous system**.[9]

In two (three) dimensions, a linear homogeneous system represents lines (planes) passing through the origin.

The augmented matrix notation for a homogeneous system with coefficient matrix **A** is $[\mathbf{A} \mid \mathbf{0}]$. Notice that a homogeneous system is always consistent since its solution(s) include $x_1 = x_2 = \cdots = x_n = 0$, the **zero solution** (or "trivial solution"). Therefore, part (ii) of Theorem 3 guarantees the following:

Corollary 1. A linear homogeneous system containing fewer equations than unknowns has an infinite number of solutions.

Exercises 1.4

In Problems 1–6, determine whether the given matrix is in row-echelon form. If so, determine its rank.

1. $\begin{bmatrix} 0 & 1 & 3 \\ 0 & 0 & 0 \\ 0 & 0 & 0 \end{bmatrix}$
   2. $\begin{bmatrix} 2 & 0 & 0 \\ 0 & 0 & 0 \\ 0 & 0 & 1 \end{bmatrix}$
   3. $\begin{bmatrix} 4 & 6 & 3 & 1 \\ 0 & 0 & 1 & 0 \\ 0 & 0 & 0 & 1 \end{bmatrix}$
   4. $\begin{bmatrix} 0 & 3 & 0 & 0 \\ 1 & 0 & 0 & 0 \\ 0 & 0 & 0 & 0 \end{bmatrix}$

5. $\begin{bmatrix} 2 & 0 & 1 & 8 \\ 0 & 1 & 2 & 3 \end{bmatrix}$
   6. $\begin{bmatrix} 3 & 0 \\ 0 & 1 \\ 0 & 0 \end{bmatrix}$

In Problems 7–12, an augmented matrix $[\mathbf{A} \mid \mathbf{b}]$ for a linear system with coefficient matrix $\mathbf{A}$ is given. Determine whether the system is consistent and, if so, whether the system has a unique solution or infinitely many solutions.

7. $\begin{bmatrix} 4 & 0 & 9 & \vdots & 1 \\ 0 & 0 & 3 & \vdots & 0 \end{bmatrix}$
   8. $\begin{bmatrix} 0 & 1 & 5 & \vdots & 0 \\ 0 & 0 & 2 & \vdots & 3 \end{bmatrix}$
   9. $\begin{bmatrix} 0 & 8 & 7 & \vdots & 4 \\ 0 & 0 & 0 & \vdots & 0 \\ 0 & 0 & 0 & \vdots & 0 \end{bmatrix}$

---

[9]Otherwise the system is said to be **nonhomogeneous**.

10. $\begin{bmatrix} 2 & 3 & 8 & \vdots & 9 \\ 0 & 1 & 5 & \vdots & 2 \\ 0 & 0 & 0 & \vdots & 4 \end{bmatrix}$

11. $\begin{bmatrix} 1 & 4 & 9 & 2 & \vdots & 0 \\ 0 & 3 & 6 & 9 & \vdots & 1 \\ 0 & 0 & 4 & 4 & \vdots & 2 \\ 0 & 0 & 0 & 1 & \vdots & 0 \end{bmatrix}$ **12.** $\begin{bmatrix} 2 & 3 & \vdots & 9 \\ 0 & 8 & \vdots & 2 \\ 0 & 0 & \vdots & 0 \\ 0 & 0 & \vdots & 0 \end{bmatrix}$

13. For what values of α will the following homogeneous systems have *nontrivial* solutions?

(a) $\begin{aligned} 2x_1 + 5x_2 + \alpha x_3 &= 0 \\ x_1 - 6x_2 + x_3 &= 0 \end{aligned}$ (b) $\begin{aligned} x_1 + x_2 + x_3 &= 0 \\ \alpha x_1 + 4x_2 + 3x_3 &= 0 \\ 4x_1 + 3x_2 + 2x_3 &= 0 \end{aligned}$

14. Show that the homogeneous system

$$a_{11}x_1 + a_{12}x_2 = 0$$
$$a_{21}x_1 + a_{22}x_2 = 0$$

has infinitely many solutions if and only if $a_{11}a_{22} - a_{12}a_{21} = 0$.

15. For the systems of equations represented by the given augmented matrices, verify that equation (2) holds.

(a) $\begin{bmatrix} 1 & 2 & 3 & \vdots & 0 \\ 0 & 5 & 6 & \vdots & 1 \\ 0 & 0 & 7 & \vdots & 2 \end{bmatrix}$ (b) $\begin{bmatrix} 3 & 5 & 0 & 0 & 0 & 1 & \vdots & 2 \\ 0 & 0 & 1 & 3 & 2 & 5 & \vdots & 0 \\ 0 & 0 & 0 & 0 & 0 & 1 & \vdots & 2 \end{bmatrix}$

(c) $\begin{bmatrix} 3 & 0 & 1 & 0 & 0 & 0 & \vdots & 2 \\ 0 & 0 & 1 & 0 & 0 & 5 & \vdots & 0 \\ 0 & 0 & 0 & 0 & 0 & 1 & \vdots & 2 \end{bmatrix}$

16. Prove that if $[\mathbf{A}^{\text{ech}}|\mathbf{b}]$ is *any* consistent system, then equation (2) holds.

Problems 17–22 involve the sum or difference of two n-by-1 column matrices as well as the scalar multiple of a column vector. Recall from calculus that

$$\begin{bmatrix} x_1 \\ x_2 \\ \vdots \\ x_n \end{bmatrix} \pm \begin{bmatrix} y_1 \\ y_2 \\ \vdots \\ y_n \end{bmatrix} = \begin{bmatrix} x_1 \pm y_1 \\ x_2 \pm y_2 \\ \vdots \\ x_n \pm y_n \end{bmatrix} \quad and \quad c \begin{bmatrix} x_1 \\ x_2 \\ \vdots \\ x_n \end{bmatrix} = \begin{bmatrix} cx_1 \\ cx_2 \\ \vdots \\ cx_n \end{bmatrix}.$$

17. If $\mathbf{A}$ is an *m*-by-*n* matrix and $\mathbf{x}$ and $\mathbf{y}$ are each *n*-by-1 column vector solutions to the homogeneous system $[\mathbf{A} \mid \mathbf{0}]$, show that $\mathbf{x} \pm \mathbf{y}$ is also a solution.

18. If **A** is an *m*-by-*n* matrix and **x** and **y** are each *n*-by-1 column vector solutions to the *non*homogeneous system [**A** | **b**] with **b** ≠ **0**, show that neither **x** ± **y** is a solution. Give an example. What system *does* **x** + **y** solve?

19. If **A** is an *m*-by-*n* matrix and **x** is an *n*-by-1 column vector solution to the homogeneous system [**A** | **0**], show that *c***x** is also a solution, for any scalar *c*.

20. If **A** is an *m*-by-*n* matrix and **x** is an *n*-by-1 column vector solution to the *non*homogeneous system [**A** | **b**] with **b** ≠ **0**, show that *c***x** is *not* a solution unless *c* = 1. Give an example.

21. If **A** is an *m*-by-*n* matrix and **x** and **y** are each *n*-by-1 column vector solutions to the *non*homogeneous system [**A** | **b**], show that **x** − **y** is a solution to the corresponding homogeneous system [**A** | **0**].

22. If **A** is an *m*-by-*n* matrix and **x** is an *n*-by-1 column vector solution to the nonhomogeneous system [**A** | **b**] and **y** is a solution to the corresponding *homogeneous system* [**A** | **0**], show that **x** + **y** is, again, a solution to [**A** | **b**].

23. (For physics and engineering students: see Group Project D: Kirchhoff's Laws).

    The 6 unknown currents I_1 through I_6 in Figure 1.3 must satisfy Kirchhoff's Laws (Kirchhoff's Current Law, **KCL**; Kirchhoff's Voltage Law, **KVL**). With hindsight, we formulate the system of equations as follows:

$$
\begin{array}{lrcl}
\textbf{KCL at the left node:} & I_1 - I_2 + I_3 & & = 0 \\
\textbf{KCL at the right node:} & I_2 \quad - I_4 - I_5 + I_6 & & = 0 \\
\textbf{KVL around loop \#1:} & 5I_3 & & = 5 \\
\textbf{KVL around loop \#2:} & 3I_4 & & = 7 - 6 = 1 \\
\textbf{KVL around loop \#3:} & & 2I_6 & = 6 \\
\textbf{KVL around loop \#4:} & 5I_3 & -2I_6 & = 0
\end{array}
$$

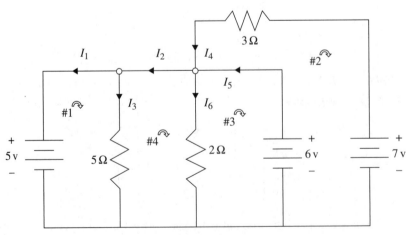

Fig. 1.3   Circuit for Problem 23.

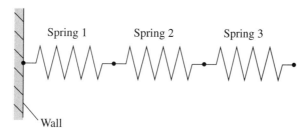

Fig. 1.4   Spring system for Problem 24.

(a) Verify these equations.

(b) Reduce this system to echelon form and show that it is inconsistent.

(c) Show that if the 5 V battery is changed to a 6 V battery, the system becomes consistent but indeterminant having an infinite number of solutions parametrized by

$$I_1 = -\frac{58}{15} + t, \; I_2 = -\frac{8}{3} + t, \; I_3 = \frac{6}{5}, \; I_4 = \frac{1}{3}, \; I_5 = t, \; I_6 = 3.$$

(d) Most experienced engineers would instantly recognize the conflict in the first circuit; the 5 V battery is wired in parallel to the 6 V battery. Changing the 5 V battery to 6 V makes one of them redundant and an indeterminant amount of current can circulate in the battery loop (I_5, I_2, and I_1). How is this revealed in the solution parametrization?

24. Each spring shown in Figure 1.4 satisfies *Hooke's Law* when it is compressed:

{compressive force (N)} = {compression (m)} *times* {spring constant (N/m)}

The spring constants are given by $k_1 = 5,000$ N/m, $k_2 = 10,000$ N/m, and $k_3 = 20,000$ N/m. If the three-spring system shown in Figure 1.4 is compressed 0.01 m, how much is each spring compressed? What force is required for this compression?

25. The three springs in Problem 24 are reconfigured as in Figure 1.5. Suppose the rigid bar is uniform and its weight exerts a force of 650 N and suppose further that the springs are 1 m apart. The compressions x_i of each spring, and the forces F_i that they exert, are as indicated.

Force equilibrium requires that the sum of the F_i equals 650 N. Torque equilibrium requires that $F_1 = F_3$. And geometric compatibility requires that x_2 equals the average of x_1 and x_3.

(a) Express these conditions in terms of x_i and solve.

(b) Now suppose that the springs are *rigid*; the spring constants are infinite, and the compressions are each zero (so that geometric compatibility is automatic). Rewrite the equilibrium equations in terms of F_i and show that there are infinitely many solutions.

Mechanical engineers say the rigidly supported system is *statically indeterminate*.

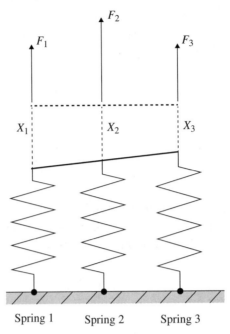

Fig. 1.5   Spring system for Problem 25.

1.5   COMPUTATIONAL CONSIDERATIONS

The Gauss elimination algorithm is ideally suited for automatic implementation. The basic computations are very simple and repetitive—multiply and add, multiply and add, ... Indeed, the calculation sequence is so routine that its human execution can be boring and prone to silly mistakes. Practically all computers sold today, and the better hand-held programmable calculators, have Gauss elimination codes available (although most computers implement Gauss elimination through computationally equivalent *matrix factorization codes*). The key advantages of the algorithm are its versatility, efficiency, and accuracy, as we shall see.

Gauss elimination and electronic computation share a rather fascinating history. Here we shall give only a brief sketch, but the interested reader can find names and dates in the reference list at the end of the section. The solution technique, of course, is quite elementary, and a form of it was known to the Chinese in 250 BC. Gauss's real contribution was essentially nothing more than the explicit, systematic, and efficient formulation of the computation sequence.

The solution of systems of simultaneous linear equations is the problem most frequently encountered in scientific computations. Consequently it was no surprise that Gauss elimination would be one of the first tasks for the "new" electronic computers of the 1940s. Soon the effects of roundoff errors began to draw increasing attention because digital computers cannot represent all numbers exactly. (If we want $1/3$, for example, we have to settle for something like 0.333333 in a decimal machine, or 0.0101010101010101 in a binary machine. Consequently we cannot obtain an exact solution to even a trivial equation like $3x = 1$; worse yet, we can't even read the equation $(1/3)x = 1$ into the computer.)

By the 1950s, John von Neumann and Herbert Goldstine published a very elaborate analysis of the cumulative effects of roundoff error and it was interpreted as stating that the accurate solution of large systems was, essentially, hopeless. This pessimistic result cast a pall of gloom on the subject until the early 1960s when John Wilkinson discovered the proper way to look at the conclusions. His *backward error analysis* showed that the Gauss elimination algorithm, coupled with the pivoting strategy which we discuss below, gives answers as accurate as one can reasonably ask.

Later interest focused on the *speed* of the algorithm. Many alternative, and faster, procedures have been developed for solving systems possessing special properties. For example, the conversion of red/blue/green levels in computer displays to the JPEG 2000 format, the computation of the discrete Fourier transform, and the solution of systems whose coefficient matrices have lots of zeros or predominantly large diagonal entries are accomplished by special algorithms tailored to the specific situation. Indeed, Gauss elimination itself is customarily coded as a computationally equivalent "matrix factorization" scheme, largely because the latter can be adapted to take advantage of symmetry in the coefficient matrix when it is present.[10] However, the basic algorithm is still regarded as the most efficient, practical, general purpose linear equation solver (if its memory requirements can be met). In the past, many mathematicians speculated that it was the best possible, but the emergence of some radical new designs and concepts in computer science undermined this conviction. However, these procedures appear too complicated to displace Gauss elimination as the method of choice.

Most of these topics are best reserved for advanced courses in numerical linear algebra or computational complexity. The textbooks by Strang, Stewart, and Watkins (listed in the references) are quite readable and suitable for self study. As we mentioned, however, there is one feature of Gauss elimination that we will consider here because of its importance in digital computer implementation: the pivoting strategy.

Consider the linear system

$$\begin{bmatrix} 0.000001 & 1 & \vdots & 1 \\ -1 & 1 & \vdots & 2 \end{bmatrix}. \tag{1}$$

Hand-calculated Gauss elimination produces

$$\begin{bmatrix} 0.000001 & 1 & \vdots & 1 \\ 0 & 1,000,001 & \vdots & 1,000,002 \end{bmatrix} \tag{2}$$

[10]See Group Project A, Part I.

and back substitution yields the (easily verified) answers

$$x_2 = \frac{1,000,002}{1,000,001} \quad (\approx 1)$$

$$x_1 = -\frac{1,000,000}{1,000,001} (\approx -1). \tag{3}$$

Now consider how a typical modern computer, which rounds all data to (say) 6 digits would solve (1).[11] Instead of (2) after the elimination, the numbers would be rounded by the machine to

$$\begin{bmatrix} 1 \times 10^{-6} & 1 & \vdots & 1 \\ 0 & 1.00000 \times 10^6 & \vdots & 1.00000 \times 10^6. \end{bmatrix} \tag{4}$$

Back substitution now yields

$$x_2 = 1$$
$$x_1 = 0 \quad (!) \tag{5}$$

This is awful! For such a simple problem, we should expect better results from a six-digit computer. But the roundoff that took place in (4) doomed the computation.

The source of the problem is the oversized factor, 10^6, that multiplied the first equation in (1) when eliminating x_1 from the second. The coefficients in the first equation were magnified so greatly that they "swamped out" the least significant digits "1" and "2" in the original second equation (2); there is no trace of these digits in the resulting system (4). In effect, *most of the information in the original second equation was simply overwhelmed by the addition of 10^6 times the first equation.*

But look what happens if we first reorder the equations. The reordered system is

$$\begin{bmatrix} -1 & 1 & \vdots & 2 \\ 0.000001 & 1 & \vdots & 1 \end{bmatrix}. \tag{6}$$

Now the multiplier is 10^{-6} and Gauss elimination results in

$$\begin{bmatrix} -1 & 1 & \vdots & 2 \\ 0 & 1.000001 & \vdots & 1.000002 \end{bmatrix}, \tag{7}$$

---

[11]Of course most computers do not use a base 10 number system, but the overall effect is the same, so we stick with familiar decimal examples.

which the computer rounds to

$$\begin{bmatrix} -1 & 1 & \vdots & 2 \\ 0 & 1 & \vdots & 1 \end{bmatrix}. \tag{8}$$

Once again the "1" and "2" in the original system (1) are overwhelmed when the second equation is combined with the first, but they are not entirely lost—they are still present in the first equation. And the resulting back substitution produces a perfectly acceptable approximation to the exact answers (3):

$$x_2 = 1$$
$$x_1 = -1. \tag{9}$$

Generalizing from this example, we can see that the occurrence of outsized multipliers in the algorithm can lead to loss of accuracy on digital machines but that a reordering of the equations can keep the multipliers within bounds. Specifically, the first equation should always be the one with the largest coefficient of x_1 (in magnitude), and at later stages the equation that is used to eliminate x_i from the remaining equations should be the one with the largest coefficient of x_i. When pivoting is used to achieve this, the technique is known among numerical analysts as "partial pivoting" (sometimes the word "partial" is merely understood. A seldom used variation called complete pivoting is described in the Problem 5.) It ensures that none of the multipliers ever exceeds 1 in magnitude. A pivoting strategy is an integral part of every respectable computer code for Gauss elimination. Clearly partial pivoting is compatible with the customary reordering necessitated by the occurrence of zeros.

For this reason, you may have difficulty checking some of your hand-calculated answers with your computer. By hand, one would probably reduce $\mathbf{A} = \begin{bmatrix} \frac{1}{2} & 1 \\ 1 & 1 \end{bmatrix}$ to row-echelon form by subtracting twice the first row from the second, resulting in $\begin{bmatrix} \frac{1}{2} & 1 \\ 0 & -1 \end{bmatrix}$. However, good software would switch the rows first, and then subtract half the (new) first row from the (new) second, producing $\begin{bmatrix} 1 & 1 \\ 0 & \frac{1}{2} \end{bmatrix}$ as the row-echelon form. After back solving, of course, the two final solutions to a system $\mathbf{A}x = \mathbf{b}$ should agree.

In coding a pivoting strategy, it is important to note that when rows of a matrix are switched, it is not necessary actually to switch the location of all the entries in computer memory (any more than it is necessary to keep recopying equations when solving systems by hand!). All that is needed is to exchange the *addresses* and this can easily be handled by intelligent coding.

Example 1. How would a computer that rounds to 12 digits implement Gauss elimination for the two systems represented by

$$
\begin{bmatrix} 2 & 4/3 & \vdots & 2 & 2 \\ 1 & 2/3 & \vdots & 1 & 2 \end{bmatrix} ?
\tag{10}
$$

Solution. Before proceeding, notice that *exact* Gauss elimination shows that the first system has an infinite number of solutions and the second has none:

$$
\begin{bmatrix} 2 & 4/3 & \vdots & 2 & 2 \\ 0 & 0 & \vdots & 0 & 1 \end{bmatrix} .
\tag{11}
$$

The original system (10), rounded to 12 digits, reads

$$
\begin{bmatrix} 2 & 1.333333333333 & \vdots & 2 & 2 \\ 1 & 0.666666666667 & \vdots & 1 & 2 \end{bmatrix} .
\tag{12}
$$

When the computer divides 1.3333333333 by 2 and retains 12 places, it gets 0.666666666665. So subtracting half the first row from the second results in the row-echelon form

$$
\begin{bmatrix} 2 & 1.333333333333 & \vdots & 2 & 2 \\ 0 & 2 \times 10^{-12} & \vdots & 0 & 1 \end{bmatrix} .
\tag{13}
$$

The computer concludes that the second equation is neither redundant nor inconsistent!

With back substitution, we readily see that this sophisticated machine finds only one of the infinite number of solutions of the consistent system and reports an absurd answer for the inconsistent system. ∎

An intelligent numerical analyst, of course, would recognize the entry 2×10^{-12} as "roundoff garbage" in (13) and reset this coefficient to zero before back substituting. For more complex systems, however, these decisions are not so transparent and a deeper understanding of the situation is necessary. Such matters are dealt with in specialized texts.

Top-grade matrix computation software will alert the user if it encounters situations like these, when the credibility of the computed answer should be scrutinized.

References on Computations

Of historical significance

1. Boyer, C.B. (1968). *A History of Mathematics*. Wiley, New York.

2. von Neumann, J. and Goldstine, H.H. (1947). Numerical inverting of matrices of high order. *Bull. Am. Math. Soc.* 53, 1021–1099.

3. Wilkinson, J.H. (1961). Error analysis of direct methods of matrix inversion. *J. Assoc. Comput. Mach.* 8, 281–330.

4. Strassen, V. (1969). Gauss elimination is not optimal. *Numerische Mathematika* 13, 354–356.

Classic textbooks on numerical linear algebra

5. Forsythe, G. and Moler, C.B. (1967). *Computer Solution of Linear Algebraic Systems.* Prentice-Hall, Upper Saddle River, NJ.

6. Householder, A.S. (1964). *The Theory of Matrices in Numerical Analysis.* Dover Publications, Mineola, NY.

7. Stewart, G.W. (1973). *Introduction to Matrix Computations.* Elsevier, Amsterdam, Netherlands.

8. Strang, G. (1980). *Linear Algebra and Its Applications* (fourth edition). Wellesley-Cambridge Press, Wellesley, MA.

9. Wilkinson, J.H. (1965). *The Algebraic Eigenvalue Problem.* Oxford University Press, London, UK.

Documentated computer codes

10. Wilkinson, J.H. and Reinsch, C., eds. (1971). *Handbook for Automatic Computation*, Vol. II, *Linear Algebra.* Springer-Verlag, New York.

11. Dongarra, B., Moer, S. (1979). *LINPACK.* IMSL, Houston, TX.

Recent textbooks

12. Golub, G.H. and van Loan, C.F. (2013). *Matrix Computations* (fourth edition). The Johns Hopkins University Press, Baltimore, MD.

13. Higham, N.J. (2013). *Accuracy and Stability of Numerical Algorithms* (second edition). Society for Industrial and Applied Mathematics, Philadelphia, PA.

14. Demmel, J. (1997). *Applied Numerical Linear Algebra.* Society for Industrial and Applied Mathematics, Philadelphia, PA.

15. Trefethen. L.N. and Bau, D. III (1997). *Numerical Linear Algebra.* Society for Industrial and Applied Mathematics, Philadelphia, PA.

16. Watkins, D.S. *Fundamentals of Matrix Computations* (third edition). Wiley-Interscience, New York, 2010.

Exercises 1.5

1. With a calculator, simulate Gauss elimination as performed on a 4-digit machine, with and without partial pivoting, for the following system. Compare both answers with the exact solution: $x_1 = 10.00$ and $x_2 = 1.000$.

$$0.003000x_1 + 59.14x_2 = 59.17$$
$$5.291x_1 - 6.130x_2 = 46.78$$

2. This problem analyzes pivoting from a theoretical, machine-free point of view.

   (a) Show that "symbolic" Gauss elimination generates the row-echelon form as shown:

$$
\begin{bmatrix} \varepsilon & 2 & \vdots & 2 \\ 1 & 1 & \vdots & 2 \end{bmatrix} \Rightarrow \begin{bmatrix} \varepsilon & 2 & \vdots & 2 \\ 0 & 1 - \dfrac{2}{\varepsilon} & \vdots & 2 - \dfrac{2}{\varepsilon} \end{bmatrix}, \tag{14}
$$

and derive the formula for the solution $x = x(\varepsilon), y = y(\varepsilon)$. Show that as $\varepsilon \to 0$ the solutions tend to $x = 1, y = 1$.

   (b) Argue that on any finite-precision machine executing the algorithm without pivoting, for sufficiently small ε, the computation indicated in (14) will be rendered as

$$
\begin{bmatrix} \varepsilon & 2 & \vdots & 2 \\ 1 & 1 & \vdots & 2 \end{bmatrix} \Rightarrow \begin{bmatrix} \varepsilon & 2 & \vdots & 2 \\ 0 & 1 - \dfrac{2}{\varepsilon} & \vdots & 2 - \dfrac{2}{\varepsilon} \end{bmatrix} \Rightarrow \begin{bmatrix} \varepsilon & 2 & \vdots & 2 \\ 0 & -\dfrac{2}{\varepsilon} & \vdots & -\dfrac{2}{\varepsilon} \end{bmatrix}, \tag{15}
$$

   resulting in the (unacceptable) solution $x = 0, y = 1$.

   (c) Argue that *with* pivoting the machine will generate

$$
\begin{bmatrix} \varepsilon & 2 & \vdots & 2 \\ 1 & 1 & \vdots & 2 \end{bmatrix} \Rightarrow \begin{bmatrix} 1 & 1 & \vdots & 2 \\ \varepsilon & 2 & \vdots & 2 \end{bmatrix} \Rightarrow \begin{bmatrix} 1 & 1 & \vdots & 2 \\ 0 & 2 - \varepsilon & \vdots & 2 - 2\varepsilon \end{bmatrix} \Rightarrow \begin{bmatrix} 1 & 1 & \vdots & 2 \\ 0 & 2 & \vdots & 2 \end{bmatrix}
$$

   and produce the (acceptable) solution $x = 1, y = 1$.

3. A healthy scotch and soda contains 290 ml of liquid, consisting of x_1 ml of lemon seltzer and x_2 ml of scotch. A careful calorimetric analysis reveals that the high-ball contains 252.8 kcal. One ml of lemon seltzer contains 0.004 kcal and 1 ml of scotch contains 2.8 kcal. Formulate two equations in x_1 and x_2 expressing (i) the calorimetric content and (ii) the volume of the highball. Using a calculator, simulate the solution of the equations via the Gauss algorithm, without pivoting, performed on a machine that retains 3 significant digits after each calculation. The repeat the simulation *with* pivoting.

   The exact solution is $x_1 = 200$ ml, $x_2 = 90$ ml. The machine retains 3 digit accuracy on a very brief calculation; yet the answer reported without pivoting is only accurate to 1 digit.

4. With a calculator, simulate Gauss elimination as performed on a 3-digit machine, with and without partial pivoting, for the following system. Compare both answers with the answer obtained using full calculator precision, then rounding.

$$\begin{bmatrix} 0.001 & 2 & 3 & \vdots & 1 \\ -1 & 3.71 & 4.62 & \vdots & 2 \\ -2 & 1.07 & 10.0 & \vdots & 3 \end{bmatrix}.$$

5. The effect of a large number "swamping out" a small number under addition, which plagued the calculation (4) when it was performed without pivoting, can still occur *with* pivoting, due to growth of the coefficients. Consider the family of systems defined by the n-row version of the augmented matrix below, where s is an integer satisfying $1 \le s < 2^{n-1}$:

$$\mathbf{M} = \begin{bmatrix} 1 & 0 & 0 & 0 & 1 & \vdots & 1 & \vdots & 0 & \vdots & 1 \\ -1 & 1 & 0 & 0 & 1 & \vdots & 1 & \vdots & 0 & \vdots & 1 \\ -1 & -1 & 1 & 0 & 1 & \vdots & 1 & \vdots & 0 & \vdots & 1 \\ -1 & -1 & -1 & 1 & 1 & \vdots & 1 & \vdots & 0 & \vdots & 1 \\ -1 & -1 & -1 & -1 & 1 & \vdots & 1 & \vdots & s & \vdots & 1+s \end{bmatrix}$$

Notice that the solution for the third right-hand side is the column sum of the solutions for the first two (so you can forego its calculation in the following).

(a) Perform the forward part of the Gauss elimination algorithm for the first two right-hand sides, in the case of $n = 5$ rows as shown. Note that the computation is very easy and that pivoting is "naturally" satisfied. Observe that the largest entry that occurs is 16 in the final (5,5) and (5,6) positions.

(b) Repeat part (a) for the version of this problem containing 6, 7, … rows until you can confidently conjecture the echelon form of the augmented matrix for the n-row case. You should see that the largest entries in the echelon form are 2^{n-1}, in positions (n, n) and $(n, n+1)$. What would be the largest entry in the third system?

(c) Back-solve the first two systems for 5, 6, … rows until you can confidently conjecture the form of the solutions for the n-row case. (Did you anticipate the obvious solution for the first system?) Observe that the $(n-1)$st entry of the answer for the third system is always $-s/2$ and is the largest entry in magnitude.

(d) Now consider solving the third system on a finite precision machine. There will be a value for n for which the computation of $2^{n-1} + s$ (in the final entry of the row-echelon form, part (b)) will "overflow;" that is, $2^{n-1} + s$ will be replaced by $2^{n-1} + s'$ for some $s' < s$. Thus, the machine will either stop and issue an error flag, or the row-echelon form will be in error, and back substitution will return the value $s'/2$ for the $(n-1)$st entry of the answer. In other words, even with the pivoting strategy, a finite-precision machine will either quit or commit an error of magnitude at least $1/2$ in the largest entry of the answer, due to unbridled growth of the data. A strategy described in the literature as *complete pivoting*, where the pivot element is chosen to be the largest-in-magnitude entry in the entire coefficient matrix (rather than the largest in the current column), prevents this. However, the growth exhibited in this contrived example is so unlikely, and the wider search strategy so time-consuming, that simple "partial pivoting" has been deemed as sufficient for most commercial software.

6. Even when pivoting is performed, sometimes a linear system simply cannot be solved accurately on a finite precision machine due to an inherent "instability" in the coefficient matrix. This instability, which magnifies the effect of rounding errors, is measured by the *condition number* μ of the matrix. The effect of the condition number is roughly as follows: if the computer retains p decimal digits of accuracy, so that the relative accuracy is about $10^{-(p+1)}$, then the relative accuracy of the solution will usually be about $\mu 10^{-(p+1)}$. In other words, you can expect to lose $\log_{10} \mu$ significant digits in the course of solving any system, simply due to the rounding errors generated while reading in the data! Although we defer a thorough explanation of this phenomenon until Section 6.4, it is amusing to perform the following experiment involving the notoriously ill-conditioned *Hilbert matrix*, where $a_{ij} = 1/(i+j-1)$. (These matrices turned up in Problem 8, Exercises 1.2.) The condition number for the 4-by-4 Hilbert matrix is approximately 15,000.

(a) Use a computer to obtain the solution to the system

$$
\left[\begin{array}{cccc:c}
1 & 1/2 & 1/3 & 1/4 & 1 \\
1/2 & 1/3 & 1/4 & 1/5 & 0 \\
1/3 & 1/4 & 1/5 & 1/6 & 0 \\
1/4 & 1/5 & 1/6 & 1/7 & 0
\end{array}\right]
\tag{16}
$$

(You may use whatever subroutines are available at your facility.)

(b) Compare the computed solution with the exact solution:

$$x_1 = 16, \quad x_2 = -120, \quad x_3 = 240, \quad x_4 = 140.$$

To how many digits is the computed answer correct? How many digits does your computer retain? Did you lose about four digits?

(c) It is important to understand that the blame for the errors is not due to the algorithm but to the problem itself. Program the computer to perform the following calculations of the left-hand side of (16):

$$
\begin{aligned}
&x_1 + (1/2)x_2 + (1/3)x_3 + (1/4)x_4, \\
&(1/2)x_1 + (1/3)x_2 + (1/4)x_3 + (1/5)x_4, \\
&(1/3)x_1 + (1/4)x_2 + (1/5)x_3 + (1/6)x_4, \\
&(1/4)x_1 + (1/5)x_2 + (1/6)x_3 + (1/7)x_4;
\end{aligned}
\tag{17}
$$

and

$$\begin{aligned}
x_1 + x_2/2 + x_3/3 + x_4/4,\\
x_1/2 + x_2/3 + x_3/4 + x_4/5,\\
x_1/3 + x_2/4 + x_3/5 + x_4/6,\\
x_1/4 + x_2/5 + x_3/6 + x_4/7.
\end{aligned} \tag{18}$$

Substitute the computed solution and the exact solution into (17). Observe that the exact solution does not "check" much better than the computed solution. (Indeed, if your facility's equation-solving subroutine is fairly sophisticated, the computed solution may seem superior.) On the other hand, substituting these solutions into (18), which avoids the rounding errors in the coefficient matrix (at least, for the exact solution), clearly distinguishes the exact solution.

(d) Repeat these computations using double precision.

1.6 SUMMARY

The tasks of analyzing data sets such as CAT scans and of fitting solution expressions for a linear differential equation to specified auxiliary conditions are instances where solving linear algebraic systems is an essential step. Because the latter are pervasive in mathematics and applications, it is important to have a methodical, foolproof algorithm for this problem that is suitable for implementation on computers. Gauss elimination fills this need.

Gauss Elimination

Gauss elimination incorporates the familiar techniques of substitution, cross-multiplication and subtraction, and solution by inspection into a systematic sequence of steps that result in explicit solutions to linear systems. The basic operations, which are guaranteed not to alter the solution set, are

(i) reordering the equations;
(ii) adding (or subtracting) a constant multiple of one equation to another and replacing the latter with the result;
(iii) multiplying an equation by a nonzero constant and replacing the original equation with the result.

Augmented Matrix Formulation

The efficiency of the Gauss elimination algorithm for solving a system of m linear equations in n unknowns is enhanced by representing the linear system in an augmented matrix format. The m-by-$(n+1)$ augmented matrix contains m rows and $n+1$ columns, in which the rows depict the (m) individual equations, the final ($[n+1]$st) column contains the nonhomogeneous terms, and the first n columns (which constitute the coefficient matrix) contain the coefficients of the unknowns.

Row-Echelon Form

In the forward part of the Gauss elimination algorithm, the basic operations are applied to the rows of the augmented matrix to reduce it to a row-echelon form, in which the first nonzero entry in each row lies to the right of the first nonzero entry in the previous row. From this form, one can identify redundant equations (a row containing all zeros) and inconsistent systems (the appearance of a row with exactly one nonzero element, in its last column).

Back Substitution

The back substitution part of the Gauss algorithm assembles the complete solution to the system by solving the equations represented by the row-echelon form in order, from last to first. Unknowns corresponding to columns that contain no nonzero leading row entries are classified as free variables and the other unknowns are expressed in terms of them through a parametric representation of the solution set.

Rank of Row-Echelon Form

The rank of a row-echelon matrix, that is the number of its nonzero rows, can be used to classify the number of solutions of the system.

   (i) If the rank of the augmented matrix equals the rank of the coefficient matrix, the system is consistent (has a solution); otherwise it is inconsistent.

  (ii) If the system is consistent and the rank of the coefficient matrix equals its number of columns (the number of unknowns), the system has a unique solution.

 (iii) If the system is consistent and the rank of the coefficient matrix is less than its number of columns (the number of unknowns), the system has an infinite number of solutions.

Linear Homogeneous Systems

A linear homogeneous system always possesses the trivial solution in which every unknown equals zero. If the number of equations is less than the number of unknowns, it has an infnite number of solutions.

Computational Considerations

Performing Gauss elimination on computers having finite precision can lead to unacceptable errors if large multiples of some equations are added to others. The effect is controlled by a partial pivoting strategy, which reorders the equations so that these multiples never exceed one, in absolute value.

2

MATRIX ALGEBRA

2.1 MATRIX MULTIPLICATION

In this section we will be studying an algebraic system based on matrices. Although "matrix algebra" is interesting in its own right, the big payoff is that it gives us a new lexicon that enables us to articulate many operations—physical (rotations, projections) and computational (the row operations)—in a compact format that facilitates understanding and utilization.

Matrix Products

The matrix product has proved to be one of the most useful inventions in mathematics. It undoubtedly was motivated by the desire to represent a system of linear algebraic equations compactly in the form $\mathbf{Ax} = \mathbf{b}$. (As one would expect, we say matrices are equal when their corresponding entries are equal.)

The system that we solved in Example 1 of Section 1.1,

$$\begin{aligned} x_1 + 2x_2 + 2x_3 &= 6 \\ 2x_1 + x_2 + x_3 &= 6 \\ x_1 + x_2 + 3x_3 &= 6 \end{aligned} \tag{1}$$

has a coefficient matrix

$$\mathbf{A} = \begin{bmatrix} 1 & 2 & 2 \\ 2 & 1 & 1 \\ 1 & 1 & 3 \end{bmatrix}.$$

Fundamentals of Matrix Analysis with Applications,
First Edition. Edward Barry Saff and Arthur David Snider.
© 2016 John Wiley & Sons, Inc. Published 2016 by John Wiley & Sons, Inc.

Therefore, introducing *column matrices* for the unknowns **x** and the right-hand side **b**,

$$\mathbf{x} = \begin{bmatrix} x_1 \\ x_2 \\ x_3 \end{bmatrix}, \quad \mathbf{b} = \begin{bmatrix} 6 \\ 6 \\ 6 \end{bmatrix},$$

we define the matrix product **Ax** so that the equation **Ax** = **b** expresses system (1):

$$\begin{bmatrix} 1 & 2 & 2 \\ 2 & 1 & 1 \\ 1 & 1 & 3 \end{bmatrix} \begin{bmatrix} x_1 \\ x_2 \\ x_3 \end{bmatrix} := \begin{bmatrix} x_1 + 2x_2 + 2x_3 \\ 2x_1 + x_2 + x_3 \\ x_1 + x_2 + 3x_3 \end{bmatrix}; \qquad (2)$$

$$\mathbf{Ax} = \begin{bmatrix} x_1 + 2x_2 + 2x_3 \\ 2x_1 + x_2 + x_3 \\ x_1 + x_2 + 3x_3 \end{bmatrix} = \begin{bmatrix} 6 \\ 6 \\ 6 \end{bmatrix} = \mathbf{b}.$$

The first entry of the product **Ax** is a mathematical combination that is familiar from vector calculus. Namely, it is the dot product of two vectors: the first row of **A**, and the unknowns.

$$x_1 + 2x_2 + 2x_3 = (1, 2, 2) \cdot (x_1, x_2, x_3). \qquad (3)$$

Similarly, the second and third members are dot products also, involving the corresponding rows of **A**.

In higher dimensions, the **product of a matrix**—that is, an *m*-by-*n* rectangular array of numbers—**and a column vector** (with *n* components) is defined to be the collection of dot products of the rows of the matrix with the vector, arranged as a column:

$$\begin{bmatrix} \text{row } \#1 \rightarrow \\ \text{row } \#2 \rightarrow \\ \vdots \\ \text{row } \#m \rightarrow \end{bmatrix} \begin{bmatrix} \\ \mathbf{x} \\ \downarrow \\ \end{bmatrix} = \begin{bmatrix} (\text{row } \#1) \cdot \mathbf{x} \\ (\text{row } \#2) \cdot \mathbf{x} \\ \vdots \\ (\text{row } \#m) \cdot \mathbf{x} \end{bmatrix}$$

where the dot product of two *n*-dimensional vectors is computed in the obvious way:

$$(a_1, a_2, \ldots, a_n) \cdot (x_1, x_2, \ldots, x_n) = a_1 x_1 + a_2 x_2 + \cdots + a_n x_n.$$

The arrangement of terms in the dot product has come to be known also as the **inner product**. We use the terms "dot product" and "inner product" interchangeably.

The fact that a linear system of equations can be written in matrix jargon as $\mathbf{Ax} = \mathbf{b}$ suggests that its solution $\mathbf{x}$ might be expressed in some form such as "$\mathbf{x} = \mathbf{b}/\mathbf{A}$". In Section 2.3 we will, in fact, fulfill this expectation.

Having defined products of matrices times column vectors, we now generalize the definition to embrace products of matrices with other matrices. The **product of two matrices A and B** is formed by taking the array of the matrix products of $\mathbf{A}$ with the individual *columns* of the second factor $\mathbf{B}$. For example, the product

$$\begin{bmatrix} 1 & 0 & 1 \\ 3 & -1 & 2 \end{bmatrix} \begin{bmatrix} 1 & 2 & x \\ -1 & -1 & y \\ 4 & 1 & z \end{bmatrix}$$

is assembled from the (matrix-vector) products

$$\begin{bmatrix} 1 & 0 & 1 \\ 3 & -1 & 2 \end{bmatrix} \begin{bmatrix} 1 \\ -1 \\ 4 \end{bmatrix} = \begin{bmatrix} 5 \\ 12 \end{bmatrix}, \quad \begin{bmatrix} 1 & 0 & 1 \\ 3 & -1 & 2 \end{bmatrix} \begin{bmatrix} 2 \\ -1 \\ 1 \end{bmatrix} = \begin{bmatrix} 3 \\ 9 \end{bmatrix},$$

$$\begin{bmatrix} 1 & 0 & 1 \\ 3 & -1 & 2 \end{bmatrix} \begin{bmatrix} x \\ y \\ z \end{bmatrix} = \begin{bmatrix} x+z \\ 3x - y + 2z \end{bmatrix}$$

to form

$$\begin{bmatrix} 1 & 0 & 1 \\ 3 & -1 & 2 \end{bmatrix} \begin{bmatrix} 1 & 2 & x \\ -1 & -1 & y \\ 4 & 1 & z \end{bmatrix} = \begin{bmatrix} 5 & 3 & x+z \\ 12 & 9 & 3x - y + 2z \end{bmatrix}.$$

More generally, if $\mathbf{A}$ is *m-by-n* and $\mathbf{B}$ is *n-by-p* with $\mathbf{b}_1, \mathbf{b}_2, \dots, \mathbf{b}_p$ denoting its columns, then

$$\mathbf{AB} = \mathbf{A} \begin{bmatrix} \mathbf{b}_1 & \mathbf{b}_2 & \cdots & \mathbf{b}_p \end{bmatrix} = \begin{bmatrix} \mathbf{Ab}_1 & \mathbf{Ab}_2 & \cdots & \mathbf{Ab}_p \end{bmatrix},$$

i.e. the *j*th column of the product is given by $\mathbf{Ab}_j$.

Caution: Note that $\mathbf{AB}$ is only defined when the number of rows of $\mathbf{B}$ matches the number of columns of $\mathbf{A}$.

One can also express matrix multiplication in terms of dot products: the (i,j)th entry of the product $\mathbf{AB}$ equals the dot product of the *i*th row of $\mathbf{A}$ with the *j*th column:

$$\begin{bmatrix} 1 & 0 & 1 \\ 3 & -1 & 2 \end{bmatrix} \begin{bmatrix} 1 & 2 & x \\ -1 & -1 & y \\ 4 & 1 & z \end{bmatrix} = \begin{bmatrix} 1+0+4 & 2+0+1 & x+0+y \\ 3+1+8 & 6+1+2 & 3x - y + 2z \end{bmatrix}$$

$$= \begin{bmatrix} 5 & 3 & x+z \\ 12 & 9 & 3x - y + 2z \end{bmatrix}.$$

The mathematical formula[1] for the product of an m-by-n matrix $\mathbf{A}$ and an n-by-p matrix $\mathbf{B}$ is

$$\mathbf{AB}\text{-: } [c_{ij}], \text{ where } c_{ij} = \sum_{k=1}^{n} a_{ik}b_{kj}. \tag{4}$$

Because $\mathbf{AB}$ is computed in terms of the *rows* of the first factor and the *columns* of the second factor, it should not be surprising that, in general, $\mathbf{AB}$ does not equal $\mathbf{BA}$ (matrix multiplication does not *commute*):

$$\begin{bmatrix} 1 & 2 \\ 3 & 4 \end{bmatrix} \begin{bmatrix} 0 & 1 \\ 1 & 0 \end{bmatrix} = \begin{bmatrix} 2 & 1 \\ 4 & 3 \end{bmatrix}, \quad \text{but} \quad \begin{bmatrix} 0 & 1 \\ 1 & 0 \end{bmatrix} \begin{bmatrix} 1 & 2 \\ 3 & 4 \end{bmatrix} = \begin{bmatrix} 3 & 4 \\ 1 & 2 \end{bmatrix}.$$

$$\begin{bmatrix} 1 & 2 & 3 \end{bmatrix} \begin{bmatrix} 4 \\ 5 \\ 6 \end{bmatrix} = \begin{bmatrix} 32 \end{bmatrix}, \quad \text{but} \quad \begin{bmatrix} 4 \\ 5 \\ 6 \end{bmatrix} \begin{bmatrix} 1 & 2 & 3 \end{bmatrix} = \begin{bmatrix} 4 & 8 & 12 \\ 5 & 10 & 15 \\ 6 & 12 & 18 \end{bmatrix}.$$

In fact, the dimensions of $\mathbf{A}$ and $\mathbf{B}$ may render one or the other of these products undefined.

$$\begin{bmatrix} 1 & 2 \\ 3 & 4 \end{bmatrix} \begin{bmatrix} 0 \\ 1 \end{bmatrix} = \begin{bmatrix} 2 \\ 4 \end{bmatrix}; \quad \begin{bmatrix} 0 \\ 1 \end{bmatrix} \begin{bmatrix} 1 & 2 \\ 3 & 4 \end{bmatrix} \quad \text{not defined.}$$

By the same token, one might not expect $(\mathbf{AB})\mathbf{C}$ to equal $\mathbf{A}(\mathbf{BC})$, since in $(\mathbf{AB})\mathbf{C}$ we take dot products with *columns* of $\mathbf{B}$, whereas in $\mathbf{A}(\mathbf{BC})$ we employ the *rows* of $\mathbf{B}$. So it is a pleasant surprise that the customary parenthesis grouping rule does indeed hold.

Example 1.   Verify the associative law for the product $\mathbf{ABC}$, where

$$\mathbf{A} = \begin{bmatrix} 1 \\ 2 \end{bmatrix}, \quad \mathbf{B} = \begin{bmatrix} 3 & 4 \end{bmatrix}, \quad \mathbf{C} = \begin{bmatrix} 1 & 2 & 3 & 4 \\ 4 & 3 & 2 & 1 \end{bmatrix}.$$

Solution.   Straightforward calculation verifies the law:

$$(\mathbf{AB})\mathbf{C} = \begin{bmatrix} 3 & 4 \\ 6 & 8 \end{bmatrix} \begin{bmatrix} 1 & 2 & 3 & 4 \\ 4 & 3 & 2 & 1 \end{bmatrix} = \begin{bmatrix} 19 & 18 & 17 & 16 \\ 38 & 36 & 34 & 32 \end{bmatrix},$$

$$\mathbf{A}(\mathbf{BC}) = \begin{bmatrix} 1 \\ 2 \end{bmatrix} \begin{bmatrix} 19 & 18 & 17 & 16 \end{bmatrix} = \begin{bmatrix} 19 & 18 & 17 & 16 \\ 38 & 36 & 34 & 32 \end{bmatrix} \quad \blacksquare$$

To fully exploit the advantages of matrix notation we "flesh out" the algebra by including matrix addition and scalar multiplication.

---

[1] The dot product of the ith row of $\mathbf{A}$ and the jth column of $\mathbf{B}$ is sometimes called the "sum of products" expression for c_{ij}.

Matrix Addition and Scalar Multiplication

The operations of matrix addition and scalar multiplication are very straightforward. Addition is performed by adding corresponding elements:

$$\begin{bmatrix} 1 & 2 & 3 \\ 4 & 5 & 6 \end{bmatrix} + \begin{bmatrix} 1 & 7 & 1 \\ 0 & 1 & 1 \end{bmatrix} = \begin{bmatrix} 2 & 9 & 4 \\ 4 & 6 & 7 \end{bmatrix}.$$

Formally, the *sum* of two *m*-by-*n* matrices is given by

$$\mathbf{A} + \mathbf{B} = \begin{bmatrix} a_{ij} \end{bmatrix} + \begin{bmatrix} b_{ij} \end{bmatrix} := \begin{bmatrix} a_{ij} + b_{ij} \end{bmatrix}.$$

The sole novelty here is that addition is undefined for two matrices having different dimensions.

   To multiply a matrix by a scalar (number), we simply multiply each element in the matrix by the number:

$$3 \begin{bmatrix} 1 & 2 & 3 \\ 4 & 5 & 6 \end{bmatrix} = \begin{bmatrix} 3 & 6 & 9 \\ 12 & 15 & 18 \end{bmatrix} = \begin{bmatrix} 1 & 2 & 3 \\ 4 & 5 & 6 \end{bmatrix} 3$$

In other words, $r\mathbf{A} = r[a_{ij}] := [ra_{ij}]$ (and $\mathbf{A}r$ *is the same as* $r\mathbf{A}$). The notation $-\mathbf{A}$ stands for $(-1)\mathbf{A}$.

Properties of Matrix Addition and Scalar Multiplication

Matrix addition and scalar multiplication are nothing more than mere bookkeeping, and the usual algebraic properties hold. If $\mathbf{A}$, $\mathbf{B}$, and $\mathbf{C}$ are *m*-by-*n* matrices, $\mathbf{0}$ is the *m*-by-*n* matrix whose entries are all zeros, and r, s are scalars, then

$$\mathbf{A} + (\mathbf{B} + \mathbf{C}) = (\mathbf{A} + \mathbf{B}) + \mathbf{C}, \quad \mathbf{A} + \mathbf{B} = \mathbf{B} + \mathbf{A},$$
$$\mathbf{A} + \mathbf{0} = \mathbf{A}, \quad \mathbf{A} + (-\mathbf{A}) = \mathbf{0},$$
$$r(\mathbf{A} + \mathbf{B}) = r\mathbf{A} + r\mathbf{B}, \quad (r + s)\mathbf{A} = r\mathbf{A} + s\mathbf{A},$$
$$r(s\mathbf{A}) = (rs)\mathbf{A} = s(r\mathbf{A}).$$

The parenthesis grouping rules for matrix algebra are then given as follows:

Properties of Matrix Algebra

$$(\mathbf{A}\mathbf{B})\mathbf{C} = \mathbf{A}(\mathbf{B}\mathbf{C}) \qquad \text{(Associativity)}$$
$$(\mathbf{A} + \mathbf{B})\mathbf{C} = \mathbf{A}\mathbf{C} + \mathbf{B}\mathbf{C} \qquad \text{(Distributivity)}$$

$$A(B+C) = AB + AC \qquad \text{(Distributivity)}$$
$$(rA)B = r(AB) = A(rB) \qquad \text{(Associativity)}$$

The proofs of these properties involve meticulous sorting of the products in (4); see Problem 16.

In general, the algebra of matrices (with compatible dimensions) proceeds much like the standard algebra of numbers, except that we must never presume that we can switch the order of matrix factors. Notice that the zero matrices act like the number zero in that

$$0 + A = A, \quad 0 \cdot B = 0$$

when the sum (product) is defined. There are also "multiplicative identity" matrices, namely, n-by-n matrices denoted I (or I_n) with ones down the main diagonal and zeros elsewhere. Multiplying I on the right or left by any other compatible matrix reproduces the latter:

$$\begin{bmatrix} 1 & 0 & 0 \\ 0 & 1 & 0 \\ 0 & 0 & 1 \end{bmatrix} \begin{bmatrix} 1 & 2 & 1 \\ 1 & 3 & 2 \\ 1 & 0 & 1 \end{bmatrix} = \begin{bmatrix} 1 & 2 & 1 \\ 1 & 3 & 2 \\ 1 & 0 & 1 \end{bmatrix} = \begin{bmatrix} 1 & 2 & 1 \\ 1 & 3 & 2 \\ 1 & 0 & 1 \end{bmatrix} \begin{bmatrix} 1 & 0 & 0 \\ 0 & 1 & 0 \\ 0 & 0 & 1 \end{bmatrix}$$

Armed with the full arsenal of matrix algebra, we turn to a simple application. The basic row operations of Gauss elimination (Definition 4, Section 1.2) can be expressed as left multiplication by appropriate matrices, the "elementary row matrix operators". Observe that the following product accomplishes an exchange of rows 2 and 3 in the matrix on the right:

$$\begin{bmatrix} 1 & 0 & 0 \\ 0 & 0 & 1 \\ 0 & 1 & 0 \end{bmatrix} \begin{bmatrix} a & b & c & d \\ e & f & g & h \\ i & j & k & l \end{bmatrix} = \begin{bmatrix} a & b & c & d \\ i & j & k & l \\ e & f & g & h \end{bmatrix}.$$

Similarly, the addition of α times row 1 to row 2 is accomplished by the product

$$\begin{bmatrix} 1 & 0 & 0 \\ \alpha & 1 & 0 \\ 0 & 0 & 1 \end{bmatrix} \begin{bmatrix} a & b & c & d \\ e & f & g & h \\ i & j & k & l \end{bmatrix} = \begin{bmatrix} a & b & c & d \\ e+\alpha a & f+\alpha b & g+\alpha c & h+\alpha d \\ i & j & k & l \end{bmatrix},$$

and the multiplication of row 2 by β is achieved by the product

$$\begin{bmatrix} 1 & 0 & 0 \\ 0 & \beta & 0 \\ 0 & 0 & 1 \end{bmatrix} \begin{bmatrix} a & b & c & d \\ e & f & g & h \\ i & j & k & l \end{bmatrix} = \begin{bmatrix} a & b & c & d \\ \beta e & \beta f & \beta g & \beta h \\ i & j & k & l \end{bmatrix}.$$

Example 2. Write out the matrix operator representation for the elementary row operations used in the forward part of the Gauss elimination algorithm depicted below:

$$
\begin{bmatrix} 2 & 2 & 2 & 2 & 2 \\ 2 & 2 & 4 & 6 & 4 \\ 0 & 0 & 2 & 3 & 1 \\ 1 & 4 & 3 & 2 & 3 \end{bmatrix}
\begin{array}{c} (-1)\rho_1 + \rho_2 \rightarrow \rho_2 \\ \\ (-1/2)\rho_1 + \rho_4 \rightarrow \rho_4 \end{array}
\begin{bmatrix} 2 & 2 & 2 & 2 & 2 \\ 0 & 0 & 2 & 4 & 2 \\ 0 & 0 & 2 & 3 & 1 \\ 0 & 3 & 2 & 1 & 2 \end{bmatrix}
$$

$$
\rho_2 \leftrightarrow \rho_4
\begin{bmatrix} 2 & 2 & 2 & 2 & 2 \\ 0 & 3 & 2 & 1 & 2 \\ 0 & 0 & 2 & 3 & 1 \\ 0 & 0 & 2 & 4 & 2 \end{bmatrix}
\begin{array}{c} \\ \\ (-1)\rho_3 + \rho_4 \rightarrow \rho_4 \end{array}
\begin{bmatrix} 2 & 2 & 2 & 2 & 2 \\ 0 & 3 & 2 & 1 & 2 \\ 0 & 0 & 2 & 3 & 1 \\ 0 & 0 & 0 & 1 & 1 \end{bmatrix}.
$$

Solution. The location of the (-1) in

$$
\mathbf{E}_1 = \begin{bmatrix} 1 & 0 & 0 & 0 \\ -1 & 1 & 0 & 0 \\ 0 & 0 & 1 & 0 \\ 0 & 0 & 0 & 1 \end{bmatrix}
$$

dictates that left multiplication of a matrix by $\mathbf{E}_1$ adds (-1) times the first row to the second row of the matrix: $(-1)\rho_1 + \rho_2 \rightarrow \rho_2$. Similarly, the $(-1/2)$ in

$$
\mathbf{E}_2 = \begin{bmatrix} 1 & 0 & 0 & 0 \\ 0 & 1 & 0 & 0 \\ 0 & 0 & 1 & 0 \\ -\frac{1}{2} & 0 & 0 & 1 \end{bmatrix}
$$

enforces, through left multiplication, the addition of $(-1/2)$ times the first row to the fourth row: $(-1/2)\rho_1 + \rho_4 \rightarrow \rho_4$. The second row of

$$
\mathbf{E}_3 = \begin{bmatrix} 1 & 0 & 0 & 0 \\ 0 & 0 & 0 & 1 \\ 0 & 0 & 1 & 0 \\ 0 & 1 & 0 & 0 \end{bmatrix}
$$

overwrites (through premultiplication) the second row of a matrix with a copy of its fourth row, while the fourth row of $\mathbf{E}_3$ overwrites the fourth row with the second; thus $\mathbf{E}_3$ exchanges the second and fourth rows of any matrix it premultiplies: $\rho_2 \leftrightarrow \rho_4$. Finally, premultiplication by

$$
\mathbf{E}_4 = \begin{bmatrix} 1 & 0 & 0 & 0 \\ 0 & 1 & 0 & 0 \\ 0 & 0 & 1 & 0 \\ 0 & 0 & -1 & 1 \end{bmatrix}.
$$

achieves $(-1)\rho_3 + \rho_4 \rightarrow \rho_4$. Since *one interprets a sequence of left multiplications by reading from right to left*, we can express the sequence of row operations in matrix operator form as follows:

$$\left(\mathbf{E}_4 \left(\mathbf{E}_3 \left(\mathbf{E}_2 \left(\mathbf{E}_1 \begin{bmatrix} 2 & 2 & 2 & 2 & 2 \\ 2 & 2 & 6 & 4 & 4 \\ 0 & 0 & 2 & 3 & 1 \\ 1 & 4 & 3 & 2 & 3 \end{bmatrix} \right) \right) \right) \right) = \begin{bmatrix} 2 & 2 & 2 & 2 & 2 \\ 0 & 3 & 2 & 1 & 2 \\ 0 & 0 & 2 & 3 & 1 \\ 0 & 0 & 0 & -2 & 0 \end{bmatrix}.$$

In fact, by regrouping the parentheses we can see that left multiplication by the single matrix (revealed after tedious computation)

$$\mathbf{E}_4 \mathbf{E}_3 \mathbf{E}_2 \mathbf{E}_1 = \begin{bmatrix} 1 & 0 & 0 & 0 \\ -1/2 & 0 & 0 & 1 \\ 0 & 0 & 1 & 0 \\ -1 & 1 & -1 & 0 \end{bmatrix}$$

puts the original matrix into echelon form.   ∎

A quick way to find the matrix operator $\mathbf{E}$ corresponding to an elementary row operation is to interpret the identity $\mathbf{E} = \mathbf{EI}$ as saying one can get $\mathbf{E}$ by simply performing the row operation on the identity matrix. Take a second to verify this for the operators in Example 2.

Exercises 2.1

1. Let $\mathbf{A} := \begin{bmatrix} 2 & 1 \\ 3 & 5 \end{bmatrix}$ and $\mathbf{B} := \begin{bmatrix} 2 & 3 \\ -1 & 0 \end{bmatrix}$

   Find: (a) $\mathbf{A} + \mathbf{B}$ (b) $3\mathbf{A} - \mathbf{B}$

2. Let $\mathbf{A} := \begin{bmatrix} 2 & 1 & 1 \\ 2 & 0 & 5 \end{bmatrix}$ and $\mathbf{B} := \begin{bmatrix} 1 & -1 & 2 \\ 0 & 3 & -2 \end{bmatrix}$

   Find: (a) $\mathbf{A} + \mathbf{B}$ (b) $7\mathbf{A} - 4\mathbf{B}$

3. Let $\mathbf{A} := \begin{bmatrix} 2 & 4 & 0 \\ 1 & 1 & 3 \\ 2 & 1 & 3 \end{bmatrix}$ and $\mathbf{B} := \begin{bmatrix} -1 & 3 & 0 \\ 5 & 2 & 1 \\ 4 & 5 & 1 \end{bmatrix}$

   Find: (a) $\mathbf{AB}$ (b) $\mathbf{BA}$ (c) $\mathbf{A}^2 = \mathbf{AA}$ (d) $\mathbf{B}^2 = \mathbf{BB}$

4. Let $\mathbf{A} := \begin{bmatrix} 2 & 1 \\ 0 & 4 \\ -1 & 3 \end{bmatrix}$ and $\mathbf{B} := \begin{bmatrix} 0 & 3 & -1 \\ 1 & 1 & 1 \end{bmatrix}$

   Find: (a) $\mathbf{AB}$ (b) $\mathbf{BA}$

5. Let $\mathbf{A} := \begin{bmatrix} 2 & -4 & 1 \\ 1 & 2 & 3 \\ -1 & 1 & 3 \end{bmatrix}$, $\mathbf{B} := \begin{bmatrix} 3 & 4 & 0 \\ 7 & 1 & 3 \\ 2 & -1 & -3 \end{bmatrix}$, and $\mathbf{C} := \begin{bmatrix} 0 & -4 & 0 \\ 1 & 1 & 5 \\ -3 & 1 & 5 \end{bmatrix}$

   Find: (a) $\mathbf{AB}$ (b) $\mathbf{AC}$ (c) $\mathbf{A(B+C)}$

6. Let $\mathbf{A} := \begin{bmatrix} 1 & 2 \\ 15 & 1 \end{bmatrix}$, $\mathbf{B} := \begin{bmatrix} 0 & 3 \\ 1 & 2 \end{bmatrix}$, and $\mathbf{C} := \begin{bmatrix} 1 & -4 \\ 4 & 6 \end{bmatrix}$

   Find: (a) $\mathbf{AB}$ (b) $\mathbf{BC}$ (c) $(\mathbf{AB})\mathbf{C}$ (d) $\mathbf{A}(\mathbf{BC})$

7. Let $\mathbf{A} := \begin{bmatrix} 2 & -1 \\ -3 & 4 \end{bmatrix}$ and $\mathbf{B} := \begin{bmatrix} 1 & 2 \\ 3 & 2 \end{bmatrix}$

   Verify that $\mathbf{AB} \neq \mathbf{BA}$.

8. Let $\mathbf{A} := \begin{bmatrix} 4 & 4 \\ 0 & 0 \\ 2 & 2 \end{bmatrix}$ and $\mathbf{B} := \begin{bmatrix} 1 & 0 & -1 \\ -1 & 0 & 1 \end{bmatrix}$

   Show that $\mathbf{AB} = \begin{bmatrix} 0 & 0 & 0 \\ 0 & 0 & 0 \\ 0 & 0 & 0 \end{bmatrix}$, but $\mathbf{BA} \neq \begin{bmatrix} 0 & 0 \\ 0 & 0 \end{bmatrix}$.

9. Construct, by trial and error, a 2-by-2 matrix $\mathbf{A}$ such that $\mathbf{A}^2 = \mathbf{0}$, but $\mathbf{A} \neq \mathbf{0}$. (Here $\mathbf{0}$ is the 2-by-2 matrix with all entries zero.)

10. Show that if two n-by-n matrices $\mathbf{A}$ and $\mathbf{B}$ *commute* (i.e., $\mathbf{AB} = \mathbf{BA}$), then $(\mathbf{A} + \mathbf{B})^2 = \mathbf{A}^2 + 2\mathbf{AB} + \mathbf{B}^2$. Then show by example that this formula can fail if $\mathbf{A}$ and $\mathbf{B}$ do not commute.

11. Construct 2-by-2 matrices $\mathbf{A}$, $\mathbf{B}$, and $\mathbf{C} \neq \mathbf{0}$ such that $\mathbf{AC} = \mathbf{BC}$, but $\mathbf{A} \neq \mathbf{B}$. [*Hint:* Start with $\mathbf{C} = \begin{bmatrix} 1 & -1 \\ 1 & -1 \end{bmatrix}$ and experiment to find suitable matrices $\mathbf{A}$ and $\mathbf{B}$.]

12. Show that if $\mathbf{A}$, $\mathbf{B}$, and $\mathbf{C}$ are n-by-n matrices such that $\mathbf{AC} = \mathbf{BC}$ and, furthermore, that there exists a matrix $\mathbf{D}$ such that $\mathbf{CD} = \mathbf{I}$, then $\mathbf{A} = \mathbf{B}$. (Compare with Problem 11.)

13. For each of the following, decide whether the statement made is always True or sometimes False.
    (a) If $\mathbf{A}$ and $\mathbf{B}$ are each n-by-n matrices and all the entries of $\mathbf{A}$ and of $\mathbf{B}$ are positive numbers, then all the entries of the product $\mathbf{AB}$ are positive numbers.
    (b) If $\mathbf{A}$ and $\mathbf{B}$ are each n-by-n matrices and $\mathbf{AB} = \mathbf{0}$, then either $\mathbf{A} = \mathbf{0}$ or $\mathbf{B} = \mathbf{0}$. (Here $\mathbf{0}$ is the n-by-n matrix with all entries equal to zero.)
    (c) If $\mathbf{A}$ and $\mathbf{B}$ are each n-by-n upper triangular matrices, then $\mathbf{AB}$ is upper triangular.
    (d) If $\mathbf{A}$ is an n-by-n diagonal matrix (i.e., a matrix whose only nonzero entries are on the diagonal), then $\mathbf{A}$ commutes with every n-by-n matrix $\mathbf{B}$ (i.e., $\mathbf{AB} = \mathbf{BA}$).

14. Write each of the following systems in matrix product form:
    (a) $\begin{array}{rcrcrcl} 8x_1 & & & - & 4x_3 & = & 2 \\ -3x_1 & + & x_2 & + & x_3 & = & -1 \\ 6x_1 & - & x_2 & & & = & \pi \end{array}$

$$
\begin{array}{rrrrrrrrr}
 & x & + & 2y & + & 6z & - & 8 & = & 0 \\
\text{(b)} \ 3 & + & y & + & 4x & + & z & & = & 0 \\
 & x & - & 4 & + & y & & & = & 0
\end{array}
$$

15. Write each of the following systems in matrix product form.

$$
\begin{array}{rrrrrrr}
 & & x_2 & + & 3x_3 & = & 0 \\
\text{(a)} \ -x_1 & + & 7x_2 & - & 9x_3 & = & -9 \\
 6x_1 & - & 4x_2 & & & = & 2
\end{array}
$$

$$
\begin{array}{rrrrrrrr}
 & & 3y & - & 2x & + & 7 & = & 0 \\
\text{(b)} \ -x & + & 2y & - & 6 & + & z & = & 0 \\
 3z & - & y & - & x & + & 3 & = & 0
\end{array}
$$

16. Verify the associativity property of multiplication, that is $(\mathbf{AB})\,\mathbf{C} = \mathbf{A}(\mathbf{BC})$, where $\mathbf{A} = [a_{ij}]$ is an m-by-n matrix, $\mathbf{B} = [b_{jk}]$ is n-by-p, and $\mathbf{C} = [c_{kr}]$ is p-by-q. [*Hint:* Let $\mathbf{D} := (\mathbf{AB})\mathbf{C}$ and $\mathbf{E} := \mathbf{A}(\mathbf{BC})$ and show, using the formula (4), that the (i, r)th entry d_{ir} of $\mathbf{D}$ is given by

$$
d_{ir} = \sum_{j=1}^{n} a_{ij} \sum_{k=1}^{p} b_{jk} c_{kr},
$$

and then interchange the order of summations to show that this is the same as the (i, r)th entry of $\mathbf{E}$.]

17. Let $\mathbf{A} := \begin{bmatrix} 2 & -4 & 1 \\ 1 & 2 & 3 \\ -1 & 1 & 3 \end{bmatrix}$. Find a matrix operator $\mathbf{B}$ such the following elementary row operations on $\mathbf{A}$ are accomplished via left multiplication by $\mathbf{B}$.

(a) $\mathbf{BA} = \begin{bmatrix} 2 & -4 & 1 \\ -1 & 1 & 3 \\ 1 & 2 & 3 \end{bmatrix}$,

(b) $\mathbf{BA} = \begin{bmatrix} 2 & -4 & 1 \\ 0 & 4 & \frac{5}{2} \\ -1 & 1 & 3 \end{bmatrix}$,

(c) $\mathbf{BA} = \begin{bmatrix} 2 & -4 & 1 \\ 2 & 4 & 6 \\ -1 & 1 & 3 \end{bmatrix}$.

18. Let $\mathbf{A} := \begin{bmatrix} 0 & 0 & -1 & 5 \\ 2 & 1 & -1 & 4 \\ 6 & -3 & 2 & 2 \end{bmatrix}$. Find a matrix operator $\mathbf{B}$ such the following elementary row operations on $\mathbf{A}$ are accomplished via left multiplication by $\mathbf{B}$.

(a) $\mathbf{BA} = \begin{bmatrix} 6 & -3 & 2 & 2 \\ 2 & 1 & -1 & 4 \\ 0 & 0 & -1 & 5 \end{bmatrix}$,

(b) $\mathbf{BA} = \begin{bmatrix} 0 & 0 & -1 & 5 \\ 2 & 1 & -1 & 4 \\ -3 & \frac{3}{2} & -1 & -1 \end{bmatrix}$,

(c) $\mathbf{BA} = \begin{bmatrix} 0 & 0 & -1 & 5 \\ 2 & 1 & -1 & 4 \\ 0 & -6 & 5 & -10 \end{bmatrix}$.

19. Transform the following matrices into row echelon form and display how this can be accomplished by a sequence of left multiplications.

(a) $\mathbf{A} = \begin{bmatrix} 0 & 0 & -1 \\ 2 & 1 & -1 \\ 6 & -3 & 2 \end{bmatrix}$,

(b) $\mathbf{B} = \begin{bmatrix} 2 & 0 & -3 & 5 \\ 2 & 1 & -1 & 4 \\ 6 & -3 & 2 & 2 \end{bmatrix}$,

(c) $\mathbf{C} = \begin{bmatrix} 1 & 2 & 1 & 1 & -1 & 0 \\ 1 & 2 & 0 & -1 & 1 & 0 \\ 0 & 0 & 1 & 2 & -2 & 1 \end{bmatrix}$.

20. Transform the following matrices into row echelon form and display how this can be accomplished by a sequence of left multiplications.

(a) $\mathbf{A} = \begin{bmatrix} 2 & -1 & 0 \\ 2 & 3 & -1 \\ 0 & -3 & 2 \end{bmatrix}$,

(b) $\mathbf{B} = \begin{bmatrix} 0 & 0 & 5 & 3 \\ 0 & -2 & -1 & 0 \\ 4 & -3 & 1 & 2 \end{bmatrix}$,

(c) $\mathbf{C} = \begin{bmatrix} 0 & 2 & 1 & 1 \\ 2 & 4 & 4 & 2 \\ 3 & 6 & 6 & 0 \\ 0 & -2 & -1 & -2 \end{bmatrix}$.

21. Years of experience have taught a fisherman that a good day of fishing is followed by another good day 75% of the time, while a bad day of fishing is followed by another bad day 40% of the time. So if Sunday is a good day, the probability that Monday is a good day is 0.75 and the probability that it is bad is 0.25.

   (a) If Sunday is a good day, what are the probabilities for Tuesday?

   (b) If $\mathbf{x}_n = \begin{bmatrix} g_n \\ b_n \end{bmatrix}$ gives the probabilities that day n is good (g_n) or bad (b_n), express the probabilities $\mathbf{x}_{n+1}$ for day $n + 1$ as a matrix product $\mathbf{x}_{n+1} = \mathbf{A}\mathbf{x}_n$.

   (c) What is the sum of the entries of the column vectors $\mathbf{x}_{n+1}$? What is the sum of the entries in each column of $\mathbf{A}$?

   (d) Suppose we refined the classification of fishing day quality to excellent, good, fair, and bad and modeled the probabilities as in (b). Would either of the answers in (c) remain the same?

22. (**Iteration of Stochastic Matrices**) Problem 21 is generalized as follows. An
    n-by-*n stochastic matrix* is a matrix for which each column vector has non-negative
    entries that add up to 1 (such vectors are called *probability vectors*).
    (a) Show that if **x** is probability vector and **A** is a stochastic matrix, then **Ax** is a
        probability vector.
    (b) Starting with a probability vector $\mathbf{x}_0$ and a stochastic matrix **A**, the sequence
        of probability vectors $\mathbf{x}_0$, $\mathbf{x}_1 := \mathbf{Ax}_0$, $\mathbf{x}_2 := \mathbf{Ax}_1$, etc. is called a *Markov chain*
        which can be represented by the iterative formula $\mathbf{x}_{k+1} := \mathbf{Ax}_k$, $k = 0, 1, \ldots$
        Explain why $\mathbf{x}_k = \mathbf{A}^k \mathbf{x}_0$, where $\mathbf{A}^0 := \mathbf{I}_n$.
    (c) Using a computer software package for multiplication, predict the long-term
        behavior (as $k \to \infty$) of the Markov chain for

$$\mathbf{A} = \begin{bmatrix} .33 & .40 & .32 & .55 \\ .11 & .15 & .41 & .03 \\ .11 & .30 & .12 & .35 \\ .45 & .15 & .15 & .07 \end{bmatrix}, \quad \text{with } \mathbf{x}_0 = \begin{bmatrix} .25 \\ .25 \\ .25 \\ .25 \end{bmatrix},$$

   that is, predict the limit of the entries in the $\mathbf{x}_k$ as $k \to \infty$. This limit vector is
   called a *steady-state probability vector*.
    (d) Experiment with different starting probability measures $\mathbf{x}_0$, to see if the
        predicted long-term behavior depends on the starting vector.

23. If **A** and **B** are *n*-by-*n* matrices and **v** is an *n*-by-1 vector, which of the following
    calculations is more efficient?
    (a) $\mathbf{A}(\mathbf{Bv})$ or $(\mathbf{AB})\mathbf{v}$?
    (b) $(\mathbf{A} + \mathbf{B})\mathbf{v}$ or $\mathbf{Av} + \mathbf{Bv}$?
    [*Hint*: How many dot products are required in each of the four calculations?]

2.2 SOME PHYSICAL APPLICATIONS OF MATRIX OPERATORS

In this section we explore several ways matrix language is used to express some of the
operations performed in applied mathematics.

We will see that, while it is important to adhere strictly to the rules of matrix algebra as listed in the previous section, some applications call for a little flexibility. For
example, sometimes the algebra requires the *x,y* coordinates of a point in the plane to
be listed as a row vector [*x y*], and other times as a column vector

$$\begin{bmatrix} x \\ y \end{bmatrix}.$$

The *matrix transpose* gives us this flexibility. When **a** is, say, a 1-by-*m* row vector, the
symbol $\mathbf{a}^T$ denotes the *m*-by-1 column vector with the same entries; and similarly when
b is a column, $\mathbf{b}^T$ is the corresponding row:

$$[1 \quad 2 \quad 3]^T = \begin{bmatrix} 1 \\ 2 \\ 3 \end{bmatrix} ; \quad \begin{bmatrix} 4 \\ 5 \\ 6 \end{bmatrix}^T = [4 \quad 5 \quad 6].$$

We'll denote the set of all m-by-1 column (1-by-m row) vectors with real entries by $\mathbf{R}_{\text{col}}^m$ ($\mathbf{R}_{\text{row}}^m$).[2]

In many mechanical systems, an object executes a rotational motion in a stationary force field. For example, communications satellites rotate around the fixed earth's gravitational field; and inside electrical motors the rotor spins in the fixed magnetic field provided by the stator (see Fig. 2.1). The equations in the next example, which link the rotator's position to the stationary coordinate system, are among the most important relations in engineering analysis.

Example 1. Determine the matrix implementing the operation of rotating a vector in the x, y-plane.

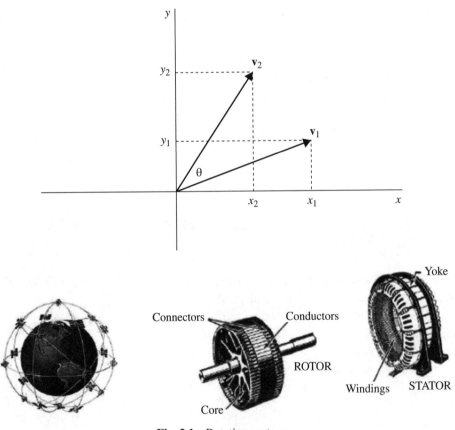

Fig. 2.1 Rotating systems.

---

[2]Our printer encourages us to use the row form whenever possible, to minimize the number of pages.

Solution. Figure 2.1 depicts the vector $\mathbf{v}_1$ with its tail at the origin and its tip at (x_1, y_1). After $\mathbf{v}_1$ is rotated through the angle θ, its tip lies at (x_2, y_2). It is a classic exercise in trigonometry to show that the relations between the tip coordinates are dot products, which can be assembled into the matrix format:

$$x_2 = x_1 \cos\theta - y_1 \sin\theta \qquad \text{or} \qquad \begin{bmatrix} x_2 \\ y_2 \end{bmatrix} = \mathbf{M}_{\text{rot}} \begin{bmatrix} x_1 \\ y_1 \end{bmatrix},$$
$$y_2 = x_1 \sin\theta + y_1 \cos\theta$$

where the matrix for the rotation operator is

$$\mathbf{M}_{\text{rot}} = \begin{bmatrix} \cos\theta & -\sin\theta \\ \sin\theta & \cos\theta \end{bmatrix}. \tag{1}$$

■

For example, the matrix $\begin{bmatrix} 0 & -1 \\ 1 & 0 \end{bmatrix}$ performs a 90-degree counterclockwise rotation, sending $[1\ 0]^T$ to $[0\ 1]^T$ and $[0\ 1]^T$ to $[-1\ 0]^T$.

When the motion of an object is mechanically constrained, as in the instance of a bead on a wire or a block sliding down an inclined plane, its acceleration is not determined by the full force of gravity; the component of gravity that opposes the constraint is nullified by the constraint itself, and only the component parallel to the allowed motion affects the dynamics. Example 2 shows how such components can be expressed in the matrix jargon.

Example 2. Express the matrix describing the operation of orthogonally projecting a vector $\mathbf{v} = \begin{bmatrix} v_1 \\ v_2 \end{bmatrix}$ onto the direction of a unit vector $\mathbf{n} = \begin{bmatrix} n_1 \\ n_2 \end{bmatrix}$ in the x, y-plane.

Solution. Figure 2.2 depicts the projection, $\mathbf{v}_n$, of $\mathbf{v}$ onto the direction of $\mathbf{n}$.

The *length* of $\mathbf{v}_n$ equals the length of $\mathbf{v}$ times the cosine of the angle θ between $\mathbf{v}$ and $\mathbf{n}$. The *direction* of $\mathbf{v}_n$ is, of course, $\mathbf{n}$.

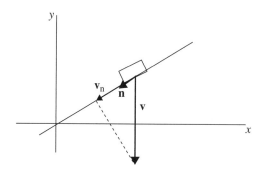

Fig. 2.2 Orthogonal projection.

Now recall that the dot product has a geometric interpretation in vector calculus; in two or three dimensions,

$$\mathbf{u} \cdot \mathbf{w} = (\text{length of } \mathbf{u}) \times (\text{length of } \mathbf{w}) \times (\text{cosine of the angle between } \mathbf{u} \text{ and } \mathbf{w}).$$

Therefore $\mathbf{n} \cdot \mathbf{v}$ equals the length of $\mathbf{v}_n$, since the length of $\mathbf{n}$ is unity. The formula for the projection is thus $\mathbf{v}_n = (\mathbf{n} \cdot \mathbf{v})\mathbf{n}$.

To rearrange this into a matrix-vector product, we exploit the associative law:

$$\mathbf{v}_n = \begin{bmatrix} v_{n,1} \\ v_{n,2} \end{bmatrix} = (\mathbf{n} \cdot \mathbf{v}) \begin{bmatrix} n_1 \\ n_2 \end{bmatrix} = \begin{bmatrix} n_1 \\ n_2 \end{bmatrix} (\mathbf{n} \cdot \mathbf{v}) = \begin{bmatrix} n_1 \\ n_2 \end{bmatrix} \left(\begin{bmatrix} n_1 & n_2 \end{bmatrix} \begin{bmatrix} v_1 \\ v_2 \end{bmatrix} \right)$$

$$= \left(\begin{bmatrix} n_1 \\ n_2 \end{bmatrix} \begin{bmatrix} n_1 & n_2 \end{bmatrix} \right) \begin{bmatrix} v_1 \\ v_2 \end{bmatrix} = \mathbf{M}_{\text{proj}} \begin{bmatrix} v_1 \\ v_2 \end{bmatrix},$$

and conclude that the matrix for the projection operator is

$$\mathbf{M}_{\text{proj}} := \begin{bmatrix} n_1 \\ n_2 \end{bmatrix} \begin{bmatrix} n_1 & n_2 \end{bmatrix} = \begin{bmatrix} n_1^2 & n_1 n_2 \\ n_1 n_2 & n_2^2 \end{bmatrix} = \mathbf{n}\mathbf{n}^T. \tag{2}$$

∎

For example, the matrix $\begin{bmatrix} 0 & 0 \\ 0 & 1 \end{bmatrix}$ projects a vector onto the y-axis having direction $\mathbf{n} = \begin{bmatrix} 0 \\ 1 \end{bmatrix}$, and $\begin{bmatrix} 1/2 & 1/2 \\ 1/2 & 1/2 \end{bmatrix}$ projects it onto the line $y = x$, which has direction $\mathbf{n} = \pm \begin{bmatrix} \sqrt{2}/2 \\ \sqrt{2}/2 \end{bmatrix}$.

The enhanced illumination of a darkened room by a candle held in front of a wall mirror can be predicted quantitatively by postulating a second, identical candle located at the position of the original's image in the mirror. This "imaging" effect is exploited in radiotelemetry by employing "ground planes" in the presence of radiating antennas. Example 3 shows how reflector matrices are used to pinpoint the mirror image.

Example 3. Express the matrix describing the operation of reflecting a vector in the x, y-plane through a mirror-line with a unit *normal* $\mathbf{n}$.

Solution. Figure 2.3 illustrates that the reflection $\hat{\mathbf{v}}$ of $\mathbf{v}$ is obtained by subtracting, from $\mathbf{v}$, two times its orthogonal projection onto the normal. Using (2), then, we exploit the identity matrix to write

$$\begin{bmatrix} \hat{v}_1 \\ \hat{v}_2 \end{bmatrix} = \begin{bmatrix} v_1 \\ v_2 \end{bmatrix} - 2 \left(\begin{bmatrix} n_1 \\ n_2 \end{bmatrix} \begin{bmatrix} n_1 & n_2 \end{bmatrix} \right) \begin{bmatrix} v_1 \\ v_2 \end{bmatrix}$$

$$= \left(\begin{bmatrix} 1 & 0 \\ 0 & 1 \end{bmatrix} - 2 \left(\begin{bmatrix} n_1 \\ n_2 \end{bmatrix} \begin{bmatrix} n_1 & n_2 \end{bmatrix} \right) \right) \begin{bmatrix} v_1 \\ v_2 \end{bmatrix} = \mathbf{M}_{\text{ref}} \begin{bmatrix} v_1 \\ v_2 \end{bmatrix}$$

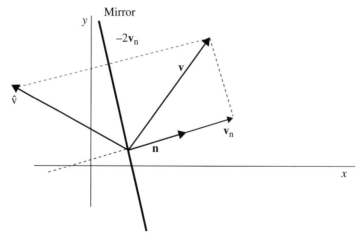

Fig. 2.3   Mirror reflector.

and the matrix for the reflection operator is

$$\mathbf{M}_{\text{ref}} := \begin{bmatrix} 1 & 0 \\ 0 & 1 \end{bmatrix} - 2 \begin{bmatrix} n_1 \\ n_2 \end{bmatrix} \begin{bmatrix} n_1 & n_2 \end{bmatrix} = \begin{bmatrix} 1 - 2n_1^2 & -2n_1 n_2 \\ -2n_1 n_2 & 1 - 2n_2^2 \end{bmatrix} = \mathbf{I} - 2\mathbf{nn}^T. \quad (3)$$

∎

For example, $\begin{bmatrix} 0 & 1 \\ 1 & 0 \end{bmatrix}$ reflects a vector through the line $y = x$, whose unit normal is $\pm \begin{bmatrix} \sqrt{2}/2 \\ -\sqrt{2}/2 \end{bmatrix}$.

The associativity property, $\mathbf{A}(\mathbf{BC}) = (\mathbf{AB})\mathbf{C}$, has a nice interpretation in this context. Consider the complicated operation of rotating vectors and then reflecting them, regarded as a single (concatenated) operation. Is this describable as a matrix product? And if so, what is the matrix that performs the operation? The composite operation is clearly prescribed by $\mathbf{y} = \mathbf{M}_{\text{ref}}(\mathbf{M}_{\text{rot}}\mathbf{x})$. Associativity equates this with $\mathbf{y} = (\mathbf{M}_{\text{ref}}\mathbf{M}_{\text{rot}})\mathbf{x}$, providing us with the single matrix $(\mathbf{M}_{\text{ref}}\mathbf{M}_{\text{rot}})$ that performs the move. For example, if we employ the 90° rotation analyzed in Example 1 followed by reflecting in the mirror $y = x$ as per Example 3, the vector $[1\ 0]^T$ is rotated to $[0\ 1]^T$ and then reflected back to $[1\ 0]^T$, while $[0\ 1]^T$ is rotated to $[-1\ 0]^T$ and reflected to $[0\ -1]^T$. The matrix accomplishing this maneuver is

$$\mathbf{M}_{\text{ref/rot}} = \begin{bmatrix} 1 & 0 \\ 0 & -1 \end{bmatrix},$$

in agreement with associativity, which stipulates the composite operator

$$\mathbf{M}_{\text{ref}}\mathbf{M}_{\text{rot}} = \begin{bmatrix} 0 & 1 \\ 1 & 0 \end{bmatrix} \begin{bmatrix} 0 & -1 \\ 1 & 0 \end{bmatrix} = \begin{bmatrix} 1 & 0 \\ 0 & -1 \end{bmatrix}$$

(Equations (1) and (3)).

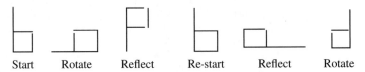

Start   Rotate   Reflect   Re-start   Reflect   Rotate

Fig. 2.4   Violation of commutativity.

This example also demonstrates the violation of commutative multiplication, because $\mathbf{M}_{\text{rot}}\,\mathbf{M}_{\text{ref}}$ equals $\begin{bmatrix} -1 & 0 \\ 0 & 1 \end{bmatrix}$. Indeed, compare the progress of the symbol in Figure 2.4, as it is first rotated/reflected, then reflected/rotated.

Example 4.   Find a 2-by-2 matrix that orthogonally projects a vector onto the line $y = 2x$, and then rotates it counterclockwise by $45°$.

Solution. A vector parallel to the line $y = 2x$ is $[1\ 2]^T$; a *unit* vector is then $\mathbf{n}_{\text{proj}} = [1\ 2]^T/\sqrt{5}$. Thus (Eq. (2))

$$\mathbf{M}_{\text{proj}} = \mathbf{n}_{\text{proj}}\mathbf{n}_{\text{proj}}^T = \frac{1}{5}\begin{bmatrix} 1 \\ 2 \end{bmatrix}[1\ 2] = \frac{1}{5}\begin{bmatrix} 1 & 2 \\ 2 & 4 \end{bmatrix}.$$

Eq. (1) gives the rotator matrix:

$$\mathbf{M}_{\text{rot}} = \begin{bmatrix} \cos 45° & -\sin 45° \\ \sin 45° & \cos 45° \end{bmatrix} = \frac{1}{\sqrt{2}}\begin{bmatrix} 1 & -1 \\ 1 & 1 \end{bmatrix}.$$

The composite action is then represented by the matrix (watch the order!)

$$\mathbf{M}_{\text{rot}}\mathbf{M}_{\text{proj}} = \frac{1}{\sqrt{2}}\begin{bmatrix} 1 & -1 \\ 1 & 1 \end{bmatrix}\frac{1}{5}\begin{bmatrix} 1 & 2 \\ 2 & 4 \end{bmatrix} = \frac{1}{5\sqrt{2}}\begin{bmatrix} -1 & -2 \\ 3 & 6 \end{bmatrix}. \qquad \blacksquare$$

We will see that these examples, which demonstrate that elementary row operations, rotations, projections, and reflections are equivalent to left multiplication by appropriate "operator" matrices, are extremely useful as theoretical tools. They provide us with an algebraic language for expressing the operations. In practice, however, they are quite inefficient. For instance, it would be foolish to execute a full matrix multiplication simply to exchange two rows in a matrix.

Exercises 2.2

In Problems 1–4 we identify points of the x,y-plane as 2-by-1 column vectors.

1. Find a 2-by-2 matrix $\mathbf{A}$ such that
    (a) $\mathbf{A}\mathbf{x}$ rotates a vector $\mathbf{x}$ by $30°$ in the counterclockwise direction;
    (b) $\mathbf{A}\mathbf{x}$ reflects a vector $\mathbf{x}$ through the line $2x + y = 0$;
    (c) $\mathbf{A}\mathbf{x}$ orthogonally projects a vector $\mathbf{x}$ onto the direction of the line $y = x$.

2. Find a 2-by-2 matrix **A** such that
   (a) **Ax** rotates a vector **x** by $45°$ in the clockwise direction;
   (b) **Ax** reflects a vector **x** through the line $x - 3y = 0$;
   (c) **Ax** orthogonally projects a vector **x** onto the direction of the line $y = 3x$.

3. Show that a 2-by-2 matrix $\mathbf{A} = \begin{bmatrix} a & b \\ c & d \end{bmatrix}$ with $ad - bc \neq 0$, regarded as a function that maps **x** into **Ax**, transforms any straight line in the plane into another straight line. Construct an example to show that if $ad - bc = 0$, then a straight line might be transformed to a single point.

4. Show that the matrix **A** in Problem 3 with $ad - bc \neq 0$ transforms the sides of any parallelogram having a vertex in the origin to the sides of another such parallelogram.

5. What is the matrix that performs the following composite operations in the x, y-plane?
   (a) Reflects a vector **x** in the line $x + 2y = 0$, and then orthogonally projects it onto the direction of the line $y = -x$.
   (b) Rotates a vector **x** counterclockwise by $120°$, then reflects it in the line $y - 2x = 0$.

6. What is the matrix that performs the following composite operations in the x, y-plane?
   (a) Orthogonally projects a vector **x** onto the direction of the line $y = 3x$, and then rotates it clockwise by $45°$.
   (b) Reflects a vector **x** in the line $y = -x$, and then orthogonally projects it onto the direction of the x-axis.

7. Construct a single 2-by-2 matrix **A** that transforms (via left multiplication by **A**) any column vector **x** in the plane to a vector one-half its length that is rotated $90°$ in the counterclockwise direction.

8. Verify, and then explain why, $\mathbf{M}_{\text{proj}}^2 = \mathbf{M}_{\text{proj}}$ in Example 2.

9. Verify, and then explain why, $\mathbf{M}_{\text{rot}}^4 = \mathbf{I}$ for a 90-degree rotation.

10. (a) Review Example 2, interpreting the vectors in Figure 2.2 as three dimensional, and argue that the formula $\mathbf{M}_{\text{proj}} = \mathbf{n}\mathbf{n}^T$ (see Eq. (2)) represents the orthogonal projection operation in three dimensions (when **n** is a unit vector in $\mathbf{R}_{\text{col}}^3$).
    (b) Confirm the following properties of the orthogonal projection operation algebraically, by manipulating the formula $\mathbf{M}_{\text{proj}} = \mathbf{n}\mathbf{n}^T$ (when **n** is a unit vector in $\mathbf{R}_{\text{col}}^3$):

    - Repeated applications of $\mathbf{M}_{\text{proj}}$ simply reproduce the result of the first projection.
    - For any vector **v**, the *residual* vector $\mathbf{v} - \mathbf{M}_{\text{proj}}\mathbf{v}$ is orthogonal to the projection $\mathbf{M}_{\text{proj}}\mathbf{v}$.

11. Construct the three dimensional orthogonal projection matrices that project a vector in $\mathbf{R}_{\text{col}}^3$
    (a) onto the z-axis and
    (b) onto the line $x = y = z$.

12. Apply the projection matrices constructed in Problem 11 to the vector $\mathbf{v} = [1\,2\,3]^T$, and confirm the properties listed in Problem 10(b).

13. (a) Review Example 3, interpreting the vectors in Figure 2.3 as *three* dimensional and the mirror as a *plane* with unit normal $\mathbf{n}$, and argue that the formula $\mathbf{M}_{\text{ref}} = \mathbf{I} - 2\mathbf{n}\mathbf{n}^T$ (see Eq. (3)) represents the mirror reflection (when $\mathbf{n}$ is a unit vector in $\mathbf{R}_{\text{col}}^3$).

    (b) Confirm the following properties of the reflection operation algebraically, by manipulating the formula $\mathbf{M}_{\text{ref}} = \mathbf{I} - 2\mathbf{n}\mathbf{n}^T$ (when $\mathbf{n}$ is a unit vector in $\mathbf{R}_{\text{col}}^3$):

    • Two applications of $\mathbf{M}_{\text{ref}}$ simply reproduce the original vector.

    • For any vector $\mathbf{v}$, the *residual* vector $\mathbf{v} - \mathbf{M}_{\text{ref}}\mathbf{v}$ is parallel to the mirror normal $\mathbf{n}$.

14. Construct the three-dimensional reflection matrices that reflect a vector in $\mathbf{R}_{\text{col}}^3$
    (a) in the mirror given by the x, y-plane and
    (b) in the plane mirror normal to the line $x = y = z$.
    (c) Apply these reflection matrices to the vector $\mathbf{v} = [1\,2\,3]^T$, and confirm the properties listed in Problem 13(b).

2.3   THE INVERSE AND THE TRANSPOSE

We have noted that the formulas $\mathbf{IA} = \mathbf{A}$ and $\mathbf{AI} = \mathbf{A}$ demonstrate that the identity matrix $\mathbf{I}$ is analogous to the scalar 1 in arithmetic. So it is natural to look for the analog to the scalar inverse. Keeping in mind that matrix multiplication is noncommutative, we tentatively call the matrix $\mathbf{B}$ a **right inverse** for $\mathbf{A}$ if $\mathbf{AB} = \mathbf{I}$, and a **left inverse** if $\mathbf{BA} = \mathbf{I}$. For example, the calculation

$$\mathbf{AB} = \begin{bmatrix} 1 & 2 & 1 \\ 1 & 3 & 2 \\ 1 & 0 & 1 \end{bmatrix} \begin{bmatrix} 3/2 & -1 & 1/2 \\ 1/2 & 0 & -1/2 \\ -3/2 & 1 & 1/2 \end{bmatrix} = \begin{bmatrix} 1 & 0 & 0 \\ 0 & 1 & 0 \\ 0 & 0 & 1 \end{bmatrix} \tag{1}$$

shows that $\mathbf{B}$ is a right inverse of $\mathbf{A}$; and, of course, that $\mathbf{A}$ is a *left* inverse of $\mathbf{B}$.

In this section we will show that if a square matrix $\mathbf{A}$ has a right inverse, it has *only* one. Furthermore this right inverse is also a left inverse, as is exemplified by the matrices in (1), because $\mathbf{B}$ is also a left inverse:

$$\mathbf{BA} = \begin{bmatrix} 3/2 & -1 & 1/2 \\ 1/2 & 0 & -1/2 \\ -3/2 & 1 & 1/2 \end{bmatrix} \begin{bmatrix} 1 & 2 & 1 \\ 1 & 3 & 2 \\ 1 & 0 & 1 \end{bmatrix} = \begin{bmatrix} 1 & 0 & 0 \\ 0 & 1 & 0 \\ 0 & 0 & 1 \end{bmatrix}.$$

Thus a square matrix either has no inverse of any kind, or it has a unique, two-sided inverse which we can designate as *the* inverse, $\mathbf{A}^{-1}$. Anticipating these results, we formulate the following definition.

Inverse of a Square Matrix

Definition 1. The **inverse** of an *n*-by-*n* matrix $\mathbf{A}$ is the unique matrix $\mathbf{A}^{-1}$ satisfying

$$\mathbf{AA}^{-1} = \mathbf{A}^{-1}\mathbf{A} = \mathbf{I}. \tag{2}$$

Square matrices that have inverses are said to be **invertible** or **nonsingular**; otherwise they are **noninvertible** or **singular**.

If a matrix has an "operator" interpretation, it may be easy to figure out its inverse.

Example 1. Derive the inverses of the rotation and reflection matrices of Section 2.2.

Solution. Equation (2) says that the inverse undoes the operation performed by the original matrix. To undo a rotation through an angle θ, we simply need to rotate through the angle $(-\theta)$. Therefore, if $\mathbf{M}_{\text{rot}}(\theta)$ denotes the rotation matrix (recall Example 1, Section 2.2), $\mathbf{M}_{\text{rot}}(-\theta)\mathbf{M}_{\text{rot}}(\theta) = \mathbf{I}$:

$$\begin{bmatrix} \cos(-\theta) & -\sin(-\theta) \\ \sin(-\theta) & \cos(-\theta) \end{bmatrix} \begin{bmatrix} \cos\theta & -\sin\theta \\ \sin\theta & \cos\theta \end{bmatrix} = \begin{bmatrix} 1 & 0 \\ 0 & 1 \end{bmatrix}, \tag{3}$$

and $\mathbf{M}_{\text{rot}}(\theta)^{-1} = \mathbf{M}_{\text{rot}}(-\theta)$. ∎

Undoing a reflection (Example 3, Section 2.2) is even simpler—we reflect again: $\mathbf{M}_{\text{ref}}\,\mathbf{M}_{\text{ref}} = \mathbf{I}$, and $\mathbf{M}_{\text{ref}}^{-1} = \mathbf{M}_{\text{ref}}$. Problem 13, Exercises 2.2, confirmed this identity directly using the formula for $\mathbf{M}_{\text{ref}}$ from Section 2.1.

The orthogonal projection operation described in Example 2 of Section 2.2 cannot be inverted; one cannot deduce, simply from the resultant projection $\mathbf{v_n}$, what the original vector $\mathbf{v}$ was. Therefore, the projection matrix has no inverse. Neither does the zero matrix $\mathbf{0}$, because the equation $\mathbf{0B} = \mathbf{I}$ can never be satisfied.

Example 2.   Derive the inverses of the elementary row operation matrices.

Solution.   To undo a row exchange, one simply repeats the exchange. To undo the addition of a multiple of one row to another, one *subtracts* the same multiple. To undo multiplication of a row by a nonzero constant, one *divides* the row by that constant. Applying these statements to the elementary row operation matrices of Section 2.1 yields

$$
\begin{bmatrix} 1 & 0 & 0 \\ 0 & 0 & 1 \\ 0 & 1 & 0 \end{bmatrix}^{-1} = \begin{bmatrix} 1 & 0 & 0 \\ 0 & 0 & 1 \\ 0 & 1 & 0 \end{bmatrix}, \quad
\begin{bmatrix} 1 & 0 & 0 \\ \alpha & 1 & 0 \\ 0 & 0 & 1 \end{bmatrix}^{-1} = \begin{bmatrix} 1 & 0 & 0 \\ -\alpha & 1 & 0 \\ 0 & 0 & 1 \end{bmatrix},
$$

$$
\begin{bmatrix} 1 & 0 & 0 \\ 0 & \beta & 0 \\ 0 & 0 & 1 \end{bmatrix}^{-1} = \begin{bmatrix} 1 & 0 & 0 \\ 0 & \frac{1}{\beta} & 0 \\ 0 & 0 & 1 \end{bmatrix}. \qquad \blacksquare
$$

Now we are going to show how to compute a square matrix's right inverse (or detect if it has none). A subsequent perusal of the calculation will reveal why the right inverse is also a left inverse and is unique.

Consider the requirement that a matrix $\mathbf{X}$ be a right inverse of the matrix $\mathbf{A}$ in (1):

$$
\mathbf{AX} = \begin{bmatrix} 1 & 2 & 1 \\ 1 & 3 & 2 \\ 1 & 0 & 1 \end{bmatrix} \begin{bmatrix} x_{11} & x_{12} & x_{13} \\ x_{21} & x_{22} & x_{23} \\ x_{31} & x_{32} & x_{33} \end{bmatrix} = \begin{bmatrix} 1 & 0 & 0 \\ 0 & 1 & 0 \\ 0 & 0 & 1 \end{bmatrix}.
$$

The individual columns of $\mathbf{X}$ satisfy

$$
\begin{bmatrix} 1 & 2 & 1 \\ 1 & 3 & 2 \\ 1 & 0 & 1 \end{bmatrix} \begin{bmatrix} x_{11} \\ x_{21} \\ x_{31} \end{bmatrix} = \begin{bmatrix} 1 \\ 0 \\ 0 \end{bmatrix}, \quad
\begin{bmatrix} 1 & 2 & 1 \\ 1 & 3 & 2 \\ 1 & 0 & 1 \end{bmatrix} \begin{bmatrix} x_{12} \\ x_{22} \\ x_{32} \end{bmatrix} = \begin{bmatrix} 0 \\ 1 \\ 0 \end{bmatrix}, \qquad (4)
$$

$$
\begin{bmatrix} 1 & 2 & 1 \\ 1 & 3 & 2 \\ 1 & 0 & 1 \end{bmatrix} \begin{bmatrix} x_{13} \\ x_{23} \\ x_{33} \end{bmatrix} = \begin{bmatrix} 0 \\ 0 \\ 1 \end{bmatrix}.
$$

Thus computing the right inverse amounts to solving a linear system with several right-hand sides, starting with the augmented matrix $[\mathbf{A}\!:\!\mathbf{I}]$. The Gauss–Jordan algorithm is best suited for this task.

Example 3.   Find the right inverse of the matrix $\mathbf{A}$ in Equation (1).

Solution.   We form the matrix $[\mathbf{A}\!:\!\mathbf{I}]$ and perform the Gauss–Jordan algorithm:

$$
\begin{bmatrix} 1 & 2 & 1 & : & 1 & 0 & 0 \\ 1 & 3 & 2 & : & 0 & 1 & 0 \\ 1 & 0 & 1 & : & 0 & 0 & 1 \end{bmatrix}
\begin{array}{l} (-1)\rho_1 + \rho_2 \to \rho_2 \\ (-1)\rho_1 + \rho_3 \to \rho_3 \end{array}
\begin{bmatrix} 1 & 2 & 1 & : & 1 & 0 & 0 \\ 0 & 1 & 1 & : & -1 & 1 & 0 \\ 0 & -2 & 0 & : & -1 & 0 & 1 \end{bmatrix}
$$

$$
\begin{array}{l} (-2)\rho_2 + \rho_1 \to \rho_1 \\[4pt] 2\rho_2 + \rho_3 \to \rho_3 \end{array}
\begin{bmatrix} 1 & 0 & 0 & : & \frac{3}{2} & -1 & \frac{1}{2} \\ 0 & 1 & 1 & : & -1 & 1 & 0 \\ 0 & 0 & 2 & : & -3 & 2 & 1 \end{bmatrix}
$$

$$(-1/2)\rho_3 \to \rho_3 \begin{bmatrix} 1 & 0 & 0 & \vdots & \frac{3}{2} & -1 & \frac{1}{2} \\ 0 & 1 & 1 & \vdots & -1 & 1 & 0 \\ 0 & 0 & 1 & \vdots & -\frac{3}{2} & 1 & \frac{1}{2} \end{bmatrix}$$

$$(-1)\rho_3 + \rho_2 \to \rho_2 \begin{bmatrix} 1 & 0 & 0 & \vdots & \frac{3}{2} & -1 & \frac{1}{2} \\ 0 & 1 & 0 & \vdots & \frac{1}{2} & 0 & -\frac{1}{2} \\ 0 & 0 & 1 & \vdots & -\frac{3}{2} & 1 & \frac{1}{2} \end{bmatrix}.$$

Now we can read the inverse from the final three columns.

$$\mathbf{A}^{-1} = \begin{bmatrix} 3/2 & -1 & 1/2 \\ 1/2 & 0 & -1/2 \\ -3/2 & 1 & 1/2 \end{bmatrix} \text{ (in agreement with Eq. (1)).} \qquad \blacksquare$$

Having employed only elementary row operations to transform $[\mathbf{A} : \mathbf{I}]$ into $[\mathbf{I} : \mathbf{B}]$, we can use a little juggling to prove that the right inverse $\mathbf{B}$ is a left inverse.

Right and Left Inverses

Theorem 1. If the square matrix $\mathbf{A}$ has a right inverse $\mathbf{B}$, then $\mathbf{B}$ is the *unique* right inverse, and $\mathbf{B}$ is also the unique left inverse.

Proof. We demonstrate that the right inverse $\mathbf{B}$ constructed in Example 3 is a left inverse by invoking the operator-matrix characterization of the row operations employed in its construction; the argument generalizes to arbitrary square matrices. Let $\mathbf{E}_1, \mathbf{E}_2, \dots, \mathbf{E}_6$ be the operator matrices that accomplish (through left multiplication) the various row operations that were executed in the example, so that we can write

$$\mathbf{E}_6 \, \mathbf{E}_5 \cdots \mathbf{E}_2 \mathbf{E}_1 \mathbf{A} = \mathbf{I}. \tag{5}$$

Right multiplication by $\mathbf{B}$ gives

$$\mathbf{E}_6 \, \mathbf{E}_5 \cdots \mathbf{E}_2 \, \mathbf{E}_1 \, \mathbf{A} \, \mathbf{B} = \mathbf{E}_6 \, \mathbf{E}_5 \cdots \mathbf{E}_2 \, \mathbf{E}_1 \mathbf{I} = \mathbf{I} \, \mathbf{B} = \mathbf{B}. \tag{6}$$

Equation (6) implies $\mathbf{B}$ equals $\mathbf{E}_6 \mathbf{E}_5 \cdots \mathbf{E}_2 \mathbf{E}_1$, and thus (5) shows that $\mathbf{B}\mathbf{A} = \mathbf{I}$. In other words, the right inverse $\mathbf{B}$ is a left inverse also.

To see that (the two-sided inverse) $\mathbf{B}$ is the only possible *right* inverse, suppose that $\mathbf{C}$ is another right inverse. Then left multiplication of the relation $\mathbf{I} = \mathbf{A}\mathbf{C}$ by $\mathbf{B}$, together with associativity, implies that $\mathbf{C}$ is the same as $\mathbf{B}$:

$$\mathbf{B}\mathbf{I} = \mathbf{B}(\mathbf{A}\mathbf{C}) = (\mathbf{B}\mathbf{A})\mathbf{C} = \mathbf{I}\mathbf{C};$$

that is, $\mathbf{B} = \mathbf{C}$. A similar argument shows that $\mathbf{B}$ is the unique left inverse of $\mathbf{A}$. $\qquad \blacksquare$

The fact that the right inverse (if it exists) is unique means that the solutions to the systems (4) are unique. In the terminology of Section 1.3, then, there are no "free variables"; thus there are no zeros on the diagonal of the row echelon form resulting from the forward part of the Gauss elimination algorithm. Moreover, when there are no such zeros, back substitution can be carried out and the computation of the inverse is unhindered.

Another way of stating this is to use the *rank* concept introduced in Section 1.4, since the rank of a square row echelon matrix with no zeros on the diagonal is equal to its number of columns; that is, the matrix has *full rank*. Noting that a matrix has full rank when and only when the homogeneous sytem $\mathbf{Ax} = \mathbf{0}$ has $\mathbf{x} = \mathbf{0}$ as its only solution, we summarize with the following.

Equivalent Conditions for the Existence of the Inverse

Theorem 2. Let $\mathbf{A}$ be a square matrix. Then the following statements are equivalent:

- $\mathbf{A}$ has an inverse $\mathbf{A}^{-1}$ (which is the unique left inverse as well as unique right inverse).
- $\mathbf{A}$ can be reduced by elementary row operations to a row echelon form with full rank.
- The homogeneous system $\mathbf{Ax} = \mathbf{0}$ has only the trivial solution $\mathbf{x} = \mathbf{0}$.
- For every column vector $\mathbf{b}$, the system $\mathbf{Ax} = \mathbf{b}$ has a unique solution.

We noted earlier that an orthogonal projection operation is noninvertible. The matrix

$$\begin{bmatrix} 1/2 & 1/2 \\ 1/2 & 1/2 \end{bmatrix}$$

projects a vector onto the line $y = x$, as we saw in Example 2 of the preceding section; and clearly subtracting the first row from the second yields a row echelon matrix of rank *one*, so its noninvertibility is consistent with Theorem 2.

Once we *know* an inverse for the coefficient matrix $\mathbf{A}$ in a system of linear equations $\mathbf{Ax} = \mathbf{b}$, the solution can be calculated directly by computing $\mathbf{x} = \mathbf{A}^{-1}\mathbf{b}$, as the following reasoning shows:

$$\mathbf{Ax} = \mathbf{b} \text{ implies } \mathbf{A}^{-1}\mathbf{Ax} = \mathbf{A}^{-1}\mathbf{b} \text{ implies } \mathbf{x} = \mathbf{A}^{-1}\mathbf{b}.$$

For the system

$$\begin{bmatrix} 1 & 2 & 1 \\ 1 & 3 & 2 \\ 1 & 0 & 1 \end{bmatrix} \begin{bmatrix} x_1 \\ x_2 \\ x_3 \end{bmatrix} = \begin{bmatrix} 1 \\ -1 \\ 0 \end{bmatrix}$$

whose coefficient matrix and its inverse appear in Equation (1),

$$
\begin{bmatrix} x_1 \\ x_2 \\ x_3 \end{bmatrix} = \begin{bmatrix} 3/2 & -1 & 1/2 \\ 1/2 & 0 & -1/2 \\ -3/2 & 1 & 1/2 \end{bmatrix} \begin{bmatrix} 1 \\ -1 \\ 0 \end{bmatrix} = \begin{bmatrix} 5/2 \\ 1/2 \\ -5/2 \end{bmatrix}.
$$

When $\mathbf{A}^{-1}$ is known, this calculation is certainly easier than applying Gauss elimination. However, the reader may correctly surmise that computing $\mathbf{A}^{-1}$ and using it to solve a system is less efficient than solving it directly by Gauss elimination, since one must perform the Gauss (or Gauss–Jordan) algorithm anyway, to get the inverse. It can be shown that even if the system has multiple right-hand sides, computing and using the inverse is less efficient than Gauss elimination. Although the inverse is an extremely useful theoretical concept (as we shall see), in practice good computists avoid its calculation if at all possible.

It is often useful in matrix analysis to switch one's focus from a matrix's rows to its columns, or vice versa—reflecting the matrix across its diagonal, as it were. For this purpose we extend the (vector) transpose introduced in the previous section.

The Matrix Transpose

Definition 2. The **transpose** $\mathbf{A}^T$ of an m-by-n matrix $\mathbf{A}$ is the n-by-m matix whose (i,j)th entry is the (j,i)th entry of $\mathbf{A}$.

For example,

$$
\begin{bmatrix} 1 & 2 & 3 \\ 4 & 5 & 6 \\ 7 & 8 & 9 \end{bmatrix}^T = \begin{bmatrix} 1 & 4 & 7 \\ 2 & 5 & 8 \\ 3 & 6 & 9 \end{bmatrix}, \quad \begin{bmatrix} 1 & 2 & 3 \\ 4 & 5 & 6 \end{bmatrix}^T = \begin{bmatrix} 1 & 4 \\ 2 & 5 \\ 3 & 6 \end{bmatrix}, \tag{7}
$$

$$
\begin{bmatrix} 1 & 2 & 3 \end{bmatrix}^T = \begin{bmatrix} 1 \\ 2 \\ 3 \end{bmatrix}, \quad \begin{bmatrix} 1 \\ 2 \\ 3 \end{bmatrix}^T = \begin{bmatrix} 1 & 2 & 3 \end{bmatrix}.
$$

Note that the rows of $\mathbf{A}$ become the columns of $\mathbf{A}^T$:

$$
\begin{bmatrix} - & \mathbf{u} & - \\ - & \mathbf{v} & - \\ - & \mathbf{w} & - \end{bmatrix}^T = \begin{bmatrix} | & | & | \\ \mathbf{u} & \mathbf{v} & \mathbf{w} \\ | & | & | \end{bmatrix}, \tag{8}
$$

and that when $\mathbf{A}^T = \mathbf{A}$ the matrix is *symmetric* (about its diagonal):

$$
\begin{bmatrix} a & b & c \\ b & d & e \\ c & e & f \end{bmatrix}^T = \begin{bmatrix} a & b & c \\ b & d & e \\ c & e & f \end{bmatrix}. \tag{9}
$$

We make two brief, elementary observations:

(a) $(\mathbf{A}^T)^T = \mathbf{A}$;

(b) the dot product $\mathbf{v} \cdot \mathbf{w}$ can be written as $\mathbf{v}^T\mathbf{w}$ when $\mathbf{v}$ and $\mathbf{w}$ are column vectors, and as $\mathbf{v}\mathbf{w}^T$ when they are rows.

Example 4. Show that the transposes of the operator matrices that perform the elementary row operations are themselves operator matrices for elementary row operations.

Solution. These matrices were illustrated in Example 2, Section 2.1. For a row-switching matrix, such as

$$\begin{bmatrix} 1 & 0 & 0 & 0 \\ 0 & 0 & 0 & 1 \\ 0 & 0 & 1 & 0 \\ 0 & 1 & 0 & 0 \end{bmatrix},$$

that exchanges rows 2 and 4, the transpose equals the original matrix; it is symmetric. The matrix

$$\begin{bmatrix} 1 & 0 & 0 & 0 \\ 0 & c & 0 & 0 \\ 0 & 0 & 1 & 0 \\ 0 & 0 & 0 & 1 \end{bmatrix},$$

multiplies row 2 by c, and is also symmetric. Finally the matrix

$$\begin{bmatrix} 1 & 0 & 0 & 0 \\ 0 & 1 & 0 & 0 \\ 0 & 0 & 1 & 0 \\ 0 & c & 0 & 1 \end{bmatrix},$$

which adds c times row 2 to row 4, has as its transpose

$$\begin{bmatrix} 1 & 0 & 0 & 0 \\ 0 & 1 & 0 & c \\ 0 & 0 & 1 & 0 \\ 0 & 0 & 0 & 1 \end{bmatrix};$$

it adds c times row 4 to row 2. The generalization to n-by-n matrices is clear. ∎

Curiously, the inverse and the transpose of a matrix product are calculated by the same rule:

Transpose and Inverse of the Matrix Product

Theorem 3. If the matrix product $\mathbf{AB}$ is defined, then we have

$$(\mathbf{AB})^T = \mathbf{B}^T \mathbf{A}^T. \tag{10}$$

Moreover, if $\mathbf{A}$ and $\mathbf{B}$ are invertible, then so is $\mathbf{AB}$ and

$$(\mathbf{AB})^{-1} = \mathbf{B}^{-1} \mathbf{A}^{-1}. \tag{11}$$

Proof. The inverse property (11) is easier to prove; we simply need to confirm that $(\mathbf{B}^{-1}\mathbf{A}^{-1})(\mathbf{AB}) = \mathbf{I}$, and this follows from manipulation of parentheses (as sanctioned by associativity): $(\mathbf{B}^{-1}\mathbf{A}^{-1})(\mathbf{AB}) = \mathbf{B}^{-1}(\mathbf{A}^{-1}\mathbf{A})\mathbf{B} = \mathbf{B}^{-1}\mathbf{B} = \mathbf{I}$.

To see (10) requires some computation. Compare the calculation of the (2,3) entry of $\mathbf{AB}$ with the (3,2) entry of $\mathbf{B}^T\mathbf{A}^T$ in the following:

$$\mathbf{AB} = \begin{bmatrix} - & - & - \\ s & t & u \end{bmatrix} \begin{bmatrix} - & - & x & - \\ - & - & y & - \\ - & - & z & - \end{bmatrix} = \begin{bmatrix} - & - & - & - \\ - & - & * & - \end{bmatrix},$$

$$\mathbf{B}^T\mathbf{A}^T = \begin{bmatrix} - & - & - \\ - & - & - \\ x & y & z \\ - & - & - \end{bmatrix} \begin{bmatrix} - & s \\ - & t \\ - & u \end{bmatrix} = \begin{bmatrix} - & - \\ - & - \\ - & * \\ - & - \end{bmatrix}.$$

Both entries contain $sx + ty + uz$, the dot product of the second row of $\mathbf{A}$ (which is the second *column* of $\mathbf{A}^T$) with the third column of $\mathbf{B}$ (i.e., the third *row* of $\mathbf{B}^T$). Generalizing, we see that the computation of the (i,j)th entry of $\mathbf{B}^T\mathbf{A}^T$ is identical to that of the (j,i)th entry of $\mathbf{AB}$, proving (10). ∎

Theorem 3 generalizes to extended products easily:

$$(\mathbf{A}_1\mathbf{A}_2 \cdots \mathbf{A}_{k-1}\mathbf{A}_k)^{-1} = \mathbf{A}_k^{-1}\mathbf{A}_{k-1}^{-1} \cdots \mathbf{A}_2^{-1}\mathbf{A}_1^{-1} \tag{12}$$
$$(\mathbf{A}_1\mathbf{A}_2 \cdots \mathbf{A}_{k-1}\mathbf{A}_k)^T = \mathbf{A}_k^T\mathbf{A}_{k-1}^T \cdots \mathbf{A}_2^T\mathbf{A}_1^T,$$

where in (12) we presume that all the inverses exist. (See Problems 21 and 22.)

Example 5. Show that the transpose of the inverse equals the inverse of the transpose; that is, for invertible matrices $(\mathbf{A}^{-1})^T = (\mathbf{A}^T)^{-1}$.

Solution. By Equation (10), the transpose of the equation $\mathbf{AA}^{-1} = \mathbf{I}$ is $(\mathbf{A}^{-1})^T\mathbf{A}^T = \mathbf{I}^T = \mathbf{I}$. so $(\mathbf{A}^{-1})^T$ is, indeed, the inverse of $\mathbf{A}^T$. ∎

Exercises 2.3

In Problems 1–6, without performing any computations, find the inverse of the given elementary row operation matrix.

1. $\begin{bmatrix} 1 & 0 & -5 \\ 0 & 1 & 0 \\ 0 & 0 & 1 \end{bmatrix}$ 2. $\begin{bmatrix} 1 & 0 & 0 \\ 0 & 1 & 0 \\ 0 & 0 & \pi \end{bmatrix}$

3. $\begin{bmatrix} 1 & 0 & 0 \\ 0 & 5 & 0 \\ 0 & 0 & 1 \end{bmatrix}$ 4. $\begin{bmatrix} 1 & 0 & 0 \\ 0 & 0 & 1 \\ 0 & 1 & 0 \end{bmatrix}$

5. $\begin{bmatrix} 1 & 0 & 0 & 0 \\ 0 & 0 & 1 & 0 \\ 0 & 1 & 0 & 0 \\ 0 & 0 & 0 & 1 \end{bmatrix}$ 6. $\begin{bmatrix} 1 & 0 & 0 & 0 \\ 0 & 1 & 0 & 0 \\ 0 & 0 & 1 & 0 \\ 0 & -18 & 0 & 1 \end{bmatrix}$

In Problems 7–12, compute the inverse of the given matrix, if it exists.

7. $\begin{bmatrix} 2 & 1 \\ -1 & 4 \end{bmatrix}$ 8. $\begin{bmatrix} 4 & 1 \\ 5 & 9 \end{bmatrix}$

9. $\begin{bmatrix} 1 & 1 & 1 \\ 1 & 2 & 1 \\ 2 & 3 & 2 \end{bmatrix}$ 10. $\begin{bmatrix} 1 & 1 & 1 \\ 1 & 2 & 3 \\ 0 & 1 & 1 \end{bmatrix}$

11. $\begin{bmatrix} 3 & 0 & 0 & 1 \\ 2 & 0 & 0 & 0 \\ 0 & 0 & 2 & 0 \\ 0 & 1 & 2 & 0 \end{bmatrix}$ 12. $\begin{bmatrix} 3 & 4 & 5 & 6 \\ 0 & 0 & 1 & 2 \\ 0 & 0 & 4 & 5 \\ 0 & 0 & 0 & 8 \end{bmatrix}$

13. Let $\mathbf{A} = \begin{bmatrix} 1 & 0 & 3 \\ 0 & 2 & 4 \\ 4 & -3 & 8 \end{bmatrix}$. Compute $\mathbf{A}^{-1}$ and use it to solve each of the following systems:

    (a) $\mathbf{Ax} = [1\ 0\ 3]^T$ (b) $\mathbf{Ax} = [-3\ 7\ 1]^T$ (c) $\mathbf{Ax} = [0\ -4\ 1]^T$

14. Let $\mathbf{A}$ be the matrix in Problem 11. Use $\mathbf{A}^{-1}$ to solve the following systems:
    (a) $\mathbf{Ax} = [-1\ 0\ 2\ 1]^T$ (b) $\mathbf{Ax} = [2\ 3\ 0\ -2]^T$

15. Let $\mathbf{A}$ be the matrix in Problem 10. Use $\mathbf{A}^{-1}$ to solve the following matrix equation for $\mathbf{X}$:

$$\mathbf{AX} = \begin{bmatrix} 1 & 0 & 3 \\ 0 & 2 & 4 \\ 4 & -3 & 8 \end{bmatrix}. \tag{13}$$

16. Prove that if $\mathbf{A}$ and $\mathbf{B}$ are both *m*-by-*n* matrices, then $(\mathbf{A}+\mathbf{B})^T = \mathbf{A}^T + \mathbf{B}^T$.

17. For each of the following, determine whether the statement made is always True or sometimes False.

    (a) If $\mathbf{A}$ and $\mathbf{B}$ are n-by-n invertible matrices, then $\mathbf{A}+\mathbf{B}$ is also invertible.

    (b) If $\mathbf{A}$ is invertible, so is $\mathbf{A}^3$.

    (c) If $\mathbf{A}$ is a symmetric matrix, then so is $\mathbf{A}^2$.

    (d) If $\mathbf{A}$ is invertible, so is $\mathbf{A}^T$.

18. Prove that if $\mathbf{A}$ is any n-by-n matrix, then the matrices $\mathbf{A} + \mathbf{A}^T$ and $\mathbf{A}\mathbf{A}^T$ are symmetric.

19. Prove that if $\mathbf{A} = \begin{bmatrix} a & b \\ c & d \end{bmatrix}$, then $\mathbf{A}$ is invertible if and only if $ad - cb \neq 0$. Show that if this condition holds, then

$$\mathbf{A}^{-1} = \frac{1}{ad - cb} \begin{bmatrix} d & -b \\ -c & a \end{bmatrix}.$$

20. Use the formula in Problem 19 to provide another proof that $\mathbf{M}_{\text{rot}}(\theta)^{-1} = \mathbf{M}_{\text{rot}}(-\theta)$ and that $\mathbf{M}_{\text{ref}}^{-1} = \mathbf{M}_{\text{ref}}$ (see Section 2.2).

21. Use Theorem 3 to prove that if $\mathbf{A}$, $\mathbf{B}$, and $\mathbf{C}$ are n-by-n matrices, then $(\mathbf{ABC})^T = \mathbf{C}^T\mathbf{B}^T\mathbf{A}^T$.

22. Prove that if $\mathbf{A}$, $\mathbf{B}$, and $\mathbf{C}$ are invertible n-by-n matrices, then $\mathbf{ABC}$ is invertible and $(\mathbf{ABC})^{-1} = \mathbf{C}^{-1}\mathbf{B}^{-1}\mathbf{A}^{-1}$.

23. Show that the transpose of the rotation operator matrix $\mathbf{M}_{\text{rot}}$ discussed in Example 1 is also the inverse of $\mathbf{M}_{\text{rot}}$.

24. If $\mathbf{A}$ is a *nilpotent matrix*; that is, a square matrix such that $\mathbf{A}^k = 0$ for some nonnegative integer k, prove that $\mathbf{I} - \mathbf{A}$ is invertible and that

$$(\mathbf{I} - \mathbf{A})^{-1} = \mathbf{I} + \mathbf{A} + \mathbf{A}^2 + \cdots + \mathbf{A}^{k-1}.$$

25. Verify the Sherman–Morrison–Woodbury formula:

$$(\mathbf{A} + \mathbf{BCD})^{-1} = \mathbf{A}^{-1} - \mathbf{A}^{-1}\mathbf{B}(\mathbf{C}^{-1} + \mathbf{DA}^{-1}\mathbf{B})^{-1}\mathbf{A}^{-1}.$$

[*Hint*: Left-multiply the right-hand member by $\mathbf{A} + \mathbf{BCD}$, and judiciously insert $\mathbf{I} = \mathbf{CC}^{-1}$.] This formula is helpful when the inverse is known for a matrix $\mathbf{A}$, but then $\mathbf{A}$ must be "updated" by adding $\mathbf{BCD}$, with an easily invertible $\mathbf{C}$. Let $\mathbf{A}$ be the 5-by-5 identity matrix, $\mathbf{B} = [0\ 0\ 0\ 0\ 1]^T$, $\mathbf{C} = [0.1]$, and $\mathbf{D} = [1\ 1\ 1\ 1\ 1]$ to obtain a quick evaluation of

$$\begin{bmatrix} 1 & 0 & 0 & 0 & 0 \\ 0 & 1 & 0 & 0 & 0 \\ 0 & 0 & 1 & 0 & 0 \\ 0 & 0 & 0 & 1 & 0 \\ 0.1 & 0.1 & 0.1 & 0.1 & 1.1 \end{bmatrix}^{-1}.$$

2.4 DETERMINANTS

The matrix concept is very powerful; a simple symbol like $\mathbf{M}$ stands for $10,000$ pieces of data, when $\mathbf{M}$ is a 100-by-100 matrix. However, this much information can be overwhelming, so it is important to have a few simple numbers that give a rough description of the matrix. For square matrices, one such descriptor is the *determinant*. The reader should be familiar with its formula for a 2-by-2 matrix:

$$\det \begin{bmatrix} a_1 & a_2 \\ b_1 & b_2 \end{bmatrix} = a_1 b_2 - a_2 b_1. \tag{1}$$

Example 1. Compute $\det \begin{bmatrix} 1 & -2 \\ 3 & 4 \end{bmatrix}$.

Solution. An easy way to remember (1) is to note that it combines products of entries along the "slopes" $\begin{bmatrix} \searrow \\ \searrow \end{bmatrix}$ and $\begin{bmatrix} \nearrow \\ \nearrow \end{bmatrix}$, signed positively for the "downhills" and negatively for the "uphills."

So $\det \begin{bmatrix} 1 & -2 \\ 3 & 4 \end{bmatrix} = (+)(1)(4) - (3)(-2) = 10.$ ∎

There is a geometric interpretation of (1). From elementary calculus, recall that the *cross product* $\mathbf{a} \times \mathbf{b}$ of two vectors $\mathbf{a}$ and $\mathbf{b}$ is defined to be the vector perpendicular to both $\mathbf{a}$ and $\mathbf{b}$, whose length equals the length of $\mathbf{a}$ times the length of $\mathbf{b}$ times the sine of the angle between $\mathbf{a}$ and $\mathbf{b}$, and whose direction is chosen so that $\mathbf{a}$, $\mathbf{b}$, and $\mathbf{a} \times \mathbf{b}$ form a right-handed triad (see Fig. 2.5).

From this definition, we see that the cross products of unit vectors $\mathbf{i}$, $\mathbf{j}$, and $\mathbf{k}$ along the x, y, and z axes, respectively, are given by the equations

$$\mathbf{i} \times \mathbf{j} = -\mathbf{j} \times \mathbf{i} = \mathbf{k}, \quad \mathbf{j} \times \mathbf{k} = -\mathbf{k} \times \mathbf{j} = \mathbf{i}, \quad \mathbf{k} \times \mathbf{i} = -\mathbf{i} \times \mathbf{k} = \mathbf{j}, \tag{2}$$

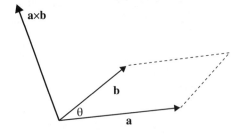

Fig. 2.5 Cross product.

and $\mathbf{a} \times \mathbf{a} = \mathbf{0}$ for every vector $\mathbf{a}$. It is also known that the cross product is distributive and anticommutative:

$$(\mathbf{a} + \mathbf{b}) \times \mathbf{c} = \mathbf{a} \times \mathbf{c} + \mathbf{b} \times \mathbf{c}, \quad \text{but } \mathbf{a} \times \mathbf{b} = -\mathbf{b} \times \mathbf{a}. \tag{3}$$

Now observe that in Figure 2.5 the pair of vectors $\mathbf{a}$ and $\mathbf{b}$ generate a parallelogram, whose base is the length of $\mathbf{a}$, and whose height is the length of $\mathbf{b}$ times the sine of the angle between $\mathbf{a}$ and $\mathbf{b}$; in other words, the area of the parallelogram is precisely the length of the cross product. If $\mathbf{a}$ and $\mathbf{b}$ lie in the x, y-plane, we can use the Equations (2) and (3) to derive a simple expression for this area in terms of the *components* of $\mathbf{a} = a_1\mathbf{i} + a_2\mathbf{j}$ and $\mathbf{b} = b_1\mathbf{i} + b_2\mathbf{j}$:

$$(a_1\mathbf{i} + a_2\mathbf{j}) \times (b_1\mathbf{i} + b_2\mathbf{j}) = [a_1b_2 - a_2b_1]\mathbf{k}. \tag{4}$$

Therefore the area of the parallelogram is given by the length of this vector, which is either $a_1b_2 - a_2b_1$ or its negative (depending on whether $\mathbf{a}, \mathbf{b}, \mathbf{k}$ form a right-handed or left-handed triad).

Note the similarity between (1) and (4). *The determinant of the 2-by-2 matrix containing the components of* $\mathbf{a}$ *in its first row and those of* $\mathbf{b}$ *in its second row equals, up to sign, the area of the parallelogram with edges* $\mathbf{a}$ *and* $\mathbf{b}$.

We can apply similar reasoning to Figure 2.6 to derive a component formula for the volume of a parallelepiped generated by three vectors $\mathbf{a}, \mathbf{b}$, and $\mathbf{c}$. Keeping in mind that the length of the cross product $\mathbf{a} \times \mathbf{b}$ equals the area of the base parallelogram, note that the *height* of the parallelepiped as depicted in the figure is given by the length of $\mathbf{c}$ times the cosine of the angle between $\mathbf{c}$ and $\mathbf{a} \times \mathbf{b}$. So the volume is the *dot product* of $\mathbf{c}$ and $\mathbf{a} \times \mathbf{b}$; it equals the length of $\mathbf{a} \times \mathbf{b}$ (i.e., the area of the base parallelogram) times the length of $\mathbf{c}$ times the cosine of this angle.

In other words, we have arrived at a vector formula for the volume of the parallelepiped—namely, $|(\mathbf{a} \times \mathbf{b}) \cdot \mathbf{c}|$. (The absolute value is necessitated by the possibility that $\mathbf{a}, \mathbf{b}, \mathbf{c}$ might comprise a left-handed triad.)

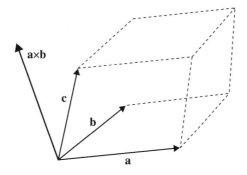

Fig. 2.6   Volume of a parallelepiped.

The component expression for $(\mathbf{a} \times \mathbf{b}) \cdot \mathbf{c}$ can be worked out by using Equations (2) again. With

$$\mathbf{a} = a_1\mathbf{i} + a_2\mathbf{j} + a_3\mathbf{k}, \quad \mathbf{b} = b_1\mathbf{i} + b_2\mathbf{j} + b_3\mathbf{k}, \quad \text{and} \quad \mathbf{c} = c_1\mathbf{i} + c_2\mathbf{j} + c_3\mathbf{k},$$

we *could* laboriously tabulate the $3 \times 3 \times 3 = 27$ terms in $(\mathbf{a} \times \mathbf{b}) \cdot \mathbf{c}$.
However, the somber student in Figure 2.7 will be gladdened to note that

(i) *all triple products containing repeated factors are zero* (e.g., $(\mathbf{i} \times \mathbf{j}) \cdot \mathbf{i} = 0$; the corresponding parallelepiped is flat and has no volume);
(ii) *all the other triple products are either +1 or −1* (e.g., $(\mathbf{i} \times \mathbf{j}) \cdot \mathbf{k} = 1$, $(\mathbf{k} \times \mathbf{j}) \cdot \mathbf{i} = -1$).

So only six terms survive, and we define the determinant of the 3-by-3 matrix containing the components of $\mathbf{a}$, $\mathbf{b}$, and $\mathbf{c}$ to be the resulting expression:

$$\det \begin{bmatrix} a_1 & a_2 & a_3 \\ b_1 & b_2 & b_3 \\ c_1 & c_2 & c_3 \end{bmatrix} = (\mathbf{a} \times \mathbf{b}) \cdot \mathbf{c} \tag{5}$$
$$= a_1 b_2 c_3 - a_1 b_3 c_2 + a_2 b_3 c_1 - a_2 b_1 c_3 + a_3 b_1 c_2 - a_3 b_2 c_1$$
$$= a_1(b_2 c_3 - b_3 c_2) + a_2(b_3 c_1 - b_1 c_3) + a_3(b_1 c_2 - b_2 c_1).$$

We display the determinant in (5) in factored form, because when we compare it with (1) we observe the appearance of 2-by-2 determinants in the formula:

Fig. 2.7   Determined Student.

$$\det \begin{bmatrix} a_1 & a_2 & a_3 \\ b_1 & b_2 & b_3 \\ c_1 & c_2 & c_3 \end{bmatrix} = a_1 \det \begin{bmatrix} b_2 & b_3 \\ c_2 & c_3 \end{bmatrix} - a_2 \det \begin{bmatrix} b_1 & b_3 \\ c_1 & c_3 \end{bmatrix} + a_3 \det \begin{bmatrix} b_1 & b_2 \\ c_1 & c_2 \end{bmatrix}. \quad (6)$$

Notice the alternating signs, which we have introduced to maintain the original ordering of the columns.

Formula (6) is easy to remember by striking out the first row and successive columns of the matrix:

$$\det \begin{bmatrix} a_1 & a_2 & a_3 \\ b_1 & b_2 & b_3 \\ c_1 & c_2 & c_3 \end{bmatrix}$$

$$= a_1 \det \begin{bmatrix} \cancel{a_1} & a_2 & a_3 \\ \cancel{b_1} & b_2 & b_3 \\ \cancel{c_1} & c_2 & c_3 \end{bmatrix} - a_2 \det \begin{bmatrix} a_1 & \cancel{a_2} & a_3 \\ b_1 & \cancel{b_2} & b_3 \\ c_1 & \cancel{c_2} & c_3 \end{bmatrix} + a_3 \det \begin{bmatrix} a_1 & a_2 & \cancel{a_3} \\ b_1 & b_2 & \cancel{b_3} \\ c_1 & c_2 & \cancel{c_3} \end{bmatrix}.$$

Example 2. Compute $\det(\mathbf{A})$, where $\mathbf{A} = \begin{bmatrix} 1 & 2 & 3 \\ 4 & 5 & 6 \\ 7 & 8 & 9 \end{bmatrix}$.

Solution. From (6), we obtain

$$\det(A) = (1) \det \begin{bmatrix} 5 & 6 \\ 8 & 9 \end{bmatrix} - 2 \det \begin{bmatrix} 4 & 6 \\ 7 & 9 \end{bmatrix} + 3 \det \begin{bmatrix} 4 & 5 \\ 7 & 8 \end{bmatrix}$$
$$= (45 - 48) - 2(36 - 42) + 3(32 - 35)$$
$$= -3 + 12 - 9 = 0. \qquad \blacksquare$$

Now observe that a reduction identity similar to (6) arises when we juggle the 2-by-2 determinant expression (1):

$$\det \begin{bmatrix} a_1 & a_2 \\ b_1 & b_2 \end{bmatrix} = a_1 b_2 - a_2 b_1 = a_1 \det \begin{bmatrix} \cancel{a_1} & \cancel{a_2} \\ \cancel{b_1} & b_2 \end{bmatrix} - a_2 \det \begin{bmatrix} \cancel{a_1} & \cancel{a_2} \\ b_1 & \cancel{b_2} \end{bmatrix} \quad (7)$$

(if we take the determinant of a 1-by-1 matrix to be simply the value of the entry).

Since (6), (7) express 3-by-3 determinants in terms of 2-by-2's and 2-by-2's in terms of 1-by-1's, we can use the same pattern to define 4-by-4 determinants in terms of 3-by-3's, and so on, to formulate a *recursive* definition.

Determinants

Definition 3. The determinant of a 1-by-1 matrix $[a_{11}]$ is simply a_{11}, and the determinant of an n-by-n matrix with $n > 1$ is defined via lower-order determinants by

$$
\det(\mathbf{A}) = \det \begin{bmatrix} a_{11} & a_{12} & a_{13} & \cdots & a_{1n} \\ a_{21} & a_{22} & a_{23} & \cdots & a_{2n} \\ a_{31} & a_{32} & a_{33} & \cdots & a_{3n} \\ \vdots & \vdots & \vdots & \ddots & \vdots \\ a_{n1} & a_{n2} & a_{n3} & \cdots & a_{nn} \end{bmatrix} \tag{8}
$$
$$
:= a_{11}\det(\mathbf{A}_{11}) - a_{12}\det(\mathbf{A}_{12}) + a_{13}\det(\mathbf{A}_{13})
$$
$$
- \cdots + (-1)^{n+1}a_{1n}\det(\mathbf{A}_{1n})
$$
$$
= \sum_{j=1}^{n}(-1)^{1+j}a_{1j}\det(\mathbf{A}_{1j}),
$$

where $\mathbf{A}_{1j}$ is the $(n-1)$-by-$(n-1)$ submatrix obtained by deleting the first row and jth column of $\mathbf{A}$.

Note that from this recursive definition it follows that *the determinant of a diagonal matrix is the product of the entries along the diagonal.* This is consistent with our original volume interpretation of the determinant; we are inclined to postulate that a rectangle in n-dimensional space having edge lengths $a_1, a_2, \ldots, a_n$ along the axes should have a volume given by $a_1 a_2 \cdots a_n$. Since its edge *vectors* would be represented by $[a_1\ 0\ \cdots\ 0]$, $[0\ a_2\ \cdots\ 0]$, $\ldots$, $[0\ 0\ \cdots\ a_n]$, the determinant does indeed give the volume:

$$
\det \begin{bmatrix} a_1 & 0 & \cdots & 0 \\ 0 & a_2 & \cdots & 0 \\ \vdots & \vdots & \ddots & \vdots \\ 0 & 0 & \cdots & a_n \end{bmatrix} = a_1 a_2 \cdots a_n.
$$

Remark. An alternative notation for the determinant utilizes vertical bars:

$$
\begin{vmatrix} a_{11} & \cdots & a_{1n} \\ \vdots & \ddots & \vdots \\ a_{n1} & \cdots & a_{nn} \end{vmatrix} := \det \begin{bmatrix} a_{11} & \cdots & a_{nn} \\ \vdots & \ddots & \vdots \\ a_{n1} & \cdots & a_{nn} \end{bmatrix}
$$

We employ this notation in the following example.

Example 3. Compute $\begin{vmatrix} 2 & 3 & 4 & 0 \\ 5 & 6 & 3 & 0 \\ 2 & 5 & 3 & 7 \\ 2 & 2 & 2 & 2 \end{vmatrix}$.

Solution. (We highlight the lower right-hand 2-by-2 submatrix for future reference.) Using the definition we reduce the computation to 3-by-3 determinants,

$$\det(\mathbf{A}) = 2 \times \begin{vmatrix} 6 & 3 & 0 \\ 5 & 3 & 7 \\ 2 & 2 & 2 \end{vmatrix} - 3 \times \begin{vmatrix} 5 & 3 & 0 \\ 2 & 3 & 7 \\ 2 & 2 & 2 \end{vmatrix} + 4 \times \begin{vmatrix} 5 & 6 & 0 \\ 2 & 5 & 7 \\ 2 & 2 & 2 \end{vmatrix} - 0 \times \begin{vmatrix} 5 & 6 & 3 \\ 2 & 5 & 3 \\ 2 & 2 & 2 \end{vmatrix}. \quad (9)$$

These are further reduced to 2-by-2's:

$$\det(\mathbf{A}) = 2 \left\{ 6 \times \begin{vmatrix} 3 & 7 \\ 2 & 2 \end{vmatrix} - 3 \times \begin{vmatrix} 5 & 7 \\ 2 & 2 \end{vmatrix} + (0) \right\} \quad (10)$$

$$- 3 \left\{ 5 \times \begin{vmatrix} 3 & 7 \\ 2 & 2 \end{vmatrix} - 3 \times \begin{vmatrix} 2 & 7 \\ 2 & 2 \end{vmatrix} + (0) \right\}$$

$$+ 4 \left\{ 5 \times \begin{vmatrix} 5 & 7 \\ 2 & 2 \end{vmatrix} - 6 \times \begin{vmatrix} 2 & 7 \\ 2 & 2 \end{vmatrix} + (0) \right\} - (0)$$

$$= (2 \times 6) \begin{vmatrix} 3 & 7 \\ 2 & 2 \end{vmatrix} - (3 \times 5) \begin{vmatrix} 3 & 7 \\ 2 & 2 \end{vmatrix} + (4 \times 5) \begin{vmatrix} 5 & 7 \\ 2 & 2 \end{vmatrix}$$

$$- (2 \times 3) \begin{vmatrix} 5 & 7 \\ 2 & 2 \end{vmatrix} + (3 \times 3 - 4 \times 6) \begin{vmatrix} 2 & 7 \\ 2 & 2 \end{vmatrix},$$

and the final answer is evaluated using (1) to calculate the 2-by-2 determinants:

$$(12 - 15)(-8) + (20 - 6)(-4) + (9 - 24)(-10) = 118. \qquad \blacksquare$$

Observe from (1) and (6) that if two rows are switched in a 3-by-3 or a 2-by-2 matrix, the determinant changes sign. (This can be traced back to the cross product appearing in (4) and (5).) To investigate this for larger matrices, consider the next example.

Example 4. Compute the determinant of the matrix in Example 3 with its first two rows switched.

Solution. If we blindly retrace the sequence of calculations of Example 3 starting with the row-switched matrix

$$\det \begin{bmatrix} 5 & 6 & 3 & 0 \\ 2 & 3 & 4 & 0 \\ 2 & 5 & 3 & 7 \\ 2 & 2 & 2 & 2 \end{bmatrix},$$

it should be clear that the lower right subdeterminant $\det \begin{bmatrix} 3 & 7 \\ 2 & 2 \end{bmatrix}$ in (9), (10) would acquire coefficients of (5×3) and $-(6 \times 2)$ instead of $(2 \times 6 - 3 \times 5)$; similarly $\det \begin{bmatrix} 5 & 7 \\ 2 & 2 \end{bmatrix}$ would acquire $(3 \times 2 - 5 \times 4)$ instead of $(4 \times 5 - 2 \times 3)$. Every term in (10), in fact, would be negated, and the determinant would turn out to be -118.   ∎

Examples 3 and 4 demonstrate that switching the first two rows of a 4-by-4 matrix changes the sign of its determinant. Moreover (9) shows that switching *any* pair of the last 3 rows also changes the sign, since a row switch would occur in each (3-by-3) determinant on the right of (9).

What happens if row 1 is switched with, say, row 3? Easy; first we switch row 3 with row 2—one sign change; then we switch (the new) row 2 with row 1—another sign change; then we switch (the new) row 2 with row 3. That's three sign changes or, equivalently, one sign change. (An alternative argument can be constructed based on the *permutation formula for the determinant* discussed in Problem 20.)

Row Interchange Property

Theorem 4.   The determinant of a square matrix changes sign if any two of its rows are switched.

Note that if the two rows to be exchanged happened to be identical, the determinant would remain unchanged; therefore, to be consistent with Theorem 4, the determinant must be zero.

Corollary 1.   If any two rows of a square matrix are equal, its determinant is zero.

As a consequence of Theorem 4, we shall show that we can use *any* row to implement the calculation of a determinant by Definition 3, if we are careful about counting the sign changes. To begin the argument, we first generalize the nomenclature.

Minors and Cofactors

Definition 4.   If the elements of a square matrix $\mathbf{A}$ are denoted by a_{ij}, the determinant of the submatrix $\mathbf{A}_{ij}$, obtained by deleting the ith row and jth column of $\mathbf{A}$, is called the (i,j)th **minor** of $\mathbf{A}$. The (i,j)th **cofactor** of $\mathbf{A}$, denoted $\mathrm{cof}(i,j)$, is the (i,j)th minor of $\mathbf{A}$ times $(-1)^{i+j}$:

$$\mathrm{cof}(i,j) := (-1)^{i+j} \det(\mathbf{A}_{ij}) = (-1)^{i+j} \times \{(i,j)\text{th minor of } \mathbf{A}\}.$$

Thus the definition of the determinant (8) can be stated in terms of cofactors by

$$\det(\mathbf{A}) = \sum_{j=1}^{n} a_{1,j} (-1)^{1+j} \det(\mathbf{A}_{1,j}) = \sum_{j=1}^{n} a_{1,j} \operatorname{cof}(1,j).$$

Now we use the row-interchange property of determinants to generalize:

Cofactor Expansion

Theorem 5. The determinant of a square matrix $\mathbf{A}$ can be calculated using the entries and cofactors of *any* row i in accordance with

$$\boxed{\det(\mathbf{A}) = \sum_{j=1}^{n} a_{i,j} \operatorname{cof}(i,j).} \tag{11}$$

Proof. To see this, simply imagine the ith row of $\mathbf{A}$ moved to the top of the matrix by successively interchanging it with the rows immediately above it, in a step-ladder fashion; this entails $(i-1)$ row interchanges. Call the resulting matrix $\mathbf{A}^{\text{new}}$, and note that due to the interchanges $\det(\mathbf{A}) = (-1)^{i-1} \det(\mathbf{A}^{\text{new}})$. The entry a_{ij} in the original matrix becomes a_{1j}^{new} in the new matrix, *but they both have the same minor*, because the same rows and columns are deleted. However, the *cofactor* of a_{1j}^{new} equals $(-1)^{1+j}$ times this minor, while the cofactor of a_{ij} is $(-1)^{i+j}$ times the minor. As a result, we can argue

$$\det(\mathbf{A}) = (-1)^{i-1} \det(\mathbf{A}^{\text{new}}) = (-1)^{i-1} \sum_{j=1}^{n} a_{1,j}^{\text{new}} (-1)^{1+j} \det(\mathbf{A}_{1j}^{\text{new}}),$$

and making the indicated identifications we get (11). ∎

From the cofactor expansion of the determinant we immediately conclude

Corollary 2. If any row of a matrix is multiplied by a scalar, the determinant is multiplied by the same scalar.

Corollary 3. If a row of $\mathbf{A}$ consists of all zeros, then $\det(\mathbf{A}) = 0$.

And with a little reasoning we also see

Corollary 4. If any multiple of one row of a matrix is added to another row, its determinant is unchanged.

This follows because if a_{ij} in (11) is replaced by, say, $a_{ij} + \beta a_{kj}$, the result of adding β times row k to row i, the new determinant would be given by

$$\sum_{j=1}^{n} [a_{ij} + \beta a_{kj}] \operatorname{cof}(i,j) = \sum_{j=1}^{n} a_{ij} \operatorname{cof}(i,j) + \beta \sum_{j=1}^{n} a_{kj} \operatorname{cof}(i,j).$$

The first sum on the right is the determinant of the original $\mathbf{A}$, and the second sum is β times the determinant of a matrix with two equal rows.

A similar argument (Problem 14) establishes the **linearity property of the determinant**:

Corollary 5. If $\mathbf{A}$ is an n-by-n matrix with rows $\mathbf{a}_1, \mathbf{a}_2, \ldots, \mathbf{a}_n$ and $\mathbf{b}$ is an 1-by-n row vector, then

$$\det \begin{bmatrix} \mathbf{a}_1 \\ \vdots \\ \mathbf{a}_{k-1} \\ (\mathbf{a}_k + \mathbf{b}) \\ \mathbf{a}_{k+1} \\ \vdots \\ \mathbf{a}_n \end{bmatrix} = \det \begin{bmatrix} \mathbf{a}_1 \\ \vdots \\ \mathbf{a}_{k-1} \\ \mathbf{a}_k \\ \mathbf{a}_{k+1} \\ \vdots \\ \mathbf{a}_n \end{bmatrix} + \det \begin{bmatrix} \mathbf{a}_1 \\ \vdots \\ \mathbf{a}_{k-1} \\ \mathbf{b} \\ \mathbf{a}_{k+1} \\ \vdots \\ \mathbf{a}_n \end{bmatrix}.$$

Example 5. Compute the determinant of the triangular matrices $\mathbf{A}$ and $\mathbf{B}$.

$$\mathbf{A} = \begin{bmatrix} \alpha & 0 & 0 & 0 & 0 \\ \beta & \gamma & 0 & 0 & 0 \\ \delta & \epsilon & \zeta & 0 & 0 \\ \eta & \theta & \sigma & \kappa & 0 \\ \lambda & \mu & \nu & \tau & \omega \end{bmatrix}, \quad \mathbf{B} = \begin{bmatrix} \alpha & \beta & \gamma & \delta & \epsilon \\ 0 & \zeta & \eta & \theta & \sigma \\ 0 & 0 & \kappa & \lambda & \mu \\ 0 & 0 & 0 & \nu & \tau \\ 0 & 0 & 0 & 0 & \omega \end{bmatrix}.$$

Solution. For the matrix $\mathbf{A}$, we use the cofactor expansion along the top rows.

$$\det(\mathbf{A}) = \alpha \det \begin{bmatrix} \gamma & 0 & 0 & 0 \\ \epsilon & \zeta & 0 & 0 \\ \theta & \sigma & \kappa & 0 \\ \mu & \nu & \tau & \omega \end{bmatrix} - 0 + 0 - 0 + 0 = \alpha\gamma \det \begin{bmatrix} \zeta & 0 & 0 \\ \sigma & \kappa & 0 \\ \nu & \tau & \omega \end{bmatrix} + (0)$$

$$= \alpha\gamma\zeta \det \begin{bmatrix} \kappa & 0 \\ \tau & \omega \end{bmatrix} + (0) = \alpha\gamma\zeta\kappa \det [\omega] + (0) = \alpha\gamma\zeta\kappa\omega.$$

For **B**, we expand along the *bottom* rows:

$$\det(\mathbf{B}) = 0 - 0 + 0 - 0 + \omega \det \begin{bmatrix} \alpha & \beta & \gamma & \delta \\ 0 & \zeta & \eta & \theta \\ 0 & 0 & \kappa & \lambda \\ 0 & 0 & 0 & \nu \end{bmatrix} = (0) + \omega\nu \det \begin{bmatrix} \alpha & \beta & \gamma \\ 0 & \zeta & \eta \\ 0 & 0 & \kappa \end{bmatrix}$$

$$\doteq (0) + \omega\nu\kappa \det \begin{bmatrix} \alpha & \beta \\ 0 & \zeta \end{bmatrix} = (0) + \omega\nu\kappa\zeta \det [\alpha] = \omega\nu\kappa\zeta\alpha.$$

In both cases, the determinant is simply the product of the elements along the diagonal.   ∎

In general, the determinant of a triangular matrix is the product of the entries along the diagonal.

Does the formula in Definition 3 give a practical way to calculate an n-by-n determinant? The answer is NO. Example 3 demonstrates that one must ultimately evaluate all products of entries drawn one from each row and column. That's $n - 1$ multiplications per term, and $n!$ terms.

In 2014, the Tianhe-2 computer developed by the National University of Defense Technology in Changsha city in Central China could do 3.386×10^{16} multiplications per second (Forbes, November 17, 2014). To evaluate a modest 25-by-25 determinant this way would take the machine hundreds of years!

But the forward stage of the Gauss elimination process reduces a matrix to upper triangular form by adding multiples of one row to another (which doesn't change the determinant) and switching rows (changing only the sign). And the determinant of an upper triangular matrix requires only $(n - 1)$ multiplications. Problem 28 of Exercises 1.1 tells us that the forward stage of Gauss elimination requires roughly $n^3/3$ multiplications. So using elementary row operations the Tianhe-2 could calculate a 25-by-25 determinant in a fraction of a picosecond.

Exercises 2.4

In Problems 1–11 use cofactor expansions to calculate the determinants of the given matrices.

1. $\begin{bmatrix} -1 & -1 \\ -1 & -1 \end{bmatrix}$ **2.** $\begin{bmatrix} 2 & 1 - 3i \\ 2 - i & 2 \end{bmatrix}$ **3.** $\begin{bmatrix} 1 & 2 & 2 \\ 2 & 1 & 1 \\ 1 & 1 & 3 \end{bmatrix}$

4. $\begin{bmatrix} 0 & 2 & 3 \\ 0 & 2 & 0 \\ 2 & 1 & 1 \end{bmatrix}$ **5.** $\begin{bmatrix} 2 & 1 & 4 + i \\ 3 & 3 & 0 \\ -1 & 4 & 2 - 3i \end{bmatrix}$ **6.** $\begin{bmatrix} 0 & 2 & 1 & 1 & 0 \\ 2 & 4 & 4 & 2 & 2 \\ 3 & 6 & 6 & 0 & 0 \\ 0 & -2 & -1 & -2 & 2 \\ 0 & 2 & 1 & 2 & 4 \end{bmatrix}$

7. $\begin{bmatrix} 0 & 0 & 0 & a \\ 0 & 0 & b & c \\ 0 & d & e & f \\ g & h & i & j \end{bmatrix}$
8. $\begin{bmatrix} 0 & 1 & 0 & 0 \\ 0 & 0 & 1 & 0 \\ 0 & 0 & 0 & 1 \\ a & b & c & d \end{bmatrix}$
9. $\begin{bmatrix} x & 1 & 0 & 0 \\ 0 & x & 1 & 0 \\ 0 & 0 & x & 1 \\ a & b & c & d \end{bmatrix}$

10. $\begin{bmatrix} a & 0 & f & 0 & 0 \\ 0 & b & g & 0 & 0 \\ 0 & 0 & c & 0 & 0 \\ 0 & 0 & h & d & 0 \\ 0 & 0 & i & 0 & e \end{bmatrix}$
11. $\begin{bmatrix} x & 1 & 1 & 1 \\ 1 & x & 1 & 1 \\ 1 & 1 & x & 1 \\ 1 & 1 & 1 & x \end{bmatrix}$

12. Generalize your answers to Problems 7–11 to n-by-n matrices having the same structures.

13. If a particular entry a_{ij} of the n-by-n matrix $\mathbf{A}$ is regarded as a variable, show that the function $f(a_{ij}) := \det(\mathbf{A})$ is either identically zero, never zero, or zero for exactly one value of a_{ij}. Construct 3-by-3 examples of each case.

14. Prove the linearity property of the determinant, Corollary 5.

15. Use determinants to compute the volume of the parallepiped determined by the vectors $\mathbf{a} = -2\mathbf{i} - 3\mathbf{j} + \mathbf{k}$, $\mathbf{b} = 2\mathbf{i} + 6\mathbf{j} - 2\mathbf{k}$, and $\mathbf{c} = 4\mathbf{i} - 4\mathbf{k}$.

16. Use determinants to compute the volume of the parallepiped determined by the vectors $\mathbf{a} = 2\mathbf{i} - 3\mathbf{j} + \mathbf{k}$, $\mathbf{b} = \mathbf{i} + 6\mathbf{j} - 2\mathbf{k}$, and $\mathbf{c} = \mathbf{i} - 4\mathbf{k}$.

17. Let $\mathbf{A}$ be the matrix

$$\mathbf{A} = \begin{bmatrix} a & b & c & d \\ e & f & g & h \\ i & j & k & l \\ m & n & o & p \end{bmatrix}.$$

Use the properties derived in this section to express the determinants of the following matrices in terms of $\det(\mathbf{A})$:

(a) $\begin{bmatrix} a+e & b+f & c+g & d+h \\ e & f & g & h \\ i & j & k & l \\ m & n & o & p \end{bmatrix}$
(b) $\begin{bmatrix} a & b & c & d \\ 3e & 3f & 3g & 3h \\ i & j & k & l \\ m & n & o & p \end{bmatrix}$

(c) $\begin{bmatrix} a & b & c & d \\ e & f & g & h \\ -i & -j & -k & -l \\ m & n & o & p \end{bmatrix}$
(d) $\begin{bmatrix} -a & -b & -c & -d \\ -e & -f & -g & -h \\ -i & -j & -k & -l \\ -m & -n & -o & -p \end{bmatrix}$

(e) $\begin{bmatrix} 3a & 3b & 3c & 3d \\ 3e & 3f & 3g & 3h \\ 3i & 3j & 3k & 3l \\ 3m & 3n & 3o & 3p \end{bmatrix}$
(f) $\begin{bmatrix} i & j & k & l \\ m & n & o & p \\ a & b & c & d \\ e & f & g & h \end{bmatrix}$

(g) $\begin{bmatrix} a & b & c & d \\ 3e-a & 3f-b & 3g-c & 3h-d \\ i & j & k & l \\ m & n & o & p \end{bmatrix}$

(h) $\begin{bmatrix} a & b & c & d \\ e-3a & f-3b & g-3c & h-3d \\ i & j & k & l \\ m & n & o & p \end{bmatrix}$

18. What are the determinants of the matrices that implement the elementary row operations by left multiplication? (Recall Example 2, Section 2.1.)

19. Reduce the following matrices to lower triangular form by subtracting appropriately chosen multiples of lower rows from upper rows, and thereby evaluate the determinants. Formulate the n-by-n generalizations.

(a) $\begin{bmatrix} 1 & x & x^2 & x^3 \\ x & 1 & x & x^2 \\ x^2 & x & 1 & x \\ x^3 & x^2 & x & 1 \end{bmatrix}$
(b) $\begin{bmatrix} 1 & 1 & 1 & 1 \\ 1 & 2 & 2 & 2 \\ 1 & 2 & 3 & 3 \\ 1 & 2 & 3 & 4 \end{bmatrix}$

20. (*Permutation formula for the determinant*)

(a) Continue performing cofactor expansions on (6) and confirm the formulation

$$\det \begin{bmatrix} a_1 & a_2 & a_3 \\ b_1 & b_2 & b_3 \\ c_1 & c_2 & c_3 \end{bmatrix} =$$

$$= a_1 b_2 \det \begin{bmatrix} 1 & 0 & 0 \\ 0 & 1 & 0 \\ 0 & 0 & c_3 \end{bmatrix} + a_1 b_3 \det \begin{bmatrix} 1 & 0 & 0 \\ 0 & 0 & 1 \\ 0 & c_2 & 0 \end{bmatrix}$$

$$+ a_2 b_1 \det \begin{bmatrix} 0 & 1 & 0 \\ 1 & 0 & 0 \\ 0 & 0 & c_3 \end{bmatrix} + \cdots$$

$$= a_1 b_2 c_3 \det \begin{bmatrix} 1 & 0 & 0 \\ 0 & 1 & 0 \\ 0 & 0 & 1 \end{bmatrix} + a_1 b_3 c_2 \det \begin{bmatrix} 1 & 0 & 0 \\ 0 & 0 & 1 \\ 0 & 1 & 0 \end{bmatrix}$$

$$+ a_2 b_1 c_3 \det \begin{bmatrix} 0 & 1 & 0 \\ 1 & 0 & 0 \\ 0 & 0 & 1 \end{bmatrix} + a_2 b_3 c_1 \det \begin{bmatrix} 0 & 1 & 0 \\ 0 & 0 & 1 \\ 1 & 0 & 0 \end{bmatrix}$$

$$+ a_3 b_1 c_2 \det \begin{bmatrix} 0 & 0 & 1 \\ 1 & 0 & 0 \\ 0 & 1 & 0 \end{bmatrix} + a_3 b_2 c_1 \det \begin{bmatrix} 0 & 0 & 1 \\ 0 & 1 & 0 \\ 1 & 0 & 0 \end{bmatrix}.$$

(b) The (valid) generalization of part (a) is the formula

$$\det(\mathbf{A}) = \sum_{i_1, i_2, \ldots, i_n} a_{1i_1} a_{2i_2} \cdots a_{ni_n} \det(\mathbf{P}_{i_1, i_2, \ldots, i_n}),$$

where the sum extends over every reordering $\{i_1, i_2, \ldots, i_n\}$ of the sequence $\{1, 2, \ldots, n\}$, and the **permutation** $\mathbf{P}_{i_1,i_2,\ldots,i_n}$ is the matrix whose jth row has a one in column i_j and zeros elsewhere (for $j = 1, 2, \ldots, n$). For example,

$$\mathbf{P}_{2,3,1} = \begin{bmatrix} 0 & 1 & 0 \\ 0 & 0 & 1 \\ 1 & 0 & 0 \end{bmatrix}.$$

Verify this formula for the 4-by-4 determinant.

(c) By applying the determinant rule for switching rows to the matrix $\mathbf{P}_{i_1,i_2,\ldots,i_n}$, argue that $\det(\mathbf{P}_{i_1,i_2,\ldots,i_n})$ is ± 1, positive if $\mathbf{P}_{i_1,i_2,\ldots,i_n}$ can be manipulated into the form of the n-by-n identity with an even number of row exchanges, and minus if an odd number of exchanges are required. You have shown that $\det(\mathbf{A})$ *is the sum of all possible products of one entry from each row and column of* $\mathbf{A}$, *with appropriate signs.*

(d) The "evenness" ($+1$) or "oddness" (-1) of the number of exchanges required to reorder the sequence $\{i_1, i_2, \ldots, i_n\}$ into the standard order $\{1, 2, \ldots, n\}$ is called the *parity* of the sequence. Note that since $\det(\mathbf{P}_{i_1,i_2,\ldots,i_n})$ is a fixed number, the parity depends only on the sequence itself (and not on the choice of exchanges).

21. Using the previous problem, identify the coefficient of x^4, x^3, and $x^0 = 1$ in the polynomial given by

$$\det \begin{bmatrix} x+a & b & c & d \\ e & x+f & g & h \\ i & j & x+k & l \\ m & n & o & x+p \end{bmatrix}.$$

Generalize to the n-by-n case.

22. A "block lower triangular" matrix has the form

$$\mathbf{M} = \begin{bmatrix} a_{11} & \cdots & a_{1i} & 0 & \cdots & 0 \\ \vdots & \cdots & \vdots & \vdots & \cdots & \vdots \\ a_{i1} & \cdots & a_{i,i} & 0 & \cdots & 0 \\ b_{i+1,1} & \cdots & b_{i+1,i} & c_{i+1,i+1} & \cdots & c_{i+1,n} \\ \vdots & \cdots & \vdots & \vdots & \cdots & \vdots \\ b_{n1} & \cdots & b_{n,i} & c_{n,i+1} & \cdots & c_{nn} \end{bmatrix} \equiv \begin{bmatrix} \mathbf{A} & \mathbf{0} \\ \mathbf{B} & \mathbf{C} \end{bmatrix}.$$

(Note that $\mathbf{M}$, $\mathbf{A}$, and $\mathbf{C}$ are square.)

(a) Show that $\det(\mathbf{M})$ does not depend on any of the entries b_{jk}. (This will be easy to see if you experiment by using the first row to expand a 4-by-4 block lower triangular matrix $\mathbf{M}$ having 2-by-2 blocks.)

(b) Show that $\det(\mathbf{M}) = \det(\mathbf{A}) \det(\mathbf{B})$.

(c) Formulate the analogous rule for block upper triangular matrices.

23. The 3-by-3 *Vandermonde* determinant is the determinant of the matrix of the form

$$\mathbf{V} = \begin{bmatrix} 1 & z & z^2 \\ 1 & x_1 & x_1{}^2 \\ 1 & x_2 & x_2{}^2 \end{bmatrix}$$

(a) For the moment, regard "z" as a variable, and consider x_1 and x_2 to be distinct constants. Without completely carrying out the cofactor expansion of $\det(\mathbf{V})$ along the first row, argue that the expansion would reveal that $\det(\mathbf{V})$ is a second-degree polynomial in z. As such, it can be factored using its roots z_1 and z_2: $a_2z^2 + a_1z + a_0 = a_2(z - z_1)(z - z_2)$.

(b) From the expansion, identify the coefficient a_2 as a 2-by-2 Vandermonde determinant. Evaluate a_2.

(c) What are the two roots? (By looking at $\mathbf{V}$ itself, it will become obvious that two values of z will render it a matrix with determinant zero.)

(d) Assemble the formula for the 3-by-3 Vandermonde determinant.

(e) Reason similarly to work out the formula for the 4-by-4 Vandermonde determinant:

$$\det \begin{bmatrix} 1 & z & z^2 & z^3 \\ 1 & x_1 & x_1^2 & x_1^3 \\ 1 & x_2 & x_2^2 & x_2^3 \\ 1 & x_3 & x_3^2 & x_3^3 \end{bmatrix}$$

[*Hint*: If p and q are two polynomials of degree 3, and they have three zeros in common, then one of them is a constant times the other.]

(f) Generalize for the n-by-n case.

24. The determinant is, of course, a function of a (square) matrix. We will study several other matrix functions in the coming chapters: the trace, eigenvalues, norms, ... In this light, it is amusing to compare the determinant with other functions sharing its properties. In the displays that follow, the $\mathbf{a}_i$'s and $\mathbf{b}$'s are matrix rows, and α and β are scalar parameters.

(a) (Recall Corollaries 2 and 5.) A matrix function $f(\mathbf{A})$ is said to be a *multilinear function of the rows of* $\mathbf{A}$ if, for all values of the parameters shown,

$$f\left(\begin{bmatrix} \mathbf{a}_1 \\ \mathbf{a}_2 \\ \vdots \\ \alpha\mathbf{a}_i + \beta\mathbf{b}_i \\ \vdots \\ \mathbf{a}_n \end{bmatrix}\right) = \alpha f\left(\begin{bmatrix} \mathbf{a}_1 \\ \mathbf{a}_2 \\ \vdots \\ \mathbf{a}_i \\ \vdots \\ \mathbf{a}_n \end{bmatrix}\right) + \beta f\left(\begin{bmatrix} \mathbf{a}_1 \\ \mathbf{a}_2 \\ \vdots \\ \mathbf{b}_i \\ \vdots \\ \mathbf{a}_n \end{bmatrix}\right).$$

Are there any other multilinear matrix row functions (other than the determinant)? [*Hint*: There is an easy way to alter the 2-by-2 determinant formula (1) while preserving "multilinearity." Generalize to formula (8).]

(b) (Recall Theorem 4.) A matrix function $f(\mathbf{A})$ is said to satisfy the *row interchange property* if, for any values of the parameters shown,

$$f\left(\begin{bmatrix} \mathbf{a}_1 \\ \vdots \\ \mathbf{a}_i \\ \vdots \\ \mathbf{a}_j \\ \vdots \\ \mathbf{a}_n \end{bmatrix}\right) = -f\left(\begin{bmatrix} \mathbf{a}_1 \\ \vdots \\ \mathbf{a}_j \\ \vdots \\ \mathbf{a}_i \\ \vdots \\ \mathbf{a}_n \end{bmatrix}\right).$$

Prove that for any multilinear matrix row function, the row interchange property implies, and is implied by, the *zero-row property*:

$$f\left(\begin{bmatrix} \mathbf{a}_1 \\ \vdots \\ \mathbf{0} \\ \vdots \\ \mathbf{a}_n \end{bmatrix}\right) = 0.$$

Are there any other multilinear matrix row functions that have the row interchange property (other than the determinant)? [*Hint*: There is a very trivial one.]

(c) Prove that if f is any multilinear matrix row function with the row interchange property, and if a forward Gauss elimination reduces a square matrix $\mathbf{A}$ to echelon form $\mathbf{E}$, then $f(\mathbf{A}) = \pm f(\mathbf{E})$: (+) if the number of row switches is even, (−) otherwise.

Furthermore, if the diagonal entries of $\mathbf{E}$ are nonzero and "back elimination" is performed to further reduce $\mathbf{E}$ to a diagonal matrix $\mathbf{D}$, then $f(\mathbf{A}) = \pm f(\mathbf{D})$.

(d) Conclude that each multilinear matrix row function with the row interchange property is determined solely by its value on the identity matrix.

2.5   THREE IMPORTANT DETERMINANT RULES

In this section, we will be deriving a few theoretical properties of determinants that greatly enhance and facilitate their utility. One of the most important of these is the rule for the determinant of the transpose:

Transpose Rule

$$\det(\mathbf{A}^{\mathrm{T}}) = \det(\mathbf{A}).\qquad(1)$$

In anticipation of the derivation, notice that (1) will immediately justify the column version of some of the "row rules" derived earlier:

  (i) If any row *or column* of $\mathbf{A}$ is multiplied by a scalar, the determinant is multiplied by the same scalar.

 (ii) If a row *or column* of $\mathbf{A}$ consists of all zeros, $\det(\mathbf{A}) = 0$.

(iii) If any two rows *or columns* of a matrix are switched, its determinant changes sign.

 (iv) If any two rows *or columns* of $\mathbf{A}$ are equal, $\det(\mathbf{A}) = 0$.

  (v) If any multiple of one row (*column*) of a matrix is added to another row (*column*), its determinant is unchanged.

 (vi) The cofactor expansion is valid for any row *and column*:

$$\det(\mathbf{A}) = \sum_{k=1}^{n} a_{ik}\mathrm{cof}(i,k) = \sum_{k=1}^{n} a_{kj}\mathrm{cof}(k,j).$$

(vii) The *linearity properties*

$$\det\begin{bmatrix} \mathbf{a}_1 & \rightarrow \\ \vdots & \\ \mathbf{a}_{k-1} & \rightarrow \\ (\mathbf{a}_k + \mathbf{b}) & \rightarrow \\ \mathbf{a}_{k+1} & \rightarrow \\ \vdots & \\ \mathbf{a}_n & \rightarrow \end{bmatrix} = \det\begin{bmatrix} \mathbf{a}_1 & \rightarrow \\ \vdots & \\ \mathbf{a}_{k-1} & \rightarrow \\ \mathbf{a}_k & \rightarrow \\ \mathbf{a}_{k+1} & \rightarrow \\ \vdots & \\ \mathbf{a}_n & \rightarrow \end{bmatrix} + \det\begin{bmatrix} \mathbf{a}_1 & \rightarrow \\ \vdots & \\ \mathbf{a}_{k-1} & \rightarrow \\ \mathbf{b} & \rightarrow \\ \mathbf{a}_{k+1} & \rightarrow \\ \vdots & \\ \mathbf{a}_n & \rightarrow \end{bmatrix},$$

$$\det\begin{bmatrix} \mathbf{a}_1 & \cdots & \mathbf{a}_{k-1} & (\mathbf{a}_k + \mathbf{b}) & \mathbf{a}_{k+1} & \cdots & \mathbf{a}_n \\ \downarrow & \vdots & \downarrow & \downarrow & \downarrow & \vdots & \downarrow \end{bmatrix}$$

$$= \det\begin{bmatrix} \mathbf{a}_1 & \cdots & \mathbf{a}_{k-1} & \mathbf{a}_k & \mathbf{a}_{k+1} & \cdots & \mathbf{a}_n \\ \downarrow & \vdots & \downarrow & \downarrow & \downarrow & \vdots & \downarrow \end{bmatrix}$$

$$+ \det\begin{bmatrix} \mathbf{a}_1 & \cdots & \mathbf{a}_{k-1} & \mathbf{b} & \mathbf{a}_{k+1} & \cdots & \mathbf{a}_n \\ \downarrow & \vdots & \downarrow & \downarrow & \downarrow & \vdots & \downarrow \end{bmatrix}.$$

Our derivation of the transpose rule hinges on another important theoretical property, the rule for matrix products:

Product Rule

$$\det(\mathbf{A}_1\mathbf{A}_2\cdots\mathbf{A}_k) = \det(\mathbf{A}_1)\det(\mathbf{A}_2)\cdots\det(\mathbf{A}_k). \tag{2}$$

We will see that the product rule leads to, among other things, a direct formula for $\mathbf{A}^{-1}$ (the "adjoint formula") and a determinant expression for the solution to a linear system (Cramer's Rule).

Equation (2) is obviously true for 1-by-1 matrices, and one can do the algebra to verify its truth for any pair of 2-by-2 matrices (Problem 16). But the algebraic manipulations required for handling n-by-n matrices quickly become unwieldy. The product rule is one of those innocent-looking propositions whose proof is infuriatingly elusive.[3] The full derivation is outlined in Problem 20; it exploits the following fact (which is also useful in other contexts).

Lemma 1. Suppose an arbitrary square matrix $\mathbf{A}$ is reduced to row echelon form $\mathbf{U}$ using the forward stage of Gauss elimination. If q row switches are performed during the reduction, then

$$\det(\mathbf{A}) = (-1)^q \det(\mathbf{U}). \tag{3}$$

Proof. Use the matrix operator representation of the elementary row operations to express the reduction with the matrix equation

$$\mathbf{U} = \mathbf{E}_p\mathbf{E}_{p-1}\cdots\mathbf{E}_2\mathbf{E}_1\mathbf{A}. \tag{4}$$

Now only row switches and additions of multiples of one row to another are involved in the forward part of the Gauss elimination algorithm. By Theorem 4 and Corollary 4 (Section 2.4), only the row switches affect the determinant, and Lemma 1 follows. ∎

This lemma establishes a very important connection between the determinant and the inverse. Suppose we try to calculate $\mathbf{A}^{-1}$ by solving the systems represented by $[\mathbf{A}\!:\!\mathbf{I}]$ (recall Example 3, Section 2.3). If we apply the forward part of the Gauss elimination algorithm, the systems $[\mathbf{A}\!:\!\mathbf{I}]$ will evolve to a form $[\mathbf{U}\!:\!\mathbf{B}]$. The particular nature of the matrix $\mathbf{B}$ is of no concern here. If each of the diagonal entries of $\mathbf{U}$ is nonzero, we will succeed in computing $\mathbf{A}^{-1}$ by back substitution. But if any of them are zero, the systems will either have multiple solutions or be inconsistent, indicating that $\mathbf{A}$ is not invertible. In the first case, $\det(\mathbf{U}) \neq \mathbf{0}$; in the second case, $\det(\mathbf{U}) = \mathbf{0}$. From (3) it then follows that

(viii) $\mathbf{A}$ is invertible if and only if $\det(\mathbf{A}) \neq 0$.

---

[3]The product rule was first proved by the great French mathematician Augustin-Louis Cauchy in 1812.

Further illumination of this connection is obtained by applying the product rule (2) (which we now take for granted) to the equation $\mathbf{A}^{-1}\mathbf{A} = \mathbf{I}$; we conclude

(ix) $\det(\mathbf{A}^{-1}) = 1/\det(\mathbf{A})$ if $\mathbf{A}$ is invertible.

We now turn to the verification of the transpose rule (1). Notice that it is immediate for symmetric matrices, and it is obvious for triangular matrices as well. Example 2, Section 1, showed that each of the elementary row operator matrices is either symmetric or triangular, so (1) holds for them too. For an arbitrary square matrix $\mathbf{A}$, we apply the reasoning of Lemma 1 and take the transpose of Equation (4) to write

$$\mathbf{U}^T = \mathbf{A}^T \mathbf{E}_1^T \mathbf{E}_2^T \cdots \mathbf{E}_{p-1}^T \mathbf{E}_p^T.$$

Applying the product rule, we conclude (recall that q is the number of row switches)

$$\det(\mathbf{U}) = \det(\mathbf{U}^T) = \det(\mathbf{A}^T)(-1)^q. \tag{5}$$

Combining this with (3) we confirm the transpose rule (1) (and its consequences (i) through (vii)). (For an alternate proof of (1), see Problem 32.)

Adjoint Matrix

A very useful fact emerges when we maneuver the cofactor expansion formula (vi) into the format of a matrix product. If we form the matrix whose (i,j)th entry is cof(i,j), the (i,j)th cofactor of $\mathbf{A}$, and then take the transpose, the result is called the *adjoint matrix* $\mathbf{A}^{\text{adj}}$.

$$\mathbf{A}^{\text{adj}} := \begin{bmatrix} \text{cof}(1,1) & \text{cof}(2,1) & \cdots & \text{cof}(n,1) \\ \text{cof}(1,2) & \text{cof}(2,2) & \cdots & \text{cof}(n,2) \\ \vdots & \vdots & \ddots & \vdots \\ \text{cof}(1,n) & \text{cof}(2,n) & \cdots & \text{cof}(n,n) \end{bmatrix}.$$

Look at what results when we multiply $\mathbf{A}$ by its adjoint:

$$\mathbf{A}\,\mathbf{A}^{\text{adj}} = \begin{bmatrix} a_{11} & a_{12} & \cdots & a_{1n} \\ a_{21} & a_{22} & \cdots & a_{2n} \\ \vdots & \vdots & \ddots & \vdots \\ a_{n1} & a_{n2} & \cdots & a_{nn} \end{bmatrix} \begin{bmatrix} \text{cof}(1,1) & \text{cof}(2,1) & \cdots & \text{cof}(n,1) \\ \text{cof}(1,2) & \text{cof}(2,2) & \cdots & \text{cof}(n,2) \\ \vdots & \vdots & \ddots & \vdots \\ \text{cof}(1,n) & \text{cof}(2,n) & \cdots & \text{cof}(n,n) \end{bmatrix}$$

On the diagonal—the (2,2) entry, for example—we get $\sum_{j=1}^{n} a_{2j} \operatorname{cof}(2,j)$, which is the expansion, by the cofactors of the second row, for $\det(\mathbf{A})$. So the diagonal of $\mathbf{A}\,\mathbf{A}^{\text{adj}}$ looks like

$$
\mathbf{A}\,\mathbf{A}^{\text{adj}} = \begin{bmatrix} \det(\mathbf{A}) & - & \cdots & - \\ - & \det(\mathbf{A}) & \cdots & - \\ \vdots & \vdots & \ddots & \vdots \\ - & - & \cdots & \det(\mathbf{A}) \end{bmatrix}.
$$

Off the diagonal—say, the (1,2) entry—we get $\sum_{j=1}^{n} a_{1j} \operatorname{cof}(2,j)$. This resembles a cofactor expansion for $\det(\mathbf{A})$ by the cofactors of the second row, *but* the entries $a_{11}, a_{22}, \ldots, a_{1n}$ are drawn from the *first* row. Therefore this is a cofactor expansion of the determinant of *the matrix* $\mathbf{A}$ *with its second row replaced by its first row*,

$$
\det \begin{bmatrix} a_{11} & a_{12} & \cdots & a_{1n} \\ a_{11} & a_{12} & \cdots & a_{1n} \\ \vdots & \vdots & \ddots & \vdots \\ a_{n1} & a_{n2} & \cdots & a_{nn} \end{bmatrix} = \sum_{j=1}^{n} a_{1j} \operatorname{cof}(2,j).
$$

This determinant is, of course, zero.

Generalizing, we conclude that

$$
\mathbf{A}\,\mathbf{A}^{\text{adj}} = \begin{bmatrix} \det(\mathbf{A}) & 0 & \cdots & 0 \\ 0 & \det(\mathbf{A}) & \cdots & 0 \\ \vdots & \vdots & \ddots & \vdots \\ 0 & 0 & \cdots & \det(\mathbf{A}) \end{bmatrix} = \det(\mathbf{A})\,\mathbf{I}, \tag{6}
$$

which yields a direct formula for $\mathbf{A}^{-1}$ if $\det(\mathbf{A}) \neq 0$.

Adjoint Formula for the Inverse

$\mathbf{A}^{-1} = \mathbf{A}^{\text{adj}} / \det(\mathbf{A})$ if $\mathbf{A}$ is invertible.

From these properties, we will be able to derive a direct formula for the solution to a linear system.

Cramer's Rule[4]

If $\mathbf{A} = [a_{ij}]$ is an n-by-n matrix and $\det(\mathbf{A}) \neq 0$, the system $\mathbf{Ax} = \mathbf{b}$ has a unique solution whose ith component is given by

$$x_i = \frac{1}{\det(\mathbf{A})} \det \begin{bmatrix} a_{11} & a_{12} & \cdots & a_{1,i-1} & b_1 & a_{1,i+1} & \cdots & a_{1n} \\ a_{21} & a_{22} & \cdots & a_{2,i-1} & b_2 & a_{2,i+1} & \cdots & a_{2n} \\ \vdots & \vdots & \ddots & \vdots & \vdots & \vdots & \ddots & \vdots \\ a_{n1} & a_{n2} & \cdots & a_{n,i-1} & b_n & a_{n,i+1} & \cdots & a_{nn} \end{bmatrix}. \qquad (7)$$

Example 1. Use Cramer's rule to solve the system

$$\begin{bmatrix} 1 & 2 & 1 \\ 1 & 3 & 2 \\ 1 & 0 & 1 \end{bmatrix} \begin{bmatrix} x_1 \\ x_2 \\ x_3 \end{bmatrix} = \begin{bmatrix} 1 \\ -1 \\ 0 \end{bmatrix}.$$

(This replicates system (2) in Section 2.1.)

Solution. Following (7), we have

$$x_1 = \frac{\det \begin{bmatrix} 1 & 2 & 1 \\ -1 & 3 & 2 \\ 0 & 0 & 1 \end{bmatrix}}{\det \begin{bmatrix} 1 & 2 & 1 \\ 1 & 3 & 2 \\ 1 & 0 & 1 \end{bmatrix}} = \frac{5}{2}, \quad x_2 = \frac{\det \begin{bmatrix} 1 & 1 & 1 \\ 1 & -1 & 2 \\ 1 & 0 & 1 \end{bmatrix}}{\det \begin{bmatrix} 1 & 2 & 1 \\ 1 & 3 & 2 \\ 1 & 0 & 1 \end{bmatrix}} = \frac{1}{2},$$

$$x_3 = \frac{\det \begin{bmatrix} 1 & 2 & 1 \\ 1 & 3 & -1 \\ 1 & 0 & 1 \end{bmatrix}}{\det \begin{bmatrix} 1 & 2 & 1 \\ 1 & 3 & 2 \\ 1 & 0 & 1 \end{bmatrix}} = \frac{-5}{2}. \qquad \blacksquare$$

Now we'll show why Cramer's rule follows from property (ix). When $\mathbf{A}$ is invertible the (unique) solution to the system $\mathbf{Ax} = \mathbf{b}$ is given by $\mathbf{x} = \mathbf{A}^{-1}\mathbf{b}$, or

$$\mathbf{x} = \begin{bmatrix} x_1 \\ x_2 \\ \vdots \\ x_n \end{bmatrix} = \frac{1}{\det(\mathbf{A})} \mathbf{A}^{\text{adj}} \mathbf{b}$$

[4]First published by the Swiss mathematician Gabriel Cramer in 1850.

$$
= \frac{1}{\det(\mathbf{A})}
\begin{bmatrix}
\text{cof}(1,1) & \text{cof}(2,1) & \cdots & \text{cof}(n,1) \\
\text{cof}(1,2) & \text{cof}(2,2) & \cdots & \text{cof}(n,2) \\
\vdots & \vdots & \ddots & \vdots \\
\text{cof}(1,n) & \text{cof}(2,n) & \cdots & \text{cof}(n,n)
\end{bmatrix}
\begin{bmatrix} b_1 \\ b_2 \\ \vdots \\ b_n \end{bmatrix}
$$

$$
= \frac{1}{\det(\mathbf{A})}
\begin{bmatrix}
b_1\text{cof}(1,1) + b_2\text{cof}(2,1) + \cdots + b_n\text{cof}(n,1) \\
b_1\text{cof}(1,2) + b_2\text{cof}(2,2) + \cdots + b_n\text{cof}(n,2) \\
\vdots \\
b_1\text{cof}(1,n) + b_2\text{cof}(2,n) + \cdots + b_n\text{cof}(n,n)
\end{bmatrix}.
\tag{8}
$$

But comparison with the cofactor expansion formula (*vi*) shows that the first entry of the final (column) matrix is the expansion, by the cofactors of the first *column*, of the determinant of the matrix $\mathbf{A}$ with its first column replaced by the vector $\mathbf{b}$:

$$
x_1 = \frac{1}{\det(\mathbf{A})} \det
\begin{bmatrix}
b_1 & a_{12} & \cdots & a_{1n} \\
b_2 & a_{22} & \cdots & a_{2n} \\
\vdots & \vdots & \ddots & \vdots \\
b_n & a_{n2} & \cdots & a_{nn}
\end{bmatrix}.
$$

The second entry of the final column matrix displayed in (8) is the cofactor expansion of

$$
\det
\begin{bmatrix}
a_{11} & b_1 & \cdots & a_{1n} \\
a_{21} & b_2 & \cdots & a_{2n} \\
\vdots & \vdots & \ddots & \vdots \\
a_{n1} & b_n & \cdots & a_{nn}
\end{bmatrix}
$$

and so on, confirming the general formula (7).

Cramer's rule is not nearly as computationally efficient as Gauss elimination, but it provides a neat *formula* for the solution to a system (which the algorithm does not). This can be very helpful in some situations. For example, suppose one is studying an engineering design governed by a linear system of equations $\mathbf{Ax} = \mathbf{b}$. The solution $\mathbf{x}$ depends, of course, on the data in the coefficient matrix $\mathbf{A}$, as well as the right-hand vector $\mathbf{b}$. It might be desirable to know the influence of a particular piece of data on a particular component in the system. Consider the following:

Example 2.   The following system contains a parameter α in its right-hand side.

$$
\begin{bmatrix}
1 & 2 & 1 \\
1 & 3 & 2 \\
1 & 0 & 1
\end{bmatrix}
\begin{bmatrix} x_1 \\ x_2 \\ x_3 \end{bmatrix}
=
\begin{bmatrix} \alpha \\ -1 \\ 0 \end{bmatrix},
$$

What is the derivative of the solution component x_3 with respect to α?

Solution. Using Cramer's rule and the cofactor expansion for the third column in the numerator, we find

$$x_3 = \frac{\det \begin{bmatrix} 1 & 2 & \alpha \\ 1 & 3 & -1 \\ 1 & 0 & 0 \end{bmatrix}}{\det \begin{bmatrix} 1 & 2 & 1 \\ 1 & 3 & 2 \\ 1 & 0 & 1 \end{bmatrix}} = \frac{\alpha \cof(1,3) - 1 \cof(2,3) + 0 \cof(3,3)}{2},$$

$$\frac{dx_3}{d\alpha} = \frac{\cof(1,3)}{2} = \frac{3}{2}. \qquad \blacksquare$$

Exercises 2.5

In Problems 1–4, use the Gauss elimination calculations performed in preceding sections to deduce the value of the determinant of the given matrix.

1. $\begin{bmatrix} 1 & 2 & 2 \\ 2 & 1 & 1 \\ 1 & 1 & 3 \end{bmatrix}$ (Example 1, Section 1.2)

2. $\begin{bmatrix} 2 & 1 & 4 \\ 3 & 3 & 0 \\ -1 & 4 & 2 \end{bmatrix}$ (Example 2, Section 1.2

3. $\begin{bmatrix} 0.202131 & 0.732543 & 0.141527 & 0.359867 \\ 0.333333 & -0.112987 & 0.412989 & 0.838838 \\ -0.486542 & 0.500000 & 0.989989 & -0.246801 \\ 0.101101 & 0.321111 & -0.444444 & 0.245542 \end{bmatrix}$ (Example 2, Section 1.1)

4. $\begin{bmatrix} 0 & 2 & 1 & 1 & 0 \\ 2 & 4 & 4 & 2 & 2 \\ 3 & 6 & 6 & 0 & 0 \\ 0 & -2 & -1 & -2 & 2 \\ 0 & 2 & 1 & 2 & 4 \end{bmatrix}$ (Example 3, Section 1.3)

In Problems 5–11, use the adjoint matrix to compute the inverse of the given matrix.

5. $\begin{bmatrix} 0 & 1 \\ 1 & 0 \end{bmatrix}$ 6. $\begin{bmatrix} \cos\theta & -\sin\theta \\ \sin\theta & \cos\theta \end{bmatrix}$ 7. $\begin{bmatrix} 1 & 0 & 0 \\ 0 & 0 & 1 \\ 0 & 1 & 0 \end{bmatrix}$

8. $\begin{bmatrix} 1 & 0 & 0 \\ \alpha & 1 & 0 \\ 0 & 0 & 1 \end{bmatrix}$ **9.** $\begin{bmatrix} 1 & 0 & 0 \\ 0 & \beta & 0 \\ 0 & 0 & 1 \end{bmatrix}$ **10.** $\begin{bmatrix} 1 & 2 & 1 \\ 1 & 3 & 2 \\ 1 & 0 & 1 \end{bmatrix}$

11. $\begin{bmatrix} 3/2 & -1 & 1/2 \\ 1/2 & 0 & -1/2 \\ -3/2 & 1 & 1/2 \end{bmatrix}$

12. Prove that if $\mathbf{A}^2 = \mathbf{A}$, the determinant of $\mathbf{A}$ is 1 or 0.

   In Problems 13–15, use Cramer's rule to solve the given system.

13. $\begin{bmatrix} 2 & 2 & \vdots & 4 \\ 3 & 1 & \vdots & 2 \end{bmatrix}$

14. $\begin{bmatrix} 1 & 2 & 2 & \vdots & 6 \\ 2 & 1 & 1 & \vdots & 6 \\ 1 & 1 & 3 & \vdots & 6 \end{bmatrix}$ (Cf. Example 1, Section 1.2)

15. $\begin{bmatrix} 2 & 1 & 4 & \vdots & -2 \\ 3 & 3 & 0 & \vdots & 6 \\ -1 & 4 & 2 & \vdots & -5 \end{bmatrix}$ (Cf. Example 2, Section 1.2)

16. For arbitrary 2-by-2 matrices $\mathbf{A}$ and $\mathbf{B}$, verify algebraically that $\det(\mathbf{AB}) = \det(\mathbf{A})\det(\mathbf{B})$

17. For the system containing the variable α,

$$\begin{bmatrix} 1 & 2 & 1 \\ 1 & 3 & 2 \\ 1 & 0 & 1 \end{bmatrix} \begin{bmatrix} x_1 \\ x_2 \\ x_3 \end{bmatrix} = \begin{bmatrix} 2 \\ \alpha \\ 0 \end{bmatrix},$$

use Cramer's rule to express the derivatives of the solution components x_j ($j = 1, 2, 3$) with respect to α in terms of determinants and cofactors.

18. Generalize the preceding problem: for the system $\mathbf{Ax} = \mathbf{b}$, show that the partial derivative of x_j with respect to b_k (holding the other b_i's constant) is given by $\mathrm{cof}(k, j) / \det(\mathbf{A})$.

19. For the system containing the variable α,

$$\begin{bmatrix} 2 & 2 & 1 \\ \alpha & 3 & 2 \\ 1 & 0 & 1 \end{bmatrix} \begin{bmatrix} x_1 \\ x_2 \\ x_3 \end{bmatrix} = \begin{bmatrix} 2 \\ 3 \\ 4 \end{bmatrix},$$

use Cramer's rule to find an expression for

(a) the derivative of the solution component x_1 with respect to α;

(b) the derivative of the solution component x_2 with respect to α.

20. Carry out the following argument to prove the product rule (2) for determinants. All matrices herein are n-by-n.

(a) Argue that $\det(\mathbf{DB}) = \det(\mathbf{D})\det(\mathbf{B})$ when $\mathbf{D}$ is diagonal (and $\mathbf{B}$ is arbitrary), by considering

$$
\begin{bmatrix}
d_{11} & \mathbf{0} & \cdots & \mathbf{0} \\
\mathbf{0} & d_{22} & \cdots & \mathbf{0} \\
\vdots & \vdots & \ddots & \vdots \\
\mathbf{0} & \mathbf{0} & \cdots & d_{nn}
\end{bmatrix}
\begin{bmatrix}
b_{11} & b_{12} & \cdots & b_{1n} \\
b_{21} & b_{22} & \cdots & b_{2n} \\
\vdots & \vdots & \ddots & \vdots \\
b_{n1} & b_{n2} & \cdots & b_{nn}
\end{bmatrix}
$$

$$
=
\begin{bmatrix}
d_{11}b_{11} & d_{11}b_{12} & \cdots & d_{11}b_{1n} \\
d_{22}b_{21} & d_{22}b_{22} & \cdots & d_{22}b_{2n} \\
\vdots & \vdots & \ddots & \vdots \\
d_{nn}b_{n1} & d_{nn}b_{n2} & \cdots & d_{nn}b_{nn}
\end{bmatrix}
=
\begin{bmatrix}
d_{11} \times [\text{row1 of }\mathbf{B}] \\
d_{22} \times [\text{row2 of }\mathbf{B}] \\
\vdots \\
d_{nn} \times [\text{row}n\text{ of }\mathbf{B}]
\end{bmatrix}
$$

and applying Corollary 2 of Section 2.4.

(b) Generalize: argue that

$$\det(\mathbf{UB}) = \det(\mathbf{U})\det(\mathbf{B}) \tag{9}$$

if $\mathbf{U}$ is an upper triangular matrix, by considering

$$
\begin{bmatrix}
\alpha & \mathbf{0} & \mathbf{0} \\
\mathbf{0} & \beta & \gamma \\
\mathbf{0} & \mathbf{0} & \delta
\end{bmatrix}
\begin{bmatrix}
b_{11} & b_{12} & b_{13} \\
b_{21} & b_{22} & b_{23} \\
b_{31} & b_{32} & b_{33}
\end{bmatrix}
=
\begin{bmatrix}
\alpha \times [\text{row1}] \\
\beta \times [\text{row2}] + \gamma \times [\text{row3}] \\
\delta \times [\text{row3}]
\end{bmatrix}.
$$

and applying Corollaries 2 and 4 of Section 2.4.

(c) Suppose as in Lemma 1 that the reduction of the (arbitrary) matrix $\mathbf{A}$ to upper triangular form $\mathbf{U}$ during the forward stage of Gauss elimination is represented by the formula (4), with q row exchange operators in the set $\{\mathbf{E}_1, \mathbf{E}_1, \cdots, \mathbf{E}_{p-1}, \mathbf{E}_p\}$. Multiply (4) by $\mathbf{B}$, on the right, and apply the inverses of the row operators to derive

$$\mathbf{E}_1^{-1}\mathbf{E}_2^{-1}\cdots\mathbf{E}_{p-1}^{-1}\mathbf{E}_p^{-1}\mathbf{UB} = \mathbf{AB}. \tag{10}$$

(d) Argue that (10) demonstrates that the matrix $(\mathbf{UB})$ evolves into the matrix $(\mathbf{AB})$ after a sequence of p elementary row operations, including q row exchanges (and no multiplications by scalars). Thus

$$\det(\mathbf{AB}) = (-1)^q \det(\mathbf{UB}). \tag{11}$$

(e) Combine (9), (11), and (3) in Lemma 1 to conclude that $\det(\mathbf{AB}) = \det(\mathbf{A})\det(\mathbf{B})$ in general. Extend to derive the product rule (2).

21. Express $\det \begin{bmatrix} p & o & n & m \\ l & k & j & i \\ h & g & f & e \\ d & c & b & a \end{bmatrix}$ in terms of $\det \begin{bmatrix} a & b & c & d \\ e & f & g & h \\ i & j & k & l \\ m & n & o & p \end{bmatrix}$. Formulate the generalization for the n-by-n case.

22. A *skew-symmetric* n-by-n matrix $\mathbf{A}$ has the property $\mathbf{A} = -\mathbf{A}^T$. For example, $\begin{bmatrix} 0 & a & b \\ -a & 0 & c \\ -b & -c & 0 \end{bmatrix}$ is skew symmetric. Show that the determinant of such a matrix is zero if n is odd.

23. Let D_n be the determinant of the n-by-n submatrix defined by the first n rows and columns of the form

$$\begin{bmatrix} a & b & 0 & 0 & 0 & \cdots \\ b & a & b & 0 & 0 & \cdots \\ 0 & b & a & b & 0 & \cdots \\ 0 & 0 & b & a & b & \cdots \\ & & & & & \\ 0 & 0 & 0 & b & a & \ddots \\ \vdots & \vdots & \vdots & \vdots & \ddots & \ddots \end{bmatrix}.$$

Evaluate D_1, D_2, and D_3.

24. (a) Use a row cofactor expansion followed by a column cofactor expansion to derive the recurrence relation $D_n = aD_{n-1} - b^2 D_{n-2}$ for the determinants defined in the preceding problem. [*Hint*: First try it out on the 5-by-5 matrix to see how it works.]

    (b) For the case $a = b = 1$, compute D_3 to D_6 and predict the value of D_{99}.

25. Let F_n be the determinant of the n-by-n submatrix defined by the first n rows and columns of the form

$$\begin{bmatrix} a & -b & 0 & 0 & 0 & \cdots \\ b & a & -b & 0 & 0 & \cdots \\ 0 & b & a & -b & 0 & \cdots \\ 0 & 0 & b & a & -b & \cdots \\ & & & & & \\ 0 & 0 & 0 & b & a & \ddots \\ \vdots & \vdots & \vdots & \vdots & \ddots & \ddots \end{bmatrix}.$$

    (a) Use a row cofactor expansion followed by a column cofactor expansion to derive the recurrence relation between F_n, F_{n-1}, and F_{n-2}. [*Hint*: Use the strategy suggested in the preceding problem.]

    (b) Write out six members of the sequence $\{F_n\}$ for the case $a = b = 1$. Do you recognize them as members of the *Fibonacci sequence*?

26. (a) Use the adjoint matrix to argue that if **A** has all integer entries and $\det(\mathbf{A}) = \pm 1$, then $\mathbf{A}^{-1}$ also has all integer entries. Construct a nondiagonal 3-by-3 integer matrix illustrating this fact.

    (b) Conversely, show that if **A** and $\mathbf{A}^{-1}$ both have all integer entries, then $\det(\mathbf{A}) = \pm 1$.

27. For each of the following decide whether the statement made is always True or sometimes False.

    (a) If **A** is a square matrix satisfying $\mathbf{A}^T\mathbf{A} = \mathbf{I}$, then $\det(\mathbf{A}) = \pm 1$.

    (b) The (i, j)th cofactor of a matrix **A** is the determinant of the matrix obtained by deleting the ith row and jth column of **A**.

    (c) If **A** and **S** are n-by-n matrices and **S** is invertible, then $\det(\mathbf{A}) = \det(\mathbf{S}^{-1}\mathbf{A}\mathbf{S})$.

    (d) If $\det(\mathbf{A}) = 0$, then there is at least one column vector $\mathbf{x} \neq \mathbf{0}$ such that $\mathbf{A}\mathbf{x} = \mathbf{0}$.

28. Suppose **A**, **B**, and **C** are square matrices of the same size whose product **ABC** is invertible. Use the properties of determinants to prove that the matrix **B** is invertible.

29. Use the adjoint matrix to construct the inverse of the following matrices:

$$
\text{(a)}\ \begin{bmatrix} 1 & 2 & 3 \\ 0 & 4 & 5 \\ 0 & 0 & 6 \end{bmatrix} \qquad
\text{(b)}\ \begin{bmatrix} 1 & 2 & 3 \\ 2 & 3 & 4 \\ 3 & 4 & 1 \end{bmatrix}
$$

30. Use Problem 29 as a guide to

    (a) show that if **A** is upper (lower) triangular and invertible, then so is $\mathbf{A}^{-1}$;

    (b) show that if **A** is symmetric and invertible, then so is $\mathbf{A}^{-1}$.

31. If **A** and **B** are n-by-n matrices and **B** is invertible, express $\det(\mathbf{B}\mathbf{A}\mathbf{B}^{-1})$ and $\det(\mathbf{B}\mathbf{A}\mathbf{B}^T)$ in terms of $\det(\mathbf{A})$ and $\det(\mathbf{B})$.

32. Use the permutation representation of the determinant, Problem 20 of Exercises 2.4, to give an alternative proof of the transpose rule (1).

2.6 SUMMARY

Matrix Algebra and Operators

Comparison of the augmented matrix representation and the customary display of a system of linear equations motivates the definition of the matrix product, which enables the system to be compactly represented as $\mathbf{A}\mathbf{x} = \mathbf{b}$. The product, coupled with the natural definition of matrix addition, gives rise to an algebra that facilitates the formulation and solution of linear systems, as well as providing a convenient instrument for analyzing the operators involved in Gauss elimination, orthogonal projections, rotations, plane mirror reflections, and many other applications that will be explored in subsequent chapters. This algebra adheres to most of the familiar rules of computation except for the complications introduced by the noncommutativity of the product and the precarious nature of the multiplicative inverse; indeed, the very existence of the latter is intimately tied to the issues of solvability of the system $\mathbf{Ax} = \mathbf{b}$. The notion of the transpose of a matrix further enriches the utility of the algebra.

Determinants

The definition of the determinant of a square matrix $\mathbf{A}$ generalizes the expression for the volume of a parallelepiped in terms of the Cartesian components of its edge vectors. The following summarizes its basic properties:

1. If any row or column of $\mathbf{A}$ is multiplied by a scalar, the deteminant is multiplied by the same scalar.
2. If a row or column of $\mathbf{A}$ consists of all zeros, $\det(\mathbf{A}) = 0$.
3. If any two rows or columns of a matrix are switched, the deteminant changes sign.
4. If any two rows or columns of $\mathbf{A}$ are equal, $\det(\mathbf{A}) = 0$.
5. If any multiple of one row (column) of a matrix is added to another row (column), the deteminant is unchanged.
6. The determinant of a triangular matrix equals the product of the entries along its diagonal.
7. Product rule: $\det(\mathbf{A}_1\mathbf{A}_2 \cdots \mathbf{A}_k) = \det(\mathbf{A}_1)\det(\mathbf{A}_2)\cdots\det(\mathbf{A}_k)$.
8. $\det(\mathbf{A}) = (-1)^q \det(\mathbf{U})$, if $\mathbf{A}$ is reduced to the upper triangular matrix $\mathbf{U}$ with q row exchanges in the forward part of the Gauss elimination algorithm.
9. $\mathbf{A}$ is invertible if and only if $\det(\mathbf{A}) \neq 0$, and $\det(\mathbf{A}^{-1}) = 1/\det(\mathbf{A})$.
10. Transpose rule: $\det(\mathbf{A}^T) = \det(\mathbf{A})$.
11. The cofactor expansion is valid for any row and column:

$$\det(\mathbf{A}) = \sum_{k=1}^{n} a_{ik}\text{cof}(i,k) = \sum_{k=1}^{n} a_{kj}\text{cof}(k,j)$$

    where $\text{cof}(i,k)$ denotes $(-1)^{i+k}\det(\mathbf{A}_{ik})$ when $\mathbf{A}_{ik}$ is obtained from $\mathbf{A}$ by deleting the ith row and kth column.
12. The adjoint matrix $\mathbf{A}^{\text{adj}}$ associated with $\mathbf{A}$ is the matrix whose (i,j)th entry is $\text{cof}(j,i)$. Furthermore, $\mathbf{A}^{-1} = \mathbf{A}^{\text{adj}}/\det(\mathbf{A})$ if $\mathbf{A}$ is invertible.
13. Cramer's rule. If $\det(\mathbf{A}) \neq 0$, the system $\mathbf{Ax} = \mathbf{b}$ has a unique solution whose ith component is given by

$$x_i = \frac{\det[\mathbf{a}_1 \, \mathbf{a}_2 \, \cdots \, \mathbf{a}_{i-1} \, \mathbf{b} \, \mathbf{a}_{i+1} \, \cdots \, \mathbf{a}_n]}{\det(\mathbf{A})}, \text{ where } \mathbf{a}_j \text{ is the } j\text{th column of } \mathbf{A}.$$

REVIEW PROBLEMS FOR PART I

1. For which values of γ does the following system possesses solutions?

$$\begin{bmatrix} -3 & 2 & 1 & \vdots & 4 \\ 1 & -1 & 2 & \vdots & 1 \\ -2 & 1 & 3 & \vdots & \gamma \end{bmatrix}$$

Are the solutions unique?

2. (a) Construct a linear system of equations for the coefficients a, b, c in the polynomial $ax^2 + bx + c$ that matches e^x at the points $x = 0, 1$, and 2.

   (b) Construct a linear system of equations for the coefficients a, b, c in the polynomial $ax^2 + bx + c$ whose value, first derivative, and second derivative match those of e^x at the point $x = 1$.

   (c) Construct a linear system of equations for the coefficients a, b, c in the polynomial $a(x - 1)^2 + b(x - 1) + c$ whose value, first derivative, and second derivative match those of e^x at the point $x = 1$.

3. Compute the following matrix products:

   (a) $\begin{bmatrix} 2 & 3 & -1 \\ 4 & 2 & 3 \end{bmatrix} \begin{bmatrix} 2 \\ 4 \\ 6 \end{bmatrix}$
   (b) $\begin{bmatrix} 2 & 3 & -1 \end{bmatrix} \begin{bmatrix} 2 & 0 & 1 \\ 4 & 2 & -1 \\ 6 & 2 & 3 \end{bmatrix}$
   (c) $\begin{bmatrix} 2 \\ 3 \\ -1 \end{bmatrix} \begin{bmatrix} 2 & 4 & 6 \end{bmatrix}$

   (d) $\begin{bmatrix} 2 & 3 & -1 \\ 4 & 2 & 3 \end{bmatrix} \begin{bmatrix} 1 & 0 & 0 \\ 0 & 1 & 0 \\ 0 & 0 & 1 \end{bmatrix}$
   (e) $\begin{bmatrix} 0 & 1 & 0 & 0 \\ 1 & 0 & 0 & 0 \\ 0 & 0 & 0 & 1 \\ 0 & 0 & 1 & 0 \end{bmatrix}^3$
   (f) $\begin{bmatrix} -1 & 0 & 1 \\ 0 & 1 & 0 \\ 0 & 0 & -1 \end{bmatrix}^{25}$

4. By experimentation, find all powers $\begin{bmatrix} 1 & 1 \\ 0 & 1 \end{bmatrix}^n$, for positive and negative integers n.

5. Construct a pair of real, nonzero 2-by-2 matrices $\mathbf{A}$ and $\mathbf{B}$ such that $\mathbf{A}^2 + \mathbf{B}^2 = \mathbf{0}$.

6. Show that if $\mathbf{A}$ is a 2-by-2 matrix satisfying $\mathbf{AB} = \mathbf{BA}$ for *every* (2-by-2) matrix $\mathbf{B}$, then $\mathbf{A}$ is a multiple of the (2-by-2) identity. [*Hint*: Pick $\mathbf{B}_1 = \begin{bmatrix} 1 & 0 \\ 0 & -1 \end{bmatrix}$ and $\mathbf{B}_2 = \begin{bmatrix} 0 & 1 \\ 1 & 0 \end{bmatrix}$.]

7. Let $\mathbf{E}$ be a matrix that performs an elementary row operation on a matrix $\mathbf{A}$ through left multiplication $\mathbf{EA}$, as described in Example 2 of Section 2.1.

   (a) Describe the form of $\mathbf{E}$ for each of the three elementary row operations corresponding to the operations listed in Theorem 1, Section 1.1.

   (b) Describe the form of $\mathbf{E}^{-1}$ for the matrices in (a).

   (c) Describe the effect of right multiplication $\mathbf{AE}$ on the *columns* of $\mathbf{A}$, for the matrices in (a).

(d) If $\mathbf{A}'$ is formed by switching two rows of $\mathbf{A}$, how does one form $(\mathbf{A}')^{-1}$ from $\mathbf{A}^{-1}$? [*Hint*: exploit the relation $(\mathbf{EA})^{-1} = \mathbf{A}^{-1}\mathbf{E}^{-1}$.]

(e) If $\mathbf{A}'$ is formed by multiplying the second row of $\mathbf{A}$ by 7, how does one form $(\mathbf{A}')^{-1}$ from $\mathbf{A}^{-1}$?

(f) If $\mathbf{A}'$ is formed by adding the first row of $\mathbf{A}$ to its second row, how does one form $(\mathbf{A}')^{-1}$ from $\mathbf{A}^{-1}$?

8. Suppose the second row of $\mathbf{A}$ equals twice its first row. Prove the same is true of any product $\mathbf{AB}$.

9. Find matrices $\mathbf{A}$ and $\mathbf{B}$ such that $\mathbf{AMB}$ equals the (i,j)th entry of $\mathbf{M}$.

10. For the system

$$2u + 3v + 7w - x + 2y = 3$$
$$-u + 2v + w + 2x + y = 2,$$

express x and y in terms of u, v, and w by first rewriting the system in the matrix form

$$\mathbf{A}\begin{bmatrix} x \\ y \end{bmatrix} + \mathbf{B}\begin{bmatrix} u \\ v \\ w \end{bmatrix} = \mathbf{C}$$

and then multiplying by $\mathbf{A}^{-1}$.

11. For each of the following, determine if the statement made is always True or sometimes False.

(a) If $\mathbf{A}$ is invertible, then so is $\mathbf{A}^2$.

(b) If $\mathbf{x}$ is an n-by-1 column vector and $\mathbf{y}$ is an 1-by-n row vector, with $n \geq 2$, then $\mathbf{xy}$ is an n-by-n singular matrix.

(c) If the n-by-n row echelon matrix $\mathbf{A}$ is invertible, then it has rank n.

(d) If $\mathbf{A}$ and $\mathbf{B}$ are n-by-n matrices satisfying $\mathbf{AB} = \mathbf{0}$, then either $\mathbf{A} = \mathbf{0}$ or $\mathbf{B} = \mathbf{0}$.

(e) If $\mathbf{x}$ is a vector in $\mathbf{R}^3_{\text{row}}$, then $\mathbf{x}$ is a linear combination of vectors $[1\ 3\ -1]$, $[4\ 6\ 1]$, and $[1\ -3\ 4]$.

(f) If $\mathbf{A}$ is an n-by-n matrix, then $\mathbf{A}$ and $\mathbf{A}^T$ commute.

12. Find necessary and sufficient conditions on the values of c and d so that the system

$$\begin{bmatrix} 1 & 0 & 1 & \vdots & b_1 \\ 0 & c & 1 & \vdots & b_2 \\ 0 & 1 & d & \vdots & b_3 \end{bmatrix}$$

always has a solution (for any b_1, b_2, b_3).

13. Prove that if $\mathbf{K}$ is *skew-symmetric* $(\mathbf{K}^T = -\mathbf{K})$, then $\mathbf{I} + \mathbf{K}$ must be nonsingular. [*Hint*: First argue that $\mathbf{v}^T \mathbf{K} \mathbf{v}$ is zero for every $\mathbf{v}$, and conclude that the homogeneous equation $(\mathbf{I} + \mathbf{K})\mathbf{x} = \mathbf{0}$ has only the trivial solution.]

14. Find the determinants using cofactor expansions:

(a) $\begin{vmatrix} 0 & 0 & 0 & 1 \\ 0 & 0 & 1 & 0 \\ 0 & 1 & 0 & 0 \\ 1 & 0 & 0 & 0 \end{vmatrix}$
(b) $\begin{vmatrix} 0 & 0 & 0 & 1 \\ 0 & 0 & 2 & 3 \\ 0 & 4 & 5 & 6 \\ 7 & 8 & 9 & 10 \end{vmatrix}$

(c) $\begin{vmatrix} 0 & 0 & 2 & 3 \\ 0 & 0 & 0 & 1 \\ 7 & 8 & 9 & 10 \\ 0 & 4 & 5 & 6 \end{vmatrix}$
(d) $\begin{vmatrix} x & 1 & 0 & 0 & 0 \\ 0 & x & 1 & 0 & 0 \\ 0 & 0 & x & 1 & 0 \\ 0 & 0 & 0 & x & 1 \\ 1 & 2 & 3 & 4 & 5 \end{vmatrix}$

15. Use Cramer's rule to determine $dx_i/d\gamma$ for $i = 1, 2, 3$, for each of the following systems involving the variable γ:

(a) $\begin{bmatrix} -3 & 2 & 1 & \vdots & 4 \\ 0 & -1 & 2 & \vdots & 1 \\ -2 & 1 & 3 & \vdots & \gamma \end{bmatrix}$
(b) $\begin{bmatrix} -3 & 2 & 1 & \vdots & 4 \\ \gamma & -1 & 2 & \vdots & 1 \\ -2 & 1 & 3 & \vdots & 0 \end{bmatrix}$.

16. Describe how to calculate $\mathbf{A}^{-1}\mathbf{B}\mathbf{v}$ and $\mathbf{B}\mathbf{A}^{-1}\mathbf{v}$ without computing $\mathbf{A}^{-1}$.

TECHNICAL WRITING EXERCISES FOR PART I

1. Describe how traces of each of the traditional solution techniques—substitution, cross-multiplication and subtraction, and inspection—are incorporated in the Gauss elimination algorithm.

2. Compare the algebraic properties of the collection of positive real numbers with those of the collection of n-by-n matrices having all positive real-number entries. Provide specific examples to justify your claims regarding, for example, multiplication and inversion.

3. Matrix multiplication has the property $\mathbf{A}(c_1\mathbf{x}+c_2\mathbf{y}) = c_1\mathbf{A}\mathbf{x}+c_2\mathbf{A}\mathbf{y}$. This property is known as *linearity* when $\mathbf{A}$ is considered as an operator or function acting on vectors. Describe some other operators, like d/dx, that possess the linearity property. Also describe some examples of nonlinear operators.

4. Explain why performing all the required row exchanges *at the outset* on a system of linear equations (if you could predict what they would be) results in a system whose Gauss elimination procedure employs exactly the same arithmetic calculations, but no pivoting.

5. Explain why the following four ways to calculate the matrix product are equivalent:

(i) (by individual entries) The (i, j)th entry of **AB** is the dot product of the ith row of **A** with the jth column of **B**.

(ii) (by individual columns) The ith column of **AB** is the linear combination of all the columns of **A** with coefficients drawn from the ith column of **B**.

(iii) (by individual rows) The ith row of **AB** is the linear combination of all the rows of **B** with coefficients drawn from the ith row of **A**.

(iv) (all at once) **AB** is sum of the products of all the columns of **A** with the corresponding rows of **B**.

GROUP PROJECTS FOR PART I

A. LU Factorization

Bearing in mind that back substitution is the quickest step in the Gauss elimination algorithm for $\mathbf{Ax} = \mathbf{b}$, consider the following proposal for solving the system in Example 1, Section 1.2, rewritten using matrix algebra as

$$\begin{bmatrix} 1 & 2 & 2 \\ 2 & 1 & 1 \\ 1 & 1 & 3 \end{bmatrix} \begin{bmatrix} x_1 \\ x_2 \\ x_3 \end{bmatrix} = \begin{bmatrix} 6 \\ 6 \\ 6 \end{bmatrix}. \tag{1}$$

(i) First perform a preliminary calculation on the coefficient matrix **A** that factors it into the product of a lower triangular matrix **L** and an upper triangular matrix **U**:

$$\mathbf{A} \equiv \begin{bmatrix} 1 & 2 & 2 \\ 2 & 1 & 1 \\ 1 & 1 & 3 \end{bmatrix} = \begin{bmatrix} l_{11} & 0 & 0 \\ l_{21} & l_{22} & 0 \\ l_{31} & l_{32} & l_{33} \end{bmatrix} \begin{bmatrix} u_{11} & u_{12} & u_3 \\ 0 & u_{22} & u_{23} \\ 0 & 0 & u_{33} \end{bmatrix} := \mathbf{LU}. \tag{2}$$

(ii) The system thus takes the form $\mathbf{LUx} = \mathbf{b}$; or equivalently, introducing the intermediate variable **y**, $\mathbf{Ly} = \mathbf{b}$, and $\mathbf{Ux} = \mathbf{y}$. Back substitute $\mathbf{Ly} = \mathbf{b}$ (well, forward substitute in the case of a lower triangular matrix!) to find **y**, and then back-substitute $\mathbf{Ux} = \mathbf{y}$ to find **x**.

Obviously the tricky part here is finding the factors **L** and **U**. Let us propose that **L** has ones on its diagonal: $l_{11} = l_{22} = l_{33} = 1$. Then the expression for the 1,1 entry of **LU** is simply u_{11}, and by equating this with the 1,1 entry of **A** we determine u_{11}; in fact, for the data in Equation (1), $u_{11} = 1$. Similarly, u_{12} and u_{13} are just as easy to determine; for Equation (1), they are each equal to 2.

Next, consider the 2,1 entry of **LU**. It only involves u_{11}, which is now known, and l_{21}; by matching this with the (2,1) entry of **A**, we thus determine l_{21}, which is $2/u_{11} = 2$ for Equation (1). The (2,2) entry of **LU** involves l_{21}, u_{12}, and l_{22}, which are known, and u_{22}. Matching with the 2,2 entry of **A** we can find u_{22}, which is -3 for the data in Equation (2).

(iii) Continue in this manner to find the complete **L** and **U** matrices for Equation (1). Then forward/back substitute to find **y** and **x**, and confirm the solution $\mathbf{x} = \begin{bmatrix} 2 & 1 & 1 \end{bmatrix}^{\mathrm{T}}$ reported in Example 1, Section 1.2.

Clearly this procedure will work on any consistent linear system (once it has been retooled to accomodate pivoting—we omit details). It may seem like idle tinkering at this point, but there are surprising payoffs. If you look over your calculations and compare them with Example 1, you can confirm the following observations:

- The final **U** matrix that you obtain by **LU** factorization is exactly the same as the upper triangular coefficient matrix that terminates the forward part of the Gauss elimination procedure (and therefore the final back-substitution calculations are identical).
- The numbers in the **L** matrix are the factors that the Gauss algorithm multiplies rows by, to eliminate variables in subsequent rows.
- In fact, all of the individual arithmetic calculations performed by **LU** factorization and by Gauss elimination are *exactly the same* (but they are not necessarily done in the same order; the final calculations for top-down solving **Ly** = **b** are performed earlier in the Gauss procedure, you will see).
- Every time an entry in the **L** matrix is computed, the corresponding entry in the **A** matrix is never used again.

Therefore the same computational effort is expended using either procedure, and **LU** factorization saves computer memory, since it can overwrite the old **A** matrix. Another advantage accrues if the data on the right-hand side vector **b** is updated; the new solution can be found using only two back substitutions, if the factorization **A** = **LU** is retained. Practically all modern scientific software systems use some form of the **LU** factorization algorithm (suitably adjusted to accomodate pivoting) for general-purpose solving of linear systems.

(b) The biggest fallout from the **LU** strategy is that it suggests other factorizations that can provide huge efficiencies. When a matrix is symmetric, for example, the triangular factors **L** and **U** can be chosen to be tranposes of each other—*so only one of them needs to be computed*! (Truthfully, the matrix **A** must also be "positive definite," a property that we'll consider in Group Project A, Part III.) Explore this by finding the so called **Cholesky factorization** of

$$\mathbf{A} = \begin{bmatrix} 4 & 2 & 0 \\ 2 & 5 & 4 \\ 0 & 4 & 8 \end{bmatrix} = \mathbf{U}^{\mathrm{T}}\mathbf{U} = \begin{bmatrix} u_{11} & 0 & 0 \\ u_{12} & u_{22} & 0 \\ u_{13} & u_{23} & u_{33} \end{bmatrix} \begin{bmatrix} u_{11} & u_{12} & u_{13} \\ 0 & u_{22} & u_{23} \\ 0 & 0 & u_{33} \end{bmatrix}.$$

(c) What difficulties will be encountered in the **LU** factorization procedure if **A** is singular? Explore this with the matrix

$$\mathbf{A} = \begin{bmatrix} 1 & 2 & 3 \\ 4 & 5 & 6 \\ 7 & 8 & 9 \end{bmatrix}.$$

B.   Two-Point Boundary Value Problem

The purpose of this project is generate approximations to the solution of the following boundary value problem:

$$\frac{d^2y}{dt^2} = -y, \quad 0 < t < 1;$$
$$y(0) = 0, \quad y(1) = 1.$$

Let $h = 1/N$ be the step size, where N is an integer. The approximations will be obtained at the points $t_k = kh$, where $k = 0, 1, \ldots, N$, using the centered difference approximation (which you can verify using L'Hopital's rule)

$$\frac{d^2y}{dt^2}(kh) \approx \frac{y([k+1]h) - 2y(kh) + y([k-1]h)}{h^2}.$$

Let y_k denote the approximation to $y(kh)$. Replace d^2y/dt^2 by its centered difference approximation in the given differential equation for $t = t_k$ to obtain a centered difference approximation to the equation at t_k.

(a) For the case $N = 5$, how many *unknown* values of y_k are there? How many centered difference approximations to the differential equation are there? Write out these centered difference approximate equations. (This is not hard, once you see the pattern.)

(b) Organize the centered difference approximate equations into a linear system represented by an augmented matrix.

(c) Conjecture the form of this system of equations for arbitrary N. Why is the coefficient matrix said to be "sparse"? Why is it called "tridiagonal"?

(d) Solve the system for $N = 5$ by hand and plot the approximate solution by connecting points. Then use software to solve the system for $N = 50$ and plot the result. Verify that the exact solution is $y(t) = \sin t$. Plot this and compare with your approximation solutions.

(e) How many multiply-and-subtract operations are required to reduce a tridiagonal augmented coefficient matrix to row echelon form? How many to solve the resulting system by back substitution?

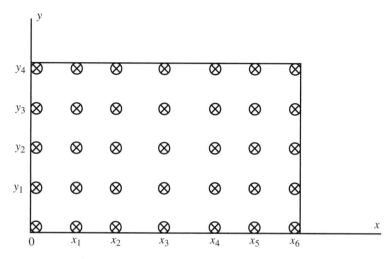

Fig. I.C.1   Voltage inside a conducting rectangle.

C.   Electrostatic Voltage

The values of the electrostatic potential $V(x, y)$ at the mesh points (x_i, y_j) inside a region bounded by a conductor, as depicted in Figure I.C.1, satisfy a "mean value property" to a high degree of accuracy. That is, each value at an interior mesh point is approximately equal to the average of the values immediately above, below, to the left, and to the right. The validity of this approximation improves as the "mesh size" $x_{i+1} - x_i \equiv y_{j+1} - y_j$ is decreased. The values on the boundary walls are prescribed; the interior values must be determined.

Note that the potential values are rather naturally organized into a matrix format $V_{ij} = V(x_i, y_j)$. However, to exploit the techniques of matrix algebra the mean value property must be expressed as a system $\mathbf{Ax} = \mathbf{b}$, where all the unknown (interior) potentials are listed as a *column*, and the known (boundary) potentials appear in the column $\mathbf{b}$.

(a) List the 15 unknown potentials from left to right, then top to bottom, and find $\mathbf{A}$ and $\mathbf{b}$ so that $\mathbf{Ax} = \mathbf{b}$ expresses the mean value conditions.

(b) The matrix $\mathbf{A}$ that you constructed consists mostly of zeros; it is *sparse*. The sparseness of a matrix can be exploited to reduce the complexity of computerized matrix computations, especially when the nonzero elements lie close to the diagonal. Experiment with a few other schemes for enumerating the unknown potential values to see if you can bring the nonzero coefficients in $\mathbf{A}$ closer to the diagonal.

(c) Suppose that the dimensions of the rectangle in Figure I.C.1 are 6-by-4, so that mesh size as depicted in the figure is unity. Potential theory tells us that the function $V(x, y) = x^3 - 3xy^2$ is a valid electrostatic potential. Evaluate this function on the edges of the rectangle to obtain the column $\mathbf{b}$; then use software to solve $\mathbf{Ax} = \mathbf{b}$ for the interior values of the potential, and compare with the true values $V(x, y)$.

(d) "Refine the mesh" in Figure I.C.1 by inserting more points so that the mesh size $x_{i+1} - x_i \equiv y_{j+1} - y_j$ is 0.5 (while the rectangle retains its original 6-by-4 size). Use software to recompute $\mathbf{b}$ and solve $\mathbf{Ax} = \mathbf{b}$ for the interior values. Compare with the true values $V(x, y)$. Repeat for another level of mesh refinement. Do you see the convergence to the true values?

(e) Formulate the three-dimensional version of this electrostatics problem, where the region is confined to a *box* and potential values directly above and below the mesh point are tallied into the averages.

D.   Kirchhoff's Laws

A typical electrical ciruit is shown in Figure I.D.2. We assume you are familiar with the physical characteristics of the circuit elements; we wish to concentrate only on the manner in which the elements are connected.

As indicated, each circuit element carries a current i_j and a voltage v_j. The reference directions for the currents are indicated by arrows, and the reference directions for the voltages by $\pm$ signs, with each current arrow pointing along the plus-to-minus direction for the corresponding voltage. The circuit elements or *branches* are numbered, as are the connection points or *nodes*.

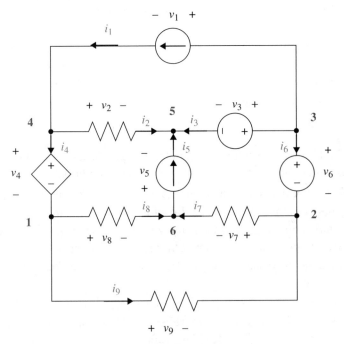

Fig. I.D.2   Electrical circuit (nodes 1−6, voltages v_1−v_9, currents i_1−i_9).

The *incidence matrix* **A** for the circuit completely specifies the connection topology, in accordance with

$$a_{ij} = 1 \text{ if current } j \text{ leaves node } i,$$
$$a_{ij} = -1 \text{ if current } j \text{ enters node } i,$$
$$a_{ij} = 0 \text{ otherwise.}$$

So for the circuit in Figure I.D.2,

$$\mathbf{A} = \begin{bmatrix} 0 & 0 & 0 & -1 & 0 & 0 & 0 & 1 & 1 \\ 0 & 0 & 0 & 0 & 0 & -1 & 1 & 0 & -1 \\ 1 & 0 & 1 & 0 & 0 & 1 & 0 & 0 & 0 \\ -1 & 1 & 0 & 1 & 0 & 0 & 0 & 0 & 0 \\ 0 & -1 & -1 & 0 & -1 & 0 & 0 & 0 & 0 \\ 0 & 0 & 0 & 0 & 1 & 0 & -1 & -1 & 0 \end{bmatrix}.$$

A has a column for every current and a row for every node in the circuit.

(a) Explain why every column of the incidence matrix always contains one $+1$ and one -1.

Kirchhoff's current law states that the sum of all currents leaving any node must equal zero.

(b) Argue that the current law is expressed by the equation $\mathbf{Ai} = \mathbf{0}$, where $\mathbf{i}$ is the column vector of branch currents $\mathbf{i} = \begin{bmatrix} i_1 & i_2 & \cdots & i_n \end{bmatrix}^{\mathrm{T}}$.

A *loop* in the circuit is a sequence of connected nodes starting and ending on the same node. For example, one loop in the circuit of Figure I.D.2 would run sequentially through nodes $1, 6, 5, 4$, and back to 1. We restrict ourselves to simple loops which visit no node, other than the initial and final, more than once. The loop can be characterized by the currents that comprise its path; for example the loop through nodes $\{1,6,5,4,1\}$ is "carried" by the currents i_8, i_5, $-i_2$, and i_4. We tagged i_2 with a minus sign to indicate that the direction of travel in the loop ran opposite to the indicated direction of i_2. In this manner any loop can be characterized by a column vector $\boldsymbol{\ell}$, with

$\ell_j = 1$ if current j is part of the loop and directed along it,
$\ell_j = -1$ if current j is part of the loop and directed oppositely, and
$\ell_j = 0$ otherwise.

Thus, the loop through nodes $\{1,6,5,4,1\}$ generates the loop vector

$$\boldsymbol{\ell} = \begin{bmatrix} 0 & -1 & 0 & 1 & 1 & 0 & 0 & 1 & 0 \end{bmatrix}^{T}.$$

Of course the loop vector does not "know" which node was the starting point; this is seldom important for a *loop*.

(c) Show that $\mathbf{A}\boldsymbol{\ell} = \mathbf{0}$ for any such loop, by interpreting $\mathbf{A}\boldsymbol{\ell}$ as a sum of (signed) columns of $\mathbf{A}$. Add the columns in the order that they appear in the loop, and argue that every partial sum contains one $+1$ and one -1 except for the final sum, which is all zeros.

Kirchhoff's voltage law states that the sum of the voltage *drops* around every loop must equal zero. The voltage drop in branch j equals v_j if the direction of the current i_j is along the direction of the loop, and equals $-v_j$ if the current direction is opposite to the loop.

(d) Argue that if $\mathbf{v}$ is the row vector of branch voltages $\mathbf{v} = \begin{bmatrix} v_1 & v_2 & \cdots & v_n \end{bmatrix}$ then the voltage law implies $\mathbf{v}\boldsymbol{\ell} = \mathbf{0}$ for any loop.

The similarity of the *topological* law $\mathbf{A}\boldsymbol{\ell} = \mathbf{0}$ and the *physical* law $\mathbf{v}\boldsymbol{\ell} = \mathbf{0}$ indicates that there is a relation between $\mathbf{A}$ and $\mathbf{v}$. To explore this, designate one of the nodes in the circuit as the reference node (or "ground" node). For example in Figure I.D.2 we take node #1 as the reference node. Then define a *node voltage* e_k at each node (k) by choosing a path along the circuit elements connecting the reference node to node #k, and adding up the branch voltage *rises* along the path. For instance, if we choose the path in Figure I.D.2 connecting node #1 to node #3 to include the nodes $\{1,6,2,3\}$ in order, then we define e_3 to be $-v_8 + v_7 + v_6$.

(e) Argue that Kirchhoff's voltage law implies that the values calculated for the node voltages are independent of the choice of paths used to evaluate them. For example, the same value for e_3 would have been assigned if we calculated $v_4 - v_2 + v_3$.

(f) Argue that if the node voltages are assembled into a row vector $\mathbf{e} = \begin{bmatrix} e_1 & e_2 & \cdots & e_m \end{bmatrix}$, then the branch voltages are given by $\mathbf{v} = \mathbf{e}\mathbf{A}$.

The relation $\mathbf{v} = \mathbf{e}\mathbf{A}$ demonstrates that the voltage law $\mathbf{v}\boldsymbol{\ell} = \mathbf{0}$ is implied by the topological law $\mathbf{A}\boldsymbol{\ell} = \mathbf{0}$. In summary, both of Kirchhoff's laws are expressible using the incidence matrix via $\mathbf{A}\mathbf{i} = \mathbf{0}$ and $\mathbf{v} = \mathbf{e}\mathbf{A}$. This enables the translation of the topology of electric circuit connections into computer code.

E. Global Positioning Systems

This project explores some the mathematical features of a GPS system.

(a) A fixed transmitter emits electromagnetic pulses that propagate in all directions at the speed of light c (taken to be a known constant). A traveler would like to be able to figure how far he/she is from the emitter, upon receiving the pulse. What kind of information can the emitter encode in the pulses, so that all travelers are accommodated?

(b) For motion in the plane, the locus of possible positions for a traveler at a known distance from an emitter is, of course, a circle. Write the equation relating the traveler's position (x, y), the emitter's position (x_1, y_1), and the distance from the emitter r_1.

(c) If the distance r_2 from a second emitter located at (x_2, y_2) is also known, how many solutions can the system of two equations describing the locus have? Why does the $(0, 1, \infty)$ rule (Theorem 3, Section 1.4) not apply?

(d) Now assume the traveler knows the distance r_3 from a third emitter located at (x_3, y_3). Reasoning geometrically, argue that all ambiguity about the traveler's position is removed, as long as the three emitters avoid certain configurations; what are these configurations?

(e) The system of three "locus" equations (in two unknowns x, y) is not linear, and is hard to solve. But luckily the differences of any two of them turn out to be linear. Formulate the three locus equations, form the differences of any two pairs, and show that Gauss elimination can be used to find x, y as long as the troublesome emitter configurations are not present.

(f) What changes need to be made to this scheme in three dimensions?

More thorough descriptions of the subtleties of global positioning systems are available on the internet; see, for example,

http://www.aero.org/education/primers/gps or
http://www.trimble.com/gps/index.shtml

F. Fixed-Point Methods

Bruce W. Atkinson, Samford University

In this project, you will be introduced to the fixed point method for approximating solutions to equations in one variable. Then you will extend this technique for approximating solutions to a linear system of equations.

(a) Let $f(x)$ be a function of one variable. A real number a is called a *fixed point* of f if $f(a) = a$. A well-known method of approximating a fixed point is *function iteration*. The basic idea is as follows: Let x_0 be an initial "guess" for a fixed point. Define a sequence of estimates (x_n), for $n \geq 1$, by letting $x_{n+1} := f(x_n)$. Thus $x_1 = f(x_0), x_2 = f(f(x_0)), x_3 = f(f(f(x_0)))$, etc. x_n is the nth *iterate* of f with initial value x_0. Now if f is continuous and $a = \lim_{n \to \infty} x_n$ exists it follows that

$$f(a) = f\left(\lim_{n \to \infty} x_n\right) = \lim_{n \to \infty} f(x_n) = \lim_{n \to \infty} x_{n+1} = a.$$

Therefore $\lim_{n \to \infty} x_n$ is a fixed point of f, and the sequence can be used to approximate the fixed point.

For example, let $f(x) = \frac{x}{2} + \frac{1}{x}$. Show that $\sqrt{2}$ is a fixed point of f. Let $x_0 = 1$. By hand, calculate the first two iterates. Then use a computer algebra system (CAS) to write the fourth iterate x_4 as an exact fraction. How well does this fraction approximate $\sqrt{2}$?

(b) The fixed point method can be used to approximate solutions to linear systems, as the following example illustrates. Consider the system:

$$4x + y + z = 17 \tag{1}$$

$$x + 4y + z = 2 \tag{2}$$

$$x + y + 4z = 11. \tag{3}$$

Solving Equation (1) for x, Equation (2) for y, and Equation (2) for z results in the equivalent system:

$$x = 4.25 - 0.25y - 0.25z \tag{4}$$

$$y = 0.5 - 0.25x - 0.25z \tag{5}$$

$$z = 2.75 - 0.25x - 0.25y. \tag{6}$$

Let

$$f\left(\begin{bmatrix} x \\ y \\ z \end{bmatrix}\right) = \mathbf{C} + \mathbf{A} \begin{bmatrix} x \\ y \\ z \end{bmatrix}, \tag{7}$$

where

$$\mathbf{C} = \begin{bmatrix} 4.25 \\ 0.5 \\ 2.75 \end{bmatrix} \text{ and } \mathbf{A} = \begin{bmatrix} 0 & -0.25 & -0.25 \\ -0.25 & 0 & -0.25 \\ -0.25 & -0.25 & 0 \end{bmatrix}$$

Thus a solution to Equations (1), (2), and (3) is a fixed point of f. It is natural to approximate the fixed point by iterating the function, starting with a specified initial vector

$$\begin{bmatrix} x_0 \\ y_0 \\ z_0 \end{bmatrix}.$$

f is an example of an *affine transformation*, that is, a linear transformation plus a constant vector. For the general affine transformation with constant vector $\mathbf{C}$ and matrix $\mathbf{A}$, as in the earlier example, it can be shown that for any initial vector the iterates converge to a unique fixed point if $\mathbf{A}^n$ converges to zero entry-wise as $n \to \infty$. Further, it can be shown $\mathbf{A}^n$ converges to zero entry-wise as $n \to \infty$ if the sum of the absolute values of the entries in each row is less than one. This is clearly the case with the above example.

In general, we are given an n-by-n linear system, with coefficient matrix $\mathbf{M}$ having non-zero diagonal entries. To generalize the technique of the earlier example, an affine function f is be constructed so that the solution to the system is a fixed point of f.

Iterating that function, starting with an initial vector, for the purpose of approximating a solution to the system, is called the *Jacobi method*.

In the example, let the initial vector be given by

$$\begin{bmatrix} 0 \\ 0 \\ 0 \end{bmatrix},$$

and denote the nth iterate of f, starting with this initial vector, by

$$\begin{bmatrix} x_n \\ y_n \\ z_n \end{bmatrix}.$$

Use a CAS to find

$$\begin{bmatrix} x_{10} \\ y_{10} \\ z_{10} \end{bmatrix}.$$

How close is that iterate to the actual solution of the system represented by Equations (1), (2), and (3)?

A matrix is said to be *diagonally dominant* if, on each row, the absolute value of the diagonal entry is greater than the sum of the absolute values of the other row entries. Use the above to prove that if $\mathbf{A}$ is diagonally dominant, the Jacobi method will converge for the system $\mathbf{Ax} = \mathbf{b}$.

(c) Again, consider the the system represented by Equations (1), (2), and (3). One can get another equivalent system which is an "improvement" of the system represented by Equations (4), (5), and (6). Take the result of Equation (4) and substitute it into the right hand side of Equation (5). The result is

$$y = -0.5625 + 0.0625y - 0.1875z. \tag{8}$$

Finally take the result of Equations (4) and (8) and substitute them into the right hand side of Equation (6). The result is

$$z = 1.828125 + 0.046875y + 0.109375z. \tag{9}$$

In summary we now have a new equivalent system:

$$x = 4.25 - 0.25y - 0.25z \tag{10}$$
$$y = -0.5625 + 0.0625y - 0.1875z \tag{11}$$
$$z = 1.828125 + 0.046875y + 0.109375z. \tag{12}$$

Let

$$g\left(\begin{bmatrix} x \\ y \\ z \end{bmatrix}\right) = \mathbf{D} + \mathbf{B}\begin{bmatrix} x \\ y \\ z \end{bmatrix}, \tag{13}$$

where

$$\mathbf{D} = \begin{bmatrix} 4.25 \\ -0.5625 \\ 1.828125 \end{bmatrix} \text{ and } \mathbf{B} = \begin{bmatrix} 0 & -0.25 & -0.25 \\ 0 & 0.0625 & -0.1875 \\ 0 & 0.046875 & 0.109375 \end{bmatrix}$$

Thus, a solution to Equations (1), (2), and (3) is a fixed point of g. One can similarly iterate g, starting with any initial vector, to solve the original system represented by Equations (1), (2), and (3).

In general, we are given an n-by-n linear system, with coefficient matrix $\mathbf{M}$ having non-zero diagonal entries. Generalizing the technique of the earlier example, an affine function g can be constructed so that the solution to the system is a fixed point of g. Iterating that function, starting with an initial vector, for the purpose of approximating a solution to the system, is called the *Gauss–Seidel method*.

For the purpose of approximating solutions to the system represented by Equations (1), (2), and (3), one can either use the Jacobi method by iterating f in Equation (7), or one can use the Gauss–Seidel method by iterating g in Equation (13). Note that the first iterate of f amounts to using the same initial values three times in providing the first "updates," x_1, y_1, and z_1. However, the first iterate of g starts with updating the x value as with f, but uses that updated x value immediately in updating the y value. Similarly, the updated z value uses the latest updated x and y values. (Sometimes, we say that the Jacobi method updates the iterations *simultaneously*, while the Gauss–Seidel method updates them *successively*.) Thus g seems to be an improvement over f because of this self-correcting feature. Numerically, the iterates of g converge to the desired solution much faster than those of f.

In the example, let the initial vector once again be given by

$$\begin{bmatrix} 0 \\ 0 \\ 0 \end{bmatrix},$$

and denote the nth iterate of g, starting with this initial vector, by

$$\begin{bmatrix} \alpha_n \\ \beta_n \\ \gamma_n \end{bmatrix}.$$

Use a CAS to find

$$\begin{bmatrix} \alpha_1 \\ \beta_1 \\ \gamma_1 \end{bmatrix}, \begin{bmatrix} \alpha_2 \\ \beta_2 \\ \gamma_2 \end{bmatrix}, \begin{bmatrix} \alpha_3 \\ \beta_3 \\ \gamma_3 \end{bmatrix}, \text{ and } \begin{bmatrix} \alpha_4 \\ \beta_4 \\ \gamma_4 \end{bmatrix}.$$

How close are these iterates to the actual solution of the system represented by Equations (1), (2), and (3)? Do these iterates support the claim that the Gauss-Seidel method is an improvement over the Jacobi method?

(d) Let $\mathbf{M}$ be the coefficient matrix for the system represented by equations (1), (2), and (3), that is

$$\mathbf{M} = \begin{bmatrix} 4 & 1 & 1 \\ 1 & 4 & 1 \\ 1 & 1 & 4 \end{bmatrix}.$$

Explain how you can use the Gauss–Seidel method to approximate $\mathbf{M}^{-1}$.

PART II

INTRODUCTION: THE STRUCTURE OF GENERAL SOLUTIONS TO LINEAR ALGEBRAIC EQUATIONS

What features do the solutions to the linear system (1) and the differential equation (2) have in common?

$$\begin{bmatrix} 1 & 2 & 1 & 1 & -1 & 0 & \vdots & 0 \\ 1 & 2 & 0 & -1 & 1 & 0 & \vdots & 0 \\ 0 & 0 & 1 & 2 & -2 & 1 & \vdots & 0 \end{bmatrix}; \qquad (1)$$

$$\frac{d^3y}{dx^3} = 0. \qquad (2)$$

Gauss elimination, applied to the homogeneous linear system (1), yields the row echelon form

$$\begin{bmatrix} 1 & 2 & 1 & 1 & -1 & 0 & \vdots & 0 \\ 0 & 0 & -1 & -2 & 2 & 0 & \vdots & 0 \\ 0 & 0 & 0 & 0 & 0 & 1 & \vdots & 0 \end{bmatrix},$$

Fundamentals of Matrix Analysis with Applications,
First Edition. Edward Barry Saff and Arthur David Snider.
© 2016 John Wiley & Sons, Inc. Published 2016 by John Wiley & Sons, Inc.

which has a three-parameter solution family

$$
\begin{aligned}
x_1 &= -2t_1 + t_2 - t_3 \\
x_2 &= t_1 \\
x_3 &= -2t_2 + 2t_3 \\
x_4 &= t_2 \\
x_5 &= t_3 \\
x_6 &= 0
\end{aligned}
\qquad \text{or} \qquad
\begin{bmatrix} x_1 \\ x_2 \\ x_3 \\ x_4 \\ x_5 \\ x_6 \end{bmatrix}
= t_1 \begin{bmatrix} -2 \\ 1 \\ 0 \\ 0 \\ 0 \\ 0 \end{bmatrix}
+ t_2 \begin{bmatrix} 1 \\ 0 \\ -2 \\ 1 \\ 0 \\ 0 \end{bmatrix}
+ t_3 \begin{bmatrix} -1 \\ 0 \\ 2 \\ 0 \\ 1 \\ 0 \end{bmatrix}. \tag{3}
$$

We do not presume that the reader has studied differential equations, but equation (2) is very simple to analyze. Indefinite integration yields, in turn,

$$
\begin{aligned}
y'' &= c_1; \\
y' &= c_1 x + c_2; \\
y &= c_1 x^2/2 + c_2 x + c_3.
\end{aligned}
\tag{4}
$$

Although the displays (3) and (4) refer to quite different mathematical objects, they are instances of a concept known as a *vector space*. By focusing attention in this chapter on the abstract notion of vector spaces, we will be able to establish general principles that can be handily applied in many diverse areas.

So let us list three apparent features that the general solutions (3) and (4) have in common. Remember that in both cases, the coefficients $\{t_i\}$, $\{c_i\}$ can take any real values; in Section 1.3, the coefficients $\{t_i\}$ were called "parameters;" of course, the coefficients $\{c_i\}$ are "constants of integration."

(i) Zero is a solution. Here we have already begun to universalize our language; "zero" for (3) means the six-component vector with all entries zero, but "zero" for (4) means the identically zero function.

(ii) The sum of any two solutions is also a solution.

(iii) Any constant multiple of a solution is also a solution.

What other mathematical structures satisfy (i)–(iii)? A familiar example is the set of vectors (directed line segments) in the x, y-plane. The sum of two vectors, as well as the multiplication of a vector by a constant, is depicted in Figure II.1.

In fact, the set of all vectors in three-dimensional space—or, all vectors along a line, or in a plane—satisfy (i)–(iii). However, any *finite* collection of such vectors would not meet the *closure* conditions (ii), (iii); such sets are not "closed" under addition and scalar multiplication. (*Exception*: The single zero vector does, trivially, satisfy (ii), (iii).) Similarly, the set of all vectors in the plane pointing into the first quadrant fails to have the closure property.

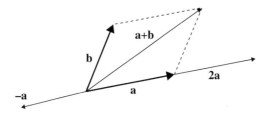

Fig. II.1 Vector addition.

In subsequent sections, we will formalize and generalize the concept of a vector space and establish general tools that can be utilized in the analysis of algebraic and differential equations, matrices, geometry, and many other areas.

3

VECTOR SPACES

3.1 GENERAL SPACES, SUBSPACES, AND SPANS

In the Introduction to Part II, we noted that solutions to linear homogeneous algebraic systems, solutions of certain linear differential equations, and geometric space vectors have some algebraic features in common. In rough terms,

- The objects themselves (call them **v**'s) enjoy an "addition" operation which satisfies the customary algebraic laws;
- These objects interact with another algebraic collection called the scalars (herein denoted by c's) through *scalar multiplication* $c\mathbf{v}$, and this operation also satisfies the expected parenthesis rules. The real numbers have played the role of the scalars in the examples we have seen so far, but in subsequent applications we shall use complex numbers. The specifics are given in the following defintion.

Then V is said to be a **vector space** over the field of scalars F.

Vector Space

Definition 1. Let V be a collection of objects which we call *vectors* (typically denoted by boldface lower case Latin letters $\mathbf{v}, \mathbf{w}, \mathbf{a}$, etc.), and let F be the set of real, or the set of complex, numbers; the elements of F will be designated as scalars (c_1, c_2, c_3, etc.). Suppose there is an operation called **vector addition**, written as $\mathbf{v}_1 + \mathbf{v}_2$, that combines pairs of vectors, and another operation called

---

Fundamentals of Matrix Analysis with Applications,
First Edition. Edward Barry Saff and Arthur David Snider.
© 2016 John Wiley & Sons, Inc. Published 2016 by John Wiley & Sons, Inc.

scalar multiplication, written $c\mathbf{v}$, that combines scalars with vectors, such that for any $\mathbf{v}, \mathbf{v}_1, \mathbf{v}_2, \mathbf{v}_3$ in V and any c, c_1, c_2 in F,

(i) $\mathbf{v}_1 + \mathbf{v}_2$ belongs to V ("V is closed under vector addition");

(ii) $c\mathbf{v}$ belongs to V ("V is closed under scalar multiplication");

(iii) V contains an additive identity vector $\mathbf{0}$ such that $\mathbf{0} + \mathbf{v} = \mathbf{v}$ for all $\mathbf{v}$ in V;

(iv) $-1\mathbf{v}$, also denoted by $-\mathbf{v}$, is the additive inverse of $\mathbf{v}$ ($-\mathbf{v} + \mathbf{v} = \mathbf{0}$);

(v) $\mathbf{v}_1 + \mathbf{v}_2 = \mathbf{v}_2 + \mathbf{v}_1$;

(vi) $\mathbf{v}_1 + (\mathbf{v}_2 + \mathbf{v}_3) = (\mathbf{v}_1 + \mathbf{v}_2) + \mathbf{v}_3$;

(vii) $c_1(c_2\mathbf{v}) = (c_1 c_2)\mathbf{v}$;

(viii) $c(\mathbf{v}_1 + \mathbf{v}_2) = c\mathbf{v}_1 + c\mathbf{v}_2$ and $(c_1 + c_2)\mathbf{v} = c_1\mathbf{v} + c_2\mathbf{v}$;

(ix) $1\mathbf{v} = \mathbf{v}$.

If the context is clear, we shorten "vector space" to, simply, "space." Note that (i) and (ii) together can be more compactly expressed as

(i$'$) $c_1\mathbf{v}_1 + c_2\mathbf{v}_2$ belongs to V.

Furthermore, it is easy to prove from the above axioms that the zero scalar times any vector equals the zero vector:

(x) $0\mathbf{v} = \mathbf{0}$ for all $\mathbf{v}$

(see Problem 10.)

The most common examples of vector spaces are the geometric vectors discussed in the Introduction (Figure II.1). Indeed, they gave rise to the nomenclature "vector space." Another ubiquitous example is the set of all column matrices with a fixed number of entries; the set of all m-by-1 column matrices with real-valued entries is a vector space over the field of real numbers, and the set of all m-by-1 column matrices with complex-valued entries is a vector space over the field of complex numbers. Note that the additive identity in these examples is the m-by-1 matrix of zeros (not the *number* zero!).

Of course row matrices also form vector spaces. In fact the set of all m-by-n rectangular matrices is a vector space over the appropriate scalar field (real matrices over the real numbers, complex matrices over the complex numbers).

A more abstract example is the set of all continuous functions on a fixed interval (a, b). Elementary calculus tells us that sums and scalar multiples of continuous functions are continuous, and the other properties are simple algebraic identities. Note that the "additive identity" in this case is the identically zero function.

The following examples of vector spaces occur frequently enough to merit special nomenclature:

Some Common Vector Spaces

- $\mathbf{R}_{\text{col}}^m$ ($\mathbf{R}_{\text{row}}^m$) (Section 2.2) is the set of all m-by-1 column (1-by-m row) matrices with real entries, over the real numbers;
- $\mathbf{C}_{\text{col}}^m$ ($\mathbf{C}_{\text{row}}^m$) denotes the set of all m-by-1 column (1-by-m row) matrices with complex entries, over the complex numbers;
- $\mathbf{R}^{m,n}$ denotes the set of all m-by-n matrices with real entries, over the real numbers;
- $\mathbf{C}^{m,n}$ denotes the set of all m-by-n matrices with complex entries, over the complex numbers;
- $C(a, b)$ denotes the set of all real-valued continuous functions on the open interval (a, b), over the real numbers (and similarly $C[a, b]$ denotes such functions on the closed interval $[a, b]$).
- $\mathbf{P}_n$ denotes the collection of all polynomial functions of degree at most n, with real coefficients. That is,

$$\mathbf{P}_n := \{a_0 + a_1 x + \cdots + a_n x^n \mid a_i\text{'s are real numbers}\}.$$

A **subspace** of a vector space is (just what you'd guess) a nonempty subset of the space which, itself, satisfies all the vector space properties. Most of these will be "inherited" from the original space; in fact, a nonempty subset of a vector space is a subspace if, and only if, it is closed under vector addition and scalar multiplication. In many cases, checking closure for a set is a simple mental exercise.

Example 1. Determine which of the following are subspaces of $\mathbf{R}_{\text{row}}^5$:

(a) All vectors of the form $[a\ b\ c\ b\ a]$;
(b) All vectors of the form $[a\ b\ 0\ -b\ -a]$;
(c) Those vectors all of whose entries are the same $[a\ a\ a\ a\ a]$;
(d) Those vectors whose second entry is zero $[a\ 0\ b\ c\ d]$;
(e) Those vectors whose middle entry is positive $[a\ b\ c(>0)\ d\ e]$;
(f) Those vectors whose entries are integers;
(g) Those vectors whose entries are nonnegative real numbers;
(h) The (single) zero vector $\mathbf{0} = [0\ 0\ 0\ 0\ 0]$.

Solution. Combinations of the form $c_1 \mathbf{v}_1 + c_2 \mathbf{v}_2$ will preserve the symmetries (a), (b), and (c) as well as the zero in (d); these subsets are subspaces. But forming the scalar

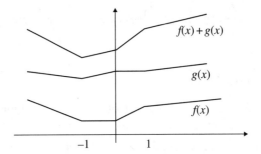

Fig. 3.1   Piecewise linear functions.

multiple $0\mathbf{v}$ would violate the nonzero restriction (e), $\mathbf{v}/2$ would destroy the integer property if any of the original integers in (f) were odd, and $(-1)\mathbf{v}$ obviously invalidates (g); so (e), (f), and (g) are not subspaces. The single vector $\mathbf{0}$ is a subspace (the only one-element subspace).   ∎

Similar reasoning shows that the following are subspaces of $C(-2,2)$:

- The set of even continuous functions $[f(x) = f(-x)]$;
- The set of odd continuous functions $[f(-x) = -f(x)]$;
- The set of continuous functions that are zero when $x = 1/2$;
- The single function $f(x) \equiv 0$;
- The set of continuous functions that are bounded as $x \to 2^{-}$;
- The set of continuous piecewise linear functions whose graphs are straight-line segments connected at the values $x = -1, 0, 1$ (Figure 3.1).

In each case, the combination $c_1\mathbf{v}_1 + c_2\mathbf{v}_2$ preserves the defining credential.

An important example of a subspace of $\mathbf{R}^m_{\text{col}}$ is the set of all solutions to a linear homogeneous algebraic system $\mathbf{A}\mathbf{x} = \mathbf{0}$; if $\mathbf{A}\mathbf{x}_1 = \mathbf{0}$ and $\mathbf{A}\mathbf{x}_2 = \mathbf{0}$, then certainly $\mathbf{A}(c_1\mathbf{x}_1 + c_2\mathbf{x}_2) = \mathbf{0}$ $(= c_1\mathbf{0} + c_2\mathbf{0})$. The collection of such solutions is called the **null space** or the **kernel** of the coefficient matrix $\mathbf{A}$. We formalize this observation in the following theorem.

Solution Space for a Linear Homogeneous Algebraic System

Theorem 1.   The set of all solutions to any linear homogeneous algebraic system $\mathbf{A}\mathbf{x} = \mathbf{0}$ (i.e., the null space or kernel of $\mathbf{A}$) comprises a vector space.

If the equation is *non*homogeneous, however, the solutions forfeit the closure property, because the sum $\mathbf{x}_1 + \mathbf{x}_2$ of two solutions of $\mathbf{A}\mathbf{x} = \mathbf{b}$ satisfies $\mathbf{A}(\mathbf{x}_1 + \mathbf{x}_2) = 2\mathbf{b}$ (which is different from $\mathbf{b}$ in the nonhomogeneous case).

Example 2. Express the subspaces in Example 1 as null spaces.

Solution. Of course, matrix notation requires that we first change these row spaces into the corresponding column spaces.

(a) A column vector has the required symmetry when its first and fifth elements are equal, and its second and fourth are also equal: $x_1 = x_5; x_2 = x_4$. These two equations in five unknowns can be expressed as $\mathbf{Ax} = \mathbf{0}$ with $\mathbf{A}$ taken to be

$$\mathbf{A} = \begin{bmatrix} 1 & 0 & 0 & 0 & -1 \\ 0 & 1 & 0 & -1 & 0 \end{bmatrix}.$$

(b) Here the first component is the negative of the fifth, the second component is the negative of the fourth, and the third component is zero. Symbolically, $x_1 = -x_5$, $x_2 = -x_4$, $x_3 = 0$, from which we get

$$\mathbf{A} = \begin{bmatrix} 1 & 0 & 0 & 0 & 1 \\ 0 & 1 & 0 & 1 & 0 \\ 0 & 0 & 1 & 0 & 0 \end{bmatrix}.$$

Similarly, subspaces (c), (d), and (h) are the null spaces of

$$\begin{bmatrix} 1 & -1 & 0 & 0 & 0 \\ 0 & 1 & -1 & 0 & 0 \\ 0 & 0 & 1 & -1 & 0 \\ 0 & 0 & 0 & 1 & -1 \end{bmatrix}, \begin{bmatrix} 0 & 1 & 0 & 0 & 0 \end{bmatrix}, \text{ and } \begin{bmatrix} 1 & 0 & 0 & 0 & 0 \\ 0 & 1 & 0 & 0 & 0 \\ 0 & 0 & 1 & 0 & 0 \\ 0 & 0 & 0 & 1 & 0 \\ 0 & 0 & 0 & 0 & 1 \end{bmatrix},$$

respectively.   ∎

Expressions like

$$t_1 \begin{bmatrix} -2 \\ 1 \\ 0 \\ 0 \\ 0 \\ 0 \end{bmatrix} + t_2 \begin{bmatrix} 1 \\ 0 \\ -2 \\ 1 \\ 0 \\ 0 \end{bmatrix} + t_3 \begin{bmatrix} -1 \\ 0 \\ 2 \\ 0 \\ 1 \\ 0 \end{bmatrix}, \quad c_1 x^2/2 + c_2 x + c_3, \text{ and } 3\mathbf{i} + 4\mathbf{j} - 2\mathbf{k}$$

are commonplace in vector space calculations. We say that any expression containing vectors and scalars in the format $c_1\mathbf{v}_1 + c_2\mathbf{v}_2 + \cdots + c_n\mathbf{v}_n$ is a *linear combination* of the vectors $\{\mathbf{v}_1, \mathbf{v}_2, \ldots, \mathbf{v}_n\}$. The following property is trivial, but so pervasive that it is convenient to spell it out explicitly, for future reference.

Linear Combination Composition Principle

Lemma 1. If $\mathbf{w}_1, \mathbf{w}_2, \ldots, \mathbf{w}_m$ are each linear combinations of the vectors $\mathbf{v}_1, \mathbf{v}_2, \ldots, \mathbf{v}_n$, then so is any linear combination of $\mathbf{w}_1, \mathbf{w}_2, \ldots, \mathbf{w}_m$.

Proof. This is a matter of simple bookkeeping. For $m = 2$ and $n = 3$, the calculation goes this way. Suppose $\mathbf{u}$ is a linear combination of $\mathbf{w}_1$ and $\mathbf{w}_2$, say

$$\mathbf{u} = d_1 \mathbf{w}_1 + d_2 \mathbf{w}_2.$$

We know that

$$\mathbf{w}_1 = c_1^{(1)} \mathbf{v}_1 + c_2^{(1)} \mathbf{v}_2 + c_3^{(1)} \mathbf{v}_3,$$
$$\mathbf{w}_2 = c_1^{(2)} \mathbf{v}_1 + c_2^{(2)} \mathbf{v}_2 + c_3^{(2)} \mathbf{v}_3.$$

Substituting these representations into the expression for $\mathbf{u}$, we get

$$\begin{aligned} \mathbf{u} &= d_1 \left(c_1^{(1)} \mathbf{v}_1 + c_2^{(1)} \mathbf{v}_2 + c_3^{(1)} \mathbf{v}_3 \right) + d_2 \left(c_1^{(2)} \mathbf{v}_1 + c_2^{(2)} \mathbf{v}_2 + c_3^{(2)} \mathbf{v}_3 \right) \\ &= \left(c_1^{(1)} d_1 + c_1^{(2)} d_2 \right) \mathbf{v}_1 + \left(c_2^{(1)} d_1 + c_2^{(2)} d_2 \right) \mathbf{v}_2 + \left(c_3^{(1)} d_1 + c_3^{(2)} d_2 \right) \mathbf{v}_3 \\ &= e_1 \mathbf{v}_1 + e_2 \mathbf{v}_2 + e_3 \mathbf{v}_3, \end{aligned}$$

as claimed. In fact, defining the matrix $\mathbf{C} = [C_{ij}]$ by $C_{ij} = c_i^{(j)}$, we can write a matrix equation for the new coefficients:

$$\begin{bmatrix} e_1 \\ e_2 \\ e_3 \end{bmatrix} = \mathbf{C} \begin{bmatrix} d_1 \\ d_2 \end{bmatrix}. \qquad \blacksquare$$

If we consider the set of all linear combinations of any set of vectors, it is easy to verify that we get a vector space. (The crucial point is, as usual, closure; if $\mathbf{u}$ and $\mathbf{w}$ are each linear combinations of $\{\mathbf{v}_1, \mathbf{v}_2, \ldots, \mathbf{v}_n\}$, then the linear combination composition principle assures us that $c_1 \mathbf{u} + c_2 \mathbf{w}$ is, also.)

Span

Definition 2. The **span**[1] of the vectors $\{\mathbf{v}_1, \mathbf{v}_2, \ldots, \mathbf{v}_n\}$ is the subspace of all linear combinations of $\{\mathbf{v}_1, \mathbf{v}_2, \ldots, \mathbf{v}_n\}$. It is abbreviated Span $\{\mathbf{v}_1, \mathbf{v}_2, \ldots, \mathbf{v}_n\}$.

---

[1] The word can also be used as a verb. We say that the vectors $\{\mathbf{v}_1, \mathbf{v}_2, \ldots, \mathbf{v}_n\}$ *span* a vector space V provided $V = \text{Span}\{\mathbf{v}_1, \mathbf{v}_2, \ldots, \mathbf{v}_n\}$.

Referring to the examples discussed in the Introduction to Part II, we can say that the vector space of solutions to $d^3y/dx^3 = 0$ is the span of $x^2/2, x$, and 1; and the solution space of

$$\left[\begin{array}{cccccc|c} 1 & 2 & 1 & 1 & -1 & 0 & 0 \\ 1 & 2 & 0 & -1 & 1 & 0 & 0 \\ 0 & 0 & 1 & 2 & -2 & 1 & 0 \end{array}\right] \tag{1}$$

is spanned by $\begin{bmatrix} -2 \\ 1 \\ 0 \\ 0 \\ 0 \\ 0 \end{bmatrix}, \begin{bmatrix} 1 \\ 0 \\ -2 \\ 1 \\ 0 \\ 0 \end{bmatrix}$, and $\begin{bmatrix} -1 \\ 0 \\ 0 \\ 2 \\ 0 \\ 1 \\ 0 \end{bmatrix}$ (see (3)) in the Introduction).

The solution space of (1) is a *subspace* of the vector space of *all* 6-by-1 column vectors $\mathbf{R}^6_{col}$, which in turn is given by the span of

$$\begin{bmatrix} 1 \\ 0 \\ 0 \\ 0 \\ 0 \\ 0 \end{bmatrix}, \begin{bmatrix} 0 \\ 1 \\ 0 \\ 0 \\ 0 \\ 0 \end{bmatrix}, \begin{bmatrix} 0 \\ 0 \\ 1 \\ 0 \\ 0 \\ 0 \end{bmatrix}, \begin{bmatrix} 0 \\ 0 \\ 0 \\ 1 \\ 0 \\ 0 \end{bmatrix}, \begin{bmatrix} 0 \\ 0 \\ 0 \\ 0 \\ 1 \\ 0 \end{bmatrix}, \text{ and } \begin{bmatrix} 0 \\ 0 \\ 0 \\ 0 \\ 0 \\ 1 \end{bmatrix}. \tag{2}$$

Example 3. Show that in $\mathbf{R}^4_{row}$,

$$\text{Span } \{[2\ 3\ 8\ 0], [1\ 0\ 0\ 1], [3\ 3\ 8\ 1]\} = \text{Span } \{[2\ 3\ 8\ 0], [1\ 0\ 0\ 1]\}.$$

Solution. Let $\mathbf{v}_1 = [2\ 3\ 8\ 0]$, $\mathbf{v}_2 = [1\ 0\ 0\ 1]$, and $\mathbf{v}_3 = [3\ 3\ 8\ 1]$. If you noticed that $\mathbf{v}_3$ is the sum of $\mathbf{v}_1$ and $\mathbf{v}_2$, you probably see why adding $\mathbf{v}_3$ into the Span "arsenal" does not increase it. Let's give a more formal argument. First note that Span$\{\mathbf{v}_1, \mathbf{v}_2\} \subseteq$ Span$\{\mathbf{v}_1, \mathbf{v}_2, \mathbf{v}_3\}$ since any linear combination of $\mathbf{v}_1$ and $\mathbf{v}_2$ can be regarded as a linear combination of $\mathbf{v}_1, \mathbf{v}_2$, and $\mathbf{v}_3$ with zero being the scalar multiplier of $\mathbf{v}_3$. On the other hand, observe that $\mathbf{v}_3 = \mathbf{v}_1 + \mathbf{v}_2$ and so, by the linear combination principle, any linear combination of $\mathbf{v}_1, \mathbf{v}_2$, and $\mathbf{v}_3$ is a linear combination of $\mathbf{v}_1$ and $\mathbf{v}_2$. Thus Span$\{\mathbf{v}_1, \mathbf{v}_2, \mathbf{v}_3\} \subseteq$ Span$\{\mathbf{v}_1, \mathbf{v}_2\}$, and so the two spans are equal. ∎

In subsequent sections, we shall delve further into the properties of the span, and we shall see that the notion of a vector space provides an organizational structure that unifies many areas of application of mathematics.

Exercises 3.1

1. Which of the following are subspaces of $\mathbf{R}_{\text{row}}^3$? Justify your answers.
   (a) All vectors $[a\ b\ c]$ with $a + b + c = 0$.
   (b) All vectors $[a\ b\ c]$ with $a^3 + b^3 + c^3 = 0$.
   (c) All vectors of the form $[0\ b\ c]$.
   (d) All vectors $[a\ b\ c]$ with $abc = 0$.

2. Which of the following subsets of $\mathbf{R}_{\text{col}}^3$ are subspaces over the real numbers? Justify your answers.
   (a) All column vectors orthogonal to $[1\ 2\ 3]^T$ (use the dot product; the zero vector is regarded as being orthogonal to all vectors).
   (b) All column vectors whose entries are rational numbers.
   (c) All column vectors of the form $[t\ 2t\ 5t]^T$, where t is any real number.
   (d) Span$\{[1\ 0\ 3]^T, [0\ 2\ 4]^T\}$.

3. Which of the following are subspaces of $C[-1, 1]$? Justify your answers.
   (a) All continuous functions f on $[-1, 1]$ with $\int_{-1}^{1} f(x)\, dx = 0$.
   (b) All functions with three continuous derivatives on $[-1, 1]$.
   (c) All continuous functions f on $[-1, 1]$ satisfying $|f(x)| \leq 1$ for all x in $[-1, 1]$.
   (d) All continuous functions f on $[-1, 1]$ with $f'(0) = 0$.

4. (a) Verify that the collection $\mathbf{P}_n$ of all polynomials of degree at most n is a vector space over the real scalars.
   (b) Is the set of all polynomials of degree exactly n (> 0) a subspace of $\mathbf{P}_n$?

5. (a) Show that $\mathbf{P}_2$ is the span of the polynomials $\{1, x - 1, (x - 1)^2\}$.
   (b) Show that $\mathbf{P}_2$ cannot be spanned by a pair of polynomials.

6. Do the polynomials of the form $(x - 5)q(x)$, with $q(x)$ in $\mathbf{P}_2$, form a subspace of $\mathbf{P}_3$?

7. Which of the following subsets of $\mathbf{R}^{3,3}$ are subspaces? Justify your answers.
   (a) All symmetric matrices.
   (b) All upper triangular matrices.
   (c) All matrices whose sum of diagonal entries is zero.
   (d) All invertible matrices.
   (e) All skew-symmetric matrices $\mathbf{A}$, that is, matrices satisfying $\mathbf{A}^T = -\mathbf{A}$.

   (f) All matrices $\mathbf{A}$ such that $\mathbf{A} \times \begin{bmatrix} 1 & 2 & 3 \\ 5 & 6 & 7 \\ 8 & 9 & 10 \end{bmatrix} = \begin{bmatrix} 0 & 0 & 0 \\ 0 & 0 & 0 \\ 0 & 0 & 0 \end{bmatrix}$.

8. Let V be the collection of all 1-by-3 vectors of the form $[f(x)\ g(x)\ h(x)]$, where f, g, and h are continuous real-valued functions on $[0, 1]$.

(a) Explain why V is a vector space over the real scalars (under the usual definitions of scalar multiplication and vector addition).

(b) What is the zero vector of this space ?

(c) Let S be the set of all vectors of the form $[ae^x \ b\sin x \ c\cos x]$, where a, b, c are real numbers. Is S a subspace of V?

9. For each of the following, determine whether the statement made is always True or sometimes False.

(a) $\mathbf{R}^2_{\text{row}} = \text{Span}\{[2\ 0], [0, \pi]\}$.

(b) If the vectors $\mathbf{v}_1$ and $\mathbf{v}_2$ are linear combinations of the vectors $\mathbf{w}_1$ and $\mathbf{w}_2$, then $\text{Span}\{\mathbf{v}_1, \mathbf{v}_2\} = \text{Span}\{\mathbf{w}_1, \mathbf{w}_2\}$.

(c) The complex numbers form a vector space over the reals (i.e., the scalars are real numbers).

(d) Any two subspaces of a vector space have a nonempty intersection.

10. Using only the axioms for a vector space (Definition 1), prove that $0\mathbf{v} = \mathbf{0}$ for all vectors $\mathbf{v}$. (*Hint*: Start by writing $0\mathbf{v} = (0+0)\mathbf{v}$.)

11. Let S be the collection of all infinite sequences $\mathbf{s} = (s_0, s_1, s_2, \ldots)$, where the s_i's are real numbers. Then S forms a vector space under the standard (elementwise) addition of sequences and scalar multiplication: $\mathbf{s} + \mathbf{t} = (s_0 + t_0, s_1 + t_1, \ldots)$, $\alpha\mathbf{s} = (\alpha s_0, \alpha s_1, \ldots)$.

(a) What is the zero vector of this space ?

(b) Is the subset of convergent sequences a subspace of S?

(c) Is the subset of all bounded sequences a subspace of S? (Recall that a sequence $\mathbf{s}$ is bounded if there is a constant M such that $|s_i| \leq M$ for all $i = 0, 1, \ldots$.)

(d) Is the subset of all sequences $\mathbf{s}$ for which $\sum_{i=0}^{\infty} s_i$ is convergent a subspace of S?

12. **Matrices as Mappings.** Let $\mathbf{A}$ be an m-by-n matrix and $\mathbf{x}$ an n-by-1 column vector. Since the product $\mathbf{y} = \mathbf{A}\mathbf{x}$ is an m-by-1 vector, $\mathbf{A}$ can be viewed as a function that maps $\mathbf{R}^n_{\text{col}}$ into $\mathbf{R}^m_{\text{col}}$; that is, $\mathbf{A}(\mathbf{x}) := \mathbf{A}\mathbf{x}$ with domain $\mathbf{R}^n_{\text{col}}$ and range a subset of $\mathbf{R}^m_{\text{col}}$. Prove that the range is a subspace of $\mathbf{R}^m_{\text{col}}$.

13. Show that the collection of all twice differentiable functions y that satisfy the homogeneous differential equation

$$x^2 y'' + e^x y' - y = 0$$

for $x > 0$ is a vector space.

14. Prove the following: If $\mathbf{A}$ is an n-by-n matrix with real number entries and r is a fixed real number, then the set of n-by-1 column vectors $\mathbf{x}$ satisfying $\mathbf{A}\mathbf{x} = r\mathbf{x}$ is a vector space (a subspace of $\mathbf{R}^n_{\text{col}}$). Furthermore, show that this vector space consists of exactly one vector if and only if $\det(\mathbf{A} - r\mathbf{I}) \neq 0$. (Values of r for which this space has more than one vector are called *eigenvalues* of $\mathbf{A}$ in Chapter 5.)

15. Prove that if S and T are subspaces of a vector space V, then the set consisting of all vectors of the form $\mathbf{s} + \mathbf{t}$, with $\mathbf{s}$ in S and $\mathbf{t}$ in T, is also a subspace of V (we denote this space by $S + T$).

16. Prove that if two nonzero vectors in the plane $\mathbf{R}^2_{\text{col}}$ are not parallel, then their span is the whole plane. (*Hint*: Show that every vector in the plane can be written as a linear combination of the two vectors.)

17. Justify the following statement: All subspaces of the plane $\mathbf{R}^2_{\text{col}}$ can be visualized either as the whole plane, a line through the origin, or just the origin itself. How could you visualize all the subspaces of $\mathbf{R}^3_{\text{col}}$?

18. Explain why $\text{Span}\{[1\ 2\ 0], [1\ 2\ 3], [0\ 0\ 6]\} = \text{Span}\{[1\ 2\ 0], [1\ 2\ 3]\}$.

19. Do the sets $\begin{bmatrix} 1 \\ 1 \\ 1 \end{bmatrix}, \begin{bmatrix} 1 \\ 0 \\ -1 \end{bmatrix}$ and $\begin{bmatrix} 1 \\ 1 \\ 1 \end{bmatrix}, \begin{bmatrix} 1 \\ 0 \\ -1 \end{bmatrix}, \begin{bmatrix} 2 \\ 1 \\ 1 \end{bmatrix}$ span the same subspace?

20. Suppose $\mathbf{v}_1$ and $\mathbf{v}_2$ are linear combinations of $\mathbf{w}_1, \mathbf{w}_2$, and $\mathbf{w}_3$. Explain why $\text{Span}\{\mathbf{v}_1, \mathbf{v}_2, \mathbf{w}_1, \mathbf{w}_2, \mathbf{w}_3\} = \text{Span}\{\mathbf{w}_1, \mathbf{w}_2, \mathbf{w}_3\}$.

21. For the vector space $C(-\infty, \infty)$, show that $f(x) = x$ is not in the span of $\{e^x, \sin x\}$. (*Hint*: What would be the consequence of $x = c_1 e^x + c_2 \sin x$ at $x = 0$?)

3.2 LINEAR DEPENDENCE

The notion of *span* gives us new insight into the analysis of systems of linear algebraic equations. Observe that the matrix product $\mathbf{Ax}$ can be identified as a linear combination of the columns of $\mathbf{A}$:

$$\mathbf{Ax} = \begin{bmatrix} 1 & 2 & 3 \\ 1 & 0 & 1 \\ 1 & 1 & 2 \\ 0 & 0 & 0 \end{bmatrix} \begin{bmatrix} x_1 \\ x_2 \\ x_3 \end{bmatrix} \equiv \begin{bmatrix} 1x_1 + 2x_2 + 3x_3 \\ 1x_1 + 0x_2 + 1x_3 \\ 1x_1 + 1x_2 + 2x_3 \\ 0x_1 + 0x_2 + 0x_3 \end{bmatrix}$$

$$= x_1 \begin{bmatrix} 1 \\ 1 \\ 1 \\ 0 \end{bmatrix} + x_2 \begin{bmatrix} 2 \\ 0 \\ 1 \\ 0 \end{bmatrix} + x_3 \begin{bmatrix} 3 \\ 1 \\ 2 \\ 0 \end{bmatrix}.$$

(1)

Therefore, $\mathbf{b}$ *belongs to the span of the columns of* $\mathbf{A}$ *if and only if the system* $\mathbf{Ax} = \mathbf{b}$ *has a solution.* This suggests that Gauss elimination could be a critical tool in settling questions about the span.

Example 1. Given the column vectors

$$\mathbf{a}_1 = \begin{bmatrix} 1 \\ 1 \\ 1 \\ 0 \end{bmatrix}, \quad \mathbf{a}_2 = \begin{bmatrix} 2 \\ 0 \\ 1 \\ 0 \end{bmatrix}, \quad \mathbf{a}_3 = \begin{bmatrix} 3 \\ 1 \\ 2 \\ 0 \end{bmatrix},$$

$$\mathbf{b}_1 = \begin{bmatrix} 6 \\ 2 \\ 4 \\ 0 \end{bmatrix}, \quad \mathbf{b}_2 = \begin{bmatrix} 1 \\ 1 \\ 2 \\ 0 \end{bmatrix}, \quad \mathbf{b}_3 = \begin{bmatrix} 1 \\ 1 \\ 1 \\ 1 \end{bmatrix},$$

which of $\mathbf{b}_1$, $\mathbf{b}_2$, $\mathbf{b}_3$ is spanned by $\{\mathbf{a}_1, \mathbf{a}_2, \mathbf{a}_3\}$?

Solution. Note that every vector in Span$\{\mathbf{a}_1, \mathbf{a}_2, \mathbf{a}_3\}$ will have a zero in its fourth row, so $\mathbf{b}_3$ can be eliminated immediately. For $\mathbf{b}_1$ and $\mathbf{b}_2$, we resort to the solution criterion for $\mathbf{Ax} = \mathbf{b}$ just established, with $\mathbf{A} = [\mathbf{a}_1 \quad \mathbf{a}_2 \quad \mathbf{a}_3]$. Gauss elimination applied to

$$[\mathbf{a}_1 \quad \mathbf{a}_2 \quad \mathbf{a}_3 \ \vdots \ \mathbf{b}_1 \quad \mathbf{b}_2] \equiv \begin{bmatrix} 1 & 2 & 3 & \vdots & 6 & 1 \\ 1 & 0 & 1 & \vdots & 2 & 1 \\ 1 & 1 & 2 & \vdots & 4 & 2 \\ 0 & 0 & 0 & \vdots & 0 & 0 \end{bmatrix}$$

results in the row echelon form

$$\begin{bmatrix} 1 & 2 & 3 & \vdots & 6 & 1 \\ 0 & -2 & -2 & \vdots & -4 & 0 \\ 0 & 0 & 0 & \vdots & 0 & 1 \\ 0 & 0 & 0 & \vdots & 0 & 0 \end{bmatrix}. \tag{2}$$

The third row in the display tells us that $\mathbf{b}_2$ is not spanned by $\{\mathbf{a}_1, \mathbf{a}_2, \mathbf{a}_3\}$; but since the solutions for $\mathbf{b}_1$ are expressible in terms of a free parameter t as $x_3 = t$, $x_2 = (-4 + 2t)/(-2) = 2 - t$, $x_1 = 6 - 2(2-t) - 3t = 2 - t$, the vector $\mathbf{b}_1$ can be expressed in terms of $\{\mathbf{a}_1, \mathbf{a}_2, \mathbf{a}_3\}$ in any number of ways. For instance by setting $t = 1$, we obtain $\mathbf{b}_1 = \mathbf{a}_1 + \mathbf{a}_2 + \mathbf{a}_3$. ∎

Sometimes, we can draw conclusions about the span of a set of vectors quite easily from elementary observations about the structure of the data. In Example 1, we saw immediately that every vector in Span$\{\mathbf{a}_1, \mathbf{a}_2, \mathbf{a}_3\}$ must have 0 in its fourth entry. If we consider the row vectors

$$\mathbf{v}_1 = \begin{bmatrix} 1 & 1 & 0 & 0 & 0 \end{bmatrix}, \mathbf{v}_2 = \begin{bmatrix} 0 & 0 & 0 & 3 & -3 \end{bmatrix}, \mathbf{v}_3 = \begin{bmatrix} 2 & 2 & 0 & -2 & 2 \end{bmatrix}, \tag{3}$$

it is clear than any vector in Span$\{\mathbf{v}_1, \mathbf{v}_2, \mathbf{v}_3\}$ will have zero in its third column, equal entries in its first two columns, and equal-but-opposite entries in its last two columns. In fact, this verbal description suffices to characterize the span completely. Usually, though, the span is more obscure and we have to resort to Gauss elimination to answer specific inquiries about it.

Example 2. Show that the sets

$$\{\mathbf{a}_1 = \begin{bmatrix} 1 & 1 & 1 & 0 \end{bmatrix}, \mathbf{a}_2 = \begin{bmatrix} 2 & 0 & 1 & 0 \end{bmatrix}, \mathbf{a}_3 = \begin{bmatrix} 3 & 1 & 2 & 0 \end{bmatrix}\}$$

and

$$\{\mathbf{c}_1 = \begin{bmatrix} 8 & 4 & 6 & 0 \end{bmatrix}, \mathbf{c}_2 = \begin{bmatrix} 5 & 1 & 3 & 0 \end{bmatrix}\}$$

span the same subspace of $\mathbf{R}_{\text{row}}^4$.

Solution. We need to show that $\mathbf{c}_1$ and $\mathbf{c}_2$ are linear combinations of $\mathbf{a}_1, \mathbf{a}_2$, and $\mathbf{a}_3$; it will then follow by the linear combination composition principle (Section 3.1) that $\text{Span}\{\mathbf{c}_1, \mathbf{c}_2\}$ is contained in $\text{Span}\{\mathbf{a}_1, \mathbf{a}_2, \mathbf{a}_3\}$. The reverse implication will follow when we show that $\mathbf{a}_1, \mathbf{a}_2$, and $\mathbf{a}_3$ are in the span of $\mathbf{c}_1$ and $\mathbf{c}_2$.

Transposing these row vectors into columns for convenience in applying Gauss elimination, we approach the first task by reducing

$$\begin{bmatrix} 1 & 2 & 3 & \vdots & 8 & 5 \\ 1 & 0 & 1 & \vdots & 4 & 1 \\ 1 & 1 & 2 & \vdots & 6 & 3 \\ 0 & 0 & 0 & \vdots & 0 & 0 \end{bmatrix}$$

to row-echelon form. The result of this calculation,

$$\begin{bmatrix} 1 & 2 & 3 & \vdots & 8 & 5 \\ 0 & -2 & -2 & \vdots & -4 & -4 \\ 0 & 0 & 0 & \vdots & 0 & 0 \\ 0 & 0 & 0 & \vdots & 0 & 0 \end{bmatrix},$$

shows that both systems are consistent, so $\mathbf{c}_1$ and $\mathbf{c}_2$ belong to $\text{Span}\{\mathbf{a}_1, \mathbf{a}_2, \mathbf{a}_3\}$.

Similarly, the second task requires reducing

$$\begin{bmatrix} 8 & 5 & \vdots & 1 & 2 & 3 \\ 4 & 1 & \vdots & 1 & 0 & 1 \\ 6 & 3 & \vdots & 1 & 1 & 2 \\ 0 & 0 & \vdots & 0 & 0 & 0 \end{bmatrix}.$$

This results in

$$\begin{bmatrix} 8 & 5 & \vdots & 1 & 2 & 3 \\ 0 & -1.5 & \vdots & 0.5 & -1 & -0.5 \\ 0 & 0 & \vdots & 0 & 0 & 0 \\ 0 & 0 & \vdots & 0 & 0 & 0 \end{bmatrix},$$

and the three systems are each consistent. Each of the subspaces $\text{Span}\{\mathbf{a}_1, \mathbf{a}_2, \mathbf{a}_3\}$ and $\text{Span}\{\mathbf{c}_1, \mathbf{c}_2\}$ is contained in the other, implying that they are equal. ∎

Observe that in Example 2, the vector $\mathbf{a}_3$ is redundant in the specification $\text{Span}\{\mathbf{a}_1, \mathbf{a}_2, \mathbf{a}_3\}$, since it is already spanned by $\mathbf{a}_1$ and $\mathbf{a}_2$; indeed, $\mathbf{a}_3 = \mathbf{a}_1 + \mathbf{a}_2$. This illustrates the important notion of *linear dependence*.

Linear Dependence and Linear Independence

Definition 3. A vector $\mathbf{v}$ is said to be **linearly dependent** on the set of vectors $\{\mathbf{v}_1, \mathbf{v}_2, \dots, \mathbf{v}_m\}$ if $\mathbf{v}$ can be expressed as a linear combination of the $\{\mathbf{v}_1, \mathbf{v}_2, \dots, \mathbf{v}_m\}$; that is, $\mathbf{v}$ belongs to $\text{Span}\{\mathbf{v}_1, \mathbf{v}_2, \dots, \mathbf{v}_m\}$. Otherwise, it is **linearly independent** of the set.

A set of distinct vectors $\{\mathbf{v}_1, \mathbf{v}_2, \dots, \mathbf{v}_n\}$ is said to be linearly dependent if at least one member of the set is linearly dependent on the rest of the set. Otherwise the set is linearly independent.

When a set of distinct vectors is linearly dependent, one often says "the vectors are dependent," with similar terminology used if the set is linearly independent. Note that a *pair* of vectors is linearly dependent if and only if one of them is a scalar multiple of the other. In $\mathbf{R}^3_{\text{col}}$, a set of three vectors is linearly dependent if and only if the vectors are coplanar (see Problem 29).

There are alternative characterizations of linear independence that are sometimes easier to test. Suppose there is a linear combination of $\{\mathbf{v}_1, \mathbf{v}_2, \cdots, \mathbf{v}_n\}$ that equals zero:

$$c_1 \mathbf{v}_1 + c_2 \mathbf{v}_2 + \cdots + c_{n-1} \mathbf{v}_{n-1} + c_n \mathbf{v}_n = \mathbf{0}.$$

This will happen, trivially, if each c_i is zero. However, if any of them (say, c_p) is *not* zero, we could divide by this coefficient and display linear dependence:

$$\mathbf{v}_p = -\frac{c_1}{c_p} \mathbf{v}_1 - \frac{c_2}{c_p} \mathbf{v}_2 - \cdots - \frac{c_{n-1}}{c_p} \mathbf{v}_{n-1} - \frac{c_n}{c_p} \mathbf{v}_n$$

($\mathbf{v}_p$ omitted in the right-hand side). Thus we have the following:

Test for Linear Independence

> **Theorem 2.** A set of distinct vectors $\{\mathbf{v}_1, \mathbf{v}_2, \ldots, \mathbf{v}_n\}$ is linearly independent if, and only if, the only linear combination of $\{\mathbf{v}_1, \mathbf{v}_2, \ldots, \mathbf{v}_n\}$ that equals zero is the trivial one (with all coefficients zero).

Remark. We cannot establish independence for a set of vectors simply by checking them in pairs; $\{[1\ 0], [0\ 1], [1\ 1]\}$ is a *pairwise* linearly independent set in $\mathbf{R}^2_{\text{row}}$, but the last vector equals the sum of the first two, so the set is linearly dependent. A valid way of establishing independence proceeds "sequentially," in the following sense. Suppose we can show that $\mathbf{v}_n$ is linearly independent of the vectors $\{\mathbf{v}_1, \mathbf{v}_2, \ldots, \mathbf{v}_{n-1}\}$, and that $\mathbf{v}_{n-1}$ is linearly independent of the remaining vectors $\{\mathbf{v}_1, \mathbf{v}_2, \ldots, \mathbf{v}_{n-2}\}$, $\mathbf{v}_{n-2}$ is linearly independent of $\{\mathbf{v}_1, \mathbf{v}_2, \ldots, \mathbf{v}_{n-3}\}$, and so on. Then we *can* be assured that $\{\mathbf{v}_1, \mathbf{v}_2, \ldots, \mathbf{v}_n\}$ is linearly independent—if not, there would a nontrivial linear combination $c_1\mathbf{v}_1 + c_2\mathbf{v}_2 + \cdots + c_{n-1}\mathbf{v}_{n-1} + c_n\mathbf{v}_n = \mathbf{0}$, but then the last $\mathbf{v}_k$ whose c_k is nonzero would be dependent on its predecessors, in violation of the assumptions.

Example 3. Let $\mathbf{A}^{\text{ech}}$ be a matrix in row echelon form, as for example

$$
\mathbf{A}^{\text{ech}} = \begin{bmatrix}
\alpha & - & - & - & - & - & - & - \\
0 & \beta & - & - & - & - & - & - \\
0 & 0 & 0 & 0 & \gamma & - & - & - \\
0 & 0 & 0 & 0 & 0 & \delta & - & - \\
& & & \vdots & & & & \\
0 & 0 & 0 & 0 & 0 & 0 & 0 & 0
\end{bmatrix}, \tag{4}
$$

where we have indicated the first nonzero entries in each row. (If $\mathbf{A}^{\text{ech}}$ happened to be the result of Gauss elimination applied to a "full" matrix, these entries would be the pivot elements discussed in Section 1.4.) Show that the columns that contain the pivot elements form a linearly independent set, while each of the other columns is linearly dependent on its predecessors. Show the nonzero rows are linearly independent.

Solution. The sixth column in (4), the final column containing a leading nonzero entry, is preceded by columns that each have zeros in their fourth row (the row of the nonzero entry). Therefore, the sixth column cannot possibly be expressed as a linear combination of the preceding columns. Similarly for the fifth, second, and first columns; each of the columns is linearly independent of the columns preceding it. Therefore, these columns form a linearly independent set.

Now if we regard the eighth column as the right-hand side of a system for which the preceding seven columns form the coefficient matrix, this system is clearly consistent. Thus column eight is expressible as a linear combination of the preceding columns. Similar considerations show that columns three, four, and seven are linearly dependent on their predecessors.

A similar analysis proves that all of the nonzero rows are linearly independent of their subsequent rows, so they are independent.   ∎

As we saw in Example 3, sometimes the data structure in a set of vectors makes it easy to determine their independence by inspection.

Example 4. Show that the set $\{\mathbf{v}_1 = \begin{bmatrix} 4 & 7 & 1 & 0 \end{bmatrix}, \mathbf{v}_2 = \begin{bmatrix} -2 & 0 & 1 & 1 \end{bmatrix}, \mathbf{v}_3 = \begin{bmatrix} 2 & 0 & 1 & 0 \end{bmatrix}$, and $\mathbf{v}_4 = \begin{bmatrix} 1 & 0 & 2 & 0 \end{bmatrix}\}$ is linearly independent.

Solution. By looking at the second and fourth columns, we see that $\mathbf{v}_1$ and $\mathbf{v}_2$ are independent of the others. And since $\mathbf{v}_3$ and $\mathbf{v}_4$ are clearly not multiples of each other, they are independent of each other. Observing $\mathbf{v}_3, \mathbf{v}_4, \mathbf{v}_1, \mathbf{v}_2$ sequentially, we conclude that the set is independent.   ∎

How can we test if an arbitrary set of vectors is linearly dependent? There's an easy test for vectors in $\mathbf{R}_{\text{col}}^n$. By Theorem 2 we need to see if we can write $c_1\mathbf{v}_1 + c_2\mathbf{v}_2 + \cdots + c_n\mathbf{v}_n = \mathbf{0}$ with at least one nonzero c_i; but as (1) demonstrates,

$$c_1\mathbf{v}_1 + c_2\mathbf{v}_2 + \cdots + c_n\mathbf{v}_n = \begin{bmatrix} \mathbf{v}_1 & \mathbf{v}_2 & \cdots & \mathbf{v}_n \end{bmatrix} \begin{bmatrix} c_1 \\ c_2 \\ \\ c_n \end{bmatrix} = 0 \tag{5}$$

amounts to a linear system of algebraic equations. It will possess nontrivial solutions (other, that is, than $c_1 = c_2 = \cdots = c_n = 0$) only if the Gauss elimination algorithm reveals the presence of free variables. This is easily detected from the final row echelon form of $\begin{bmatrix} \mathbf{v}_1 & \mathbf{v}_2 & \cdots & \mathbf{v}_n \end{bmatrix}$.

Example 5. Use Gauss elimination to verify that the vectors $\mathbf{a}_1, \mathbf{a}_2$, and $\mathbf{a}_3$ in Example 1 are linearly dependent.

Solution. A row echelon form of $\begin{bmatrix} \mathbf{a}_1 & \mathbf{a}_2 & \mathbf{a}_3 \end{bmatrix}$ was displayed earlier in the left portion of (2); it shows that they are dependent, because the related system

$$\begin{bmatrix} 1 & 2 & 3 & \vdots & 0 \\ 0 & -2 & -2 & \vdots & 0 \\ 0 & 0 & 0 & \vdots & 0 \\ 0 & 0 & 0 & \vdots & 0 \end{bmatrix}$$

has nontrivial solutions (in fact, an infinity of solutions).   ∎

It should be clear that any linearly independent set of vectors in $\mathbf{R}_{\text{col}}^n$ can contain no more than n vectors; the row echelon form of a matrix with more columns than rows would have to indicate free parameters.

Example 6. Show that in the vector space $\mathbf{R}^{3,2}$ of 3-by-2 matrices, the following matrices are independent:

$$\mathbf{A} = \begin{bmatrix} 6 & 3 \\ 5 & 2 \\ 4 & 1 \end{bmatrix}, \; \mathbf{B} = \begin{bmatrix} 1 & 0 \\ 0 & 0 \\ 0 & 0 \end{bmatrix}, \; \mathbf{C} = \begin{bmatrix} 1 & 4 \\ 2 & 5 \\ 3 & 6 \end{bmatrix}.$$

Solution. The test for dependence involves checking the expression $c_1\mathbf{A}+c_2\mathbf{B}+c_3\mathbf{C} = \mathbf{0}$. Although the matrices $\mathbf{A}$, $\mathbf{B}$, and $\mathbf{C}$ are presented in a rectangular format, this is a term-by-term comparison, so we can regard each matrix as a column of data: the crucial question becomes whether

$$c_1 \begin{bmatrix} 6 \\ 5 \\ 4 \\ 3 \\ 2 \\ 1 \end{bmatrix} + c_2 \begin{bmatrix} 1 \\ 0 \\ 0 \\ 0 \\ 0 \\ 0 \end{bmatrix} + c_3 \begin{bmatrix} 1 \\ 2 \\ 3 \\ 4 \\ 5 \\ 6 \end{bmatrix} = \mathbf{0}$$

requires that each $c_i = 0$. Gauss elimination reduces the system

$$\begin{bmatrix} 6 & 1 & 1 & \vdots & 0 \\ 5 & 0 & 2 & \vdots & 0 \\ 4 & 0 & 3 & \vdots & 0 \\ 3 & 0 & 4 & \vdots & 0 \\ 2 & 0 & 5 & \vdots & 0 \\ 1 & 0 & 6 & \vdots & 0 \end{bmatrix} \quad \text{to the row echelon form} \quad \begin{bmatrix} 6 & 1 & 1 & \vdots & 0 \\ 0 & -\frac{5}{6} & \frac{7}{6} & \vdots & 0 \\ 0 & 0 & \frac{28}{5} & \vdots & 0 \\ 0 & 0 & 0 & \vdots & 0 \\ 0 & 0 & 0 & \vdots & 0 \\ 0 & 0 & 0 & \vdots & 0 \end{bmatrix},$$

whose only solution is $c_1 = c_2 = c_3 = 0$. The matrices are linearly independent. ∎

Since independence reduces to a question of solvability, surely we can find an application for the determinant somewhere. So we conclude this section with

Linear Independence of the Rows and Columns of a Square Matrix

Theorem 3. The rows (or the columns) of a square matrix $\mathbf{A}$ are linearly independent if and only if $\det(\mathbf{A}) \neq 0$.

Proof. If $\det(\mathbf{A}) \neq 0$, the matrix $\mathbf{A}$ has an inverse so the only solution to $\mathbf{AX} = \mathbf{0}$ is $\mathbf{x} = 0$, that is the only linear combination of the columns of $\mathbf{A}$ that equals zero is the trivial one; hence the columns are independent. Since $\det(\mathbf{A}^T) = \det(\mathbf{A})$, the same is true of the rows.

If $\det(\mathbf{A}) = 0$, then (by the discussion in Section 2.5) $\mathbf{A}$ can be reduced by elementary row operations to an upper triangular matrix $\mathbf{U}$ in row echelon form whose determinant is also zero. But a square upper triangular matrix in row echelon form with zero determinant must have all zeros in its last row. Therefore, the corresponding row in the original matrix $\mathbf{A}$ was reduced to zero by subtracting multiples of other rows; in other words, it was equal to a linear combination of the other rows. Hence the rows of A are linearly dependent. Again, applying this reasoning to $\mathbf{A}^T$ demonstrates that $\mathbf{A}$'s columns are also dependent. ■

In Example 3 of Section 2.4, we computed the determinant of the matrix

$$\mathbf{A} = \begin{bmatrix} 2 & 3 & 4 & 0 \\ 5 & 6 & 3 & 0 \\ 2 & 5 & 3 & 7 \\ 2 & 2 & 2 & 2 \end{bmatrix}$$

to be 118. Therefore, we know its columns and its rows are linearly independent.

The issue of linear independence for vector spaces of *functions* is subtler than that for $\mathbf{R}_{\text{col}}^m$ or $\mathbf{R}^{m,n}$. Section 4.4 will address this topic.

Exercises 3.2

1. Which of the following vectors in $\mathbf{R}_{\text{row}}^3$ lie in the span of the vectors $\mathbf{a}_1 = [1 \ -1 \ 2]$, $\mathbf{a}_2 = [2 \ 5 \ 6]$?
   (a) $[1 \ 6 \ 7]$ (b) $[3 \ 4 \ 8]$ (c) $[1 \ -1 \ 2]$ (d) $[0 \ 0 \ 0]$

2. Which of the following vectors in $\mathbf{R}_{\text{row}}^3$ lie in the span of the vectors $\mathbf{a}_1 = [1 \ 0 \ 1]$, $\mathbf{a}_2 = [2 \ 0 \ 1]$?
   (a) $[1 \ 0 \ -3]$ (b) $[0 \ -1 \ 2]$ (c) $[1 \ 0 \ 2]$ (d) $[0 \ 0 \ 0]$

3. Given the row vectors $\mathbf{a}_1 = [1 \ 0 \ 2 \ 0]$, $\mathbf{a}_2 = [1 \ 1 \ 1 \ 1]$, $\mathbf{a}_3 = [1 \ 0 \ 2 \ 1]$, and $\mathbf{b}_1 = [1 \ -1 \ 2 \ 1]$, $\mathbf{b}_2 = [-1 \ -1 \ -1 \ -1]$, $\mathbf{b}_3 = [1 \ 0 \ 1 \ 0]$, which of $\mathbf{b}_1, \mathbf{b}_2, \mathbf{b}_3$ is spanned by $\{\mathbf{a}_1, \mathbf{a}_2, \mathbf{a}_3\}$?

4. Given the row vectors $\mathbf{a}_1 = [1 \ 1 \ 0 \ 3 \ 1]$, $\mathbf{a}_2 = [0 \ 1 \ 1 \ 3 \ 2]$, $\mathbf{a}_3 = [1 \ 1 \ 1 \ 6 \ 2]$, and $\mathbf{b}_1 = [-1 \ -1 \ 0 \ -3 \ -1]$, $\mathbf{b}_2 = [0 \ 1 \ 1 \ 3 \ 1]$, $\mathbf{b}_3 = [2 \ 3 \ 2 \ 12 \ 5]$, which of $\mathbf{b}_1, \mathbf{b}_2, \mathbf{b}_3$ is spanned by $\{\mathbf{a}_1, \mathbf{a}_2, \mathbf{a}_3\}$?

5. Show that the sets $\{[1 \ 0 \ 0 \ 1], [0 \ 1 \ 1 \ 3], [2 \ -3 \ -3 \ -7]\}$ and $\{[2 \ -2 \ -2 \ -4], [0 \ -1 \ -1 \ -3]\}$ span the same subspace of $\mathbf{R}_{\text{row}}^4$.

6. Show that the sets $\{[1 \ 1 \ 0 \ -1], [0 \ 1 \ 2 \ 3], [-1 \ 1 \ 1 \ 6]\}$ and $\{[1 \ 3 \ 4 \ 5], [7 \ 7 \ 6 \ -5], [0 \ -1 \ -2 \ -3]\}$ span the same subspace of $\mathbf{R}_{\text{row}}^4$.

   In Problems 7–16 determine whether the given set of vectors is linearly independent or linearly dependent in the prescribed vector space. If the vectors are linearly dependent, find a non-trivial linear combination of them that equals the zero vector.

7. $[0 \ 3 \ -1], [0 \ 3 \ 1]$ in $\mathbf{R}_{\text{row}}^3$.

8. $[\pi\ 9]$, $[7\ e]$, $[-1\ 800]$ in $\mathbf{R}^2_{row}$.

9. $[-2\ 1\ -1\ 1]$, $[4\ 0\ 6\ 0]$, $[5\ 4\ -2\ 1]$ in $\mathbf{R}^4_{row}$.

10. $[2\ -4\ 3\ -6]^T$, $[2\ 0\ 0\ 0]^T$, $[-5\ 0\ 5\ 0]^T$, $[7\ -3\ 8\ -5]^T$ in $\mathbf{R}^4_{col}$.

11. $[2+3i\ \ 1+i\ \ 4+i]$, $[6-2i\ \ 2-i\ \ 2-4i]$, $[-6+13i\ \ -1+5i\ \ 8+11i]$ in $\mathbf{C}^3_{row}$.

12. $[1\ 2\ 1\ 0]^T$, $[2\ -2\ 0\ 2]^T$, $[3\pi\ 0\ \pi\ 5\pi]^T$ in $\mathbf{R}^4_{col}$.

13. The rows of the matrix $\begin{bmatrix} 5 & -1 & 2 & 3 \\ 0 & 3 & 6 & -2 \\ 0 & 0 & 0 & 9 \end{bmatrix}$ in $\mathbf{R}^4_{row}$.

14. The columns of the matrix $\begin{bmatrix} -4 & 5 & 2 \\ 0 & 0 & 1 \\ 0 & 2 & 2 \\ 0 & 0 & 4 \end{bmatrix}$ in $\mathbf{R}^4_{col}$.

15. $\begin{bmatrix} 2 & 3 \\ -3 & 1 \end{bmatrix}$, $\begin{bmatrix} -8 & -9 \\ 0 & -4 \end{bmatrix}$, $\begin{bmatrix} 8 & 10 \\ -2 & 6 \end{bmatrix}$ in $\mathbf{R}^{2,2}$.

16. $\begin{bmatrix} 1 & 2 & 3 \\ 3 & 2 & 1 \end{bmatrix}$, $\begin{bmatrix} 2 & 4 & 6 \\ 6 & 4 & 2 \end{bmatrix}$ in $\mathbf{R}^{2,3}$.

17. Find necessary and sufficient conditions on the values of b and c so that the columns of $\begin{bmatrix} 1 & 0 & 1 \\ 0 & b & 1 \\ 0 & 1 & c \end{bmatrix}$ span $\mathbf{R}^3_{col}$.

18. Find necessary and sufficient conditions on the values of b and c so that the rows of $\begin{bmatrix} 1 & b & 1 \\ 0 & b & 1 \\ 1 & 1 & c \end{bmatrix}$ span $\mathbf{R}^3_{row}$.

19. For each of the following, decide whether the statement made is always True or sometimes False.

(a) Any set of vectors that contains the zero vector is linearly dependent.

(b) If a set of three vectors is linearly dependent, then one of the vectors must be a scalar multiple of one of the others.

(c) Every subset of a set of linearly dependent vectors is linearly dependent.

(d) If the columns of a square matrix $\mathbf{A}$ are linearly dependent, then $\det(\mathbf{A}) = 0$.

(e) The rows of any elementary row operation matrix are linearly independent.

20. If $\mathbf{A}$ is a 6-by-6 matrix whose second row is a scalar multiple of its fifth row, explain why the columns of $\mathbf{A}$ are linearly dependent.

21. Explain why the following is true: If $\mathbf{v}_1$, $\mathbf{v}_2$, and $\mathbf{v}_3$ are linearly independent and $\mathbf{v}_4$ is not in $\text{Span}\{\mathbf{v}_1, \mathbf{v}_2, \mathbf{v}_3\}$, then the set $\{\mathbf{v}_1, \mathbf{v}_2, \mathbf{v}_3, \mathbf{v}_4\}$ is linearly independent.

22. Prove that any four vectors in $\mathbf{R}^3_{col}$ must be linearly dependent. What about $n + 1$ vectors in $\mathbf{R}^n_{col}$?

23. Prove that if $\mathbf{A}$, $\mathbf{B}$, and $\mathbf{C}$ are linearly independent matrices in $\mathbf{R}^{3,3}$ and $\mathbf{D}$ is an invertible 3-by-3 matrix, then $\mathbf{AD}$, $\mathbf{BD}$, and $\mathbf{CD}$ are linearly independent matrices.

24. Prove that if $\mathbf{v}_1, \mathbf{v}_2, \mathbf{v}_3$ are the columns of an n-by-3 matrix $\mathbf{A}$ and $\mathbf{B}$ is an invertible 3-by-3 matrix, then the span of the columns of $\mathbf{AB}$ is identical to $\mathrm{Span}\{\mathbf{v}_1, \mathbf{v}_2, \mathbf{v}_3\}$.

25. Let $\mathbf{A}$ be an n-by-n matrix and $\mathbf{x}$ be an n-by-1 column vector. Explain why the columns of $\mathbf{Ax}, \mathbf{A}^2\mathbf{x}, \mathbf{A}^3\mathbf{x}, \ldots$ all lie in the span of the columns of $\mathbf{A}$.

26. Let $\mathbf{A}$ be an n-by-n matrix and $\mathbf{B}$ be an invertible n-by-n matrix. Prove that if the column vectors of $\mathbf{A}$ are linearly independent, then so are the columns vectors of $\mathbf{AB}$.

27. Given that the vectors $\mathbf{v}_1, \mathbf{v}_2, \mathbf{v}_3$ are linearly independent, show that the following vectors are also linearly independent:

(a) $\mathbf{w}_1 = 2\mathbf{v}_1 - 3\mathbf{v}_2$, $\mathbf{w}_2 = 4\mathbf{v}_1 - 3\mathbf{v}_2$

(b) $\mathbf{w}_1 = \mathbf{v}_1 - 3\mathbf{v}_2$, $\mathbf{w}_2 = 4\mathbf{v}_1 - 3\mathbf{v}_3$, $\mathbf{w}_3 = 2\mathbf{v}_2 - 3\mathbf{v}_3$.

28. Prove that an n-by-n matrix $\mathbf{A}$ is invertible if and only if for any set of n linearly independent vectors $\mathbf{v}_1, \mathbf{v}_2, \ldots, \mathbf{v}_n$ in $\mathbf{R}_{\mathrm{col}}^n$, the set $\mathbf{Av}_1, \mathbf{Av}_2, \ldots, \mathbf{Av}_n$ is linearly independent.

29. Prove that three vectors in $\mathbf{R}_{\mathrm{col}}^3$ are linearly dependent if and only if they are coplanar. (*Hint*: If $\mathbf{v}_1, \mathbf{v}_2$, and $\mathbf{v}_3$ are coplanar, and $\mathbf{n}$ is a normal to the plane in which they lie, then the dot products $\mathbf{v}_i \cdot \mathbf{n}$ are zero for $i = 1, 2, 3$. What does this imply about the determinant of the matrix $[\mathbf{v}_1 \ \mathbf{v}_2 \ \mathbf{v}_3]$?)

3.3 BASES, DIMENSION, AND RANK

In Section 3.1 we noted that the span of any set of vectors constitutes a subspace. We observed that the solution space of the homogeneous system

$$\begin{bmatrix} 1 & 2 & 1 & 1 & -1 & 0 & \vdots & 0 \\ 1 & 2 & 0 & -1 & 1 & 0 & \vdots & 0 \\ 0 & 0 & 1 & 2 & -2 & 1 & \vdots & 0 \end{bmatrix} \tag{1}$$

(the "null space" of its coefficient matrix) is given by

$$\mathrm{Span}\{\begin{bmatrix} -2 & 1 & 0 & 0 & 0 & 0 \end{bmatrix}^T, \begin{bmatrix} 1 & 0 & -2 & 1 & 0 & 0 \end{bmatrix}^T, \begin{bmatrix} -1 & 0 & 2 & 0 & 1 & 0 \end{bmatrix}^T\}, \tag{2}$$

and the solution space of the homogeneous differential equation

$$d^3y/dx^3 = 0$$

is given by $\mathrm{Span}\{x^2/2, \ x, \ 1\}$.

Now consider the subspace of $\mathbf{R}_{\mathrm{row}}^5$ spanned by

$$\mathbf{v}_1 = \begin{bmatrix} 1 & 1 & 0 & 0 & 0 \end{bmatrix}, \ \mathbf{v}_2 = \begin{bmatrix} 0 & 0 & 0 & 3 & -3 \end{bmatrix}, \ \mathbf{v}_3 = \begin{bmatrix} 2 & 2 & 0 & -2 & 2 \end{bmatrix}, \tag{3}$$

which could be described as the subspace in which the first two entries were equal, the final two negatives of each other, and the third entry zero. Observe that v_3 can be generated by taking $2v_1 - 2v_2/3$. This means that v_3 is redundant in Span$\{v_1, v_2, v_3\}$, since (by the linear combination composition principle) any vector expressible as a linear combination of v_1, v_2, and v_3 can be rearranged into a linear combination of v_1 and v_2 alone. So it would be economical to drop v_3 and regard the subspace as Span$\{v_1, v_2\}$. But we can't economize any further, since v_1 and v_2 are independent of each other (since v_2 is not a scalar times v_1). We formalize this idea:

Basis

Definition 4. A **basis** for a space (or subspace) is a linearly independent set of vectors that spans the (sub)space.

A basis for a (sub)space is therefore a "minimal spanning set," in the sense that if any vector is dropped from the basis, it will no longer span the space (the dropped vector itself, for example, cannot be expressed as a linear combination of the others). It can also be characterized as a "maximal linearly independent set," in the sense that there are no other vectors in the subspace that are independent of the basis (otherwise, the basis wouldn't have spanned the space).

An obvious basis for, say, $\mathbf{R}_{col}^4$ (or, for that matter, $\mathbf{C}_{col}^4$) would the "canonical basis."

$$\begin{bmatrix} 1 \\ 0 \\ 0 \\ 0 \end{bmatrix}, \begin{bmatrix} 0 \\ 1 \\ 0 \\ 0 \end{bmatrix}, \begin{bmatrix} 0 \\ 0 \\ 1 \\ 0 \end{bmatrix}, \text{ and } \begin{bmatrix} 0 \\ 0 \\ 0 \\ 1 \end{bmatrix}.$$

However the columns of *any* nonsingular square n-by-n matrix comprise a basis for $\mathbf{R}_{col}^n$; indeed, since the determinant of the matrix is nonzero, Cramer's rule implies both linear independence and the spanning property.

Example 1. Construct bases for the subspaces of $\mathbf{R}_{row}^5$ considered in Example 1 of Section 3.1:

(a) All "symmetric" vectors of the form $\begin{bmatrix} a & b & c & b & a \end{bmatrix}$;

(b) All "antisymmetric" vectors of the form $\begin{bmatrix} a & b & 0 & -b & -a \end{bmatrix}$;

(c) Those vectors all of whose entries are the same $\begin{bmatrix} a & a & a & a & a \end{bmatrix}$;

(d) Those vectors whose second entry is zero $\begin{bmatrix} a & 0 & b & c & d \end{bmatrix}$.

Solution. (a) Two obvious vectors in the subspace are $v_1 = \begin{bmatrix} 1 & 0 & 0 & 0 & 1 \end{bmatrix}$ and $v_2 = \begin{bmatrix} 0 & 1 & 0 & 1 & 0 \end{bmatrix}$. However, they don't span the subspace, since everything in Span$\{v_1, v_2\}$ will have zero in its third entry. So we throw in $v_3 = \begin{bmatrix} 0 & 0 & 1 & 0 & 0 \end{bmatrix}$.

The set $\{\mathbf{v}_1, \mathbf{v}_2, \mathbf{v}_3\}$ is clearly independent because of the location of the zeros, and it spans the space, as witnessed by the display

$$\begin{bmatrix} a & b & c & b & a \end{bmatrix} = a\mathbf{v}_1 + b\mathbf{v}_2 + c\mathbf{v}_3.$$

(b) Proceeding similarly to (a), we find the basis $\begin{bmatrix} 1 & 0 & 0 & 0 & -1 \end{bmatrix}$ and $\begin{bmatrix} 0 & 1 & 0 & -1 & 0 \end{bmatrix}$.

(c) The single vector $\begin{bmatrix} 1 & 1 & 1 & 1 & 1 \end{bmatrix}$ spans the subspace.

(d) This space is spanned by the independent vectors $\begin{bmatrix} 1 & 0 & 0 & 0 & 0 \end{bmatrix}$, $\begin{bmatrix} 0 & 0 & 1 & 0 & 0 \end{bmatrix}, \begin{bmatrix} 0 & 0 & 0 & 1 & 0 \end{bmatrix}$, and $\begin{bmatrix} 0 & 0 & 0 & 0 & 1 \end{bmatrix}$. ∎

Example 2. Find a basis for the subspace of $\mathbf{R}^4_{\text{col}}$ of vectors whose entries add up to zero.

Solution. This subspace is the null space for the matrix $\mathbf{A} = \begin{bmatrix} 1 & 1 & 1 & 1 \end{bmatrix}$. Since $\mathbf{A}$ is already in row echelon form, the solutions to $\mathbf{Ax} = \mathbf{0}$ are parametrized by back-solving the augmented system $\begin{bmatrix} 1 & 1 & 1 & 1 & \vdots & 0 \end{bmatrix}$, yielding $x_4 = t_1$, $x_3 = t_2$, $x_2 = t_3$, $x_1 = -t_1 - t_2 - t_3$, or

$$\mathbf{x} = \begin{bmatrix} x_1 \\ x_2 \\ x_3 \\ x_4 \end{bmatrix} = t_1 \begin{bmatrix} -1 \\ 0 \\ 0 \\ 1 \end{bmatrix} + t_2 \begin{bmatrix} -1 \\ 0 \\ 1 \\ 0 \end{bmatrix} + t_3 \begin{bmatrix} -1 \\ 1 \\ 0 \\ 0 \end{bmatrix}, \tag{4}$$

and the vectors in this display form a basis. ∎

Change of Coordinates

In Example 2 of Section 3.2, we saw how to test whether a vector $\mathbf{v}$ in $\mathbf{R}^n_{\text{col}}$ lies in the subspace spanned by $\{\mathbf{w}_1, \mathbf{w}_2, \ldots, \mathbf{w}_k\}$; indeed, the possibility of a relation

$$\mathbf{v} = c_1\mathbf{w}_1 + c_2\mathbf{w}_2 + \cdots + c_k\mathbf{w}_k \tag{5}$$

is easily resolved by Gauss elimination, since (5) amounts to a linear system of n equations in k unknowns:

$$\begin{bmatrix} \\ \mathbf{v} \\ \\ \end{bmatrix} = c_1 \begin{bmatrix} \\ \mathbf{w}_1 \\ \\ \end{bmatrix} + c_2 \begin{bmatrix} \\ \mathbf{w}_2 \\ \\ \end{bmatrix} + \cdots + c_k \begin{bmatrix} \\ \mathbf{w}_k \\ \\ \end{bmatrix} = \begin{bmatrix} \mathbf{w}_1 & \mathbf{w}_2 & \cdots & \mathbf{w}_k \end{bmatrix} \begin{bmatrix} c_1 \\ c_2 \\ \vdots \\ c_k \end{bmatrix}$$

$$\begin{bmatrix} \\ \mathbf{v} \\ \\ \end{bmatrix} = \begin{bmatrix} \mathbf{w}_1 & \mathbf{w}_2 & \cdots & \mathbf{w}_k \end{bmatrix} \begin{bmatrix} \\ \mathbf{c} \\ \\ \end{bmatrix}. \tag{6}$$

Gauss elimination decides the issue and supplies the values of the coefficients c_j when appropriate. For example, to express $\begin{bmatrix} 2 & 4 \end{bmatrix}^T$ in terms of the vectors $\{[1 \ 1]^T, \ [1\,{-}1]^T\}$ (which are independent and form a basis), we solve

$$\begin{bmatrix} 2 \\ 4 \end{bmatrix} = \begin{bmatrix} 1 & 1 \\ 1 & -1 \end{bmatrix} \begin{bmatrix} c_1 \\ c_2 \end{bmatrix} \tag{7}$$

and calculate $[c_1 \, c_2]^T = [3 \, -1]^T$.

Let's focus on the relationship between $\mathbf{v} = [2 \, 4]^T$ and $\mathbf{c} = [3 \, -1]^T$ in equations (5–7). They are different vectors, of course; $[3 \, -1]^T$ is not the same, numerically, as $[2 \, 4]^T$. But (5) states that $[3 \, -1]^T$ *represents* $[2 \, 4]^T$ *with respect to the basis* $\{[1 \, 1]^T, [1 \, -1]^T\}$. We say that $\mathbf{c} = [3 \, -1]^T$ gives the *coordinates* of $\mathbf{v} = [2 \, 4]^T$ *in the basis* $\{\mathbf{w}_1, \mathbf{w}_2\} = \{[1 \, 1]^T, [1 \, -1]^T\}$.

In fact, the "original" vector $[2 \, 4]^T$ can be interpreted as giving the coordinates of $\mathbf{v}$ in the canonical basis $\{[1 \, 0]^T, [0 \, 1]^T\}$. With this jargon, we interpret (6) as a *coordinate transformation* from the $\{\mathbf{w}_j\}$ basis to the canonical basis, with $\begin{bmatrix} \mathbf{w}_1 & \mathbf{w}_2 & \cdots & \mathbf{w}_k \end{bmatrix}$ playing the role of a *transition matrix*.

If the $\{\mathbf{w}_j\}$ basis spans the whole space $\mathbf{R}^n_{\text{col}}$, that is, if $k = n$, (6) implies that $\begin{bmatrix} \mathbf{w}_1 & \mathbf{w}_2 & \cdots & \mathbf{w}_n \end{bmatrix}$ is invertible, and that its inverse is a transition matrix from canonical coordinates to $\{\mathbf{w}_j\}$ coordinates. The Gauss elimination that we performed on (7) was, of course, equivalent to multiplying by the inverse.

Now let's address the issue of a *general* change of coordinates. That is, we have two different (non)canonical bases, $\{\mathbf{w}_1, \mathbf{w}_2, \ldots, \mathbf{w}_p\}$ and $\{\mathbf{u}_1, \mathbf{u}_2, \ldots, \mathbf{u}_q\}$ for the same subspace of $\mathbf{R}^n_{\text{col}}$, and a vector $\mathbf{v}$ lying in the subspace. Thus

$$\mathbf{v} = c_1 \mathbf{w}_1 + c_2 \mathbf{w}_2 + \cdots + c_p \mathbf{w}_p = d_1 \mathbf{u}_1 + d_2 \mathbf{u}_2 + \cdots + d_q \mathbf{u}_q. \tag{8}$$

How are its coordinates $\mathbf{c}$ in the $\{\mathbf{w}_j\}$ basis related to its coordinates $\mathbf{d}$ in the $\{\mathbf{u}_j\}$ basis? Again, matrix algebra resolves the problem easily; (8) is equivalent to

$$\begin{bmatrix} \mathbf{w}_1 & \mathbf{w}_2 & \cdots & \mathbf{w}_p \end{bmatrix} \begin{bmatrix} \mathbf{c} \end{bmatrix} = \begin{bmatrix} \mathbf{u}_1 & \mathbf{u}_2 & \cdots & \mathbf{u}_q \end{bmatrix} \begin{bmatrix} \mathbf{d} \end{bmatrix}. \tag{9}$$

The right-hand side is given, so Gauss elimination answers our question.

Example 3. The subspaces of $\mathbf{R}^3_{\text{col}}$ spanned by $\{[1 \, 1 \, 0]^T, [0 \, 1 \, 1]^T\}$ and by $\{[1 \, 2 \, 1]^T, [1 \, 0 \, -1]^T\}$ can be shown to be the same by the method of Example 2, Section 3.2. Find the coordinates in the first basis for the vector whose coordinates in the second basis are $[1 \, 2]^T$.

Solution. Inserting our data into equation (9) gives

$$\begin{bmatrix} 1 & 0 \\ 1 & 1 \\ 0 & 1 \end{bmatrix} \begin{bmatrix} c_1 \\ c_2 \end{bmatrix} = \begin{bmatrix} 1 & 1 \\ 2 & 0 \\ 1 & -1 \end{bmatrix} \begin{bmatrix} 1 \\ 2 \end{bmatrix} = \begin{bmatrix} 3 \\ 2 \\ -1 \end{bmatrix} \tag{10}$$

or

$$\begin{bmatrix} 1 & 0 & \vdots & 3 \\ 1 & 1 & \vdots & 2 \\ 0 & 1 & \vdots & -1 \end{bmatrix}, \tag{11}$$

which row reduces to

$$\begin{bmatrix} 1 & 0 & \vdots & 3 \\ 0 & 1 & \vdots & -1 \\ 0 & 0 & \vdots & 0 \end{bmatrix}$$

Therefore, the coordinates are $c_1 = 3$, $c_2 = -1$. This can be verified:

$$3 \begin{bmatrix} 1 \\ 1 \\ 0 \end{bmatrix} + (-1) \begin{bmatrix} 0 \\ 1 \\ 1 \end{bmatrix} = 1 \begin{bmatrix} 1 \\ 2 \\ 1 \end{bmatrix} + 2 \begin{bmatrix} 1 \\ 0 \\ -1 \end{bmatrix} \left(= \begin{bmatrix} 3 \\ 2 \\ -1 \end{bmatrix} \right). \qquad \blacksquare$$

If $p = q = n$, the matrices in (9) are square and nonsingular, and the change-of-coordinates rule can be expressed

$$\begin{bmatrix} c \end{bmatrix} = \begin{bmatrix} w_1 & w_2 & \cdots & w_k \end{bmatrix}^{-1} \begin{bmatrix} u_1 & u_2 & \cdots & u_k \end{bmatrix} \begin{bmatrix} d \end{bmatrix}. \tag{12}$$

Formula (12) is elegant, but we have seen that Gauss elimination is more efficient than using inverses, as well as universally applicable. Problem 44 discusses an alternative formulation.

Of course a (sub)space will typically have more than one basis. (For example, the columns of *any* nonsingular 3-by-3 matrix form a basis for $\mathbf{R}^3_{col}$.) But it is not possible for it to have one basis of, say, 3 vectors and another basis with 4 vectors, for the following reason:

Cardinality of Bases

Theorem 4. If the vector space V has a basis consisting of n vectors, then every collection of more than n vectors in the space must be linearly dependent. Consequently, every basis for a given space contains the same number of elements.

Proof. We take n to be 3 and prove that every set of four vectors is linearly dependent. The generalization will then be easy to see. Let $\{v_1, v_2, v_3\}$ be the basis, and $\{\mathbf{w}_1, \mathbf{w}_2, \mathbf{w}_3, \mathbf{w}_4\}$ be any four vectors in the space. We shall devise a nontrivial linear combination $d_1\mathbf{w}_1 + d_2\mathbf{w}_2 + d_3\mathbf{w}_3 + d_4\mathbf{w}_4$ that equals the zero vector. First express the $\mathbf{w}_i$'s in terms of the $\mathbf{v}_j$'s (exactly as in the proof of the linear combination composition principle of Section 3.1):

$$
\begin{aligned}
\mathbf{0} &= d_1\mathbf{w}_1 + d_2\mathbf{w}_2 + d_3\mathbf{w}_3 + d_4\mathbf{w}_4 \\
&= d_1(c_1^{(1)}\mathbf{v}_1 + c_2^{(1)}\mathbf{v}_2 + c_3^{(1)}\mathbf{v}_3) + d_2(c_1^{(2)}\mathbf{v}_1 + c_2^{(2)}\mathbf{v}_2 + c_3^{(2)}\mathbf{v}_3) \\
&\quad + d_3(c_1^{(3)}\mathbf{v}_1 + c_2^{(3)}\mathbf{v}_2 + c_3^{(3)}\mathbf{v}_3) + d_4(c_1^{(4)}\mathbf{v}_1 + c_2^{(4)}\mathbf{v}_2 + c_3^{(4)}\mathbf{v}_3) \\
&= (c_1^{(1)}d_1 + c_1^{(2)}d_2 + c_1^{(3)}d_3 + c_1^{(4)}d_4)\mathbf{v}_1 + (c_2^{(1)}d_1 + c_2^{(2)}d_2 + c_2^{(3)}d_3 + c_2^{(4)}d_4)\mathbf{v}_2 \\
&\quad + (c_3^{(1)}d_1 + c_3^{(2)}d_2 + c_3^{(3)}d_3 + c_3^{(4)}d_4)\mathbf{v}_3.
\end{aligned}
$$

Can we find d_j's not all zero such that

$$
\begin{aligned}
c_1^{(1)}d_1 + c_1^{(2)}d_2 + c_1^{(3)}d_3 + c_1^{(4)}d_4 &= 0 \\
c_2^{(1)}d_1 + c_2^{(2)}d_2 + c_2^{(3)}d_3 + c_2^{(4)}d_4 &= 0 \\
c_3^{(1)}d_1 + c_3^{(2)}d_2 + c_3^{(3)}d_3 + c_3^{(4)}d_4 &= 0 \ ?
\end{aligned}
\tag{13}
$$

Of course! As noted at the end of Section 1.4, a system like (13)—homogeneous, with fewer equations than unknowns—has an infinite number of solutions.

Theorem 4 motivates a new definition.

Dimension

Definition 5. The *dimension* of a vector space is the number of vectors in its bases.

Therefore, the dimension of $\mathbf{R}_{\text{col}}^n$, or $\mathbf{R}_{\text{row}}^n$, is n. The dimension of $\mathbf{R}^{m,n}$ is mn. The dimensions of the subspaces of $\mathbf{R}_{\text{row}}^5$ in Example 1 are 3 for symmetric vectors, 2 for antisymmetric vectors, 1 for vectors with all components equal, and 4 for vectors with zero in the second entry.

Now we are going to turn to a discussion of the dimension of some subspaces that are associated with a matrix.

Recall that the *null space* of a matrix $\mathbf{A}$ is the set of solutions to the system $\mathbf{A}\mathbf{x} = \mathbf{0}$; one could call it the set of vectors "annihilated" by $\mathbf{A}$. Its dimension is called the **nullity** of $\mathbf{A}$.

The **range** of a matrix $\mathbf{A}$ is the set of vectors $\mathbf{b}$ for which the system $\mathbf{A}\mathbf{x} = \mathbf{b}$ has solutions. As we have seen, the range is the span of the columns of $\mathbf{A}$ (and hence is a vector space). A basis for the range would be any maximal linearly independent subset of the columns of **A**, and the dimension of the range is the number of columns in such a basis. Similar considerations apply to the *row space* of **A**, the span of **A**'s rows.

These spaces are very easy to analyze if the matrix **A** is in row echelon form, as for example

$$\mathbf{A}^{\text{ech}} = \begin{bmatrix} \alpha & - & - & - & - & - & - \\ 0 & \beta & - & - & - & - & - \\ 0 & 0 & 0 & 0 & \gamma & - & - \\ 0 & 0 & 0 & 0 & 0 & \delta & - \\ 0 & 0 & 0 & 0 & 0 & 0 & 0 \end{bmatrix}. \tag{14}$$

Recalling the discussion of a similar matrix in the preceding section, we see that the bold entries (the leading nonzero elements in each row) "flag" a maximal set of independent columns *and* a maximal set of independent rows. So these sets are bases for the range and the row space respectively, whose common dimension is the rank of **A**^{ech}. (Recall from Section 1.4 that the rank of a row echelon matrix was defined to be the number of nonzero rows). If we contemplate back-solving **A**^{ech}**x** = **0**, the structure of the solution formula would look like

$$x_7 = t_1, \; x_6 = At_1, \; x_5 = Bt_1, \; x_4 = t_2, \; x_3 = t_3,$$
$$x_2 = Ct_1 + Dt_2 + Et_3, \; x_1 = Ft_1 + Gt_2 + Ht_3$$

or

$$\mathbf{x} = t_1 \begin{bmatrix} F \\ C \\ 0 \\ 0 \\ B \\ A \\ 1 \end{bmatrix} + t_2 \begin{bmatrix} G \\ D \\ 0 \\ 1 \\ 0 \\ 0 \\ 0 \end{bmatrix} + t_3 \begin{bmatrix} H \\ E \\ 1 \\ 0 \\ 0 \\ 0 \\ 0 \end{bmatrix}. \tag{15}$$

Every non-flagged column introduces a free parameter and an independent column vector into the solution expression, so (15) demonstrates that the dimension of the null space equals the number of unflagged columns. Therefore the total number of columns of a row echelon matrix equals the dimension of the range plus the dimension of the null space (the nullity). In our example, rank = 4, nullity = 3, and they sum to 7, the number of columns.

What are the dimensions of these spaces for a "full" matrix, not in row echelon form? We are going to prove that the row operations of Gauss elimination preserve the dimensions of a matrix's range and row space (!). This will show that we can find the dimensions of these subspaces by using the forward portion of the Gauss elimination algorithm to reduce the matrix to row echelon form, and reading off the dimensions as demonstrated above. Our proof, then, will justify the following generalizations:

Range, Row Space, Null Space, and Rank

Theorem 5. The dimension of the range (or column space) of any matrix equals the dimension of its row space.[2]

(You may be surprised by Theorem 5. Even if a matrix has 1000 rows and 5 columns, the maximal number of linearly independent rows is limited by the number of independent columns.)

Definition 6. The **rank** of a matrix is the dimension of its row space (or column space). A matrix is said to be of **full rank** if its rank equals its number of columns or rows.

(The following theorem generalizes equation (2) of Section 1.4.)

Theorem 6. The rank plus the nullity of any matrix equals the total number of columns.

We turn to the proof that Gauss elimination preserves column and row dependencies. We begin with the column analysis. Recall that the test for linear dependence or independence of a set of vectors is whether or not **0** can be written as a nontrivial linear combination. The next theorem says the test can be applied either to the columns of **A** or those of $\mathbf{A}^{\text{ech}}$.

Elementary Row Operations Preserve Column Dependence

Theorem 7. Suppose the matrix **A** is reduced to a row echelon form $\mathbf{A}^{\text{ech}}$ by the elementary row operations of Gauss elimination. Then if a linear combination of the columns of **A** equals **0**, the same linear combination of the columns of $\mathbf{A}^{\text{ech}}$ equals **0** also. Conversely if a linear combination of the columns of $\mathbf{A}^{\text{ech}}$ equals **0**, the same linear combination of the columns of **A** equals **0** also. Consequently, the dimension of the column space of **A** equals the dimension of the column space of $\mathbf{A}^{\text{ech}}$.

Proof. In Example 2 of Section 2.1 we saw how $\mathbf{A}^{\text{ech}}$ can be represented as

$$\mathbf{A}^{\text{ech}} = \mathbf{E}_p \mathbf{E}_{p-1} \cdots \mathbf{E}_2 \mathbf{E}_1 \mathbf{A}$$

where the **E**... matrices express elementary row operations (and thus are invertible). It follows that **Ac** is zero if, and only if, $\mathbf{A}^{\text{ech}}\mathbf{c}$ is zero. ∎

[2]Some authors say "the row rank equals the column rank." Also the wording "the number of linearly independent columns is the same as the number of independent rows" sometimes appears. The latter is hazardous, however; a statement like "**A** has two independent rows" invites the meaningless question "Which rows are they?".

The effect of elementary row operations on the rows is much simpler, because the rows of a matrix after a row operation are linear combinations of the rows before the operation (since we merely add a multiple of one row to another, multiply a row by a nonzero scalar, or switch two rows). By the linear combination composition principal (Section 3.1), then, the succession of operations that reduce $\mathbf{A}$ to $\mathbf{A}^{ech}$ render the rows of $\mathbf{A}^{ech}$ as linear combinations of the rows of $\mathbf{A}$. Likewise, the rows of $\mathbf{A}$ are linear combinations of the rows of $\mathbf{A}^{ech}$, since the elementary row operations are reversible. The row spaces of $\mathbf{A}$ and $\mathbf{A}^{ech}$ thus coincide and we have

Elementary Row Operations Preserve Row Dependence

Theorem 8. Suppose the matrix $\mathbf{A}$ is reduced to a row echelon form $\mathbf{A}^{ech}$ by the elementary row operations of Gauss elimination. Then the row space of $\mathbf{A}$ equals the row space of $\mathbf{A}^{ech}$. (In particular they have the same dimension.)

Example 4. Find the row rank, column rank, and nullity of the matrix

$$\mathbf{A} = \begin{bmatrix} 2 & 1 & 0 & 6 & 4 \\ 0 & 4 & 0 & 2 & 2 \\ 2 & 1 & 0 & 3 & 1 \end{bmatrix}.$$

Also find bases for its row space, column space, and null space.

Solution. The forward portion of the Gauss elimination algorithm reduces $\mathbf{A}$ to the row echelon form

$$\mathbf{A}^{ech} = \begin{bmatrix} 2 & 1 & 0 & 6 & 4 \\ 0 & 4 & 0 & 2 & 2 \\ 0 & 0 & 0 & -3 & -3 \end{bmatrix},$$

with no row exchanges. All three rows of $\mathbf{A}^{ech}$ are independent, so the row rank and column rank are 3. The original rows of $\mathbf{A}$, then, are independent, and they form a basis for its row space. The first, second, and fourth (or fifth) columns of $\mathbf{A}^{ech}$ are independent, so the corresponding columns of $\mathbf{A}$ form a basis for its column space. The nullity is 5 minus 3, or 2. Back substitution shows the solutions of $\mathbf{Ax} = \mathbf{0}$ to be

$$\mathbf{x} = t_1 \begin{bmatrix} 1 \\ 0 \\ 0 \\ -1 \\ 1 \end{bmatrix} + t_2 \begin{bmatrix} 0 \\ 0 \\ 1 \\ 0 \\ 0 \end{bmatrix},$$

which displays a basis for the null space.   ■

In closing, we wish to make three remarks: two on the practical aspects of the calculations we have discussed, and one of little practical value at all.

Remark (i). The rank of the matrix $\begin{bmatrix} \frac{1}{3} & 1 \\ 1 & 3 \end{bmatrix}$ is obviously 1; the second column equals three times the first. However if the matrix is entered into a 12-digit calculator or computer, its data are truncated to $\begin{bmatrix} 0.333333333333 & 1 \\ 1 & 3 \end{bmatrix}$, and the machine will determine its rank to be 2 (its determinant is -10^{-12}, its columns are linearly independent). And of course the same effect occurs in binary computers. An enlightened matrix software package must issue a warning to the user, when the data renders decisions involving linear independence or rank to be questionable.

Remark (ii). Although from a theoretical point of view all bases for a vector space are equivalent, some are better than others in practice. For instance, the following two sets are each bases for $\mathbf{R}^2_{col}$:

$$\mathbf{v}_1 = \begin{bmatrix} 3 \\ 1 \end{bmatrix} \text{ and } \mathbf{v}_2 = \begin{bmatrix} 1 \\ 3 \end{bmatrix} ; \quad \mathbf{w}_1 = \begin{bmatrix} 0.333333333333 \\ 1 \end{bmatrix} \text{ and } \mathbf{w}_2 = \begin{bmatrix} 1 \\ 3 \end{bmatrix} .$$

However if we try to write the simple vector $\begin{bmatrix} 2 \\ 0 \end{bmatrix}$ as linear combinations using these two bases,

$$\begin{bmatrix} 2 \\ 0 \end{bmatrix} = \frac{3}{4} \begin{bmatrix} 3 \\ 1 \end{bmatrix} - \frac{1}{4} \begin{bmatrix} 1 \\ 3 \end{bmatrix} ,$$

$$\begin{bmatrix} 2 \\ 0 \end{bmatrix} = -6,000,798,970,513.65 \begin{bmatrix} 0.333333333333 \\ 1 \end{bmatrix} + 2,000,266,323,504.55 \begin{bmatrix} 1 \\ 3 \end{bmatrix} ,$$

it becomes clear that in many applications the $\{\mathbf{v}_1, \mathbf{v}_2\}$ basis would be superior.

Remark (iii). In any vector space, the zero vector—alone—qualifies as a subspace. Does it have a basis? What is its dimension? Strictly speaking, there is no linearly independent set of vectors that span $\mathbf{0}$, so it has no basis and no dimension. A diligent review of our theorems shows that we should define its dimension to be zero.

Exercises 3.3

1. Find a basis for the vectors in $\mathbf{R}^6_{row}$ whose first three entries sum to zero and whose last three entries sum to zero.

2. Find a basis for the subspace of $\mathbf{R}^6_{row}$ whose vectors have the form $\begin{bmatrix} a & a & b & b & c & c \end{bmatrix}$.

3. Find a basis for the subspace of $\mathbf{R}^6_{row}$ whose vectors have the form $\begin{bmatrix} a & -a & b & -b & c & -c \end{bmatrix}$.

In Problems 4–12, find bases for the row space, column space, and null space of the given matrix. Some of these matrices were previously reduced to row echelon form in the cited sections of the book. (Don't forget to account for the row exchanges.)

4. $\begin{bmatrix} 1 & 2 & 2 \\ 2 & 1 & 1 \\ 1 & 1 & 3 \end{bmatrix}$ (Section 1.1)   **5.** $\begin{bmatrix} 2 & 1 & 4 \\ 3 & 3 & 0 \\ -1 & 4 & 2 \end{bmatrix}$ (Section 1.2)

6. $\begin{bmatrix} 0.202131 & 0.732543 & 0.141527 & 0.359867 \\ 0.333333 & -0.112987 & 0.412989 & 0.838838 \\ -0.486542 & 0.500000 & 0.989989 & -0.246801 \\ 0.101101 & 0.321111 & -0.444444 & 0.245542 \end{bmatrix}$ (Section 1.1)

7. $\begin{bmatrix} 0 & 2 & 1 & 1 & 0 \\ 2 & 4 & 4 & 2 & 2 \\ 3 & 6 & 6 & 0 & 0 \\ 0 & -2 & -1 & -2 & 2 \\ 0 & 2 & 1 & 2 & 4 \end{bmatrix}$ (Section 1.3)   **8.** $\begin{bmatrix} 1 & 0 & 1 \\ 1 & 1 & 1 \\ 0 & 1 & 1 \\ 3 & 3 & 6 \\ 1 & 2 & 2 \end{bmatrix}$ (Section 1.3)

9. $\begin{bmatrix} 1 & 0 & 1 \\ 1 & 1 & 1 \\ 0 & 1 & 1 \\ 3 & 3 & 6 \end{bmatrix}$ (Section 1.3)

10. $\begin{bmatrix} 1 & 2 & 1 & 1 & -1 & 0 \\ 1 & 2 & 0 & -1 & 1 & 0 \\ 0 & 0 & 1 & 2 & -2 & 1 \end{bmatrix}$ (Section 1.3)

11. $\begin{bmatrix} 2 & 2 & 2 & 2 & 2 \\ 2 & 2 & 4 & 6 & 4 \\ 0 & 0 & 2 & 3 & 1 \\ 1 & 4 & 3 & 2 & 3 \end{bmatrix}$ **12.** $\begin{bmatrix} 1 & 2 & 1 \\ 1 & 3 & 2 \\ 1 & 0 & 1 \end{bmatrix}$

13. Suppose the matrix **B** is formed by deleting some rows and columns from the matrix **A** ("**B** is a submatrix of **A**"). Can the rank of **B** exceed the rank of **A**?

14. Prove that if $\{\mathbf{v}_1, \mathbf{v}_2, \ldots, \mathbf{v}_n\}$ constitutes a basis for $\mathbf{R}^n_{\text{col}}$ and $\det(\mathbf{A}) \neq 0$, then $\{\mathbf{A}\mathbf{v}_1, \mathbf{A}\mathbf{v}_2, \ldots, \mathbf{A}\mathbf{v}_n\}$ also constitutes a basis for $\mathbf{R}^n_{\text{col}}$.

In Problems 15–20 find the dimension of, and a basis for, the indicated subspace of $\mathbf{R}^{5,5}$.

15. The subspace of upper triangular matrices.

16. The subspace of symmetric matrices.

17. The subspace of antisymmetric (or skew-symmetric: $\mathbf{A} = -\mathbf{A}^T$) matrices.

18. The subspace of matrices all of whose row sums are zero.

19. The subspace of matrices all of whose column sums are zero.

20. The subspace of matrices all of whose row sums and column sums are zero.

21. What is the rank of a matrix whose entries are all ones?

22. What is the rank of a "checkerboard" matrix, whose (i,j)th entry is $(-1)^{i+j}$?

23. What is the rank of the rotation matrix of Section 2.2? The orthogonal projection matrix? The mirror reflection matrix?

24. Given k linearly independent vectors in $\mathbf{R}^n_{\text{col}}$ with $k < n$, describe a procedure for finding $(n - k)$ additional vectors that, together with the original k vectors, form a basis for $\mathbf{R}^n_{\text{col}}$. In fact, show that each of the new vectors can be chosen from the canonical basis (only one nonzero entry). (*Hint:* Form the matrix whose first k columns are the given vectors, and append n more columns formed from the canonical basis vectors; then reduce to row echelon form.)

25. Find a basis for $\mathbf{R}^3_{\text{row}}$ containing $\begin{bmatrix} 1 & 2 & 3 \end{bmatrix}$ and two canonical basis vectors.

26. Show that if $\mathbf{v}$ and $\mathbf{w}$ are nonzero vectors in $\mathbf{R}^n_{\text{col}}$, then $\mathbf{A} := \mathbf{v}\mathbf{w}^T$ has rank one. Is the converse true?

27. Find a basis for $\mathbf{R}^4_{\text{row}}$ containing $\begin{bmatrix} 1 & 2 & 3 & 4 \end{bmatrix}$ and $\begin{bmatrix} 0 & 1 & 2 & 2 \end{bmatrix}$ and two canonical basis vectors.

28. If the rank of an m-by-n matrix $\mathbf{A}$ is r, describe how you could factor $\mathbf{A}$ into an m-by-r matrix $\mathbf{B}$ and an r-by-n matrix $\mathbf{C}$: $\mathbf{A} = \mathbf{BC}$. (Hint: Let $\mathbf{B}$ equal a matrix of r independent columns of $\mathbf{A}$. Then each of $\mathbf{A}$'s columns can be expressed as $\mathbf{Bc}$. Assemble $\mathbf{C}$ out of the vectors $\mathbf{c}$.) What is the factored form of a rank 1 matrix?

29. Carry out the factorization in Problem 28 for the matrix $\begin{bmatrix} 1 & 2 & 1 & 1 & -1 & 0 \\ 1 & 2 & 0 & -1 & 1 & 0 \\ 0 & 0 & 1 & 2 & -2 & 0 \end{bmatrix}$.

30. Carry out the factorization in Problem 28 for the matrix $\begin{bmatrix} 1 & 2 & 1 & 1 & -1 & 0 \\ 1 & 2 & 0 & -1 & 1 & 0 \\ 0 & 0 & 1 & 2 & -2 & 1 \end{bmatrix}$.

31. For what values of α (if any) do the vectors $[0\ 1\ \alpha]$, $[\alpha\ 0\ 1]$, and $[\alpha\ 1\ 1+\alpha]$ form a basis for $\mathbf{R}^3_{\text{row}}$?

   For Problems 32–35 it may be helpful to refer to the displays of typical row echelon forms (a)–(f) *in Section* 1.4.

32. What is the relation between the rank of the m-by-n matrix $\mathbf{A}$, the dimension of the null space of $\mathbf{A}^T$, and m?

33. If $\mathbf{A}$ is 23-by-14 and its rank is 10, what are the dimensions of the null spaces of $\mathbf{A}$ and $\mathbf{A}^T$?

34. If the rank of an m-by-n matrix $\mathbf{A}$ is 7 and its null space has dimension 4, what can you say about m and n?

35. Construct a two-dimensional subspace of $\mathbf{R}^3_{\text{row}}$ that contains none of the vectors $\begin{bmatrix} 1 & 0 & 0 \end{bmatrix}$, $\begin{bmatrix} 0 & 1 & 0 \end{bmatrix}$, or $\begin{bmatrix} 0 & 0 & 1 \end{bmatrix}$.

36. If $\mathbf{A}$ is a 7-by-8 matrix and the dimension of its null space is 1, describe its column space.

37. Let $\mathbf{A}$ be an m-by-n matrix and $\mathbf{B}$ be n-by-p. Show that rank$(\mathbf{AB}) \le \min\{\text{rank}(\mathbf{A}), \text{rank}(\mathbf{B})\}$.

38. Let **A** and **B** be matrices with the same number of rows but possibly different numbers of columns. Let **C** be the augmented matrix **C** $= [\mathbf{A}|\mathbf{B}]$. Prove that rank$(\mathbf{C}) \leq$ rank$(\mathbf{A}) +$ rank$(\mathbf{B})$.

39. Do you recognize the patterns of numbers in the following matrices, if they are read left-to-right and top-to-bottom ("sinistrodextrally")?

$$\mathbf{G} = \begin{bmatrix} 2 & 4 \\ 8 & 16 \\ 32 & 64 \end{bmatrix} ; \quad \mathbf{A} = \begin{bmatrix} 1 & 2 & 3 & 4 \\ 5 & 6 & 7 & 8 \\ 9 & 10 & 11 & 12 \end{bmatrix} ;$$

$$\mathbf{F} = \begin{bmatrix} 1 & 1 & 2 & 3 & 5 & 8 \\ 13 & 21 & 34 & 55 & 89 & 144 \\ 233 & 377 & 610 & 987 & 1597 & 2584 \end{bmatrix} ?$$

The data in **G** constitute a *geometric sequence* $a, a^2, a^3, a^4, \ldots$. Those in **A** form an *arithmetic sequence* $a, a+r, a+2r, a+3r, a+4r, \ldots$. The numbers in **F** are drawn from the *Fibonacci sequence*, where each entry is the sum of the two preceding entries. For $m, n > 1$,

(a) what is the rank of the m-by-n matrix constructed from a geometric sequence?

(b) what is the rank of the m-by-n matrix constructed from an arithmetic sequence?

(c) what is the rank of the m-by-n matrix constructed from the Fibonacci sequence?

40. The coordinates of a vector $\mathbf{v}$ in $\mathbf{R}^3_{col}$ are $[2\ 2\ 2]$ with respect to the basis $\{[1\ 1\ 1]^T, [1\ 1\ 0]^T, [1\ 0\ 0]^T\}$. Find its coordinates with respect to the basis $\{[1\ 1\ 0]^T, [0\ 1\ 1]^T, [1\ 0\ 1]^T\}$.

41. The coordinates of a vector $\mathbf{v}$ in $\mathbf{R}^3_{col}$ are $[1\ 2\ -1]$ with respect to the basis $\{[1\ 1\ 0]^T, [1\ 0\ 1]^T, [0\ 1\ 1]^T\}$. Find its coordinates with respect to the basis $\{[1\ 1\ 1]^T, [1\ 1\ 0]^T, [1\ 0\ 0]^T\}$.

42. The deliberations of Example 2, Section 3.2 and the subsequent discussion showed that the bases $\{[1\ 1\ 1\ 0], [2\ 0\ 1\ 0]\}$ and $\{[8\ 4\ 6\ 0], [5\ 1\ 3\ 0]\}$ span the same subspace of $\mathbf{R}^4_{row}$. What are coordinates, in the first basis, of the vector whose coordinates in the second basis are $[2\ 0]$?

43. The deliberations of Example 2, Section 3.2 and the subsequent discussion showed that the bases $\{[1\ 1\ 1\ 0], [2\ 0\ 1\ 0]\}$ and $\{[8\ 4\ 6\ 0], [5\ 1\ 3\ 0]\}$ span the same subspace of $\mathbf{R}^4_{row}$. What are coordinates, in the second basis, of the vector whose coordinates in the first basis are $[2\ 2]$?

44. An alternative formalism for changing coordinates with respect to two bases goes as follows. Let us return to equation (9), where $\{\mathbf{w}_1, \mathbf{w}_2, \ldots, \mathbf{w}_p\}$ and $\{\mathbf{u}_1, \mathbf{u}_2, \ldots, \mathbf{u}_q\}$ are bases for the same subspace of $\mathbf{R}^n_{col}$, and $\mathbf{v}$ has unknown coordinates $\mathbf{c}$ in the first subspace and known coordinates $\mathbf{d}$ in the second.

(a) Rewrite (9) as $\mathbf{Wc} = \mathbf{Ud}$, with the obvious definitions of **W** and **U**, and multiply on the left by $\mathbf{W}^T$ to obtain $\mathbf{W}^T\mathbf{Wc} = \mathbf{W}^T\mathbf{Ud}$. The matrix $\mathbf{W}^T\mathbf{W}$ is the "Gram matrix" for the vectors $\mathbf{w}_j$. Show that the (i, j)th entry of $\mathbf{W}^T\mathbf{W}$ is the dot product of $\mathbf{w}_i$ and $\mathbf{w}_j$, and that the (i,j)th entry of $\mathbf{W}^T\mathbf{U}$ is the dot product of $\mathbf{w}_i$ and $\mathbf{u}_j$.

(b) Show that the Gram matrix of a set of linearly independent vectors is always nonsingular. (*Hint*: if $\mathbf{W}^T\mathbf{W}\mathbf{x} = \mathbf{0}$ then $\mathbf{x}^T\mathbf{W}^T\mathbf{W}\mathbf{x} = (\mathbf{W}\mathbf{x}) \cdot (\mathbf{W}\mathbf{x}) = 0$; consider the implications when $\{\mathbf{w}_1, \mathbf{w}_2, \ldots, \mathbf{w}_p\}$ are independent.)

(c) Derive the rule for changing coordinates $\mathbf{c} = (\mathbf{W}^T\mathbf{W})^{-1}\mathbf{W}^T\mathbf{U}\mathbf{d}$.

3.4 SUMMARY

The structural similarities shared by solution sets of linear algebraic equations, general solutions of linear differential equations, and vectors in two and three dimensions give rise to the concept of an abstract vector space, characterized by the operations of addition of vectors to vectors and multiplication of vectors by scalars. Indeed, vector space theory can be summarized in a nutshell as the study of the varieties and properties of the linear combinations $c_1\mathbf{v}_1 + c_2\mathbf{v}_2 + \cdots + c_n\mathbf{v}_n$, where the c_i's are scalars and the $\mathbf{v}_i$'s are vectors. A subset of a vector space is a subspace if it contains all of its linear combinations. The set of all linear combinations of a given set of vectors is a subspace known as the span of the set.

Linear Independence

A set of distinct vectors is said to be linearly independent if none of the vectors can be expressed as a linear combination of the others; or, equivalently, if the only linear combination that equals zero is the trivial one (with all coefficients zero). Gauss elimination can be used to decide the independence of a set of column vectors. If a matrix is in row echelon form, all of its nonzero rows are independent, as are all of its columns that contain pivot elements. The rows and the columns of a square matrix are independent if and only if its determinant is nonzero.

Bases, Dimension, and Rank

A basis for a (sub)space is a linearly independent set that spans the (sub)space. All bases for a particular (sub)space contain the same number of elements, and this number is the dimension of the (sub)space. The dimension of the span of the columns of any matrix equals the dimension of the span of its rows. The rank of the matrix, which is this common dimension, is preserved by the elementary row operations. The nullity of a matrix $\mathbf{A}$ is the dimension of the subspace of vectors $\mathbf{x}$ such that $\mathbf{A}\mathbf{x} = \mathbf{0}$. The number of columns of $\mathbf{A}$ equals the sum of its rank and its nullity.

4

ORTHOGONALITY

4.1 ORTHOGONAL VECTORS AND THE GRAM–SCHMIDT ALGORITHM

In this section, we shall extend the notions of length and orthogonality to vector spaces of arbitrary dimension.

Recall that the basic computation in forming a matrix product is the *dot product*, identified in Section 2.1 via the formula

$$\mathbf{v} \cdot \mathbf{w} = \begin{bmatrix} v_1 & v_2 & \cdots & v_n \end{bmatrix} \cdot \begin{bmatrix} w_1 & w_2 & \cdots & w_n \end{bmatrix} = v_1 w_1 + v_2 w_2 + \cdots + v_n w_n.$$

The matrix notation for this combination is $\mathbf{v}\mathbf{w}^T$ if the vectors are in $\mathbf{R}_{\text{row}}^n$, and $\mathbf{v}^T\mathbf{w}$ if the vectors are in $\mathbf{R}_{\text{col}}^n$.

From calculus, recall that the dot product has a geometric interpretation. In two or three dimensions, if the angle between $\mathbf{v}$ and $\mathbf{w}$ is denoted by θ (as in Figure 4.1), and the lengths of the vectors are denoted by $\|\mathbf{v}\|$ and $\|\mathbf{w}\|$, then

$$\mathbf{v} \cdot \mathbf{w} = \|\mathbf{v}\|\|\mathbf{w}\| \cos\theta. \tag{1}$$

In particular, $\mathbf{v}$ and $\mathbf{w}$ are perpendicular or *orthogonal* if $\mathbf{v} \cdot \mathbf{w} = 0$. The length of $\mathbf{v}$ (in $\mathbf{R}^3$) is given, via the Pythagorean theorem, by

$$\|\mathbf{v}\| = (\mathbf{v} \cdot \mathbf{v})^{1/2} = (v_1^2 + v_2^2 + v_3^2)^{1/2},$$

Fundamentals of Matrix Analysis with Applications,
First Edition. Edward Barry Saff and Arthur David Snider.
© 2016 John Wiley & Sons, Inc. Published 2016 by John Wiley & Sons, Inc.

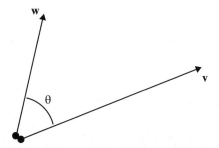

Fig. 4.1   Constituents of the dot product.

and if $\mathbf{v} \neq \mathbf{0}$ the vector $\mathbf{v}/\|\mathbf{v}\|$ is a unit vector (its length is unity). In Example 2 of Section 2.2, we pointed out that the orthogonal projection of $\mathbf{v}$ onto the direction of a unit vector $\mathbf{n}$ is given by the formula $(\mathbf{n} \cdot \mathbf{v})\mathbf{n}$ (see Figure 2.2).

We now extend the notions of length and orthogonality to higher dimensions by using the inner product.

Norm and Orthogonality in $\mathbf{R}^n_{\text{col}}$

Definition 1. The *Euclidean norm* (or simply "norm") of a vector $\mathbf{v} = [v_1 \ v_2 \ \cdots \ v_n]$ in $\mathbf{R}^n_{\text{row}}$ is given by

$$\|\mathbf{v}\| := (\mathbf{v} \cdot \mathbf{v})^{1/2} = (v_1^2 + v_2^2 + \cdots + v_n^2)^{1/2},$$

and similarly for $\mathbf{v}$ in $\mathbf{R}^n_{\text{col}}$.

Two vectors $\mathbf{v}$ and $\mathbf{w}$ are *orthogonal* if $\mathbf{v} \cdot \mathbf{w} = 0$. An *orthogonal set* of vectors is a collection of mutually orthogonal vectors. If, in addition, each vector in an orthogonal set is a unit vector, the set is said to be *orthonormal*.

The *orthogonal projection* of a vector $\mathbf{v}$ onto a unit vector $\mathbf{n}$ in $\mathbf{R}^n_{\text{col}}$ or $\mathbf{R}^n_{\text{row}}$ is the vector $(\mathbf{v} \cdot \mathbf{n})\mathbf{n}$.

Note the following:

- By convention, we have dictated that the zero vector is orthogonal to all other vectors in the space.
- Any nonzero vector can be rescaled into a unit vector by dividing by its norm: $\mathbf{v}^{\text{unit}} := \mathbf{v}/\|\mathbf{v}\|$.
- After $\mathbf{v}$ has been orthogonally projected onto the unit vector $\mathbf{n}$, the residual $\mathbf{v} - (\mathbf{v} \cdot \mathbf{n})\mathbf{n}$ is orthogonal to $\mathbf{n}$:

$$\mathbf{n} \cdot [\mathbf{v} - (\mathbf{v} \cdot \mathbf{n})\mathbf{n}] = \mathbf{n} \cdot \mathbf{v} - (\mathbf{v} \cdot \mathbf{n})\mathbf{n} \cdot \mathbf{n} = \mathbf{n} \cdot \mathbf{v} - (\mathbf{v} \cdot \mathbf{n})(1) = 0.$$

Linear combinations of orthogonal vectors are convenient to deal with because the coefficients are easily "isolated"; to extract, say, c_1 from $c_1\mathbf{v}_1 + c_2\mathbf{v}_2 + \cdots + c_n\mathbf{v}_n$, simply take the dot product with $\mathbf{v}_1$; the result is $c_1\mathbf{v}_1 \cdot \mathbf{v}_1 + c_2\mathbf{v}_1 \cdot \mathbf{v}_2 + \cdots + c_n\mathbf{v}_1 \cdot \mathbf{v}_n = c_1\|\mathbf{v}_1\|^2 + (0)$. The other c_j's are annihilated.

This makes it easy to show that orthogonality of a set of (nonzero) vectors is a stronger property than independence. To see why only the trivial linear combination gives zero,

$$c_1\mathbf{v}_1 + c_2\mathbf{v}_2 + \cdots + c_n\mathbf{v}_n = \mathbf{0}, \tag{2}$$

take the dot product of (2) with each $\mathbf{v}_j$ in turn; the result is $c_j\|\mathbf{v}_j\|^2 = 0$, and since the *vectors* are nonzero, each coefficient c_j must be zero.

It follows that an orthogonal set of nonzero vectors is an *orthogonal basis* for the space it spans. An orthogonal basis is a very valuable computational tool, because the "isolation trick" makes it very easy to express any vector $\mathbf{w}$ as a linear combination of the vectors in the basis.

Example 1. Let

$$\mathbf{v}_1 = \begin{bmatrix} 1 \\ 1 \\ 1 \\ 1 \end{bmatrix}, \quad \mathbf{v}_2 = \begin{bmatrix} 1 \\ 1 \\ -1 \\ -1 \end{bmatrix}, \quad \mathbf{v}_3 = \begin{bmatrix} 1 \\ -1 \\ 1 \\ -1 \end{bmatrix}, \quad \mathbf{v}_4 = \begin{bmatrix} 1 \\ -1 \\ -1 \\ 1 \end{bmatrix}, \text{ and } \mathbf{w} = \begin{bmatrix} 1 \\ 2 \\ 3 \\ 4 \end{bmatrix}.$$

The (nonzero) vectors $\mathbf{v}_1, \mathbf{v}_2, \mathbf{v}_3, \mathbf{v}_4$ are mutually orthogonal (mentally verify this), and since there are four of them they form an orthogonal basis for $\mathbf{R}^4_{\text{col}}$. Express $\mathbf{w}$ as a linear combination of $\mathbf{v}_1, \mathbf{v}_2, \mathbf{v}_3, \mathbf{v}_4$.

Solution. Normally to express one vector in terms of others we would assemble a matrix out of the columns,

$$\mathbf{A} = \begin{bmatrix} 1 & 1 & 1 & 1 \\ 1 & 1 & -1 & -1 \\ 1 & -1 & 1 & -1 \\ 1 & -1 & -1 & 1 \end{bmatrix},$$

and solve $\mathbf{Ac} = \mathbf{w}$ for the coefficients c_j. But look how easy it is to extract the c_j from

$$\mathbf{w} = c_1\mathbf{v}_1 + c_2\mathbf{v}_2 + c_3\mathbf{v}_3 + c_4\mathbf{v}_4$$

when we have orthogonality; premultiplying this equation by $\mathbf{v}_1^T, \mathbf{v}_2^T, \mathbf{v}_3^T, \mathbf{v}_4^T$ in turn we find

$$c_1 = \mathbf{v}_1^T\mathbf{w}/\mathbf{v}_1^T\mathbf{v}_1 = \frac{1+2+3+4}{1+1+1+1} = 2.5, \quad c_2 = \mathbf{v}_2^T\mathbf{w}/\mathbf{v}_2^T\mathbf{v}_2 = \frac{1+2-3-4}{1+1+1+1} = -1,$$

$$c_3 = \mathbf{v}_3^T\mathbf{w}/\mathbf{v}_3^T\mathbf{v}_3 = \frac{1-2+3-4}{1+1+1+1} = -0.5, \quad c_4 = \mathbf{v}_4^T\mathbf{w}/\mathbf{v}_4^T\mathbf{v}_4 = \frac{1-2-3+4}{1+1+1+1} = 0. \tag{3}$$

Thus,

$$\mathbf{w} = 2.5\mathbf{v}_1 - \mathbf{v}_2 - 0.5\mathbf{v}_3.$$ ∎

This task is made even easier if we first rescale $\{\mathbf{v}_1, \mathbf{v}_2, \mathbf{v}_3, \mathbf{v}_4\}$ to be unit vectors.

Example 2. Express $\mathbf{w}$ in Example 1 as a linear combination of the orthonormal set

$$\mathbf{v}_1^{\text{unit}} = \frac{1}{\sqrt{4}}\begin{bmatrix} 1 \\ 1 \\ 1 \\ 1 \end{bmatrix} = \begin{bmatrix} 0.5 \\ 0.5 \\ 0.5 \\ 0.5 \end{bmatrix}, \quad \mathbf{v}_2^{\text{unit}} = \begin{bmatrix} 0.5 \\ 0.5 \\ -0.5 \\ -0.5 \end{bmatrix},$$

$$\mathbf{v}_3^{\text{unit}} = \begin{bmatrix} 0.5 \\ -0.5 \\ 0.5 \\ -0.5 \end{bmatrix}, \quad \mathbf{v}_4^{\text{unit}} = \begin{bmatrix} 0.5 \\ -0.5 \\ -0.5 \\ 0.5 \end{bmatrix}.$$

Solution. Isolating the coefficients in

$$\mathbf{w} = c_1\mathbf{v}_1^{\text{unit}} + c_2\mathbf{v}_2^{\text{unit}} + c_3\mathbf{v}_3^{\text{unit}} + c_4\mathbf{v}_4^{\text{unit}} \tag{4}$$

we obtain (note the denominators that appeared in (3) now become ones)

$$c_1 = \left(\mathbf{v}_1^{\text{unit}}\right)^T \mathbf{w}/(1) = 5, \ c_2 = \left(\mathbf{v}_2^{\text{unit}}\right)^T \mathbf{w}/(1) = -2, \tag{5}$$
$$c_3 = \left(\mathbf{v}_3^{\text{unit}}\right)^T \mathbf{w}/(1) = -1, \ c_4 = \left(\mathbf{v}_4^{\text{unit}}\right)^T \mathbf{w}/(1) = 0.$$

Hence $\mathbf{w} = 5\mathbf{v}_1^{\text{unit}} - 2\mathbf{v}_2^{\text{unit}} - \mathbf{v}_3^{\text{unit}}$. ∎

For future reference, we note the generalizations of formulas (3)–(5):

Orthogonal (Orthonormal) Basis Expansions

Theorem 1. Let $\{\mathbf{v}_1, \mathbf{v}_2, \ldots, \mathbf{v}_m\}$ be an orthogonal basis for the vector space V, and let $\{\mathbf{v}_1^{\text{unit}}, \mathbf{v}_2^{\text{unit}}, \ldots, \mathbf{v}_m^{\text{unit}}\}$ be an orthonormal basis. Then any vector $\mathbf{w}$ in V can be expressed as

$$\mathbf{w} = \frac{\mathbf{w} \cdot \mathbf{v}_1}{\mathbf{v}_1 \cdot \mathbf{v}_1}\mathbf{v}_1 + \frac{\mathbf{w} \cdot \mathbf{v}_2}{\mathbf{v}_2 \cdot \mathbf{v}_2}\mathbf{v}_2 + \cdots + \frac{\mathbf{w} \cdot \mathbf{v}_m}{\mathbf{v}_m \cdot \mathbf{v}_m}\mathbf{v}_m \tag{6}$$

or

$$\mathbf{w} = \left(\mathbf{w} \cdot \mathbf{v}_1^{\text{unit}}\right)\mathbf{v}_1^{\text{unit}} + \left(\mathbf{w} \cdot \mathbf{v}_2^{\text{unit}}\right)\mathbf{v}_2^{\text{unit}} + \cdots + \left(\mathbf{w} \cdot \mathbf{v}_m^{\text{unit}}\right)\mathbf{v}_m^{\text{unit}}. \tag{7}$$

Moreover,

$$\|\mathbf{w}\|^2 = \left(\mathbf{w} \cdot \mathbf{v}_1^{\text{unit}}\right)^2 + \left(\mathbf{w} \cdot \mathbf{v}_2^{\text{unit}}\right)^2 + \cdots + \left(\mathbf{w} \cdot \mathbf{v}_m^{\text{unit}}\right)^2. \tag{8}$$

(Identity (8) follows easily by evaluating $\mathbf{w} \cdot \mathbf{w}$ using (7).)

Stimulated by these examples, we are led to ask how do you find an orthogonal basis? One answer is given by the *Gram–Schmidt algorithm.*

Gram–Schmidt Algorithm

To replace a collection of (nonzero) vectors $\{\mathbf{v}_1, \mathbf{v}_2, \mathbf{v}_3, \ldots, \mathbf{v}_m\}$ by an orthogonal set with the same span, first set

$$\mathbf{w}_1 := \mathbf{v}_1. \tag{9}$$

Then subtract off, from $\mathbf{v}_2$, its orthogonal projection onto $\mathbf{w}_1^{\text{unit}}$:

$$\mathbf{w}_2 := \mathbf{v}_2 - \left(\mathbf{v}_2 \cdot \mathbf{w}_1^{\text{unit}}\right)\mathbf{w}_1^{\text{unit}} = \mathbf{v}_2 - \left(\mathbf{v}_2 \cdot \mathbf{w}_1\right)\mathbf{w}_1/\left(\mathbf{w}_1 \cdot \mathbf{w}_1\right). \tag{10}$$

(As previously noted, the residual $\mathbf{w}_2$ of the projection of $\mathbf{v}_2$ onto $\mathbf{w}_1^{\text{unit}}$ is orthogonal to $\mathbf{w}_1^{\text{unit}}$ and hence to $\mathbf{w}_1$; See Figure 4.2(a).) Continue to subtract off in succession, from each $\mathbf{v}_j$, its orthogonal projection onto the previous *nonzero* $\mathbf{w}_k^{\text{unit}}$'s:

$$\mathbf{w}_3 := \mathbf{v}_3 - \left(\mathbf{v}_3 \cdot \mathbf{w}_1^{\text{unit}}\right)\mathbf{w}_1^{\text{unit}} - \left(\mathbf{v}_3 \cdot \mathbf{w}_2^{\text{unit}}\right)\mathbf{w}_2^{\text{unit}} \;[\text{Figure 4.2(b)}]$$

$$\vdots \tag{11}$$

$$\mathbf{w}_m := \mathbf{v}_m - \left(\mathbf{v}_m \cdot \mathbf{w}_1^{\text{unit}}\right)\mathbf{w}_1^{\text{unit}} - \left(\mathbf{v}_m \cdot \mathbf{w}_2^{\text{unit}}\right)\mathbf{w}_2^{\text{unit}}$$

$$- \left(\mathbf{v}_m \cdot \mathbf{w}_3^{\text{unit}}\right)\mathbf{w}_3^{\text{unit}} - \cdots - \left(\mathbf{v}_m \cdot \mathbf{w}_{m-1}^{\text{unit}}\right)\mathbf{w}_{m-1}^{\text{unit}}.$$

We ignore any vectors $\mathbf{w}_k$ that turn out to be zero. The set $\{\mathbf{w}_1, \mathbf{w}_2, \mathbf{w}_3, \ldots, \mathbf{w}_m\}$, with the zero vectors deleted, will then be an orthogonal basis spanning the same subspace as $\{\mathbf{v}_1, \mathbf{v}_2, \mathbf{v}_3, \ldots, \mathbf{v}_m\}$.

To verify the claim in the algorithm, note the following:

- each $\mathbf{w}_k$, $k \geqslant 2$, is orthogonal to $\mathbf{w}_1, \mathbf{w}_2, \ldots, \mathbf{w}_{k-1}$, and so $\{\mathbf{w}_1, \mathbf{w}_2, \ldots, \mathbf{w}_m\}$ are all mutually orthogonal;

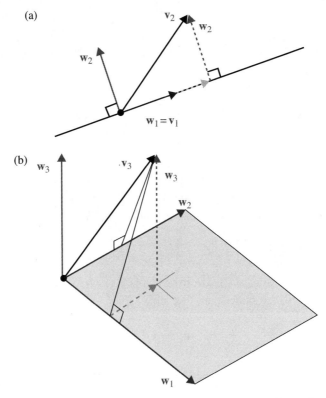

Fig. 4.2 Gram–Schmidt algorithm. (a) Two dimensions. (b) Three dimensions.

- each $\mathbf{w}_k$ is a linear combination of the $\mathbf{v}_j$'s, so the span of the $\mathbf{w}_k$'s is contained in the span of the $\mathbf{v}_k$'s;
- if a particular $\mathbf{w}_k$ turns out to be zero, the corresponding $\mathbf{v}_k$ is linearly dependent on the preceding $\mathbf{v}_j$'s, so the maximal number of independent $\mathbf{v}_j$'s equals the number of nonzero $\mathbf{w}_j$'s.

Thus the number of nonzero $\mathbf{w}_j$'s equals the dimension of $\mathrm{Span}\{\mathbf{v}_1, \mathbf{v}_2, \ldots, \mathbf{v}_m\}$, and the nonzero $\mathbf{w}_j$'s are independent. Consequently, they are a basis for $\mathrm{Span}\{\mathbf{v}_1, \mathbf{v}_2, \ldots, \mathbf{v}_m\}$.

Example 3. Use the Gram–Schmidt algorithm to find an orthonomal basis for the span of the following vectors:

$$\mathbf{v}_1 = \begin{bmatrix} 2 \\ 2 \\ 1 \\ 0 \end{bmatrix}, \quad \mathbf{v}_2 = \begin{bmatrix} 6 \\ -2 \\ 1 \\ -2 \end{bmatrix}, \quad \mathbf{v}_3 = \begin{bmatrix} 2 \\ -2 \\ 0 \\ 1 \end{bmatrix}, \quad \mathbf{v}_4 = \begin{bmatrix} 0 \\ 0 \\ 0 \\ 1 \end{bmatrix}.$$

Solution. We take $\mathbf{w}_1 = \mathbf{v}_1$ as in (9). Since $\mathbf{w}_1 \cdot \mathbf{w}_1 = \|\mathbf{w}_1\|^2 = 9$, we get from (10),

$$\mathbf{w}_1^{\text{unit}} = \frac{1}{\sqrt{9}}\mathbf{w}_1 = \begin{bmatrix} 2/3 \\ 2/3 \\ 1/3 \\ 0 \end{bmatrix},$$

$$\mathbf{w}_2 = \begin{bmatrix} 6 \\ -2 \\ 1 \\ 2 \end{bmatrix} - \left\{ \begin{bmatrix} 6 & -2 & 1 & 2 \end{bmatrix} \begin{bmatrix} 2/3 \\ 2/3 \\ 1/3 \\ 0 \end{bmatrix} \right\} \begin{bmatrix} 2/3 \\ 2/3 \\ 1/3 \\ 0 \end{bmatrix}$$

$$= \begin{bmatrix} 6 \\ -2 \\ 1 \\ 2 \end{bmatrix} - \{3\} \begin{bmatrix} 2/3 \\ 2/3 \\ 1/3 \\ 0 \end{bmatrix} = \begin{bmatrix} 4 \\ -4 \\ 0 \\ 2 \end{bmatrix},$$

$$\mathbf{w}_2^{\text{unit}} = \frac{1}{\sqrt{36}} \begin{bmatrix} 4 \\ -4 \\ 0 \\ 2 \end{bmatrix} = \begin{bmatrix} 2/3 \\ -2/3 \\ 0 \\ 1/3 \end{bmatrix}.$$

Similarly,

$$\mathbf{w}_3 = \begin{bmatrix} 2 \\ -2 \\ 0 \\ 1 \end{bmatrix} - \begin{bmatrix} 2 & -2 & 0 & 1 \end{bmatrix} \begin{bmatrix} 2/3 \\ 2/3 \\ 1/3 \\ 0 \end{bmatrix} \begin{bmatrix} 2/3 \\ 2/3 \\ 1/3 \\ 0 \end{bmatrix} - \begin{bmatrix} 2 & -2 & 0 & 1 \end{bmatrix} \begin{bmatrix} 2/3 \\ -2/3 \\ 0 \\ 1/3 \end{bmatrix} \begin{bmatrix} 2/3 \\ -2/3 \\ 0 \\ 1/3 \end{bmatrix}$$

$$= \begin{bmatrix} 0 \\ 0 \\ 0 \\ 0 \end{bmatrix}$$

(implying that $\mathbf{v}_3$ is *not* linearly independent of $\mathbf{v}_1$ and $\mathbf{v}_2$);

$$\mathbf{w}_4 = \cdots = \begin{bmatrix} -2/9 \\ 2/9 \\ 0 \\ 8/9 \end{bmatrix}, \quad \mathbf{w}_4^{\text{unit}} = \begin{bmatrix} -\sqrt{2}/6 \\ \sqrt{2}/6 \\ 0 \\ 2\sqrt{2}/3 \end{bmatrix}.$$

The vectors $\mathbf{w}_1^{\text{unit}}, \mathbf{w}_2^{\text{unit}}$, and $\mathbf{w}_4^{\text{unit}}$ form an orthonormal basis for Span$\{\mathbf{v}_1, \mathbf{v}_2, \mathbf{v}_3, \mathbf{v}_4\}$. ∎

In Section 4.2, we will see how the introduction of a new type of matrix facilitates the incorporation of the concepts of the present section into our matrix language.

Exercises 4.1

1. Verify that the vectors $\{[3/5 \quad 4/5], [-4/5 \quad 3/5]\}$ form an orthonormal set and express $[5 \quad 10]$ as a linear combination of them.

2. Verify that the vectors $\{[1/\sqrt{2} \quad 1/\sqrt{2}], [1/\sqrt{2} \quad -1/\sqrt{2}]\}$ form an orthonormal set and express $[2 \quad 4]$ as a linear combination of them.

3. Verify that the vectors

$$\left\{ [1/3 \quad 2/3 \quad 2/3], [-2/3 \quad -1/3 \quad 2/3], [-2/3 \quad 2/3 \quad -1/3] \right\}$$

form an orthonormal set and express $[3 \quad 6 \quad 9]$ as a linear combination of them.

4. Verify that the vectors

$$\left\{ [1/2 \quad 1/2 \quad 1/2 \quad 1/2], [1/2 \quad -1/2 \quad 1/2 \quad -1/2] \right.$$
$$\left. [-1/2 \quad -1/2 \quad 1/2 \quad 1/2], [1/2 \quad -1/2 \quad -1/2 \quad 1/2] \right\}$$

form an orthonormal set and express $[1 \quad 2 \quad 3 \quad 4]$ as a linear combination of them.

In Problems 5–10, use the Gram–Schmidt algorithm to construct a set of orthonormal vectors spanning the same space as the given set.

5. $[-8 \quad 6]$ and $[-1 \quad 7]$.

6. $[1 \quad 2 \quad 2]$ and $[2 \quad 3 \quad 4]$.

7. $[1 \quad -1 \quad 1 \quad -1]$ and $[3 \quad -1 \quad 3 \quad -1]$.

8. $[1 \quad 1 \quad 1 \quad 1], [1 \quad 0 \quad 1 \quad 0],$ and $[4 \quad 0 \quad 2 \quad -2]$.

9. $[2 \quad 2 \quad -2 \quad -2], [3 \quad 1 \quad -1 \quad -3],$ and $[2 \quad 0 \quad -2 \quad -4]$.

10. $[-1 \quad 1 \quad -1 \quad 1], [2 \quad 0 \quad 2 \quad 0],$ and $[3 \quad 1 \quad 1 \quad -1]$.

11. If you just purchased an **iGram**—a device that rapidly performs the Gram–Schmidt algorithm on any set of vector inputs—what inputs would you send it to solve the following?

    (a) Find a vector orthogonal to [5 7].

    (b) Find two vectors orthogonal to [1 2 3] and to each other.

In Problems 12–15 append the canonical basis to the given family and apply the Gram–Schmidt algorithm to find the requested vectors.

12. Find a nonzero vector orthogonal to the family in Problem 6.

13. Find a nonzero vector orthogonal to the family in Problem 8.

14. Find a nonzero vector orthogonal to the family in Problem 9.

15. Find two nonzero vectors orthogonal to the family in Problem 7 and to each other.

16. Find an orthonormal basis for the subspace of $\mathbf{R}^3_{\text{row}}$ formed by vectors $[x\ y\ z]$ in the plane $3x - y + 2z = 0$.

17. Find an orthonormal basis for the subspace of $\mathbf{R}^4_{\text{row}}$ formed by vectors $[x\ y\ z\ w]$ in the hyperplane $2x - y + z + 3w = 0$.

18. Find an orthonormal basis for the null space in $\mathbf{R}^3_{\text{col}}$ of the matrix

$$\mathbf{A} = \begin{bmatrix} 1 & -3 & 0 \\ -2 & 6 & 0 \\ 1 & 3 & 0 \end{bmatrix}.$$

19. Find an orthonormal basis for the null space in $\mathbf{R}^4_{\text{col}}$ of the matrix

$$\mathbf{A} = \begin{bmatrix} 2 & 3 & 1 & 0 \\ 1 & 1 & 1 & 1 \end{bmatrix}.$$

20. The polar coordinate description of a two-dimensional vector $\mathbf{v}$ is given by $\mathbf{v} = \begin{bmatrix} x \\ y \end{bmatrix} = \begin{bmatrix} r\cos\theta \\ r\sin\theta \end{bmatrix}$. Use the characterization $\mathbf{v}_1 \cdot \mathbf{v}_2 = 0$ to demonstrate that $\mathbf{v}_1$ and $\mathbf{v}_2$ are orthogonal if their polar angles θ_1 and θ_2 differ by $90°$.

21. The spherical coordinate description of a three-dimensional vector $\mathbf{v}$ is given by

$$\mathbf{v} = \begin{bmatrix} x \\ y \\ z \end{bmatrix} = \begin{bmatrix} r\sin\phi\cos\theta \\ r\sin\phi\sin\theta \\ r\cos\phi \end{bmatrix}.$$

Using the characterization $\mathbf{v}_1 \cdot \mathbf{v}_2 = 0$, derive conditions on the spherical angles $\theta_1, \phi_1, \theta_2, \phi_2$ guaranteeing that $\mathbf{v}_1$ and $\mathbf{v}_2$ are orthogonal.

22. The calculus identity $\mathbf{v}_1 \cdot \mathbf{v}_2 = \|\mathbf{v}\| \|\mathbf{w}\| \cos\theta$ implies that $|\mathbf{v}_1 \cdot \mathbf{v}_2| \leq \|\mathbf{v}\| \|\mathbf{w}\|$ in 2 or 3 dimensions. The *Cauchy–Schwarz inequality* extends this to higher dimensions:

$$|\mathbf{v}^T \mathbf{w}| \leq \|\mathbf{v}\| \|\mathbf{w}\| \text{ for all vectors in } \mathbf{R}^n_{\text{col}}.$$

Prove the inequality by completing the following steps:
(a) Express $(\mathbf{v} + x\mathbf{w})^T (\mathbf{v} + x\mathbf{w})$ as a quadratic polynomial in x.
(b) Use the quadratic formula to express the roots of this polynomial.
(c) What does the fact $(\mathbf{v} + x\mathbf{w})^T (\mathbf{v} + x\mathbf{w}) \geq 0$ tell you about the roots?
(d) Use these observations to prove the inequality.

23. Prove that equality holds in the Cauchy–Schwarz inequality (Problem 22) if and only if $\mathbf{v}$ and $\mathbf{w}$ are linearly dependent.

24. Use the Cauchy–Schwarz inequality (Problem 22) to give a proof of the *triangle inequality* $\|\mathbf{v}+\mathbf{w}\| \leq \|\mathbf{v}\|+\|\mathbf{w}\|$ for any $\mathbf{v}, \mathbf{w}$ in $\mathbf{R}^n_{\text{col}}$. (*Hint*: Consider $(\mathbf{v}+\mathbf{w}) \cdot (\mathbf{v}+\mathbf{w})$.)

25. Prove the parallelogram law: $\|\mathbf{v} + \mathbf{w}\|^2 + \|\mathbf{v} - \mathbf{w}\|^2 = 2\|\mathbf{v}\|^2 + 2\|\mathbf{w}\|$. (Draw a two-dimensional sketch to see why this identity is so named.)

4.2   ORTHOGONAL MATRICES

The following theorem motivates the definition of a new, extremely useful, class of matrices.

Characterization of Norm-Preserving Matrices

> **Theorem 2.** If $\mathbf{Q}$ is a square matrix with real number entries, the following properties are all equivalent:
>
> (i) The inverse of $\mathbf{Q}$ is its transpose (i.e., $\mathbf{QQ}^T = \mathbf{Q}^T\mathbf{Q} = \mathbf{I}$);
>
> (ii) The rows of $\mathbf{Q}$ are orthonormal;
>
> (iii) The columns of $\mathbf{Q}$ are orthonormal;
>
> (iv) $\mathbf{Q}$ preserves inner products: $\mathbf{v}^T\mathbf{w} = (\mathbf{Qv})^T(\mathbf{Qw})$ for any $\mathbf{v}, \mathbf{w}$;
>
> (v) $\mathbf{Q}$ preserves Euclidean norms: $\|\mathbf{v}\| = \|\mathbf{Qv}\|$ for any $\mathbf{v}$.

Orthogonal Matrices

> **Definition 2.**   A real, square matrix satisfying any of the properties in Theorem 2 is said to be an **orthogonal** matrix.

From Example 2 of the preceding section, we conclude that the matrix

$$\begin{bmatrix} 0.5 & 0.5 & 0.5 & 0.5 \\ 0.5 & 0.5 & -0.5 & -0.5 \\ 0.5 & -0.5 & 0.5 & -0.5 \\ 0.5 & -0.5 & -0.5 & 0.5 \end{bmatrix}$$

is orthogonal. Because of (ii) and (iii), a better name for this class would be "orthonormal matrices," but tradition dictates the stated nomenclature.

Proof of Theorem 2.   If we depict $\mathbf{Q}$ by its rows, then the expression of the statement $\mathbf{QQ}^T = \mathbf{I}$ looks like

$$\mathbf{QQ}^T = \begin{bmatrix} -\text{row } 1- \\ -\text{row } 2- \\ \vdots \\ -\text{row } n- \end{bmatrix} \begin{bmatrix} | & | & & | \\ [\text{row } 1]^T & [\text{row } 2]^T & \cdots & [\text{row } n]^T \\ | & | & & | \end{bmatrix}$$

$$= \mathbf{I} = \begin{bmatrix} 1 & 0 & \cdots & 0 \\ 0 & 1 & \cdots & 0 \\ \vdots & \vdots & \ddots & \vdots \\ 0 & 0 & \cdots & 1 \end{bmatrix},$$

Thus the condition $\mathbf{QQ}^T = \mathbf{I}$ says no more, and no less, than that the rows of $\mathbf{Q}$ are orthonormal. Similarly, $\mathbf{Q}^T\mathbf{Q} = \mathbf{I}$ states that the columns of $\mathbf{Q}$ are orthonormal; thus (i), (ii). and (iii) are all equivalent.

Statements (iv) and (v) are equivalent, because the norm and the inner product can be expressed in terms of each other:

$$\|\mathbf{v}\| = (\mathbf{v} \cdot \mathbf{v})^{1/2} \text{ and } 2\mathbf{v} \cdot \mathbf{w} = \|\mathbf{v} + \mathbf{w}\|^2 - \|\mathbf{v}\|^2 - \|\mathbf{w}\|^2$$

(the latter is easily verified by writing the norms as inner products). So our proof will be complete if we can show, for instance, that (i) implies (iv) and (iv) implies (iii).

The right-hand member in (iv) equals $\mathbf{v}^T\mathbf{Q}^T\mathbf{Qw}$, so (iv) holds whenever $\mathbf{Q}^T\mathbf{Q} = \mathbf{I}$; that is, (i) implies (iv). And if $\mathbf{e}_i$ denotes the ith column of the identity matrix, then $\mathbf{Qe}_i$ is the ith column of $\mathbf{Q}$; so applying (iv) when $\mathbf{v} = \mathbf{e}_i$ and $\mathbf{w} = \mathbf{e}_j$, we conclude that the columns of $\mathbf{Q}$ have the same orthonormality properties as the columns of $\mathbf{I}$ when (iv) holds. Thus (iv) implies (iii). ∎

Note that Theorem 2 implies that inverses, transposes, and products of orthogonal matrices are, again, orthogonal.

Property (v) is the key to many of the applications of orthogonal matrices. For example, since rotations and mirror reflections preserve lengths, we can immediately state that the corresponding matrix operators are orthogonal. In fact, it can be shown that in two or three dimensions the most general orthogonal matrix can be considered as a sequence of rotations and reflections (See Group Project $\mathbf{A}$ at the end of Part II).

Example 1. Verify that the transpose of the reflection operator in Example 3 of Section 2.2 is equal to its inverse.

Solution. If $\mathbf{n}$ denotes the column matrix representing the unit normal to the mirror, formula (3) in Section 2.2 for the reflector operator states

$$\mathbf{M}_{\text{ref}} = \mathbf{I} - 2\mathbf{nn}^T. \tag{1}$$

Then we can verify

$$\mathbf{M}_{\text{ref}}^T\mathbf{M}_{\text{ref}} = \{\mathbf{I} - 2\mathbf{nn}^T\}^T\{\mathbf{I} - 2\mathbf{nn}^T\} = \{\mathbf{I}^T - 2\mathbf{n}^{TT}\mathbf{n}^T\}\{\mathbf{I} - 2\mathbf{nn}^T\} \tag{2}$$

$$= \{\mathbf{I} - 2\mathbf{nn}^T\}\{\mathbf{I} - 2\mathbf{nn}^T\} = \mathbf{I} - 2\mathbf{nn}^T - 2\mathbf{nn}^T + 4\mathbf{nn}^T\mathbf{nn}^T$$

$$= \mathbf{I} - 2\mathbf{nn}^T - 2\mathbf{nn}^T + 4\mathbf{n}(1)\mathbf{n}^T = \mathbf{I} - 4\mathbf{nn}^T + 4\mathbf{nn}^T = \mathbf{I}. \quad ∎$$

Note that although the formula (1) was derived for reflections in two dimensions, the calculation is independent of the number of dimensions. Accordingly, we say that

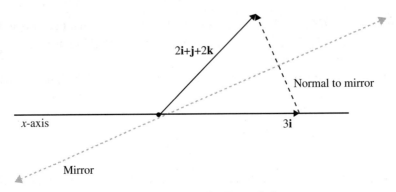

Fig. 4.3   Mirror for Example 2.

matrices of the form (1) generate reflections in $\mathbf{R}^n_{\text{col}}$ whenever $\mathbf{n}$ is a n-by-1 unit vector. (And such matrices continue to satisfy $\mathbf{M}^T_{\text{ref}} = \mathbf{M}^{-1}_{\text{ref}} = \mathbf{M}_{\text{ref}}$.)

Example 2. Construct a reflector matrix that carries the three-dimensional vector $2\mathbf{i} + \mathbf{j} + 2\mathbf{k}$ into a vector along the x-axis.

Solution. We will use the formula (1), once we figure out the unit normal to the mirror. Since reflections preserve length (norm), the final vector must be (plus or minus) $\sqrt{2^2 + 1^2 + 2^2}\mathbf{i} = \pm 3\mathbf{i}$. The normal to the mirror must then be parallel to the vector connecting the tips of (say) $+3\mathbf{i}$ and $(2\mathbf{i}+\mathbf{j}+2\mathbf{k})$; that is, parallel to their difference (see Figure 4.3). Therefore,

$$\mathbf{n}^{\text{unit}} = \begin{bmatrix} 3-2 \\ -1 \\ -2 \end{bmatrix} \frac{1}{\sqrt{(3-2)^2 + (-1)^2 + (-2)^2}} = \frac{1}{\sqrt{6}} \begin{bmatrix} 1 \\ -1 \\ -2 \end{bmatrix},$$

and the reflector matrix is

$$\mathbf{M}_{\text{ref}} = \begin{bmatrix} 1 & 0 & 0 \\ 0 & 1 & 0 \\ 0 & 0 & 1 \end{bmatrix} - 2\frac{1}{\sqrt{6}} \begin{bmatrix} 1 \\ -1 \\ -2 \end{bmatrix} \frac{1}{\sqrt{6}} \begin{bmatrix} 1 & -1 & -2 \end{bmatrix} = \begin{bmatrix} 2/3 & 1/3 & 2/3 \\ 1/3 & 2/3 & -2/3 \\ 2/3 & -2/3 & -1/3 \end{bmatrix}.$$

■

Matrices constructed as in Example 2 are known as *Householder reflectors*. Clearly, they can be designed to zero out the lower entries in the first column of any matrix, as depicted:

$$\mathbf{M}_{\text{ref}} \begin{bmatrix} \text{col } \#1 & \text{col } \#2 & \cdots & \text{col } \#n \\ \downarrow & \downarrow & & \downarrow \end{bmatrix} = \begin{bmatrix} \|\text{col } \#1\| & X & \cdots & X \\ 0 & X & \cdots & X \\ 0 & X & \cdots & X \\ \vdots & \vdots & \ddots & \vdots \\ 0 & X & \cdots & X \end{bmatrix}. \tag{3}$$

A little thought (Group Project B at the end of Part II) will reveal that a sequence of well-chosen reflectors can render the original matrix into *row echelon form*, producing a competitor to the Gauss elimination algorithm! (And pivoting is unnecessary.) In fact, Householder reflector schemes can also replace the Gram–Schmidt algorithm.

The use of orthogonal matrices in matrix computations has very important advantages with regard to computational accuracy, because property (v) in Theorem 2 implies that multiplication by orthogonal matrices does not increase errors. For example, if we were to solve a system $\mathbf{Ax} = \mathbf{b}$ using exact arithmetic (so that $\mathbf{x} = \mathbf{A}^{-1}\mathbf{b}$), but we later learned that the vector $\mathbf{b}$ contained a measurement error $\delta\mathbf{b}$, then our answer would be in error by $\delta\mathbf{x} = \mathbf{A}^{-1}(\mathbf{b} + \delta\mathbf{b}) - \mathbf{A}^{-1}\mathbf{x} = \mathbf{A}^{-1}\delta\mathbf{b}$. If $\mathbf{A}^{-1}$ is orthogonal, we have $\|\delta\mathbf{x}\| = \|\mathbf{A}^{-1}\delta\mathbf{b}\| = \|\delta\mathbf{b}\|$, showing that the error norm in the solution is no bigger than the error norm in the data.

Most enlightened matrix software packages contain options for some computations that use *only* orthogonal matrix multiplications. They usually require more arithmetic operations, but the increased "numerical stability" may warrant the effort for a particular application.[1]

Example 3. Calculate the matrix expressing the result of rotating a three-dimensional vector through an angle θ around the z-axis, followed by a rotation through ϕ around the x-axis.

Solution. In Example 1 of Section 2.2, we pointed out that the (x, y) coordinates of a vector $\mathbf{v}_1$ are changed when it is rotated to the position $\mathbf{v}_2$ as in Figure 4.4a, according to

$$x_2 = x_1 \cos\theta - y_1 \sin\theta \tag{4}$$
$$y_2 = x_1 \sin\theta + y_1 \cos\theta.$$

Since the z-coordinate is unchanged, as a three-dimensional rotation (4) has the matrix formulation

$$\begin{bmatrix} x_2 \\ y_2 \\ z_2 \end{bmatrix} = \begin{bmatrix} \cos\theta & -\sin\theta & 0 \\ \sin\theta & \cos\theta & 0 \\ 0 & 0 & 1 \end{bmatrix} \begin{bmatrix} x_1 \\ y_1 \\ z_1 \end{bmatrix}. \tag{5}$$

To modify these equations for rotations around the x-axis, we redraw the diagram with the x-axis coming out of the page, as in Figure 4.4b. We conclude $x_3 = x_2$ and simply copy (4) with the variables appropriately renamed:

$$y_3 = y_2 \cos\phi - z_2 \sin\phi, \ z_3 = y_2 \sin\phi + z_2 \cos\phi;$$

---

[1]In 1977, one of the authors was directed to program a reflector matrix algorithm to replace Gauss elimination for onboard guidance control calculations in the space shuttle, to ensure reliability.

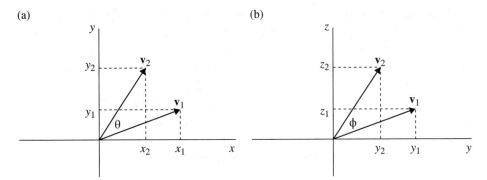

Fig. 4.4 Rotation sequence.

that is,

$$\begin{bmatrix} x_3 \\ y_3 \\ z_3 \end{bmatrix} = \begin{bmatrix} 1 & 0 & 0 \\ 0 & \cos\phi & -\sin\phi \\ 0 & \sin\phi & \cos\phi \end{bmatrix} \begin{bmatrix} x_2 \\ y_2 \\ z_2 \end{bmatrix}. \tag{6}$$

Concatenating the two operations, we obtain

$$\begin{aligned} \begin{bmatrix} x_3 \\ y_3 \\ z_3 \end{bmatrix} &= \begin{bmatrix} 1 & 0 & 0 \\ 0 & \cos\phi & -\sin\phi \\ 0 & \sin\phi & \cos\phi \end{bmatrix} \begin{bmatrix} x_2 \\ y_2 \\ z_2 \end{bmatrix} \\ &= \begin{bmatrix} 1 & 0 & 0 \\ 0 & \cos\phi & -\sin\phi \\ 0 & \sin\phi & \cos\phi \end{bmatrix} \begin{bmatrix} \cos\theta & -\sin\theta & 0 \\ \sin\theta & \cos\theta & 0 \\ 0 & 0 & 1 \end{bmatrix} \begin{bmatrix} x_1 \\ y_1 \\ z_1 \end{bmatrix} \\ &= \begin{bmatrix} \cos\theta & -\sin\theta & 0 \\ \cos\phi\sin\theta & \cos\phi\cos\theta & -\sin\phi \\ \sin\phi\sin\theta & \sin\phi\cos\theta & \cos\phi \end{bmatrix} \begin{bmatrix} x_1 \\ y_1 \\ z_1 \end{bmatrix}. \end{aligned} \tag{7}$$

(Of course the product of two orthogonal matrices is, itself, orthogonal. Indeed, mental calculation confirms the orthonormality of the columns above.) ■

Exercises 4.2

1. What values can the determinant of an orthogonal matrix take?

2. For the matrix M_{ref} in Example 2, verify each of the properties (i) through (v) in Theorem 2.

3. Find all 2-by-2 orthogonal matrices of the form

(a) $\begin{bmatrix} x & y \\ y & x \end{bmatrix}$ (b) $\begin{bmatrix} x & -y \\ y & x \end{bmatrix}$.

4. Show that if $\mathbf{Q}$ is orthogonal and symmetric, then $\mathbf{Q}^2 = \mathbf{I}$.

5. Find the 3-by-3 rotation matrix that rotates every vector through ψ radians about the y-axis.

6. Determine all 2-by-2 orthogonal matrices all of whose entries are either zero or one.

7. Determine all 3-by-3 orthogonal matrices all of whose entries are either zero or one.

8. Prove that the jth column of *any* square invertible matrix $\mathbf{A}$ is orthogonal to every row of $\mathbf{A}^{-1}$, except the jth.

9. Construct a reflector matrix that carries the 3-dimensional vector $5\mathbf{i} + 12\mathbf{j} + 13\mathbf{k}$ into a vector along the x-axis.

10. Construct a reflector matrix that carries the 3-dimensional vector $5\mathbf{i} + 12\mathbf{j} + 13\mathbf{k}$ into a vector along the y-axis.

11. Construct a reflector matrix that carries the 4-dimensional vector $\begin{bmatrix} 1 & 1 & 1 & 1 \end{bmatrix}$ into a vector having all entries zero except the first.

12. Find a 3-by-3 Householder reflector matrix $\mathbf{M}_{\text{ref}}$ that, through left multiplication, zeros out all entries below the first entry, in the first column of

$$\mathbf{A} = \begin{bmatrix} 1 & 0 & 2 \\ 2 & 1 & 0 \\ 3 & 0 & -1 \end{bmatrix}.$$

13. Find a Householder reflector matrix $\mathbf{M}_{\text{ref}}$ that, through left multiplication, zeros out all entries below the first entry, in the *second* column of

$$\mathbf{A} = \begin{bmatrix} 4 & 2 & 0 \\ 0 & 1 & 1 \\ 0 & 3 & 0 \end{bmatrix}. \tag{8}$$

Is the zero pattern in the first column preserved?

14. Find a Householder reflector matrix $\mathbf{M}_{\text{ref}}$ that, through left multiplication, zeros out all entries *above and below* the *second* entry, in the *second* column of the matrix in (8). (In other words, all nondiagonal entries in the second column will become 0.) Is the zero pattern in the first column preserved? (*Hint*: Bearing in mind the reflector preserves norms, decide to which vector the second column will be reflected, and then figure out the mirror normal.)

15. Find a Householder reflector matrix $\mathbf{M}_{\text{ref}}$ that, through left multiplication, zeros out all entries *below* the *second* entry, in the *second* column of the matrix in (8), and at the same time leaves the *first* entry in the second column intact. Is the zero pattern in the first column preserved? (*Hint*: Bearing in mind the reflector preserves norms, decide to which vector the second column will be reflected, and then figure out the mirror normal.)

16. (*An alternative to Gauss elimination*) Describe how a sequence of n Householder reflectors can be constructed to render, through left multiplication, an n-by-n matrix into upper triangular form. (*Hint*: See Problem 15.)

17. Is *reflection* matrix multiplication commutative ($\mathbf{M}_{ref1}\mathbf{M}_{ref2} = \mathbf{M}_{ref2}\mathbf{M}_{ref1}$)? Resolve this question in two ways:

    (a) Express the reflections using formula (1) in two dimensions and carry out the multiplications.

    (b) Compare a succession of reflections through the line $y = x$ and through the y-axis, in both orders.

18. Is *rotation* matrix multiplication in three dimensions commutative ($\mathbf{M}_{rot1}\mathbf{M}_{rot2} = \mathbf{M}_{rot2}\mathbf{M}_{rot1}$)? Resolve this question in two ways:

    (a) Reevaluate the final matrix in Example 3 if the x rotation is performed first;

    (b) Execute a succession of $90°$ rotations of your textbook about a set of z, x-axes, in both orders.

19. Show that if $\mathbf{A} = -\mathbf{A}^T$, then

    (a) $(\mathbf{I} - \mathbf{A})$ is invertible. (*Hint*: First prove $||(\mathbf{I} - \mathbf{A})\mathbf{v}||^2 = ||\mathbf{v}||^2 + ||\mathbf{A}\mathbf{v}||^2$.)

    (b) $(\mathbf{I}-\mathbf{A})^{-1}(\mathbf{I}+\mathbf{A})$ is orthogonal. (*Hint*: Multiply the matrix by its transpose and use the identity $(\mathbf{M}^{-1})^T = (\mathbf{M}^T)^{-1}$ (Example 5, Section 2.3).)

20. In modern algebra, a set G with a binary operation $*$ is said to be a *group* if it satisfies the following properties:

    (i) Closure: If a, b belong to G, so does $a * b$;

    (ii) Associativity: If a, b, c belong to G then $(a * b) * c = a * (b * c)$;

    (iii) Identity: There is an element e in G such that $e * a = a * e$ for every a in G;

    (iv) Inverse: For each a in G there is a b in G such that $a * b = b * a = e$;

    Which of the following sets forms a group under the operation of matrix multiplication?

    (a) All n-by-n matrices;

    (b) All invertible n-by-n matrices;

    (c) All orthogonal n-by-n matrices.

4.3 LEAST SQUARES

The formula $\mathbf{v}_n = \mathbf{n}(\mathbf{n} \cdot \mathbf{v})$ orthogonally projects $\mathbf{v}$ onto the direction specified by the unit vector $\mathbf{n}$; that is, onto the one-dimensional subspace spanned by $\mathbf{n}$. And indeed, the residual $(\mathbf{v} - \mathbf{v}_n)$ is orthogonal to every vector in the subspace (Figure 4.5a). Now we generalize this by employing the Gram–Schmidt algorithm to project $\mathbf{v}$ onto higher dimensional subspaces (Figure 4.5b).

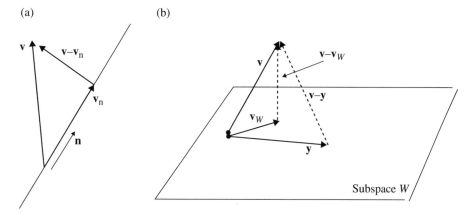

Fig. 4.5 Orthogonal projections. (a) One-dimensional subspace. (b) Two-dimensional subspace.

Beginning with a (finite-dimensional) vector space V and a subspace W onto which we will project the vectors in V, the first step is to determine bases for these spaces—say, $\{\mathbf{v}_1, \mathbf{v}_2, \ldots, \mathbf{v}_n\}$ for V and $\{\mathbf{w}_1, \mathbf{w}_2, \ldots, \mathbf{w}_m\}$ for W. Now we can obtain an *orthonormal* basis for W by performing the Gram–Schmidt algorithm on $\{\mathbf{w}_1, \mathbf{w}_2, \ldots, \mathbf{w}_m\}$ and rescaling to obtain unit vectors. Actually, we propose to apply this strategy to the extended set $\{\mathbf{w}_1, \mathbf{w}_2, \ldots, \mathbf{w}_m, \mathbf{v}_1, \mathbf{v}_2, \ldots, \mathbf{v}_n\}$. The resulting vectors are independent and span V; they, too, form a basis. The first m vectors extracted, $\{\mathbf{w}'_1, \mathbf{w}'_2, \ldots, \mathbf{w}'_m\}$ will be the orthonormal basis for W.

The remaining $(n-m)$ vectors, $\{\mathbf{v}'_{m+1}, \mathbf{v}'_{m+2}, \ldots, \mathbf{v}'_n\}$, span a subspace of V which we denote by $W^\perp$ ("W perp"), because every vector in $W^\perp$ is orthogonal (perpendicular) to every vector in W; indeed,

$$\left(c_1\mathbf{w}'_1 + c_2\mathbf{w}'_2 + \cdots + c_m\mathbf{w}'_m\right) \cdot \left(c_{m+1}\mathbf{v}'_{m+1} + c_{m+2}\mathbf{v}'_{m+2} + \cdots + c_n\mathbf{v}'_n\right) \tag{1}$$
$$= c_1 c_{m+1}\mathbf{w}'_1 \cdot \mathbf{v}'_{m+1} + c_1 c_{m+2}\mathbf{w}'_1 \cdot \mathbf{v}'_{m+2} + \cdots$$
$$= \mathbf{0}.$$

Thus W and $W^\perp$ are **mutually orthogonal subspaces**. Notice also that the set $\{\mathbf{w}'_1, \mathbf{w}'_2, \ldots, \mathbf{w}'_m, \mathbf{v}'_{m+1}, \mathbf{v}'_{m+2}, \ldots, \mathbf{v}'_n\}$ is a full orthonormal basis for V.

Next we use the orthonormal basis expansion formula of Theorem 1 (Section 4.1) to express an arbitrary vector $\mathbf{v}$ in V as a linear combination of elements of this basis:

$$\mathbf{v} = (\mathbf{v} \cdot \mathbf{w}'_1)\mathbf{w}'_1 + (\mathbf{v} \cdot \mathbf{w}'_2)\mathbf{w}'_2 + \cdots + (\mathbf{v} \cdot \mathbf{w}'_m)\mathbf{w}'_m \tag{2}$$
$$+ (\mathbf{v} \cdot \mathbf{v}'_{m+1})\mathbf{v}'_{m+1} + (\mathbf{v} \cdot \mathbf{v}'_{m+2})\mathbf{v}'_{m+2} + \cdots + (\mathbf{v} \cdot \mathbf{v}'_{n+2})\mathbf{v}'_n$$
$$= \mathbf{v}_W + \mathbf{v}_{W\perp}.$$

This construction—orthogonalizing a basis for W and extending it to an orthonormal basis for V, spawning identity (2)—demonstrates the decomposition of a vector space into orthogonal subspaces. To summarize:

Orthogonal Subspace Decomposition

When W is a subspace of the vector space V, the notation

$$V = W \oplus W^\perp, \qquad (3)$$

signifies the following:

- Every vector in V can be decomposed into a sum of a vector in W plus one in $W^\perp$;
- W and $W^\perp$ are mutually orthogonal subspaces.

Moreover, it is easy to prove that the decomposition is unique, and that the zero vector is the only vector that W and $W^\perp$ have in common; see Problem 22. Note that if $\mathbf{v} = \mathbf{v}_W + \mathbf{v}_{W^\perp}$ as in (2), then the orthonormal basis expansion given in equation (8) of Theorem 1, Section 4.1, implies that

$$\|\mathbf{v}\|^2 = \|\mathbf{v}_W\|^2 + \|\mathbf{v}_{W^\perp}\|^2. \qquad (4)$$

Of course we have discussed the orthogonal projection of a vector onto another *vector* many times. Now we can call the "W-part" of (2), $\mathbf{v}_W = c_1\mathbf{w}_1' + c_2\mathbf{w}_2' + \cdots + c_m\mathbf{w}_m'$, the *orthogonal projection of* $\mathbf{v}$ *onto the subspace* W. This orthogonal projection has a remarkable property; *it's the nearest vector to* $\mathbf{v}$ *among all the vectors in the subspace* W:

Best Approximation

Theorem 3. Let $V = W \oplus W^\perp$ and let $\mathbf{v}_W$ be the orthogonal projection of the vector $\mathbf{v}$ into the subspace W. If $\mathbf{y}$ is any other vector in W, then

$$\|\mathbf{v} - \mathbf{v}_W\| < \|\mathbf{v} - \mathbf{y}\|.$$

In other words, of all vectors in W the orthogonal projection $\mathbf{v}_W$ is the best approximation to the vector $\mathbf{v}$ (as measured by the Euclidean norm). As illustrated in Figure 4.5, this is just a generalization of the familiar fact that the shortest distance from a point to a plane is the length of the perpendicular from the point to the plane.

Proof of Theorem 3.   First observe that for any y in W, we can write

$$\mathbf{v} - \mathbf{y} = (\mathbf{v} - \mathbf{v}_W) + (\mathbf{v}_W - \mathbf{y}). \tag{5}$$

The vector $(\mathbf{v} - \mathbf{v}_W)$ lies in the subspace $W^\perp$ (in fact, it *is* $\mathbf{v}_{W^\perp}$), while the vector $(\mathbf{v}_W - \mathbf{y})$ lies in W. So applying (4) to (5) we learn that

$$\|\mathbf{v} - \mathbf{y}\|^2 = \|\mathbf{v} - \mathbf{v}_W\|^2 + \|\mathbf{v}_W - \mathbf{y}\|^2,$$

demonstrating that $\|\mathbf{v} - \mathbf{y}\|$ exceeds $\|\mathbf{v} - \mathbf{v}_W\|$ unless $\mathbf{y}$ is, in fact, $\mathbf{v}_W$.   ∎

Theorem 3 is used in many areas of applied mathematics. For our purposes, it provides an immediate answer to the dilemma of what to do about inconsistent systems.

Suppose an engineering design has led to a system $\mathbf{Ax} = \mathbf{b}$ that turns out to have no solutions. (For instance, it may have more equations than unknowns.) Rather than abandon the design, the engineer may be willing to accept a "solution" vector $\mathbf{x}$ that brings $\mathbf{Ax}$ as close to $\mathbf{b}$ as possible. We have seen that the product $\mathbf{Ax}$ can be interpreted as a linear combination of the columns of $\mathbf{A}$, with coefficients given by $\mathbf{x}$. So our "compromise solution" will be to find the linear combination of columns of $\mathbf{A}$ that best approximates $\mathbf{b}$. In other words, we have the subspace W, given by the span of the columns of $\mathbf{A}$, and we seek the vector in that subspace that is closest to $\mathbf{b}$. Theorem 3 says that the best approximation to $\mathbf{b}$ will be the orthogonal projection $\mathbf{b}_W$ of $\mathbf{b}$ onto W. So we tentatively formulate the following procedure:

Least-Squares Solution to $\mathbf{Ax} = \mathbf{b}$

To find a vector $\mathbf{x}$ providing the best approximation to a solution of the inconsistent system expressed as $\mathbf{Ax} = \mathbf{b}$,[2]

- (i) Compute an orthonormal basis for the subspace spanned by the columns of $\mathbf{A}$ (using, for instance, the Gram–Schmidt algorithm);
- (ii) Find the orthogonal projection $\mathbf{b}_{\text{colA}}$ of $\mathbf{b}$ onto this subspace (using, for instance, (2));
- (iii) Solve the (consistent) system $\mathbf{Ax} = \mathbf{b}_{\text{colA}}$.

Then the Euclidean norm of $(\mathbf{b} - \mathbf{Ax})$ will be less than or equal to $\|\mathbf{b} - \mathbf{Ay}\|$ for any vector $\mathbf{y}$.

Note that the square of the Euclidean norm of the vector $\mathbf{Ax} - \mathbf{b}$ is given by the sum of the squares of its entries, which is precisely the sum of the squares of the differences between the left- and right-hand sides of the "equations" represented by $\mathbf{Ax} = \mathbf{b}$. This

---

[2]Rigorously speaking, since the system is inconsistent, we honestly should refrain from writing "$\mathbf{Ax} = \mathbf{b}$".

is commonly known as the *sum of squares of errors*, from which the procedure derives its name, "Least Squares."

Rather than display an example of this procedure as formulated, we turn directly to a short cut. Since $\mathbf{b}_{\text{colA}}$ is the orthogonal projection of $\mathbf{b}$ onto the subspace spanned by the columns of $\mathbf{A}$, the residual $\mathbf{b} - \mathbf{b}_{\text{colA}}$ is orthogonal to this subspace, and hence to every column of $\mathbf{A}$. There's an easy way to express this orthogonality using the transpose:

$$
\mathbf{A} = \begin{bmatrix} | & | & | & | \\ \text{col 1} & \text{col 2} & \cdots & \text{col } n \\ | & | & | & | \end{bmatrix}, \quad \mathbf{A}^T = \begin{bmatrix} - [\text{col 1}]^T - \\ - [\text{col 2}]^T - \\ \vdots \\ -[\text{col } n]^T - \end{bmatrix}
$$

and therefore

$$
\mathbf{A}^T (\mathbf{b} - \mathbf{A}\mathbf{x}) = \begin{bmatrix} - [\text{col 1}]^T - \\ - [\text{col 2}]^T - \\ \vdots \\ -[\text{col } n]^T - \end{bmatrix} (\mathbf{b} - \mathbf{A}\mathbf{x}) = \mathbf{0}
$$

expresses the orthogonality.

Theorem 4 summarizes these observations. We leave it to the reader (Problem 21) to provide the details of the proof.

Characterization of Least Squares Approximation

Theorem 4. The vector $\mathbf{x}$ provides a least squares approximation for the (m-by-n) system $\mathbf{A}\mathbf{x} = \mathbf{b}$ if and only if $\mathbf{A}^T\{\mathbf{b} - \mathbf{A}\mathbf{x}\} = \mathbf{0}$ or, equivalently,

$$\mathbf{A}^T \mathbf{A}\mathbf{x} = \mathbf{A}^T \mathbf{b}. \tag{6}$$

Equations (6) are called the *normal equations* for the system represented by $\mathbf{A}\mathbf{x} = \mathbf{b}$. The normal equations provide an alternative to the least squares procedure described earlier.

Example 1.   Find a vector $\mathbf{x}$ providing the least-squares approximation to the system expressed by

$$
\begin{bmatrix} 1 & -1 & 1 & \vdots & 1 \\ 1 & 1 & 3 & \vdots & 0 \\ 1 & 1 & 3 & \vdots & 1 \\ 1 & 1 & 3 & \vdots & 2 \end{bmatrix}. \tag{7}
$$

Solution. Note that the last three equations in (7) are blatantly inconsistent. The normal equations for this system are

$$\mathbf{A}^T\mathbf{A}\mathbf{x} = \begin{bmatrix} 1 & 1 & 1 & 1 \\ -1 & 1 & 1 & 1 \\ 1 & 3 & 3 & 3 \end{bmatrix} \begin{bmatrix} 1 & -1 & 1 \\ 1 & 1 & 3 \\ 1 & 1 & 3 \\ 1 & 1 & 3 \end{bmatrix} \begin{bmatrix} x_1 \\ x_2 \\ x_3 \end{bmatrix} = \mathbf{A}^T\mathbf{b} = \begin{bmatrix} 1 & 1 & 1 & 1 \\ -1 & 1 & 1 & 1 \\ 1 & 3 & 3 & 3 \end{bmatrix} \begin{bmatrix} 1 \\ 0 \\ 1 \\ 2 \end{bmatrix},$$

which, after multiplication, take the following augmented matrix form:

$$\begin{bmatrix} 4 & 2 & 10 & \vdots & 4 \\ 2 & 4 & 8 & \vdots & 2 \\ 10 & 8 & 28 & \vdots & 10 \end{bmatrix}.$$

Gauss elimination reveals the solutions to this system to be

$$\begin{bmatrix} x_1 \\ x_2 \\ x_3 \end{bmatrix} = \begin{bmatrix} 1 - 2t \\ -t \\ t \end{bmatrix} = \begin{bmatrix} 1 \\ 0 \\ 0 \end{bmatrix} + t\begin{bmatrix} -2 \\ -1 \\ 1 \end{bmatrix}.$$

The quality of the solution is exhibited by the "error" or "residual" $\mathbf{b} - \mathbf{A}\mathbf{x}$:

$$\mathbf{b} = \begin{bmatrix} 1 \\ 0 \\ 1 \\ 2 \end{bmatrix}, \quad \mathbf{b}_{\text{colA}} = \mathbf{A}\mathbf{x} = \begin{bmatrix} 1 \\ 1 \\ 1 \\ 1 \end{bmatrix}, \quad \text{error} = \mathbf{b} - \mathbf{A}\mathbf{x} = \begin{bmatrix} 0 \\ -1 \\ 0 \\ 1 \end{bmatrix}. \qquad \blacksquare$$

Example 1 highlights several noteworthy points about least squares approximations:

- The error $\mathbf{b} - \mathbf{A}\mathbf{x}$ is always orthogonal to the best approximation $\mathbf{A}\mathbf{x}$;
- The coefficient matrix $\mathbf{A}^T\mathbf{A}$ for the normal equations is always square and symmetric;
- The normal equations are always consistent;
- The best approximation to $\mathbf{b}$, namely $\mathbf{b}_{\text{colA}} = \mathbf{A}\mathbf{x}$, is unique. However the "solution" vector $\mathbf{x}$, that *provides* the best approximation to $\mathbf{A}\mathbf{x} = \mathbf{b}$, might not be unique.[3]

In Problem 15, it is demonstrated that the rank of $\mathbf{A}^T\mathbf{A}$ equals the rank of $\mathbf{A}$. Therefore *if the columns of* $\mathbf{A}$ *are linearly independent,* $\mathbf{A}^T\mathbf{A}$ *is invertible and the least*

---

[3]The "pseudoinverse" of $\mathbf{A}$, described in Section 6.4, provides the solution vector $\mathbf{x}$ of *minimal norm*. For Example 1, this would be $[1/3 \ -1/3 \ 1/3]^T$ (for $t = 1/3$).

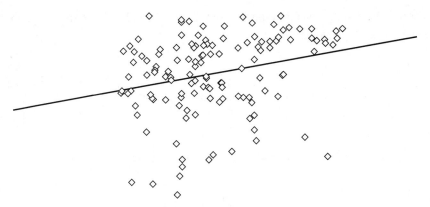

Fig. 4.6 Scatter plot.

squares problem $\mathbf{Ax} = \mathbf{b}$ *has a unique solution given by*

$$\mathbf{x} = (\mathbf{A}^T\mathbf{A})^{-1}\mathbf{A}^T\mathbf{b}. \tag{8}$$

Our final example addresses a common situation. One has a "scatter plot," that is, a collection of measured data points in the plane, (x_i, y_i). The task is to find the best straight line $y = mx + b$ passing through the data. See Figure 4.6.

Example 2. ("*Least Squares Linear Fit*") Given a collection of pairs of numbers $\{x_i, y_i\}_{i=1}^{N}$, derive equations for the slope m and intercept b of the straight line $y = mx + b$ that minimizes the sum of squares of errors

$$\sum_{i=1}^{N} [y_i - (mx_i + b)]^2.$$

Solution. We need to cast our assignment as a least squares problem for some system $\mathbf{Ax} = \mathbf{b}$. The trick is to note that our unknowns are m and b (and not the known data x_i). We are trying to match up, as best possible, the left and right-hand sides of

$$
\begin{aligned}
mx_1 + b &= y_1, \\
mx_2 + b &= y_2, \\
&\vdots \\
mx_N + b &= y_N.
\end{aligned}
\tag{9}
$$

The system (9) can be expressed in matrix form as

$$
\begin{bmatrix} x_1 & 1 \\ x_2 & 1 \\ \vdots & \vdots \\ x_N & 1 \end{bmatrix}
\begin{bmatrix} m \\ b \end{bmatrix}
=
\begin{bmatrix} y_1 \\ y_2 \\ \vdots \\ y_N \end{bmatrix}.
\tag{10}
$$

The normal equations for (10) are

$$
\begin{bmatrix} x_1 & x_2 & \cdots & x_N \\ 1 & 1 & \cdots & 1 \end{bmatrix}
\begin{bmatrix} x_1 & 1 \\ x_2 & 1 \\ \vdots & \vdots \\ x_N & 1 \end{bmatrix}
\begin{bmatrix} m \\ b \end{bmatrix}
=
\begin{bmatrix} x_1 & x_2 & \cdots & x_N \\ 1 & 1 & \cdots & 1 \end{bmatrix}
\begin{bmatrix} y_1 \\ y_2 \\ \vdots \\ y_N \end{bmatrix},
$$

or

$$
\begin{bmatrix} \sum_{i=1}^{N} x_i^2 & \sum_{i=1}^{N} x_i \\ \sum_{i=1}^{N} x_i & N \end{bmatrix}
\begin{bmatrix} m \\ b \end{bmatrix}
=
\begin{bmatrix} \sum_{i=1}^{N} x_i y_i \\ \sum_{i=1}^{N} y_i \end{bmatrix}.
$$

Cramer's rule then gives the solution

$$
m = \frac{N S_{xy} - S_x S_y}{N S_{x^2} - S_x^2}, \quad b = \frac{S_{x^2} S_y - S_x S_{xy}}{N S_{x^2} - S_x^2}, \tag{11}
$$

where the sums are abbreviated

$$
S_x := \sum_{i=1}^{N} x_i, \; S_y := \sum_{i=1}^{N} y_i, \; S_{xy} := \sum_{i=1}^{N} x_i y_i, \; S_{x^2} := \sum_{i=1}^{N} x_i^2. \quad \blacksquare
$$

In practice, the numerical implementation of the normal equations approach to solving least squares is more sensitive to errors than the orthogonal subspace procedure described earlier. Most enlightened matrix software will automatically opt for the latter, despite its extra complexity.

Exercises 4.3

1. Use the three-step procedure described after Theorem 3 to find all vectors **x** providing the least squares solution to the inconsistent system

$$
\begin{bmatrix} 1 & 3 & \vdots & 1 \\ -1 & -1 & \vdots & 0 \\ 1 & 3 & \vdots & 0 \\ -1 & -1 & \vdots & 1 \end{bmatrix}.
$$

(Save work by using Problem 7, Exercises 4.1.) What is the error for this least squares approximation?

2. Use the three-step procedure described after Theorem 3 to find all vectors **x** providing the least squares solution to the inconsistent system

$$\left[\begin{array}{ccc:c} 2 & 3 & 2 & 0 \\ 2 & 1 & 0 & 0 \\ -2 & -1 & -2 & 0 \\ -2 & -3 & -4 & 1 \end{array}\right].$$

(Save work by using Problem 9, Exercises 4.1.) What is the error for this least squares approximation?

3. Write out and solve the normal equations for the inconsistent system in Problem 2.

4. Write out and solve the normal equations for the inconsistent system

$$\left[\begin{array}{ccc:c} -1 & 2 & 3 & 1 \\ 1 & 0 & 1 & 1 \\ -1 & 2 & 1 & 1 \\ 1 & 0 & -1 & 0 \end{array}\right].$$

5. Use the three-step procedure described after Theorem 3 to find all vectors **x** providing the least squares solution to the inconsistent system in Problem 4. (See Problem 10, Exercises 4.1.) What is the error for this least squares approximation?

6. Show that the least squares solutions to the system $x + y = 1$, $x + y = 3$ are the solutions to $x + y = 2$.

7. Modify Example 2 to find the slope m for the straight line *through the origin* $y = mx$ that best fits the data $\{(x_i, y_i)\}_{i=1}^{N}$.

8. Modify Example 2 to find the constant b for the *horizontal* line $y = b$ that best fits the data $\{(x_i, y_i)\}_{i=1}^{N}$.

9. Modify Example 2 to find the slope m for the straight line *with a prescribed intercept b_0*, $y = mx + b_0$, that best fits the data $\{(x_i, y_i)\}_{i=1}^{N}$.

10. Modify Example 2 to find the intercept b for the straight line *with a prescribed slope m_0*, $y = m_0x + b$, that best fits the data $\{(x_i, y_i)\}_{i=1}^{N}$.

11. Given the data $\{(x_i, y_i)\}_{i=1}^{N}$, take the averages $\bar{x} := \frac{1}{N}\sum_{i=1}^{N} x_i$, $\bar{y} := \frac{1}{N}\sum_{i=1}^{N} y_i$. Show that the parameters m and b in Example 2 satisfy $\bar{y} = m\bar{x} + b$.

12. Suppose that in the inconsistent system $A\mathbf{x} = \mathbf{b}$, the error in the ith equation is weighted by the factor w_i, and that our task is to minimize the *weighted sum of squares of errors*. This would occur, for example, if some of the equations were expressed in dollars, others in euros, pounds, etc.

(a) Argue that the quantity to be minimized is $\|\mathbf{W}(\mathbf{b} - \mathbf{Ax})\|^2$, where $\mathbf{W}$ is the diagonal matrix with the weights w_i on its diagonal.

(b) Derive the normal equations for the weighted least squares problem.

13. Example 2, Section 2.2, derived the formula $\mathbf{Mb} = \mathbf{nn}^T\mathbf{b}$
    for the matrix orthogonally projecting a two-dimensional column vector $\mathbf{b}$ onto the direction of a unit vector $\mathbf{n}$.
    (a) Show that if $\mathbf{v}$ is not a unit vector, the formula for projecting $\mathbf{b}$ onto the direction of $\mathbf{v}$ (which must be nonzero, of course) is $\mathbf{vv}^T\mathbf{b}/\mathbf{v}^T\mathbf{v}$.
    (b) In higher dimensions, this problem would be expressed as finding the closest vector of the form $x\mathbf{v}$ to the vector $\mathbf{b}$. Express this as a least squares system, and show that the normal equations imply that the same formula $\mathbf{vv}^T\mathbf{b}/\mathbf{v}^T\mathbf{v}$ holds.

14. Generalize the formula in Problem 13; if $\{\mathbf{v}_1, \mathbf{v}_2, \ldots, \mathbf{v}_m\}$ forms an orthonormal basis for the subspace W of $\mathbf{R}^n_{\text{col}}$, derive a formula for the matrix operator that orthogonally projects $\mathbf{b}$ into the subspace W.

15. Prove that the rank of $\mathbf{A}^T\mathbf{A}$ equals the rank of $\mathbf{A}$ by reasoning as follows:
    (a) Show that the null space of $\mathbf{A}$ is contained in the null space of $\mathbf{A}^T\mathbf{A}$.
    (b) Show that the null space of $\mathbf{A}^T\mathbf{A}$ is contained in the null space of $\mathbf{A}$ (and thus by part (a) these two null spaces are identical). (*Hint*: if $\mathbf{A}^T\mathbf{Ax} = \mathbf{0}$, then $\mathbf{x}^T\mathbf{A}^T\mathbf{Ax} \equiv \|\mathbf{Ax}\|^2 = 0$.)
    (c) Apply Theorem 6, Section 3.3, to complete the argument.

16. If the columns of $\mathbf{A}$ are independent, show that $(\mathbf{A}^T\mathbf{A})^{-1}\mathbf{A}^T$ is a left inverse of $\mathbf{A}$. (See Problem 15.)

17. Suppose that one wishes to find a good fit to a set of data $\{(x_i, y_i)\}_{i=1}^N$, with each $y_i > 0$, using a function of the form $y = ce^{bx}$. Here both b and c are adjustible. By taking logarithms, show that this can be reduced to the situation in Example 2. What are the optimal values of c and b? What, precisely, is the quantity being minimized?

18. Suppose that one wishes to find the least squares fit to a set of data $\{x_i, y_i\}_{i=1}^N$ using a polynomial $y = c_0 + c_1 x + c_2 x^2 + \cdots + c_k x^k$. What is the form of the normal equations for the coefficients $\{c_0, c_1, c_2, \ldots, c_k\}$?

19. Suppose that one wishes to find the least squares fit to a set of data $\{x_i, y_i\}_{i=1}^N$ using a linear combination of exponentials $y = c_0 + c_1 e^{-x} + c_2 e^{-2x} + \cdots + c_k e^{-kx}$. What is the form of the normal equations for the coefficients $\{c_0, c_1, c_2, \ldots, c_k\}$?

20. The *Fredholm alternative* states that $\mathbf{Ax} = \mathbf{b}$ has a solution if and only if $\mathbf{b}$ is orthogonal to every solution of $\mathbf{A}^T\mathbf{y} = \mathbf{0}$.
    (a) Argue that $[\text{Span}\{\text{rows of } \mathbf{A}\}]^\perp = \text{null space of } \mathbf{A}$.
    (b) Argue that $\text{Span}\{\text{columns of } \mathbf{A}\} = [\text{null space of } \mathbf{A}^T]^\perp$ by applying (a) to $\mathbf{A}^T$.
    (c) From these considerations, assemble a proof of the Fredholm alternative.

21. Prove Theorem 4 as follows:
    (a) The sum of squares of errors is given by

    $$\text{SSE} = \|\mathbf{b} - \mathbf{Ax}\|^2 = (\mathbf{b} - \mathbf{Ax})^T(\mathbf{b} - \mathbf{Ax}).$$

Expand this formula in terms of the entries of $\mathbf{A}$, $\mathbf{b}$, and $\mathbf{x}$ to obtain

$$\text{SSE} = \sum_{i=1}^{n} \sum_{j=1}^{n} \sum_{k=1}^{n} A_{ji}A_{jk}x_ix_k - 2\sum_{j=1}^{n} \sum_{k=1}^{n} A_{jk}b_jx_k + \sum_{j=1}^{n} b_j^2.$$

(b) SSE is a smooth function of $\mathbf{x}$ that is bounded below (by zero). Therefore at its minimum point, each of its partials (with respect to x_p) must be zero. Carefully compute this partial to derive

$$\frac{\partial SSE}{\partial x_p} = \sum_{j=1}^{n} \sum_{k=1}^{n} A_{jp}A_{jk}x_k + \sum_{i=1}^{n} \sum_{j=1}^{n} A_{ji}A_{jp}x_i - 2\sum_{j=1}^{n} A_{jp}b_j = 0.$$

[*Hint*: This might be easier to see if you first work it out explicitly for the cases $n = 2$ and $n = 3$.]

(c) Argue that the equation in (b) is the pth component of the system $2\mathbf{A}^T\mathbf{A}\mathbf{x} - 2\mathbf{A}^T\mathbf{b} = 0$.

22. Show that if W, W_1, and W_2 are three subspaces of V such that $V = W \oplus W_1$ and $V = W \oplus W_2$ are orthogonal decompositions as described by equation (3), then $W_1 = W_2$. Show also that the only vector that W and W_1 have in common is $\mathbf{0}$.

4.4   FUNCTION SPACES

Although in Section 3.1 we identified some examples of vector spaces whose elements are functions, our subsequent examples and theorems have focused on the standard spaces $\mathbf{R}_{\text{col}}^n$ and $\mathbf{R}_{\text{row}}^n$. We discussed how Gauss elimination can be used to determine independence and find bases, and how the dot product facilitates the exploitation of orthogonality.

For more exotic vector spaces, these notions can be more subtle. For instance, consider the task of determining the independence of three elements $y_1(x)$, $y_2(x)$, $y_3(x)$ of the space of continuous functions on, say, the interval $[0, 1]$. We need to decide if the requirement $c_1y_1(x) + c_2y_2(x) + c_3y_3(x) = 0$ for every x in $[0, 1]$ necessitates that each c_i be zero. That's an infinite number of conditions at our disposal!

We could contemplate proceeding by picking 3 points $\{x_1, x_2, x_3\}$ in $[0, 1]$ and reducing the equations

$$c_1y_1(x_1) + c_2y_2(x_1) + c_3y_3(x_1) = 0$$
$$c_1y_1(x_2) + c_2y_2(x_2) + c_3y_3(x_2) = 0$$
$$c_1y_1(x_3) + c_2y_2(x_3) + c_3y_3(x_3) = 0$$

or

$$\begin{bmatrix} y_1(x_1) & y_2(x_1) & y_3(x_1) \\ y_1(x_2) & y_2(x_2) & y_3(x_2) \\ y_1(x_3) & y_2(x_3) & y_3(x_3) \end{bmatrix} \begin{bmatrix} c_1 \\ c_2 \\ c_3 \end{bmatrix} = \begin{bmatrix} 0 \\ 0 \\ 0 \end{bmatrix} \qquad (1)$$

to row echelon form. If this reveals that (1) has only the zero solution, we will have shown that the only way $c_1 y_1(x) + c_2 y_2(x) + c_3 y_3(x) \equiv 0$ can hold is if $c_1 = c_2 = c_3 = 0$, and thus the functions must be independent. However, if nontrivial solutions do exist, we could not conclude that the functions are dependent, because we could never be sure that a different selection of $\{x_1, x_2, x_3\}$ might have established independence. So Gauss elimination *may* prove *in*dependence, if we're lucky in our choice of test values for x, but it can never establish dependence.

Example 1. Show that the functions 1, x^2, and x^4 are linearly independent on $(-\infty, \infty)$.

Solution. If we test for independence at the points $\{-1, 0, 1\}$, the system (1) becomes

$$\begin{bmatrix} 1 & 1 & 1 \\ 1 & 0 & 0 \\ 1 & 1 & 1 \end{bmatrix} \begin{bmatrix} c_1 \\ c_2 \\ c_3 \end{bmatrix} = \begin{bmatrix} 0 \\ 0 \\ 0 \end{bmatrix},$$

which has nontrivial solutions: for example, $c_1 = 0$, $c_2 = -1$, $c_3 = 1$. So the test is inconclusive. But if we had chosen $\{0, 1, 2\}$, the condition would have been

$$\begin{bmatrix} 1 & 0 & 0 \\ 1 & 1 & 1 \\ 1 & 4 & 16 \end{bmatrix} \begin{bmatrix} c_1 \\ c_2 \\ c_3 \end{bmatrix} = \begin{bmatrix} 0 \\ 0 \\ 0 \end{bmatrix};$$

the determinant of the coefficient matrix is 12 (nonzero), so only the trivial solution is possible for $\{c_1, c_2, c_3\}$, implying the functions *are* independent. ∎

The basic villain here is that many function spaces are infinite dimensional. They have no finite bases; no one who has taken calculus will believe that you can generate *every* continuous function on $(-\infty, \infty)$ by taking linear combinations of a paltry collection of, say, only 7, or 8, or 20 billion, of them. One of the aims of the discipline known as *functional analysis* is to establish bases for the function spaces, but since such bases contain an infinite number of elements, the notion of convergence is a prerequisite to any analysis involving linear combinations.

This feature—the infinite dimensionality—is made more concrete by considering tabulations of continuous functions on, say, a spread sheet:

$$\begin{bmatrix} t & \cos t & \sin t \\ 0 & 1.000 & 0 \\ \pi/6 & 0.866 & 0.500 \\ 2\pi/6 & 0.500 & 0.866 \\ 3\pi/6 & 0 & 1.000 \\ 4\pi/6 & -0.500 & 0.866 \\ 5\pi/6 & -0.866 & 0.500 \\ \pi & -1.000 & 0 \\ 7\pi/6 & -0.866 & -0.500 \\ 8\pi/6 & -0.500 & -0.866 \\ 9\pi/6 & 0 & -1.000 \\ 10\pi/6 & 0.500 & -0.866 \\ 11\pi/6 & 0.866 & -0.500 \end{bmatrix} \tag{2}$$

In computational practice, each function is reduced to a column vector. Indeed we would tabulate $3\cos t + 4\sin t$ by forming linear combinations of the second and third columns of (2). But we cannot completely trust any mathematical conclusion concerning $\cos t$ and $\sin t$ based on only a finite number of sampled values. Therefore we need to employ other tools to decide, rigorously, the independence of sets of functions.

In some cases, we can use the growth rate. For example, the linear independence of 1, x^2, and x^4 on $(-\infty, \infty)$ is easily established as follows. If $c_1 1 + c_2 x^2 + c_3 x^4 = 0$ for all x, we can conclude that $c_3 = 0$ by dividing by x^4 and letting $x \to \infty$;

$$c_3 = \lim_{x \to \infty} \left[-\frac{c_1}{x^4} - \frac{c_2}{x^2} \right] = 0.$$

Then we apply the same reasoning to $c_1 1 + c_2 x^2$ to get $c_2 = 0$, and so on. This argument is easily extended to show that any finite set of polynomials of differing degrees is linearly independent on $(-\infty, \infty)$.

Example 2.   Verify the claim that any finite set of exponential functions with different decay times is linearly independent on $(-\infty, \infty)$.

Solution. We need to show that the equality

$$c_1 e^{\alpha_1 x} + c_2 e^{\alpha_2 x} + \cdots + c_n e^{\alpha_n x} = 0 \tag{3}$$

(all α_i distinct) can only hold if each $c_i = 0$. We exploit the growth rates again. Suppose α_j is the largest (in the algebraic sense) of the α_i's. Dividing (3) by $e^{\alpha_j x}$ and letting $x \to \infty$, we see that c_j must be zero. Now carry out the same argument with the second largest of the α_i's, and so on, to conclude that every c_i is zero. ∎

We remark that the growth rate argument will not work in analyzing linear independence over a *finite* interval. Indeed, it is possible for a collection of functions to be linearly independent over $(-\infty, \infty)$, but linearly *dependent* over some subinterval (a, b). See Problem 20.

For future reference, it is useful to note that the following sets of functions are linearly independent on *every* open interval (a, b):

- $\{1, x, x^2, \ldots, x^n\}$
- $\{1, \cos x, \sin x, \cos 2x, \sin 2x, \ldots, \cos nx, \sin nx\}$
- $\{e^{\alpha_1 x}, e^{\alpha_2 x}, \ldots, e^{\alpha_n x}\}$ (for distinct constants α_i)

(see Problems 4, 8, and 9).

Although Gauss elimination is ineffectual as a tool for analyzing function spaces, the notion of orthogonality and its attendant features carry over quite nicely. How do we form a dot product of the functions $\cos t$ and $\sin t$ (over the interval $[0, 2\pi]$)? Again taking a computational point of view, we could start by forming the dot product of the second and third columns in (2), and contemplate the limit of such dot products as the number of tabulations increases:

$$(\cos t) \cdot (\sin t) \overset{?}{=} \lim_{N \to \infty} \sum_{i=1}^{N} \cos t_i \sin t_i. \tag{4}$$

Sums of the form (4) diverge as $N \to \infty$, but with the insertion of a constant factor $\Delta t := t_{i+1} - t_i = 2\pi/N$ we observe the emergence of an old friend, the Riemann sum:

$$(\cos t) \cdot (\sin t) = \lim_{N \to \infty} \sum_{i=1}^{N} \cos t_i \sin t_i \, \Delta t = \int_0^{2\pi} \cos t \sin t \, dt. \tag{5}$$

Accordingly, one defines the *inner product* of two continuous functions on $[a, b]$ to be

$$<f, g> := \int_a^b f(t)\, g(t)\, dt \tag{6}$$

with the associated norm

$$\|f\| := \left[\int_a^b f(t)^2 \, dt \right]^{1/2}. \tag{7}$$

Through the inner product such concepts as the Gram–Schmidt algorithm, orthogonal projection, and least squares approximations are enabled in function spaces. Fourier series and eigenfunction expansions of solutions of partial differential equations are some of the advanced techniques based on these considerations.[4]

---

[4]The term *Hilbert Spaces* refers to function spaces endowed with inner products (and a few other properties).

Example 3.   Show that the (infinite) collection $\{\sin x, \sin 2x, \sin 3x, \ldots\}$ is an orthogonal family of functions in the interval $(0, \pi)$.

Solution.   The inner products (6) for this family are

$$\int_0^\pi \sin mx \sin nx \, dx = \left[\frac{\sin(m-n)x}{2(m-n)} - \frac{\sin(m+n)x}{2(m+n)}\right]_0^\pi = 0 \text{ for } m \neq n. \qquad \blacksquare$$

Example 4.   Apply the Gram–Schmidt algorithm to the collection of functions $\{1, x, x^2\}$ on the interval $(-1, 1)$.

Solution.   Executing the procedure described in Section 4.1 with the substitution of (6) for the dot product $\mathbf{v} \cdot \mathbf{w}$, we calculate

$$\mathbf{w}_1 = \mathbf{v}_1 = 1.$$

$$\mathbf{w}_2 = \mathbf{v}_2 - (\mathbf{v}_2 \cdot \mathbf{w}_1)\mathbf{w}_1/\mathbf{w}_1 \cdot \mathbf{w}_1 = x - \left[\int_{-1}^1 (x)(1) \, dx\right] 1/\left[\int_{-1}^1 (1)(1) \, dx\right] = x.$$

$$\mathbf{w}_3 := \mathbf{v}_3 - (\mathbf{v}_3 \cdot \mathbf{w}_1)\mathbf{w}_1/\mathbf{w}_1 \cdot \mathbf{w}_1 - (\mathbf{v}_3 \cdot \mathbf{w}_2)\mathbf{w}_2/\mathbf{w}_2 \cdot \mathbf{w}_2$$

$$= x^2 - \left[\int_{-1}^1 (x^2)(1) \, dx\right] 1/\left[\int_{-1}^1 (1)(1) \, dx\right] - \left[\int_{-1}^1 (x^2)(x) \, dx\right] x/\left[\int_{-1}^1 (x)(x) \, dx\right]$$

$$= x^2 - \frac{1}{3}.$$

If these functions were rescaled to take the value one when $x = 1$, they would be the first three *Legendre polynomials*. $\qquad \blacksquare$

Exercises 4.4

1. Decide if the following sets of functions are linearly independent on the indicated intervals.
   (a) $1, \sin^2 x, \cos^2 x$ on $[0, 2\pi]$.   (d) $x, x^2, 1 - 2x^2$ on $[-2, 2]$.
   (b) $x, |x|$ on $[-1, 1]$.   (e) $x, xe^x, 1$ on $(-\infty, \infty)$.
   (c) $x, |x|$ on $[-1, 0]$.   (f) $1/x, \sqrt{x}, x$ on $(0, \infty)$.

2. Perform the Gram–Schmidt algorithm on the set $1, e^x, e^{-x}$ to construct 3 functions that are orthogonal on the interval $(0, 1)$.

3. Perform the Gram–Schmidt algorithm on the set $1, x, x^2$ to construct 3 polynomials that are orthogonal on the interval $(0, 1)$.

4. Prove that the set of functions $\{1, x, x^2, \ldots, x^n\}$ is linearly independent over every open interval (a, b). (*Hint*: If $c_0 + c_1 x + \cdots + c_n x^n = 0$ in an interval, differentiate n times to show $c_n = 0$.)

5. Use the strategy of Problem 4 to argue that $\{1, (x-1), (x-1)^2, \ldots, (x-1)^n\}$ is a basis for $\mathbf{P}_n$ (the space of all polynomials of degree at most n).

6. Express the polynomial x^n as a linear combination of the basis elements in Problem 5. (*Hint*: Consider the Taylor series.)

7. (a) Rescale the functions computed in Example 4 so that they take the value one at $x = 1$.

   (b) Extend the computations in Example 4 to find the Legendre polynomial of degree 3.

8. Prove that the set of functions $\{1, \cos x, \sin x, \cos 2x, \sin 2x, \ldots, \cos nx, \sin nx\}$ is linearly independent over every open interval (a,b). (*Hint*: If $c_0 + c_1 \cos x + d_1 \sin x + \cdots + c_n \cos nx + d_n \sin nx = 0$ in an interval, differentiate $2n$ times and write the resulting $2n+1$ equations (regarding the c_j's and d_j's as unknowns) in matrix form. Row operations will reduce the matrix to one with 2-by-2 blocks along the diagonal and zeros below; the determinant will be seen to be positive.)

9. Prove that the set of functions $\{e^{\alpha_1 x}, e^{\alpha_2 x}, \ldots, e^{\alpha_n x}\}$ (for distinct constants α_i) is linearly independent over every open interval (a,b). (*Hint*: If $c_1 e^{\alpha_1 x} + c_2 e^{\alpha_2 x} + \cdots + c_n e^{\alpha_n x} = 0$ in an interval, differentiate $n-1$ times and write the resulting n equations (regarding the c_j's as unknowns) in matrix form. Factor as

$$\begin{bmatrix} 1 & 1 & \cdots & 1 \\ \alpha_1 & \alpha_2 & \cdots & \alpha_n \\ \vdots & \vdots & \ddots & \vdots \\ \alpha_1^{n-1} & \alpha_2^{n-1} & \cdots & \alpha_n^{n-1} \end{bmatrix} \begin{bmatrix} e^{\alpha_1 x} & 0 & \cdots & 0 \\ 0 & e^{\alpha_2 x} & \cdots & 0 \\ \vdots & \vdots & \ddots & \vdots \\ 0 & 0 & \cdots & e^{\alpha_n x} \end{bmatrix} \begin{bmatrix} c_1 \\ c_2 \\ \vdots \\ c_n \end{bmatrix} = \mathbf{0}$$

and consider determinants (recall Problem 23, Exercises 2.4).)

10. Prove that x, $|x|$, and x^2 are linearly independent on $(-1, 1)$.

11. Utilize the growth rates to show the following sets of functions are linearly independent on $(0, \infty)$:
    (a) $\{1, \ln x, e^x\}$
    (b) $\{e^x, xe^x, x^2 e^x\}$

12. Consider the problem of approximating a function $f(x)$ over an interval (a, b) by a linear combination of other functions $c_1 y_1(x) + c_2 y_2(x) + \cdots + c_n y_n(x)$, so as to minimize the integrated squared error

$$\int_a^b \{f(x) - [c_1 y_1(x) + c_2 y_2(x) + \cdots + c_n y_n(x)]\}^2 \, dx.$$

(a) Recast this task as the problem of orthogonally projecting $f(x)$ into $\mathrm{Span}\{y_1(x), y_2(x), \ldots, y_n(x)\}$ using the inner product (6).

(b) Retrace the discussion in Section 4.1 to argue that the remainder $\{f(x) - [c_1 y_1(x) + c_2 y_2(x) + \cdots + c_n y_n(x)]\}$ must be orthogonal to each of the functions $\{y_1(x), y_2(x), \ldots, y_n(x)\}$.

(c) Express the system of n equations in the n unknowns $\{c_1, c_2, \ldots, c_n\}$ enforcing (b), in matrix form. This is the analog of the normal equations for function spaces. The coefficient matrix is called a *Gram matrix*.

13. Use the normal equations derived in Problem 12 to find the least squares approximation of $\cos^3 x$ by a combination of $\sin x$ and $\cos x$, over the interval $(0, 2\pi)$. Sketch the graphs of $\cos^3 x$ and the approximation.

14. Repeat Problem 13 using a combination of the functions 1 and x.

15. (a) Express the normal equations derived in Problem 12 to find the least squares approximation of $f(x)$ by a straight line $mx+b$, over the interval (a, b). Compare with Example 2 in Section 4.3.

    (b) Find the straight line that provides the least squares approximation to x^2 over the interval $(0, 1)$.

    (c) The *linear Chebyshev* approximation to x^2 over the interval $(0, 1)$ is the straight line $y = mx + b$ whose maximal vertical deviation from $y = x^2$ over the interval is least (a "minimax" approximation). Find this Chebyshev approximation. (*Hint*: Do it graphically; place a ruler on the graph of $y = x^2$ and adjust it until the criterion is met.)

    (d) Plot x^2, its least squares approximation, and its Chebyshev approximation over $(0, 1)$ on the same graph.

16. Formulate the normal equations derived in Problem 12 for the task of finding the least squares approximation to $f(x)$ by a combination of $1, x, x^2, \ldots, x^n$. The coefficient matrix was identified as the *Hilbert matrix* in Problem 6, Exercises 1.5.

17. Show that the functions $\{1, \cos x, \sin x, \cos 2x, \sin 2x, \ldots, \cos nx, \sin nx\}$ are orthogonal on the interval $(0, 2\pi)$.

18. Show that any sequence of elementary row operations performed on the column vector $[1, x, x^2, \ldots, x^n]^T$ yields a vector whose entries constitute a basis for $\mathbf{P}_n$.

19. By Problem 4 the set $\{1, x, x^2, \ldots, x^n\}$ is a basis for the space $\mathbf{P}_n$ of all polynomials of degree less than or equal to n.

    (a) Find another basis for $\mathbf{P}_n$ whose elements are all polynomials of the same degree.

    (b) Find a basis for the subspace of $\mathbf{P}_n$ consisting of all polynomials taking the value zero at $x = 0$.

    (c) Find a basis for the subspace of $\mathbf{P}_n$ consisting of all polynomials taking the value zero at $x = 1$.

20. Argue that the functions

$$f_1(x) = x; \quad f_2(x) = \begin{cases} x & \text{if } -1 < x < 1 \\ x^3 & \text{otherwise} \end{cases};$$

$$f_3(x) = \begin{cases} x^3 & \text{if } -1 < x < 1 \\ x & \text{otherwise} \end{cases}$$

are linearly dependent on $[-1, 1]$ but linearly independent on $(-\infty, \infty)$.

21. Prove that *any* $n + 1$ polynomials p_k, $k = 0, 1, \ldots, n$, with degree of each p_k precisely k, forms a basis for $\mathbf{P}_n$.

4.5   SUMMARY

Orthogonality

The familiar notion of orthogonality in two and three dimensions is extended to higher dimensions via the dot product criterion, which for real column vectors takes the form $\mathbf{v}^T \mathbf{w} = 0$. Similarly, the notion of length is generalized as the Euclidean norm $\|\mathbf{v}\| = (\mathbf{v}^T \mathbf{v})^{1/2}$, and mutually orthogonal vectors having unit norm are called orthonormal. A mutually orthogonal set of nonzero vectors must be linearly independent, and indeed the coefficients in a linear combination of these, $\mathbf{w} = c_1 \mathbf{v}_1 + c_2 \mathbf{v}_2 + \cdots + c_n \mathbf{v}_n$, are easy to compute: $c_i = \mathbf{v}_i^T \mathbf{w} / \mathbf{v}_i^T \mathbf{v}_i$. The Gram–Schmidt algorithm is used to replace a given set of vectors by a linearly independent mutually orthogonal set that is "equivalent," in the sense that it spans the same subspace.

Orthogonal matrices are characterized by the following equivalent properties:

(i)   The inverse equals the transpose;

(ii) The rows are orthonormal;

(iii) The columns are orthonormal;

(iv) Multiplication by the matrix preserves inner products;

(v)   Multiplication by the matrix preserves Euclidean norms.

Least Squares

A least squares solution to an inconsistent linear system $\mathbf{Ax} = \mathbf{b}$ is a vector $\mathbf{x}$ that minimizes the sum of the squares of the differences between its left and right hand members. The left-hand side $\mathbf{Ax}$ is a best approximation, among the vectors in the span of the columns of $\mathbf{A}$, to the vector $\mathbf{b}$, in the Euclidean norm. In general, a best (Euclidean) approximation $\mathbf{w}$ to a vector $\mathbf{b}$ among all vectors in a given subspace W is characterized by the property that the residual $(\mathbf{b} - \mathbf{w})$ is orthogonal to all the vectors in the subspace.

An orthogonal decomposition $V = W \oplus W^\perp$ of a vector space V gives a unique representation of every vector $\mathbf{v}$ in V as a sum of mutually orthogonal vectors $\mathbf{w}$ in the subspace W and $\mathbf{w}^\perp$ in the suspace $W^\perp$. Then $\mathbf{w}$ is the best approximation in W to the vector $\mathbf{v}$ and the residual $\mathbf{w}^\perp$ is orthogonal to W.

It follows that a least squares solution $\mathbf{x}$ to an inconsistent m-by-n linear system $\mathbf{Ax} = \mathbf{b}$ will be a (genuine) solution to the (consistent) normal equations $\mathbf{A}^T \mathbf{Ax} = \mathbf{A}^T \mathbf{b}$.

Function Spaces

Some collections of functions may possess the properties of a vector space. Various classes of polynomials, for example, are closed under addition and scalar multiplication. We can then exploit the attendant notions of independence, spans, and bases to study them. Most function spaces are infinite-dimensional, however, and such concepts are complicated by issues of convergence. Independence can sometimes be established by considerations of growth rates, for functions defined on an infinite domain.

The similarity between the dot product of vectors in high-dimensional spaces and the Riemann sum for functional integrals suggests the extension of the idea of orthogonality to function spaces via the inner product $< f,g >:= \int_a^b f(t)g(t)dt$ and the associated norm $\|f\|^2 := \int_a^b f^2(t)dt$.

REVIEW PROBLEMS FOR PART II

1. (a) Write a matrix equation expressing the condition that all the row sums of the matrix $\mathbf{A}$ are equal; use only the symbol $\mathbf{A}$ (and not its entries a_{ij}).

   (b) Write a matrix equation expressing the conditions that all the column sums of the matrix $\mathbf{A}$ are equal.

   (c) What further properties are exhibited by the "magic square" in Albrecht Durer's Melancolia (Figure 4.7) (http://en.wikipedia.org/wiki/Melencolia_I)?

2. Is $[1\ 2\ 3]$ spanned by $[3\ 2\ 1]$ and $[1\ 1\ 1]$?

3. Find a nonzero vector that lies in both the span of $[1\ 1\ 2]$ and $[1\ 0\ 1]$, and the span of $[2\ 13\ 23]$ and $[2\ 1\ 33]$.

4. Which of these subsets of $\mathbf{R}_{\text{row}}^4$ are subspaces?

   (a) The vectors whose entries all have the same sign.

   (b) The vectors orthogonal to $[1\ 1\ 1\ 1]$.

   (c) The vectors [a b c d] satisfying ab = 0.

   (d) The vectors [a b c d] satisfying a + 2b + 3c = d.

   (e) The intersection of two arbitrary subspaces (i.e., the set of vectors belonging to both subspaces);

   (f) The union of two arbitrary subspaces (i.e., the set of vectors belonging to one subspace or the other).

16	3	2	13
5	10	11	8
9	6	7	12
4	15	14	1

Fig. 4.7   Magic Square in Albrecht Durer's Melancolia.

5. Which of these subsets of $\mathbf{R}^{4,4}$ are subspaces?
   (a) The orthogonal matrices;
   (b) The diagonal matrices;
   (c) The reflection matrices;
   (d) The Gauss-elimination row operators;
   (e) The nonsingular matrices.

6. What are the dimensions of these subspaces of $\mathbf{R}^{4,4}$? Find bases for each.
   (a) The symmetric matrices;
   (b) The skew-symmetric matrices;
   (c) The upper triangular matrices.

7. Find 2 vectors that, together with $\begin{bmatrix} 1 & 2 & 3 & 4 \end{bmatrix}$ and $\begin{bmatrix} 4 & 3 & 2 & 1 \end{bmatrix}$, form a basis for $\mathbf{R}^4_{\text{row}}$.

8. If $\mathbf{v}_1, \mathbf{v}_2, \ldots, \mathbf{v}_n$ are linearly independent, which of the following are independent:
   (a) The sums $\mathbf{v}_1 + \mathbf{v}_2,\, \mathbf{v}_2 + \mathbf{v}_3,\, \ldots,\, \mathbf{v}_{n-1} + \mathbf{v}_n,\, \mathbf{v}_1 + \mathbf{v}_n$;
   (b) The differences $\mathbf{v}_1 - \mathbf{v}_2,\, \mathbf{v}_2 - \mathbf{v}_3,\, \ldots,\, \mathbf{v}_{n-1} - \mathbf{v}_n,\, \mathbf{v}_1 - \mathbf{v}_n$?

9. Prove that any two three-dimensional subspaces of $\mathbf{R}^5_{\text{col}}$ must have a nonzero vector in common.

10. If V is the vector space of all polynomials of degree two or less, find a subset of $\{1 + 2x + 3x^2,\, 4 - x^2,\, 1 + 2x,\, 2 - x,\, 1\}$ that is a basis for V.

11. Find a basis for the space spanned by $\sin 2t, \cos 2t, \sin^2 t, \cos^2 t$, for $-\infty < t < \infty$.

12. In the Introduction to Part II, we observed that the differential equation $d^3y/dx^3 = 0$ has a set of solutions $y = c_1 x^2/2 + c_2 x + c_3$ comprising a three-dimensional vector space. Find a set of three solutions that are *orthogonal* on $[0,1]$.

13. Prove that if $\mathbf{A}$ is 5-by-3 and $\mathbf{B}$ is 3-by-5, then $\mathbf{AB}$ is singular.

14. Prove that if the rank of $\mathbf{A}$ equals one, then $\mathbf{A}^2$ is a scalar multiple of $\mathbf{A}$.

15. Is the following a basis for $\mathbf{R}^{2,2}$?

$$\begin{bmatrix} 1 & 1 \\ 1 & 0 \end{bmatrix}, \begin{bmatrix} 2 & 0 \\ 3 & 1 \end{bmatrix}, \begin{bmatrix} 0 & 0 \\ 2 & -1 \end{bmatrix}, \begin{bmatrix} -2 & 0 \\ 3 & 0 \end{bmatrix}.$$

16. (**Controllability**) Many control systems can be described by the equation $\mathbf{x}_{p+1} = \mathbf{A}\mathbf{x}_p + \mathbf{B}\mathbf{u}_p$, where the vector $\mathbf{x}_p$ describes the state of the system at time p (so the entries of $\mathbf{x}_p$ may be positions, angles, temperatures, currents, etc.); $\mathbf{A}$ and $\mathbf{B}$ are appropriately dimensioned matrices characterizing the particular system, and $\mathbf{u}_p$ is the *input vector* (at time p) selected by the system operator.
   (a) Show that any state vector $\mathbf{x}$ lying in the span of the columns of $[\mathbf{A}\mathbf{x}_0,\ \mathbf{B},\ \mathbf{AB},\ \mathbf{A}^2\mathbf{B},\ \ldots,\ \mathbf{A}^k\mathbf{B}]$ can be reached within $k + 1$ time steps by

proper choices of inputs $[\mathbf{u}_0,\ \mathbf{u}_1,\ \mathbf{u}_2,\ \ldots,\ \mathbf{u}_k]$. (*Hint*: Derive, extend, and exploit the relation $\mathbf{x}_2 = \mathbf{A}^2\mathbf{x}_0 + \mathbf{A}\mathbf{B}\mathbf{u}_0 + \mathbf{B}\mathbf{u}_1$.)

(b) If $\mathbf{A}$ is $n - by - n$, show that the system is "controllable," that is, can be driven into any state $\mathbf{x}$ from any initial state $\mathbf{x}_0$, if the rank of

$$[\mathbf{B},\ \mathbf{AB},\ \mathbf{A}^2\mathbf{B},\ \ldots,\ \mathbf{A}^s\mathbf{B}] \text{ equals } n \text{ for some } s.$$

(c) Find a sequence of inputs $\mathbf{u}_0, \mathbf{u}_1, \mathbf{u}_2$ that will drive the initial state $\mathbf{x}_0 = [1\ 1\ 1]^T$ to the state $\mathbf{x} = [1\ 0\ -1]^T$ in three time steps for the system characterized by

$$\mathbf{A} = \begin{bmatrix} 0 & 0 & 0 \\ 1 & 0 & 0 \\ 0 & 1 & 0 \end{bmatrix}, \quad \mathbf{B} = \begin{bmatrix} 1 \\ 0 \\ 0 \end{bmatrix}.$$

17. Prove that if $\mathbf{A}$ is symmetric and $\mathbf{u}^T\mathbf{A}\mathbf{u} = \mathbf{u}^T\mathbf{u}$ for every vector $\mathbf{u}$, then $\mathbf{A}$ is the identity matrix. (*Hint*: First apply the equation with vectors $\mathbf{u}$ having all zero entries except for a single "one"; then repeat, for vectors having two "ones.")

18. (a) Describe how to express an arbitrary square matrix as a sum of a symmetric matrix $(\mathbf{A} = \mathbf{A}^T)$ and a skew-symmetric matrix $(\mathbf{A} = -\mathbf{A}^T)$.

(b) Describe how to express an arbitrary square matrix as a sum of a symmetric matrix and an upper triangular matrix.

(c) Is either of these decompositions unique?

19. What is $\mathbf{u} \cdot \mathbf{v}$ if $\|\mathbf{u}\| = \|\mathbf{v}\| = \|\mathbf{u} + \mathbf{v}\| = 2$?

20. Find the least squares best approximation over the interval $[-1, 1]$ to x^3 in the form $a_1 + a_2x + a_3x^2$. (*Hint*: Use the results of Problem 12, Exercises 4.4.) How is your answer the Legendre polynomial of degree three?

TECHNICAL WRITING EXERCISES FOR PART II

1. Discuss the fact that the intersection of two subspaces of a vector space is also a subspace, but their union need not be a subspace. Illustrate with planes and lines.

2. Most mainframe, desktop, and laptop computers are binary machines; they perform arithmetic in the base-2 system, using finite-precision binary approximations to fractional numbers. However most calculators are binary-coded decimal machines; they perform decimal arithmetic using finite-precision decimal approximations, although the decimal digits are encoded in binary form. (See http://en.wikipedia.org/wiki/Binary-coded_decimal for example.) Discuss the nature of the numbers that are encoded *exactly* in each of the two systems, and describe a matrix whose rank might be reported differently by two such machines.

GROUP PROJECTS FOR PART II

A. Rotations and Reflections

Your objective in this project is to demonstrate that every 3-by-3 orthogonal matrix **Q** can be expressed as a product of rotation and reflection matrices. The following observations will help you construct the argument.

(a) The orthonormal triad $\mathbf{i} = [1\ 0\ 0]^T$, $\mathbf{j} = [0\ 1\ 0]^T$, and $\mathbf{k} = [0\ 0\ 1]^T$ represent unit vectors along each of the three axes.

(b) The three columns $\mathbf{i_Q}$, $\mathbf{j_Q}$, and $\mathbf{k_Q}$ of **Q** are the images, under multiplication by **Q**, of these three unit vectors. They also form an orthonormal triad, so $\mathbf{k_Q} = (\pm)\,\mathbf{i_Q} \times \mathbf{j_Q}$. For our purposes, one can take $\mathbf{k_Q} = \mathbf{i_Q} \times \mathbf{j_Q}$ without loss of generality by introducing, if necessary, a reflection through the xy plane. Thus the resulting orthogonal matrix is completely determined by its first two columns.

(c) Take a basketball or soccerball and imagine that the origin of the xyz axis system lies at its center. Draw dots on the ball's surface at the points where you estimate **i** and **j** penetrate the ball; also draw dots where $\mathbf{i_Q}$ and $\mathbf{j_Q}$ penetrate the ball.

(d) Let $\mathbf{M}_1$ denote any rotation of the ball that carries **i** to $\mathbf{i_Q}$. Let $\mathbf{j}'$ be the image of **j** under this rotation.

(e) Argue that it must be possible to carry $\mathbf{j}'$ to $\mathbf{j_Q}$ by an additional rotation of the ball *about the axis through* $\mathbf{i_Q}$. Call this rotation $\mathbf{M}_2$.

(f) Argue that the concatenation of these two rotations and the reflection (if needed) qualifies as an orthogonal transformation because of Theorem 2, and that the first two columns of its matrix are $\mathbf{i_Q}$ and $\mathbf{j_Q}$; thus it must equal the original orthogonal matrix.

B. Householder Reflectors

In this project, you will see how an extension of the reasoning employed in Example 2 of Section 4.2 generates alternative algorithms for solving linear systems, performing Gram–Schmidt reductions, and finding least squares solutions.

(a) Show that a second Householder reflector $\mathbf{M}'_{\mathrm{ref}}$ can be constructed to reduce the second column of the matrix in Equation (3), Section 4.2, to the form shown:

$$\mathbf{M}'_{\mathrm{ref}}\mathbf{M}_{\mathrm{ref}}\begin{bmatrix} \text{col}\ \#1 & \text{col}\ \#2 & \cdots & \text{col}\ \#n \\ \downarrow & \downarrow & & \downarrow \end{bmatrix} = \mathbf{M}'_{\mathrm{ref}}\begin{bmatrix} \|\text{col}\ \#1\| & a_{12} & \cdots & X \\ 0 & a_{22} & \cdots & X \\ 0 & a_{32} & \cdots & X \\ \vdots & \vdots & \ddots & \vdots \\ 0 & a_{n2} & \cdots & X \end{bmatrix},$$

$$\mathbf{M}'_{\text{ref}} \begin{bmatrix} \|\text{col}\,\#1\| & a_{12} & \cdots & X \\ 0 & a_{22} & \cdots & X \\ 0 & a_{32} & \cdots & X \\ \vdots & \vdots & \ddots & \vdots \\ 0 & a_{n2} & \cdots & X \end{bmatrix} = \begin{bmatrix} \|\text{col}\,\#1\| & a_{12} & X & \cdots & X \\ 0 & \beta & X & \cdots & X \\ 0 & 0 & X & \cdots & X \\ 0 & 0 & X & \cdots & X \\ 0 & 0 & X & \cdots & X \end{bmatrix}.$$

(*Hint*: You need a reflector that leaves the first component of every vector intact, but reflects the remaining components to the second axis.) Argue that the normal to the corresponding mirror is given by $[0\ \beta\ 0\ \cdots\ 0]^T - [0\ a_{22}\ a_{32}\ \cdots\ a_{n2}]^T$, where $\beta = \pm (a_{22}^2 + a_{32}^2 + \cdots + a_{n2}^2)^{1/2}$.

(b) Extend this to argue that any matrix $\mathbf{A}$ can be reduced to a row-echelon form $\mathbf{R}$ by premultiplication by a sequence of Householder reflectors $\mathbf{R} = \mathbf{M}_{\text{ref}}^{(m)} \mathbf{M}_{\text{ref}}^{(m-1)} \cdots \mathbf{M}_{\text{ref}}^{(1)} \mathbf{A}$.

(c) Argue that this shows any matrix $\mathbf{A}$ admits a factorization into an orthogonal matrix $\mathbf{Q}$ times a matrix in row echelon form $\mathbf{R}$. $\mathbf{A} = \mathbf{QR}$ is known as "the $\mathbf{QR}$ factorization."

(d) Explain how the $\mathbf{QR}$ factorization spawns an alternative for the forward part of the Gauss elimination algorithm for solving a linear system $\mathbf{Ax} = \mathbf{b}$, replacing the elementary row operations by premultiplication by Householder reflectors. (*Hint*: $\mathbf{Ax} = \mathbf{b}$ is equivalent to $\mathbf{Rx} = \mathbf{Q}^{-1}\mathbf{b}$.)

(e) Explain why pivoting is unnecessary in this procedure.

(f) For inconsistent systems, explain why the (Euclidean) norm $\|\mathbf{Ax} - \mathbf{b}\|$ is the same as $\|\mathbf{Rx} - \mathbf{Q}^{-1}\mathbf{b}\|$, and that a vector $\mathbf{x}$ minimizing the latter is obvious. Thus state an alternative procedure for finding least squares solutions.

C.   Infinite Dimensional Matrices

In Section 4.4 we pointed out that function spaces typically are infinite-dimensional, and we shied away from any reference to matrices in that context. Another, less formidable, infinite-dimensional vector space is the set of all doubly infinite sequences $\mathbf{v} = [\cdots,\ v_{-2},\ v_{-1},\ \underline{v_0},\ v_1,\ v_2,\ \cdots]$, manipulated formally like row vectors. We keep track of the addresses of the elements by underlining the entry in column #0.

The *Haar wavelet transform* extracts (backward) sums and differences from a sequence according to

$$\mathbf{sum} = [\cdots,\ v_{-4}+v_{-5},\ v_{-2}+v_{-3},\ \underline{v_0+v_{-1}},\ v_2+v_1,\ \cdots],$$

$$\mathbf{diff} = [\cdots,\ v_{-4}-v_{-5},\ v_{-2}-v_{-3},\ \underline{v_0-v_{-1}},\ v_2-v_1,\ \cdots].$$

It is amusing to express the equations for the Haar transform in matrix notation.

(a) Show that the backward sums of consecutive terms in the sequence $\mathbf{v}$ can be represented by the matrix product

$$
\mathbf{M}_+\mathbf{v}^T = \text{row } \#0 \rightarrow
\begin{bmatrix}
\ddots & \vdots & \vdots & \vdots & \vdots & \vdots & \vdots & \\
\cdots & 1 & 0 & 0 & 0 & 0 & 0 & \cdots \\
\cdots & 1 & 1 & 0 & 0 & 0 & 0 & \cdots \\
\cdots & 0 & 1 & \underline{1} & 0 & 0 & 0 & \cdots \\
\cdots & 0 & 0 & 1 & 1 & 0 & 0 & \cdots \\
\cdots & 0 & 0 & 0 & 1 & 1 & 0 & \cdots \\
\cdots & 0 & 0 & 0 & 0 & 1 & 1 & \cdots \\
& \vdots & \vdots & \vdots & \vdots & \vdots & \vdots & \ddots
\end{bmatrix}
\begin{bmatrix}
\vdots \\ v_{-2} \\ v_{-1} \\ v_0 \\ v_1 \\ v_2 \\ v_3 \\ \vdots
\end{bmatrix}
$$

$$
= \begin{bmatrix}
\vdots \\
v_{-2} + v_{-3} \\
v_{-1} + v_{-2} \\
\underline{v_0 + v_{-1}} \\
v_1 + v_0 \\
v_2 + v_1 \\
v_3 + v_2 \\
\vdots
\end{bmatrix}.
$$

(col #0 marks the ↓ column at the underlined $(0,0)$ entry)

(Note that we have underlined the $(0,0)$ entry of the matrix and the entry in row #0 of the column vector.)

(b) Construct the matrix $\mathbf{M}_-$ that forms backward differences of consecutive terms.

(c) Of course the Haar sequences **sum** and **diff** are *downsampled* versions of $\mathbf{M}_+\mathbf{v}^T$ and $\mathbf{M}_-\mathbf{v}^T$; they only retain every other term. Construct the downsampling matrix $\mathbf{M}_\downarrow$ that forms (the transposes of)

$$\mathbf{v}_\downarrow = [\cdots, v_{-2}, \underline{v_0}, v_2, \cdots] \text{ from } \mathbf{v} = [\cdots, v_{-2}, v_{-1}, \underline{v_0}, v_1, v_2, \cdots].$$

(d) Form the matrix products $\mathbf{M}_\downarrow\mathbf{M}_+$ and $\mathbf{M}_\downarrow\mathbf{M}_-$ and verify that they directly execute the Haar operations resulting in **sum** and **diff**.

(e) The upsampling matrix $\mathbf{M}_\uparrow$ inserts zeros into a sequence, forming (the transposes of)

$$\mathbf{v}_\uparrow = [\cdots, v_{-2}, 0, v_{-1}, 0, \underline{v_0}, 0, v_1, 0, v_2, \cdots] \text{ from}$$
$$\mathbf{v} = [\cdots, v_{-2}, v_{-1}, \underline{v_0}, v_1, v_2, \cdots].$$

Construct $\mathbf{M}_\uparrow$.

(f) The left-shift matrix $\mathbf{M}_{\leftarrow}$ forms $\mathbf{v}_{\leftarrow} = [\cdots, v_{-1}, v_0, \underline{v_1}, v_2, v_3, \cdots]$ from $\mathbf{v} = [\cdots, v_{-2}, v_{-1}, \underline{v_0}, v_1, v_2, \cdots]$. Construct $\mathbf{M}_{\leftarrow}$.

(g) An important property of the Haar transform is that it is easily inverted; the sequences $\frac{\text{sum}+\text{diff}}{2}$ and $\frac{\text{sum}-\text{diff}}{2}$, when properly interleaved, restore the original sequence. Devise a combination of $\mathbf{M}_{\uparrow}, \mathbf{M}_{\leftarrow}, \frac{\text{sum}+\text{diff}}{2}$ and $\frac{\text{sum}-\text{diff}}{2}$, that equals $\mathbf{v}$. Verify that the identity matrix results when $\mathbf{M}_{+}, \mathbf{M}_{-}, \mathbf{M}_{\uparrow}$, and $\mathbf{M}_{\leftarrow}$ are properly combined.

(h) In signal processing, the left-shift operation corresponds to a *time advance* and is physically impossible. Define the time delay matrix $\mathbf{M}_{\rightarrow}$ and argue that a combination of $\mathbf{M}_{\uparrow}, \mathbf{M}_{\rightarrow}, \frac{\text{sum}+\text{diff}}{2}$ and $\frac{\text{sum}-\text{diff}}{2}$ can be constructed to yield a time-delayed version of the original sequence $\mathbf{v}$.

PART III

INTRODUCTION: REFLECT ON THIS

A mirror passes through the origin of the x, y-plane. It reflects the unit vector $[1\ 0]^T$ into the vector $[0.8\ 0.6]^T$ (Fig. III.1a), and the vector $[0\ 1]^T$ is reflected to $[0.6\ -0.8]^T$ (Fig. III.1b). Find a vector normal to the mirror.

The specifications of this problem give us sufficient information to write down the reflector matrix $\mathbf{M}_{\text{ref}}$ (or simply $\mathbf{M}$): since $\mathbf{M}\begin{bmatrix}1\\0\end{bmatrix}$ equals the first column of $\mathbf{M}$ and $\mathbf{M}\begin{bmatrix}0\\1\end{bmatrix}$ equals the second,

$$\mathbf{M} = \begin{bmatrix} 0.8 & 0.6 \\ 0.6 & -0.8 \end{bmatrix}. \tag{1}$$

In this section our strategy for finding a normal to the mirror is based on the observation that the mirror reflection of a normal $\mathbf{n}$ is $-\mathbf{n}$:

$$\mathbf{Mn} = -\mathbf{n} \tag{2}$$

(Fig. III.1c).

Equation (2) can be rewritten, using the identity matrix $\mathbf{I}$, as an equation for a null vector of $\mathbf{M} + \mathbf{I}$:

$$\mathbf{Mn} + \mathbf{n} = (\mathbf{M} + \mathbf{I})\mathbf{n} = \mathbf{0}. \tag{3}$$

Fundamentals of Matrix Analysis with Applications,
First Edition. Edward Barry Saff and Arthur David Snider.
© 2016 John Wiley & Sons, Inc. Published 2016 by John Wiley & Sons, Inc.

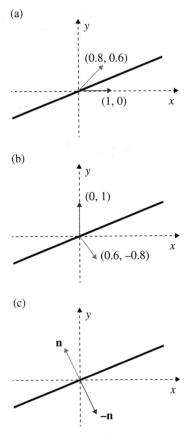

Fig. III.1   (a) Mirror image of [1, 0], (b) mirror image of [0, 1], and (c) the normal.

For (3) to have a nontrivial solution, the matrix $\mathbf{M} + \mathbf{I}$ must be singular, and the Gauss elimination algorithm will reveal its null vectors. The steps go as follows:

$$
\left[\mathbf{M} + \mathbf{I} \vdots \mathbf{0}\right] = \begin{bmatrix} 0.8 + 1 & 0.6 & \vdots 0 \\ 0.6 & -0.8 + 1 & \vdots 0 \end{bmatrix}
$$

$$
= \begin{bmatrix} 1.8 & 0.6 & \vdots 0 \\ 0.6 & 0.2 & \vdots 0 \end{bmatrix} \left(-\frac{1}{3}\rho_1 + \rho_2 \to \rho_2 \right) \begin{bmatrix} 1.8 & 0.6 & \vdots 0 \\ 0 & 0 & \vdots 0 \end{bmatrix},
$$

so that

$$
\mathbf{n} = t \begin{bmatrix} -1/3 \\ 1 \end{bmatrix},
$$

where t is any scalar. A quick calculation confirms (2).

In fact, insertion of the *unit* normal $[-1, 3]^T \frac{1}{\sqrt{10}}$ into Equation (3) of Section 2.2 reproduces the matrix $\mathbf{M}$ in (1), confirming that it is, indeed, a reflector operator.

Now we generalize. Whenever the *direction* of a vector is unchanged (or reversed) by a matrix multiplication, we say that the vector is an **eigenvector** of the matrix. More specifically, if $\mathbf{v}$ is an eigenvector, $\mathbf{Mv}$ is simply a scalar multiple of $\mathbf{v}$:

$$\mathbf{Mv} = r\mathbf{v}, \tag{4}$$

and the scalar r is called the **eigenvalue** corresponding to $\mathbf{v}$. In (4), we allow the possibility that the eigenvalue r can be negative or zero or even complex. But the zero *vector*, since it has no "direction," is excluded from the family of eigen*vectors*, although it trivially satisfies (4) (for *every* scalar r). The mirror's normal $\mathbf{n}$ is an eigenvector corresponding to the eigenvalue -1 because of Equation (2).

In fact, any vector *parallel* to the mirror in Figure III.1 is also an eigenvector, since it is *unchanged* by reflection; its eigenvalue is $+1$.

It is extremely profitable to know a matrix's eigenvectors, mainly because an n-by-n matrix—despite all its n^2 pieces of data—simply acts like a scalar when multiplying an eigenvector, and this can be exploited in a myriad of applications (including differential equations). In this chapter, we will examine the eigenvector–eigenvalue theory for matrices.

5

EIGENVECTORS AND EIGENVALUES

5.1   EIGENVECTOR BASICS

We formalize the deliberations of the introduction to Part III as follows:

Eigenvectors and Eigenvalues

> **Definition 1.** Let $\mathbf{A} = [a_{ij}]$ be a square matrix. The **eigenvalues** of $\mathbf{A}$ are those real or complex numbers r for which the equation
>
> $$\mathbf{A}\mathbf{u} = r\mathbf{u} \text{ or, equivalently, } (\mathbf{A} - r\mathbf{I})\,\mathbf{u} = \mathbf{0}, \tag{1}$$
>
> has at least one nontrivial (nonzero) solution $\mathbf{u}$. The nontrivial solutions are called the **eigenvectors** of $\mathbf{A}$ associated with r.

We have seen that the vectors normal and parallel to a mirror are eigenvectors of the reflector matrix. The normal vectors share the eigenvalue -1 and form a one-dimensional subspace, and the vectors parallel to the mirror share the eigenvalue $+1$ and also form a one-dimensional subspace (in two dimensions). (In three-dimensional space the vectors parallel to the mirror form, of course, a two-dimensional subspace.)

Here are some other examples of eigenvectors and eigenvalues.

(i) A matrix that performs a rotation in three-dimensional space has, as eigenvectors, all vectors parallel to the axis of rotation (what is the eigenvalue?).

Fundamentals of Matrix Analysis with Applications,
First Edition. Edward Barry Saff and Arthur David Snider.

But a rotation in *two* dimensions changes every vector's direction (unless the rotation is through an integer multiple of 180°). In both of these cases, we shall see that rotation matrices *do* have more eigenvectors, but they are complex (Example 4, Section 5.2).

(ii) Every singular matrix has the eigenvalue zero; its null vectors **u** satisfy **Au** = 0**u**.

(iii) Each number on the diagonal of a diagonal matrix is an eigenvalue, with eigenvectors exemplified by the following display:

$$
\begin{bmatrix}
d_{11} & 0 & 0 & \cdots & 0 \\
0 & d_{22} & 0 & \cdots & 0 \\
0 & 0 & d_{33} & \cdots & 0 \\
\vdots & & & \ddots & \vdots \\
0 & 0 & 0 & \cdots & d_{nn}
\end{bmatrix}
\begin{bmatrix} 0 \\ 0 \\ 1 \\ \vdots \\ 0 \end{bmatrix}
= d_{33}
\begin{bmatrix} 0 \\ 0 \\ 1 \\ \vdots \\ 0 \end{bmatrix}
$$

More generally, if any *single* column of a square matrix has this "diagonal structure," the matrix will have an eigenvector as indicated:

$$
\begin{bmatrix}
x & x & 0 & x & \cdots & x \\
x & x & 0 & x & \cdots & x \\
x & x & d & x & \cdots & x \\
x & x & 0 & x & \cdots & x \\
\vdots & \vdots & \vdots & \vdots & & \vdots \\
x & x & 0 & x & \cdots & x
\end{bmatrix}
\begin{bmatrix} 0 \\ 0 \\ 1 \\ 0 \\ \vdots \\ 0 \end{bmatrix}
= d
\begin{bmatrix} 0 \\ 0 \\ 1 \\ 0 \\ \vdots \\ 0 \end{bmatrix}
$$

(iv) The matrices that perform Gauss elimination's elementary row operations also have some obvious eigenvectors. A matrix that, through multiplication, adds a multiple of row j to row k of a column vector has, as eigenvectors, all nontrivial vectors with a zero in row j; a matrix that switches two rows of a column vector will have, as eigenvectors, any nontrivial vector with identical entries in those rows:

$$
\begin{bmatrix}
1 & 0 & 0 \\
0 & 1 & 0 \\
c & 0 & 1
\end{bmatrix}
\begin{bmatrix} 0 \\ s \\ t \end{bmatrix}
= (1)
\begin{bmatrix} 0 \\ s \\ t \end{bmatrix} \; ;
\qquad
\begin{bmatrix}
0 & 0 & 1 \\
0 & 1 & 0 \\
1 & 0 & 0
\end{bmatrix}
\begin{bmatrix} s \\ t \\ s \end{bmatrix}
= (1)
\begin{bmatrix} s \\ t \\ s \end{bmatrix} .
$$

(v) The product **Mu** with $\mathbf{u} = [1\ 1\ 1 \ldots 1]^T$ tabulates the row sums of **M**. Thus if all the row sums of **M** are equal, **u** is an eigenvector.

(vi) An eigenvector of **A** is also an eigenvector of $\mathbf{A} + s\mathbf{I}$, because $\mathbf{Au} = r\mathbf{u}$ implies $(\mathbf{A} + s\mathbf{I})\mathbf{u} = r\mathbf{u} + s\mathbf{u} = (r+s)\mathbf{u}$. The eigenvalue has been *shifted* by s.

(vii) An eigenvector **u** of **A** is also an eigenvector of $\mathbf{A}^n$ for any positive integer n. For example, $\mathbf{Au} = r\mathbf{u}$ implies $\mathbf{A}^2\mathbf{u} = \mathbf{A}(\mathbf{Au}) = \mathbf{A}(r\mathbf{u}) = r\mathbf{Au} = r^2\mathbf{u}$, and analogous calculations hold for $\mathbf{A}^n$; the eigenvalue is r^n. In fact if $\mathbf{A}$ is invertible, it follows easily that $\mathbf{A}^{-1}\mathbf{u} = r^{-1}\mathbf{u}$; and $\mathbf{u}$ is an eigenvector of $\mathbf{A}^{-n}$ also.

(viii) Similarly,

$$\left(a_n\mathbf{A}^n + a_{n-1}\mathbf{A}^{n-1} + \cdots + a_1\mathbf{A} + a_0\mathbf{I}\right)\mathbf{u} = \left(a_n r^n + a_{n-1}r^{n-1} + \cdots + a_1 r + a_0\right)\mathbf{u}$$

for an eigenvector $\mathbf{u}$. Thus for any polynomial $p(t)$, $\mathbf{u}$ is an eigenvector of $p(\mathbf{A})$ with eigenvalue $p(r)$.

It is important to note that any (nonzero) multiple of an eigenvector is also an eigenvector, with the same eigenvalue. (So eigenvectors are really "eigendirections," but this nomenclature has never caught on.) In fact if $\mathbf{u}_1, \mathbf{u}_2, \ldots, \mathbf{u}_k$ each satisfy $\mathbf{A}\mathbf{u} = r\mathbf{u}$ with the same eigenvalue r, so does any linear combination:

$$\begin{aligned}
\mathbf{A}\left(c_1\mathbf{u}_1 + c_2\mathbf{u}_2 + \cdots + c_k\mathbf{u}_k\right) &= c_1\mathbf{A}\mathbf{u}_1 + c_2\mathbf{A}\mathbf{u}_2 + \cdots + c_k\mathbf{A}\mathbf{u}_k \\
&= c_1 r\mathbf{u}_1 + c_2 r\mathbf{u}_2 + \cdots + c_k r\mathbf{u}_k \\
&= r\left(c_1\mathbf{u}_1 + c_2\mathbf{u}_2 + \cdots + c_k\mathbf{u}_k\right).
\end{aligned}$$

Therefore the set of all eigenvectors *with the same eigenvalue*, together with the zero vector (which we have excluded as an "eigenvector"), form a subspace—an "**eigenspace**."

On the other hand, eigenvectors with *different* eigenvalues are linearly independent. To see this, suppose that we have a collection of eigenvectors $\{\mathbf{u}_i\}$, each having a distinct eigenvalue (r_i). Choose any two; they will be independent, because if one were a multiple of the other they would have the same eigenvalue. Now keep adding, one by one, more eigenvectors from the collection *as long they are independent of the set already accumulated*, until you have a maximal independent set $\mathbf{u}_1, \mathbf{u}_2, \ldots, \mathbf{u}_k$. If there is one left over, $\mathbf{u}_{k+1}$, it would be linearly dependent on the set and expressible as

$$\mathbf{u}_{k+1} = c_1\mathbf{u}_1 + c_2\mathbf{u}_2 + \cdots + c_k\mathbf{u}_k \tag{2}$$

with not all the c_i's zero (since $\mathbf{u}_{k+1}$ is not the zero vector). But multiplication of (2) by $\mathbf{A} - r_{k+1}\mathbf{I}$ would yield

$$\left(r_{k+1} - r_{k+1}\right)\mathbf{u}_{k+1} = c_1\left(r_1 - r_{k+1}\right)\mathbf{u}_1 + c_2\left(r_2 - r_{k+1}\right)\mathbf{u}_2 + \cdots + c_k\left(r_k - r_{k+1}\right)\mathbf{u}_k. \tag{3}$$

Since (3) displays a nontrivial linear combination of $\mathbf{u}_1$ through $\mathbf{u}_k$ that equals $\mathbf{0}$, the postulated independence of $\mathbf{u}_1$ through $\mathbf{u}_k$ would be contradicted. We summarize our findings in the following theorems.

Spans of Sets of Eigenvectors

Theorem 1.   If **A** is any square matrix, the set of all eigenvectors of **A** with the same eigenvalue, together with the zero vector, forms a subspace.

Theorem 2.   Any collection of eigenvectors of **A** that correspond to distinct eigenvalues is linearly independent.

The exercises explore other examples of eigenvector–eigenvalue pairs that can be visualized. A formal procedure for calculating eigenvectors will be described in the next section.

Exercises 5.1

1. What are the eigenvalues and eigenvectors of an orthogonal projector matrix? (Recall Example 2, Section 2.2.)

2. If we regard the three-dimensional vector **m** as fixed, the cross-product $\mathbf{w} = \mathbf{m} \times \mathbf{v}$ is a linear operation on the vector **v**, and can be expressed as a matrix product $\mathbf{w} = \mathbf{M}\mathbf{v}$ (with **w** and **v** as column vectors). *Before* you work out the specific form of the matrix **M** (in terms of the components of **m**), identify an eigenvector–eigenvalue pair for the cross product. Then write out **M** and verify your eigenvector equation.

   Problems 3 and 4 explore the eigenvector–eigenvalue structure of the matrices

$$\mathbf{F} = \begin{bmatrix} 0 & 1 & 1 & 1 & 1 & 1 & 1 \\ 1 & 0 & 1 & 1 & 1 & 1 & 1 \\ 1 & 1 & 0 & 1 & 1 & 1 & 1 \\ 1 & 1 & 1 & 0 & 1 & 1 & 1 \\ 1 & 1 & 1 & 1 & 0 & 1 & 1 \\ 1 & 1 & 1 & 1 & 1 & 0 & 1 \\ 1 & 1 & 1 & 1 & 1 & 1 & 0 \end{bmatrix}, \qquad \mathbf{G} = \begin{bmatrix} a & b & b & b & b & b & b \\ b & a & b & b & b & b & b \\ b & b & a & b & b & b & b \\ b & b & b & a & b & b & b \\ b & b & b & b & a & b & b \\ b & b & b & b & b & a & b \\ b & b & b & b & b & b & a \end{bmatrix}.$$

3. Because volleyball is an extremely team-coordinated activity, it is difficult to assess the value of a player in isolation. But it is noted that the Gauss Eliminators' play suffers when Saff sits on the bench and improves when Snider is sitting, while rotating the other five members does not seem to make a difference—as long as Saff and Snider are both playing. Thus the *relative* effectiveness of each player can roughly be expressed by the vector

	Saff	Snider	Aniston	Jolie	Berry	Kardashian	Gaga
v =	+1	−1	0	0	0	0	0

Note that the jth entry of the product $\mathbf{F}\mathbf{v}^T$ (with **F** given above) measures the effectiveness of the team when the jth player is sitting. From these observations, construct six linearly independent eigenvectors of **F** with the eigenvalue -1. What is the remaining eigenvalue (and a corresponding eigenvector)?

4. What are the eigenvectors and eigenvalues of **G**? [*Hint*: Express **G** in terms of **F** and **I**.]

5. What are the eigenvectors of the Gauss-elimination matrix that multiplies one row of a matrix vector by a constant?

6. In this problem you will prove that *all of the diagonal entries of a triangular matrix are eigenvalues*. To fix our ideas let's look at the 4-by-4 upper triangular case:

$$\mathbf{U} = \begin{bmatrix} u_{11} & u_{12} & u_{13} & u_{14} \\ 0 & u_{22} & u_{23} & u_{24} \\ 0 & 0 & u_{33} & u_{34} \\ 0 & 0 & 0 & u_{44} \end{bmatrix}$$

(a) Find an eigenvector with u_{11} as eigenvalue. [*Hint*: This is easy; look for an eigenvector with only one nonzero entry.]

(b) Try to find an eigenvector corresponding to u_{22}, of the form $\mathbf{v} = \begin{bmatrix} x & 1 & 0 & 0 \end{bmatrix}^T$. Show that the equation $\mathbf{Uv} = u_{22}\mathbf{v}$ only requires $u_{11}x + u_{12} = u_{22}x$, or $x = u_{12}/(u_{22} - u_{11})$ if $u_{22} \neq u_{11}$. And if u_{22} *does* equal u_{11}, we have already proved it is an eigenvalue.

(c) Show that if u_{33} is not a duplicate of one of the earlier diagonal entries, then it has an eigenvector of the form $\begin{bmatrix} x & y & 1 & 0 \end{bmatrix}$. And so on.

7. Following the strategy of the preceding problem, show that

$$\mathbf{U} = \begin{bmatrix} 1 & 1 & 1 & 1 \\ 0 & 2 & \alpha & 1 \\ 0 & 0 & 2 & 1 \\ 0 & 0 & 0 & 4 \end{bmatrix}$$

has an eigenvector corresponding to the eigenvalue 2, of the form $\begin{bmatrix} x & 1 & 0 & 0 \end{bmatrix}^T$; but it only has an eigenvector of the form $\begin{bmatrix} y & z & 1 & 0 \end{bmatrix}^T$ if $\alpha = 0$.

The entries of the complex vector $\mathbf{u}_4 = \begin{bmatrix} i & -1 & -i & 1 \end{bmatrix}^T$, *depicted in Figure 5.1, are the* 4th *roots of unity,* $\sqrt[4]{1}$. *If they are regarded cyclically counter-clockwise (so that i "follows" 1), they have some properties that can be expressed as eigenvector equations. Refer to Figure 5.1 to assist in answering Problems 8–11.*

8. From Figure 5.1, note that each root equals i times the previous root. Interpret this as an eigenvector statement for the matrix

$$\mathbf{A} = \begin{bmatrix} 0 & 1 & 0 & 0 \\ 0 & 0 & 1 & 0 \\ 0 & 0 & 0 & 1 \\ 1 & 0 & 0 & 0 \end{bmatrix}.$$

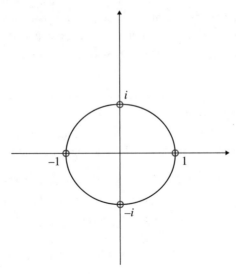

Fig. 5.1   The fourth roots of unity.

9. From Figure 5.1, note that the difference of the successor and the predecessor of each root in the list equals $2i$ times the root. Interpret this as an eigenvector statement for the matrix

$$\mathbf{B} = \begin{bmatrix} 0 & 1 & 0 & -1 \\ -1 & 0 & 1 & 0 \\ 0 & -1 & 0 & 1 \\ 1 & 0 & -1 & 0 \end{bmatrix}.$$

10. From Figure 5.1, note that the sum of the successor and predecessor of each root equals zero. Interpret this as an eigenvector statement for the matrix

$$\mathbf{C} = \begin{bmatrix} 0 & 1 & 0 & 1 \\ 1 & 0 & 1 & 0 \\ 0 & 1 & 0 & 1 \\ 1 & 0 & 1 & 0 \end{bmatrix}.$$

11. (a) The fourth roots of unity can also be written as the consecutive powers of $i \equiv e^{i2\pi/4}$, the "primitive" fourth root:

$$\mathbf{u}_4 = \begin{bmatrix} i & i^2 & i^3 & i^4 \end{bmatrix}^T.$$

If we take the vector consisting of the four consecutive powers of the *second* root in $\mathbf{u}_4$, namely $i^2 = -1$, we get another sequence

$$\begin{bmatrix} i^2 & i^4 & i^6 & i^8 \end{bmatrix}^T = \begin{bmatrix} -1 & 1 & -1 & 1 \end{bmatrix}^T$$

that enjoys neighbor relations similar to those described in Problems 8 through 10; specifically, each entry is a fixed multiple of the previous entry, and the sums/differences of each entry's successor and predecessor are fixed multiples of the entry. Hence the vector is another eigenvector of **A**, **B**, and **C**. What are the corresponding eigenvalues?

(b) The consecutive powers of the third, and even the fourth, entries in $\mathbf{u}_4$:

$$\begin{bmatrix} i^3 & i^6 & i^9 & i^{12} \end{bmatrix}^T = \begin{bmatrix} -i & -1 & i & 1 \end{bmatrix}^T \text{ and}$$
$$\begin{bmatrix} i^4 & i^8 & i^{12} & i^{16} \end{bmatrix}^T = \begin{bmatrix} 1 & 1 & 1 & 1 \end{bmatrix}^T,$$

continue to satisfy similar neighbor relations. What are the corresponding eigenvector statements for **A**, **B**, and **C**?

12. The matrices **A**, **B**, and **C** in Problems 8–11 are special instances of the general 4-by-4 **circulant** matrix:

$$\mathbf{D} = \begin{bmatrix} d_1 & d_2 & d_3 & d_4 \\ d_4 & d_1 & d_2 & d_3 \\ d_3 & d_4 & d_1 & d_2 \\ d_2 & d_3 & d_4 & d_1 \end{bmatrix},$$

wherein each row is a cyclic right-shift of the preceding row. Show that $\mathbf{u}_4$ and the three vectors described in Problem 11 are common eigenvectors of every circulant matrix. What are the corresponding eigenvalues? [*Hint*: Figure 5.1 will probably not be of any help; this problem should be tackled algebraically.]

13. Repeat the analyses of Problems 8–12 for the vector $\mathbf{u}_6$ constructed from the powers of the primitive sixth root of unity $\omega = e^{i\pi/3} = \sqrt[6]{1}$;

$$\mathbf{u}_6 = \begin{bmatrix} e^{i\pi/3} & e^{i2\pi/3} & e^{i3\pi/3} & e^{i4\pi/3} & e^{i5\pi/3} & e^{i6\pi/3} \end{bmatrix}^T$$
$$= \begin{bmatrix} \omega & \omega^2 & \omega^3 & \omega^4 & \omega^5 & \omega^6 \end{bmatrix}^T.$$

(See Fig. 5.2.) In particular, construct six vectors that are eigenvectors for each of the 6-by-6 circulant matrices corresponding to **A**, **B**, **C**, and **D**.

14. Generalize the circulant matrix in Problem 12 to its n-by-n form and formulate the corresponding eigenvector statements.[1]

----

[1] The vectors

$$\begin{bmatrix} \omega & \omega^2 & \omega^3 & \cdots & \omega^n \end{bmatrix}^T, \begin{bmatrix} \omega^2 & \omega^4 & \omega^6 & \cdots & \omega^{2n} \end{bmatrix}^T, \ldots, \begin{bmatrix} 1 & 1 & 1 & \cdots \end{bmatrix}^T,$$
$$\omega = e^{i2\pi/n} \quad (\omega^n = 1)$$

are discrete analogs of the "finite Fourier series." The fact that they are eigenvectors of every circulant matrix is crucial in the science of digital signal processing. See Group Project C at the end of Part III.

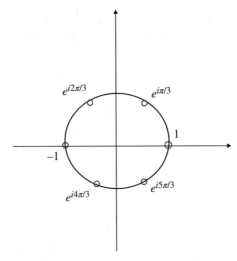

Fig. 5.2 The sixth roots of unity.

15. Show that if **u** is an eigenvector of (**AB**), then (**Bu**) is an eigenvector of **BA**, and that **AB** and **BA** have the same eigenvalues.

16. If $\mathbf{Au} = r\mathbf{u}$, what is $(3\mathbf{A}^3 - 4\mathbf{A} + 2\mathbf{I})\mathbf{u}$?

17. Show that if none of the eigenvalues of **A** equals one, then $\mathbf{I} - \mathbf{A}$ is invertible.

18. If $\mathbf{Au} = r_1\mathbf{u}$ and $\mathbf{Bu} = r_2\mathbf{u}$ (and $\mathbf{u} \neq \mathbf{0}$), show that **u** is an eigenvector of $\mathbf{A} + \mathbf{B}$ and **AB**. What are the corresponding eigenvalues?

19. Property (vi) in the text points out that changing the matrix **A** to $\mathbf{A} + s\mathbf{I}$ shifts every eigenvalue of **A** by s. In numerical analysis, it is convenient to be able to shift only one eigenvalue, leaving the others intact. If the eigenvalue is replaced by 0, the matrix is said to be *deflated*. In this problem, you will accomplish this with a "rank-one correction"—that is, adding a matrix of rank one to **A**. (Recall Problem 28, Exercises 3.3.) Suppose **A** is n-by-n and has the eigenvalue–eigenvector pairs $\mathbf{Au}_1 = r_1\mathbf{u}_1$, $\mathbf{Au}_2 = r_2\mathbf{u}_2$, ..., $\mathbf{Au}_m = r_m\mathbf{u}_m$.

    (a) Let **b** be any nonzero n-by-1 vector. Verify that $\mathbf{u}_1$ is an eigenvector of $\mathbf{A} + \mathbf{u}_1\mathbf{b}^T$ with corresponding eigenvalue

    $$r_1' = r_1 + \mathbf{b}^T\mathbf{u}_1.$$

    (b) Show that every *other* eigenvalue $r_2, \ldots, r_m$ of **A** is still an eigenvalue of $\mathbf{A} + \mathbf{u}_1\mathbf{b}^T$. [*Hint*: if $r_2 = r_1'$, part (a) has already shown that it is an eigenvalue. Otherwise right-multiply

    $$\mathbf{A} + \mathbf{u}_1\mathbf{b}^T \text{ by } \mathbf{u}_2 + \left[\mathbf{b}^T\mathbf{u}_2/(r_2 - r_1')\right]\mathbf{u}_1$$

    and observe the eigenvector equation for r_2.]

20. Show that every n-by-n matrix having $\mathbf{u}$ as an eigenvector with r as the corresponding eigenvalue can be constructed by assembling n vectors orthogonal to $\mathbf{u}$ into the rows of a matrix, and adding r to each diagonal entry.

5.2 CALCULATING EIGENVALUES AND EIGENVECTORS

How can we find eigenvalues and eigenvectors? The complete story[2] is contained in the relation $[\mathbf{A} - r\mathbf{I}]\mathbf{u} = \mathbf{0}$:

(i) the eigenvalue r must be a number that renders $\mathbf{A} - r\mathbf{I}$ singular—that is, $\det(\mathbf{A} - r\mathbf{I}) = 0$;

(ii) with r specified, the eigenvectors $\mathbf{u}$ are found by solving $[\mathbf{A} - r\mathbf{I}]\mathbf{u} = \mathbf{0}$ through Gauss elimination.

As seen from the cofactor expansion theorem (Theorem 5 of Section 2.4), the determinant of a matrix is a sum/difference of terms that are products of its entries, exactly one from each row and column. So for an n-by-n matrix $\mathbf{A}$, the term containing the highest number of occurrences of the parameter r in $\det(\mathbf{A} - r\mathbf{I})$ is the product of the diagonal entries, generating $(-r)^n$. Therefore, $\det(\mathbf{A} - r\mathbf{I})$ is an nth degree polynomial in r, and finding the eigenvalues of a matrix is equivalent to finding the zeros of its **characteristic polynomial** $p(r)$,

$$p(r) := \det(\mathbf{A} - r\mathbf{I}) = (-r)^n + c_{n-1}r^{n-1} + \cdots + c_0$$
$$= (-1)^n(r - r_1)\cdots(r - r_n). \tag{1}$$

The equation $p(r) = 0$ known as the **characteristic equation** of $\mathbf{A}$.

Problem 6 in Exercises 5.1 argued that the eigenvalues of a triangular matrix are its diagonal entries. This is readily confirmed by observing the characteristic polynomial. For an upper triangular matrix,

$$\det \begin{bmatrix} a_{11} - r & a_{12} & a_{13} & \cdots & a_{1n} \\ 0 & a_{22} - r & a_{23} & \cdots & a_{2n} \\ & & & \vdots & \\ 0 & 0 & 0 & & a_{nn} - r \end{bmatrix} = (a_{11} - r)(a_{22} - r)\ldots(a_{nn} - r).$$

The following examples of eigenvector calculations illustrate the basic procedure and the complications that result when the characteristic equation has complex or multiple roots.

---

[2]Section 6.5 addresses some important practical computational issues.

Example 1. Find the eigenvalues and eigenvectors of the matrix

$$\mathbf{A} := \begin{bmatrix} 2 & 1 \\ -3 & -2 \end{bmatrix}.$$

Solution. The characteristic equation for $\mathbf{A}$ is

$$|\mathbf{A} - r\mathbf{I}| = \begin{bmatrix} 2 - r & 1 \\ -3 & -2 - r \end{bmatrix} = (2 - r)(-2 - r) + 3 = r^2 - 1 = 0 .$$

Hence the eigenvalues of $\mathbf{A}$ are $r_1 = 1$, $r_2 = -1$. To find the eigenvectors corresponding to $r_1 = 1$, we must solve $(\mathbf{A} - r_1\mathbf{I})\mathbf{u} = \mathbf{0}$. Substituting for $\mathbf{A}$ and r_1 gives

$$\begin{bmatrix} 1 & 1 \\ -3 & -3 \end{bmatrix} \begin{bmatrix} u_1 \\ u_2 \end{bmatrix} = \begin{bmatrix} 0 \\ 0 \end{bmatrix}. \qquad (2)$$

Gauss elimination (or mental arithmetic) shows the solutions to (2) are given by $u_2 = s$, $u_1 = -s$. Consequently, the eigenvectors associated with $r_1 = 1$ can be expressed as

$$\mathbf{u}_1 = s \begin{bmatrix} -1 \\ 1 \end{bmatrix}. \qquad (3)$$

For $r_2 = -1$, the equation $(\mathbf{A} - r_2\mathbf{I})\mathbf{u} = \mathbf{0}$ becomes

$$\begin{bmatrix} 3 & 1 \\ -3 & -1 \end{bmatrix} \begin{bmatrix} u_1 \\ u_2 \end{bmatrix} = \begin{bmatrix} 0 \\ 0 \end{bmatrix}.$$

Solving, we obtain $u_1 = -t$ and $u_2 = 3t$, with t arbitrary. Therefore, the eigenvectors associated with the eigenvalue $r_2 = -1$ are

$$\mathbf{u}_2 = t \begin{bmatrix} -1 \\ 3 \end{bmatrix}. \qquad (4)$$

■

Example 2. Find the eigenvalues and eigenvectors of the matrix

$$\mathbf{A} := \begin{bmatrix} 1 & 2 & -1 \\ 1 & 0 & 1 \\ 4 & -4 & 5 \end{bmatrix}.$$

Solution. The characteristic equation for $\mathbf{A}$ is

$$|\mathbf{A} - r\mathbf{I}| = \begin{bmatrix} 1-r & 2 & -1 \\ 1 & -r & 1 \\ 4 & -4 & 5-r \end{bmatrix} = 0,$$

which simplifies to $(r-1)(r-2)(r-3) = 0$. Hence, the eigenvalues of $\mathbf{A}$ are $r_1 = 1$, $r_2 = 2$, and $r_3 = 3$. To find the eigenvectors corresponding to $r_1 = 1$, we set $r = 1$ in $(\mathbf{A} - r\mathbf{I})\mathbf{u} = \mathbf{0}$. This gives

$$\begin{bmatrix} 0 & 2 & -1 \\ 1 & -1 & 1 \\ 4 & -4 & 4 \end{bmatrix} \begin{bmatrix} u_1 \\ u_2 \\ u_3 \end{bmatrix} = \begin{bmatrix} 0 \\ 0 \\ 0 \end{bmatrix}. \tag{5}$$

Gauss elimination readily provides a parametrization of the solution set for (5). However the system is so transparent that, again, mental arithmetic suffices. The third equation is clearly redundant since it's a multiple of the second equation. And forward substitution reveals the solutions to the first two equations to be

$$u_2 = s,$$
$$u_3 = 2s, \quad u_1 = -s.$$

Hence, the eigenvectors associated with $r_1 = 1$ are

$$\mathbf{u}_1 = s \begin{bmatrix} -1 \\ 1 \\ 2 \end{bmatrix}. \tag{6}$$

For $r_2 = 2$ we solve

$$\begin{bmatrix} -1 & 2 & -1 \\ 1 & -2 & 1 \\ 4 & -4 & 3 \end{bmatrix} \begin{bmatrix} u_1 \\ u_2 \\ u_3 \end{bmatrix} = \begin{bmatrix} 0 \\ 0 \\ 0 \end{bmatrix}$$

in a similar fashion to obtain the eigenvectors

$$\mathbf{u}_2 = s \begin{bmatrix} -2 \\ 1 \\ 4 \end{bmatrix}. \tag{7}$$

Finally, for $r_3 = 3$ we solve

$$\begin{bmatrix} -2 & 2 & -1 \\ 1 & -3 & 1 \\ 4 & -4 & 2 \end{bmatrix} \begin{bmatrix} u_1 \\ u_2 \\ u_3 \end{bmatrix} = \begin{bmatrix} 0 \\ 0 \\ 0 \end{bmatrix}$$

and get the eigenvectors

$$\mathbf{u}_3 = s \begin{bmatrix} -1 \\ 1 \\ 4 \end{bmatrix}. \tag{8}$$

∎

Example 3. Find the eigenvalues and eigenvectors for

$$\mathbf{A} = \begin{bmatrix} 1 & 2 & 0 \\ 0 & 1 & -2 \\ 2 & 2 & -1 \end{bmatrix}.$$

Solution. The characteristic equation,

$$|\mathbf{A} - r\mathbf{I}| = \begin{vmatrix} 1 - r & 2 & 0 \\ 0 & 1 - r & -2 \\ 2 & 2 & -1 - r \end{vmatrix} = -r^3 + r^2 - 3r - 5 = 0,$$

simplifies to

$$(r + 1)\ (r^2 - 2r + 5) = (r + 1)(r - [1 + 2i])(r - [1 - 2i]) = 0.$$

Thus the eigenvectors satisfy $(\mathbf{A} - r\mathbf{I})\mathbf{u} = \mathbf{0}$ for each of the eigenvalues -1, $1 + 2i$, and $1 - 2i$:

$$\begin{bmatrix} 1 - (-1) & 2 & 0 \\ 0 & 1 - (-1) & -2 \\ 2 & 2 & -1 - (-1) \end{bmatrix} \begin{bmatrix} u_1 \\ u_2 \\ u_3 \end{bmatrix} = \begin{bmatrix} 0 \\ 0 \\ 0 \end{bmatrix},$$

$$\begin{bmatrix} 1 - (1 + 2i) & 2 & 0 \\ 0 & 1 - (1 + 2i) & -2 \\ 2 & 2 & -1 - (1 + 2i) \end{bmatrix} \begin{bmatrix} u_1 \\ u_2 \\ u_3 \end{bmatrix} = \begin{bmatrix} 0 \\ 0 \\ 0 \end{bmatrix},$$

$$\begin{bmatrix} 1 - (1 - 2i) & 2 & 0 \\ 0 & 1 - (1 - 2i) & -2 \\ 2 & 2 & -1 - (1 - 2i) \end{bmatrix} \begin{bmatrix} u_1 \\ u_2 \\ u_3 \end{bmatrix} = \begin{bmatrix} 0 \\ 0 \\ 0 \end{bmatrix}.$$

The first system has solutions $t[-1 \quad 1 \quad 1]^T$. The second and third systems have coefficient matrices that are complex conjugates of each other, and Gauss elimination (with complex numbers) produces the conjugate solutions $t[1 \quad (\pm i) \quad 1]^T$. Quick mental calculation confirms that

$$\begin{bmatrix} 1 & 2 & 0 \\ 0 & 1 & -2 \\ 2 & 2 & -1 \end{bmatrix} \begin{bmatrix} -1 \\ 1 \\ 1 \end{bmatrix} = (-1) \begin{bmatrix} -1 \\ 1 \\ 1 \end{bmatrix}, \quad \begin{bmatrix} 1 & 2 & 0 \\ 0 & 1 & -2 \\ 2 & 2 & -1 \end{bmatrix} \begin{bmatrix} 1 \\ i \\ 1 \end{bmatrix} = (1+2i) \begin{bmatrix} 1 \\ i \\ 1 \end{bmatrix},$$

$$\begin{bmatrix} 1 & 2 & 0 \\ 0 & 1 & -2 \\ 2 & 2 & -1 \end{bmatrix} \begin{bmatrix} 1 \\ -i \\ 1 \end{bmatrix} = (1-2i) \begin{bmatrix} 1 \\ -i \\ 1 \end{bmatrix}. \qquad\blacksquare$$

Example 4. Find the eigenvalues and eigenvectors for the rotation matrices

$$\mathbf{A} = \begin{bmatrix} \cos\theta & -\sin\theta \\ \sin\theta & \cos\theta \end{bmatrix} \quad \text{and} \quad \mathbf{B} = \begin{bmatrix} 0 & 0 & 1 \\ 1 & 0 & 0 \\ 0 & 1 & 0 \end{bmatrix}$$

Solution. The form of the two-dimensional rotation matrix $\mathbf{A}$ was derived in Example 1 of Section 2.2. Its characteristic equation,

$$|\mathbf{A} - r\mathbf{I}| = \begin{vmatrix} \cos\theta - r & -\sin\theta \\ \sin\theta & \cos\theta - r \end{vmatrix} = (\cos\theta - r)^2 + \sin^2\theta = r^2 - 2r\cos\theta + 1 = 0,$$

has the roots

$$r = \frac{2\cos\theta \pm \sqrt{4\cos^2\theta - 4}}{2} = \cos\theta \pm i\sin\theta = e^{\pm i\theta},$$

which are complex unless $\theta = 0$ or π (in which case $\mathbf{A} = \pm\mathbf{I}$). Complex eigenvectors corresponding to $e^{\pm i\theta}$ are easily seen to be $[1 \ \ \mp i]^T$.

In Figure 5.3, the matrix $\mathbf{B}$ takes the position vector $[x \ \ y \ \ z]^T$ into $[z \ \ x \ \ y]^T$; it cycles the axes $(x \to y \to z \to x)$, and thus it is a rotation about the direction making equal angles with the axes: namely, $[1 \ \ 1 \ \ 1]^T$. The characteristic equation,

$$|\mathbf{B} - r\mathbf{I}| = \begin{vmatrix} -r & 0 & 1 \\ 1 & -r & 0 \\ 0 & 1 & -r \end{vmatrix} = (-r)^3 + 1 = 0,$$

shows that the eigenvalues are the cube roots of unity: namely, $r_1 = 1$, $r_2 = e^{i2\pi/3}$, and $r_3 = e^{-i2\pi/3}$. The direction of the axis of rotation $[1 \ \ 1 \ \ 1]^T$ is easily seen to be an eigenvector corresponding to $r_1 = 1$, as predicted in Section 5.1. The complex eigenvectors are of little interest to us at this point. $\qquad\blacksquare$

The next example demonstrates some curious aberrations that we will explore later in this book.

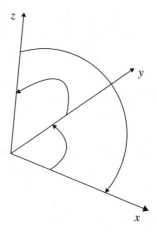

Fig. 5.3   Rotation in three dimensions.

Example 5.   Find the eigenvalues and eigenvectors for

$$\mathbf{A} = \begin{bmatrix} 1 & -2 & 2 \\ -2 & 1 & 2 \\ 2 & 2 & 1 \end{bmatrix} \quad \text{and} \quad \mathbf{B} = \begin{bmatrix} 1 & 0 \\ c & 1 \end{bmatrix} \quad (c \neq 0).$$

Solution.   The characteristic equation for **A** is

$$|\mathbf{A} - r\mathbf{I}| = \begin{bmatrix} 1-r & -2 & 2 \\ -2 & 1-r & 2 \\ 2 & 2 & 1-r \end{bmatrix}$$
$$= -r^3 + 3r^2 + 9r - 27 = -(r+3)(r-3)^2 = 0.$$

The eigenvectors for the eigenvalue $r = -3$ are found by straightforward Gauss elimination to be $t\,[-1 \quad -1 \quad 1]^T$. But when we substitute the double root $r = +3$ into the homogeneous system for the eigenvectors, Gauss elimination produces

$$\mathbf{A} - 3\mathbf{I} \mapsto \begin{bmatrix} 1-3 & -2 & 2 & \vdots & 0 \\ -2 & 1-3 & 2 & \vdots & 0 \\ 2 & 2 & 1-3 & \vdots & 0 \end{bmatrix}$$

$$= \begin{bmatrix} -2 & -2 & 2 & \vdots & 0 \\ -2 & -2 & 2 & \vdots & 0 \\ 2 & 2 & -2 & \vdots & 0 \end{bmatrix} \begin{pmatrix} -\rho_1 + \rho_2 \to \rho_2 \\ \rho_1 + \rho_3 \to \rho_3 \end{pmatrix} \begin{bmatrix} -2 & -2 & 2 & \vdots & 0 \\ 0 & 0 & 0 & \vdots & 0 \\ 0 & 0 & 0 & \vdots & 0 \end{bmatrix}$$

indicating that the double eigenvalue has a two-dimensional eigenspace,

$$t_1 \begin{bmatrix} 1 \\ 0 \\ 1 \end{bmatrix} + t_2 \begin{bmatrix} -1 \\ 1 \\ 0 \end{bmatrix}.$$

For **B** the characteristic equation also has a double root:

$$|\mathbf{B} - r\mathbf{I}| = \begin{bmatrix} 1 - r & 0 \\ c & 1 - r \end{bmatrix} = (1 - r)^2, \tag{9}$$

but its eigenspace is only one-dimensional:

$$\mathbf{B} - (1)\mathbf{I} \mapsto \begin{bmatrix} 1 - 1 & 0 & \vdots & 0 \\ c & 1 - 1 & \vdots & 0 \end{bmatrix} \begin{array}{c} \rho_1 \leftrightarrow \rho_2 \end{array} \begin{bmatrix} c & 0 & \vdots & 0 \\ 0 & 0 & \vdots & 0 \end{bmatrix}$$

has $t\begin{bmatrix} 0 & 1 \end{bmatrix}^T$ as its only solutions.[3]   ∎

To obtain some perspective for our further study, let's summarize the observations we have made so far.

- Every n-by-n matrix has at least one and at most n (possibly complex) eigenvalues. Why? Because they are the roots of a polynomial of degree n (the characteristic polynomial, Equation (1)).
- Every eigenvalue has at least one eigenvector—or, better, at least a one-dimensional eigenspace. Why? Because the eigenvectors are solutions of a singular, homogeneous system.
- If an eigenvalue of a real matrix is real, its eigenvectors can be taken to be real as well. Why? Because they can be computed using the Gauss elimination algorithm (which does not need complex numbers to reduce a real matrix). (Since any nonzero multiple of an eigenvector is also an eigenvector, we could (maliciously) make a real eigenvector complex by multiplying it by i (or any other complex number); the eigenvalue, though, would remain real.)
- Multiple (repeated) eigenvalues may, *or may not*, have eigenspaces of dimension higher than one.
- Eigenvectors corresponding to distinct eigenvalues are linearly independent. Thus if there are no multiple eigenvalues, we can create a full *basis* of n eigenvectors—one for each eigenvalue. Or, if each multiple eigenvalue has an eigenspace whose

---

[3]Did you recognize **B** as an elementary row operator matrix?

dimension is equal to its multiplicity (as a root of the characteristic equation), again we can create a basis of eigenvectors.

- The eigenvalues of a triangular matrix are its diagonal entries.

We will show later that real, symmetric matrices have only real eigenvalues, and each multiple eigenvalue has its "full set" of eigenvectors—that is, the dimension of its eigenspace equals the multiplicity of the eigenvalue. But if a matrix is **defective**, not having a full set of eigenvectors for each eigenvalue, we will see that we have to settle for a basis consisting partially of eigenvectors, and partially of *generalized eigenvectors* satisfying a "watered-down" version of the eigenvector equation (namely, $(\mathbf{A} - r\mathbf{I})^k \mathbf{u} = \mathbf{0}$ for some $k > 1$).

Exercises 5.2

The matrices in Problems 1 through 8 have distinct eigenvalues. Find these eigenvalues and corresponding eigenvectors.

1. $\begin{bmatrix} 9 & -3 \\ -2 & 4 \end{bmatrix}$ 2. $\begin{bmatrix} -1 & -3 \\ 4 & 7 \end{bmatrix}$ 3. $\begin{bmatrix} 1 & -2 \\ -2 & 1 \end{bmatrix}$ 4. $\begin{bmatrix} -3 & 2 \\ -5 & 4 \end{bmatrix}$ 5. $\begin{bmatrix} 1 & 1 \\ 1 & 1 \end{bmatrix}$

6. $\begin{bmatrix} 1 & 0 & 2 \\ 3 & 4 & 2 \\ 0 & 0 & 0 \end{bmatrix}$ 7. $\begin{bmatrix} 4 & -1 & -2 \\ 2 & 1 & -2 \\ 1 & -1 & 1 \end{bmatrix}$ 8. $\begin{bmatrix} 2 & -1 & -1 \\ -1 & 3 & 0 \\ -1 & 0 & 3 \end{bmatrix}$

The matrices in Problems 9–12 have complex eigenvalues. Find the eigenvalues and eigenvectors.

9. $\begin{bmatrix} 0 & -1 \\ 1 & 0 \end{bmatrix}$ 10. $\begin{bmatrix} 0 & 1 \\ -1+i & -1 \end{bmatrix}$ 11. $\begin{bmatrix} 1 & 2i \\ -1 & 3+i \end{bmatrix}$ 12. $\begin{bmatrix} -6 & 2 & 16 \\ -3 & 1 & 11 \\ -3 & 1 & 7 \end{bmatrix}$

The matrices in Problems 13–20 have multiple eigenvalues. Some are defective. Find the eigenvalues and eigenvectors.

13. $\begin{bmatrix} 10 & -4 \\ 4 & 2 \end{bmatrix}$ 14. $\begin{bmatrix} 1 & -1 \\ 1 & 3 \end{bmatrix}$ 15. $\begin{bmatrix} 3 & 2 \\ -2 & 7 \end{bmatrix}$ 16. $\begin{bmatrix} 0 & -i \\ i & 2i \end{bmatrix}$

17. $\begin{bmatrix} 0 & 1 & 1 \\ 1 & 0 & 1 \\ 1 & 1 & 0 \end{bmatrix}$ 18. $\begin{bmatrix} 1 & 2 & 6 \\ -1 & 4 & 6 \\ -1 & 2 & 8 \end{bmatrix}$ 19. $\begin{bmatrix} 1 & 0 & 0 \\ 0 & 2 & 3 \\ 1 & 0 & 2 \end{bmatrix}$ 20. $\begin{bmatrix} 1 & 2 & 2 \\ 2 & 4 & 3 \\ -2 & -1 & 0 \end{bmatrix}$

21. **Deflation of a matrix.** A matrix is said to be *deflated* when one of its eigenvalues is replaced by zero, while the others are left intact. Calculate the eigenvalues of each of the following matrices, deflate the matrix using the technique described in Problem 19 of Exercises 5.1 to replace the highest eigenvalue by zero, and then recalculate the eigenvalues to check your work.

(a) $\begin{bmatrix} 9 & -3 \\ -2 & 4 \end{bmatrix}$ (see Problem 1)

(b) $\begin{bmatrix} 1 & -2 & 1 \\ -1 & 2 & -1 \\ 0 & 1 & 1 \end{bmatrix}$

22. If $\mathbf{A}$ is invertible, show how one can construct the characteristic polynomial of $\mathbf{A}^{-1}$ from the characteristic polynomial of $\mathbf{A}$ (without calculating $\mathbf{A}^{-1}$).

23. (a) If $\mathbf{A}$ is a square matrix show that $\mathbf{A}$ and $\mathbf{A}^T$ have the same eigenvalues.

(b) Show that eigenvectors $\mathbf{v}_1$ and $\mathbf{v}_2$ of $\mathbf{A}^T$ and $\mathbf{A}$ respectively satisfy $\mathbf{v}_2^T \mathbf{v}_1 = 0$, if they correspond to different eigenvalues. [*Hint*: Start with the identity $\mathbf{v}_1^T(\mathbf{A}\mathbf{v}_2) = (\mathbf{A}^T\mathbf{v}_1)^T\mathbf{v}_2$.]

(c) Verify (a) and (b) for the matrices

(i) $\begin{bmatrix} 3 & 2 \\ -1 & 0 \end{bmatrix}$ (ii) $\begin{bmatrix} -3 & -2 & 2 \\ 2 & 5 & -4 \\ 1 & 5 & -4 \end{bmatrix}$.

24. Why does every real *n*-by-*n* matrix have at least one real eigenvalue if *n* is odd? Construct a real 2-by-2 matrix with *no* real eigenvalues.

25. Show that the coefficient c_0 in the characteristic polynomial (1) for $\mathbf{A}$ equals $\det(\mathbf{A})$.

26. **Trace.** Show that $(-1)^n$ times the coefficient c_{n-1} in the characteristic polynomial (1) for $\mathbf{A}$ equals the sum of the diagonal elements of $\mathbf{A}$ (otherwise known as the **trace** of $\mathbf{A}$). (It is also the sum of the eigenvalues; why?)

27. What are the eigenvalues of a 2-by-2 matrix with trace equal to 2 and determinant equal to -8?

28. Show that if $\mathbf{A}$ and $\mathbf{B}$ are both *n*-by-*n* matrices then $\mathbf{AB}$ and $\mathbf{BA}$ have the same trace, determinant, and eigenvalues. [*Hint*: compare the characteristic polynomials of $\mathbf{AB}$ and $\mathbf{BA}$.]

5.3 SYMMETRIC AND HERMITIAN MATRICES

The results of the eigenvector calculations for the symmetric matrix $\mathbf{A}$ in Example 5 of the previous section are summarized in the following text:

$$A = \begin{bmatrix} 1 & -2 & 2 \\ -2 & 1 & 2 \\ 2 & 2 & 1 \end{bmatrix};$$

$$A \begin{bmatrix} 1 \\ 0 \\ 1 \end{bmatrix} = 3 \begin{bmatrix} 1 \\ 0 \\ 1 \end{bmatrix}, \quad A \begin{bmatrix} -1 \\ 1 \\ 0 \end{bmatrix} = 3 \begin{bmatrix} -1 \\ 1 \\ 0 \end{bmatrix}, \quad A \begin{bmatrix} -1 \\ -1 \\ 1 \end{bmatrix} = -3 \begin{bmatrix} -1 \\ -1 \\ 1 \end{bmatrix}. \tag{1}$$

This display exhibits some general properties of real symmetric matrices that make them "ideal" for eigenvalue analysis.

Eigenvalues and Eigenvectors of Real Symmetric Matrices

Theorem 3. If A is a real symmetric matrix,

  (i) its eigenvalues are all real (and consequently its eigenvectors can be taken to be real);

  (ii) its eigenvectors corresponding to distinct eigenvalues are orthogonal; and

  (iii) its multiple eigenvalues have eigenspaces of dimension equal to the multiplicity of the corresponding eigenvalues.

Note how Equation (1) manifests these conclusions. The eigenvalues are the real numbers 3 (*twice*) and -3, the eigenspace for the repeated eigenvalue is spanned by the *two* independent eigenvectors $[1 \quad 0 \quad 1]^T$ and $[-1 \quad 1 \quad 0]^T$, and each of these is orthogonal to the remaining eigenvector $[-1 \quad -1 \quad 1]^T$.

The proofs of (i) and (ii) simply involve juggling the basic eigenvalue equation $Au = ru$ and the symmetry condition $A^T = A$.

Proof of (i) and (ii). Suppose $Au_1 = r_1 u_1$ and $Au_2 = r_2 u_2$. Then the relation

$$r_2 u_1^T u_2 = r_1 u_1^T u_2 \tag{2}$$

can be derived by processing $u_1^T A u_2$ in two ways:

$$u_1^T A u_2 = u_1^T (A u_2) = u_1^T (r_2 u_2) = r_2 u_1^T u_2,$$
$$\|\|\|$$
$$u_1^T A u_2 = \left(u_1^T A\right) u_2 = \left(A^T u_1\right)^T u_2 = (A u_1)^T u_2 = (r_1 u_1)^T u_2 = r_1 u_1^T u_2. \tag{3}$$
$$\uparrow$$
(Keep in mind that $A^T = A$.)

In the following, we use the overbar to denote the complex conjugate of a scalar or vector. Thus $\overline{a + ib} = a - ib$ and $\overline{v + iw} = v - iw$ (for real a, b, v, and w).

To show (2) implies (i), note that if $\mathbf{Au} = r\mathbf{u}$ and $\mathbf{A}$ is real, then by taking complex conjugates we derive $\mathbf{A\bar{u}} = \bar{r}\mathbf{\bar{u}}$, so we invoke (2) with the identifications $r_1 = r$, $r_2 = \bar{r}$, $\mathbf{u}_1 = \mathbf{u}$, $\mathbf{u}_2 = \mathbf{\bar{u}}$, to conclude $\bar{r}\mathbf{u}^T\mathbf{\bar{u}} = r\mathbf{u}^T\mathbf{\bar{u}}$. But now $\mathbf{u}^T\mathbf{\bar{u}}$ is a sum of squared magnitudes and thus is positive (since $\mathbf{u} \neq \mathbf{0}$), so $\bar{r}$ has to equal r; the eigenvalues are real. (And therefore the eigenvectors can be taken to be real also.)

To see that (2) implies (ii) we simply note that $r_1 \neq r_2$ in (2) implies orthogonality: $\mathbf{u}_1^T\mathbf{u}_2 = 0$.

The proof of (iii) is more complicated and is deferred to Section 6.3.   ■

Although Theorem 3 only guarantees that eigenvectors corresponding to *distinct* eigenvalues are orthogonal, remember that eigenvectors corresponding to the same eigenvalue form a subspace—and therefore an *orthogonal* basis of eigenvectors can be constructed for the subspace (using, say, the Gram–Schmidt algorithm of Section 4.1).

Example 1.   Form an orthogonal basis from the eigenvectors of the symmetric matrix in display (1).

Solution.   As noted, the eigenvectors $\mathbf{u}_1 = \begin{bmatrix} 1 & 0 & 1 \end{bmatrix}^T$ and $\mathbf{u}_2 = \begin{bmatrix} -1 & 1 & 0 \end{bmatrix}^T$, corresponding the repeated eigenvalue 3, are each orthogonal to the remaining eigenvector $\mathbf{u}_3 = \begin{bmatrix} -1 & -1 & 1 \end{bmatrix}^T$; but we are required to alter them so that they are also orthogonal to each other. Applying the Gram–Schmidt algorithm to $\mathbf{u}_1$ and $\mathbf{u}_2$, we retain $\mathbf{u}_1$ and replace $\mathbf{u}_2$ by

$$\mathbf{u} = \mathbf{u}_2 - \frac{\mathbf{u}_1^T\mathbf{u}_2}{\mathbf{u}_1^T\mathbf{u}_1}\mathbf{u}_1 = \begin{bmatrix} -\dfrac{1}{2} & 1 & \dfrac{1}{2} \end{bmatrix}^T,$$

revising (1) to display the *orthogonal* eigenvectors

$$\mathbf{A}\begin{bmatrix} 1 \\ 0 \\ 1 \end{bmatrix} = 3\begin{bmatrix} 1 \\ 0 \\ 1 \end{bmatrix}, \quad \mathbf{A}\begin{bmatrix} -1/2 \\ 1 \\ 1/2 \end{bmatrix} = 3\begin{bmatrix} -1/2 \\ 1 \\ 1/2 \end{bmatrix}, \quad \mathbf{A}\begin{bmatrix} -1 \\ -1 \\ 1 \end{bmatrix} = -3\begin{bmatrix} -1 \\ -1 \\ 1 \end{bmatrix}. \quad (4)$$

■

This construction allows us to refine statement (ii) of Theorem 3 and obtain a neater characterization for eigenvectors of real symmetric matrices.

Corollary 1.   A real symmetric n-by-n matrix has n orthogonal eigenvectors.

Our examples have shown that we can't afford the luxury of real arithmetic when we study eigenvalue theory; eigenvalues are roots of polynomial equations, and they can be complex even when the matrices are real.

Luckily, the extension of our deliberations to complex matrices and vectors is not too painful. The biggest concession we must make is in the inner product. Recall that if $\mathbf{v}$ and $\mathbf{w}$ are two vectors in $\mathbf{R}^3_{col}$, their *dot product* $\mathbf{v} \cdot \mathbf{w}$ is given by the matrix product $\mathbf{v}^T\mathbf{w}$, and we have employed this concept in two ways:

(i) the *norm*, or Euclidean length, of $\mathbf{v}$ equals $\sqrt{\mathbf{v} \cdot \mathbf{v}}$, and

(ii) $\mathbf{v}$ and $\mathbf{w}$ are *orthogonal* when $\mathbf{v} \cdot \mathbf{w}$ equals zero.

Now if $\mathbf{v}$ is a *complex* vector, there is little value in considering the sum of squares of its components. Indeed, the "length" of $[i \quad 1]^T$ would be $i^2 + 1^2$, or zero; so this concept of length would be useless. What we need, to measure the intensity of a complex vector $\mathbf{v}$, is the square root of the sum of squares of the *magnitudes* of its components,

$$\sqrt{|v_1|^2 + |v_2|^2} \equiv \sqrt{\overline{v_1}\,v_1 + \overline{v_2}\,v_2},$$

which is conveniently written in matrix notation as $\left(\overline{\mathbf{v}}^T\mathbf{v}\right)^{1/2}$. Thus the intensity of $[i \quad 1]^T$ becomes $\sqrt{(-i)i + 1^2} = \sqrt{2}$. Accordingly, we generalize the concepts of norm, symmetry, and orthogonality as follows.

Norm and Orthogonality of Complex Vectors; Hermitian Matrices

Definition 2. (a) The conjugate-transpose of any matrix $\mathbf{A}$ is denoted $\mathbf{A}^H$ ["**A**-Hermitian," named after the French mathematician Charles Hermite (1822–1901).]

$$\mathbf{A}^H := (\overline{\mathbf{A}})^T, \quad a_{ij}^H = \overline{a_{ji}}.$$

(b) The **inner product** $< \mathbf{v}|\mathbf{w} >$ of two complex column vectors $\mathbf{v}$ and $\mathbf{w}$ equals $\mathbf{v}^H\mathbf{w}$; for row vectors it is $\overline{\mathbf{v}}\mathbf{w}^T$. (We always conjugate the *first* vector.)[4]

(c) Complex vectors are said to be **orthogonal** when their inner product $< \mathbf{v}|\mathbf{w} >$ is zero.

(d) The **norm** of a complex vector $\mathbf{v}$ equals $< \mathbf{v}|\mathbf{v} >^{1/2}$.

(e) A matrix that is equal to its conjugate-transpose is said to be **Hermitian** or **self-adjoint**: $\mathbf{A} = \mathbf{A}^H$.

(f) A matrix $\mathbf{Q}$ whose conjugate-transpose equals its inverse is said to be **unitary**: $\mathbf{Q}^H = \mathbf{Q}^{-1}$. (So if $\mathbf{Q}$ is real, it is orthogonal.)

So "Hermitian matrix" generalizes the terminology "symmetric matrix" and "unitary matrix" generalizes "orthogonal matrix," and the nomenclature for "norm" and "orthogonal" is retained but upgraded.

---

[4]Some authors elect to conjugate the second vector.

The rule for rearranging factors when taking conjugate-transposes is quite predictable:

$$(\mathbf{ABC})^H = \overline{(\mathbf{ABC})}^T = (\overline{\mathbf{A}}\,\overline{\mathbf{B}}\,\overline{\mathbf{C}})^T = \overline{\mathbf{C}}^T\overline{\mathbf{B}}^T\overline{\mathbf{A}}^T = \mathbf{C}^H\mathbf{B}^H\mathbf{A}^H, \tag{5}$$

and it follows as usual that the product of unitary matrices is unitary:

$$(\mathbf{Q}_1\mathbf{Q}_2\mathbf{Q}_3)^{-1} = \mathbf{Q}_3^{-1}\mathbf{Q}_2^{-1}\mathbf{Q}_1^{-1} = \mathbf{Q}_3^H\mathbf{Q}_2^H\mathbf{Q}_1^H = (\mathbf{Q}_1\mathbf{Q}_2\mathbf{Q}_3)^H. \tag{6}$$

The proof of Theorem 3, asserting that eigenvectors corresponding to distinct eigenvalues of a symmetric matrix are orthogonal, is easily extended to the Hermitian case. The basic equality

$$r_2\mathbf{u}_1^H\mathbf{u}_2 = \overline{r_1}\,\mathbf{u}_1^H\mathbf{u}_2 \tag{7}$$

(analogous to (2)) is derived by adapting (3) for a Hermitian matrix as shown:

$$\mathbf{u}_1^H\mathbf{A}\mathbf{u}_2 = \mathbf{u}_1^H(\mathbf{A}\mathbf{u}_2) = \mathbf{u}_1^H(r_2\mathbf{u}_2) = r_2\mathbf{u}_1^H\mathbf{u}_2,$$
$$\text{|||}$$
$$\mathbf{u}_1^H\mathbf{A}\mathbf{u}_2 = (\overline{\mathbf{u}_1}^T\mathbf{A})\mathbf{u}_2 = (\mathbf{A}^T\overline{\mathbf{u}_1})^T\mathbf{u}_2 = \overline{(\overline{\mathbf{A}}^T\mathbf{u}_1)}^T\mathbf{u}_2 = (\overline{\mathbf{A}\mathbf{u}_1})^T\mathbf{u}_2 = (\overline{r_1\mathbf{u}_1})^T\mathbf{u}_2 = \overline{r_1}\mathbf{u}_1^H\mathbf{u}_2.$$
$$\underbrace{(\overline{\mathbf{A}}^T = \mathbf{A}^H = \mathbf{A})}.$$

So if $\mathbf{u}_1 = \mathbf{u}_2$ in (7), we conclude that every eigenvalue r_1 is real, while if $r_1 \neq r_2$ we conclude that the eigenvectors are orthogonal. As with real symmetric matrices, we have the following.

Eigenvalues and Eigenvectors of Hermitian Matrices

Theorem 4. If $\mathbf{A}$ is a Hermitian matrix,

  (i) its eigenvalues are all real [note that this implies the coefficients in its characteristic polynomial, which can be displayed by multiplying out

$$p(r) = |\mathbf{A} - r\mathbf{I}| = (r_1 - r)(r_2 - r)\cdots(r_n - r),$$

  are also real—although $\mathbf{A}$ itself may be complex.]
 (ii) its eigenvectors corresponding to distinct eigenvalues are orthogonal; and
(iii) (to be proved in Section 6.3) its multiple eigenvalues have eigenspaces of dimension equal to the multiplicity of the corresponding eigenvalues.

Finally, we note that the Gram–Schmidt algorithm of Section 4.1 for extracting orthogonal vectors is easily adapted to the complex case. One simply changes the matrix transposes in Equations (9–11) of that section to conjugate transposes (See Problem 14).

Example 2. Find three orthogonal eigenvectors for the Hermitian matrix.

$$\mathbf{A} = \begin{bmatrix} 0 & 0 & 0 \\ 0 & 1 & -1+i \\ 0 & -1-i & 2 \end{bmatrix}.$$

Solution. The eigenvalues are the roots of the characteristic polynomial

$$p(r) = |\mathbf{A} - r\mathbf{I}| = -r^3 + 3r^2 = -r^2(r-3).$$

The eigenvectors corresponding to $r = 0$ and $r = 3$ satisfy, respectively,

$$\mathbf{Au} = \mathbf{0} \quad \text{or} \quad \begin{bmatrix} 0 & 0 & 0 & \vdots & 0 \\ 0 & 1 & -1+i & \vdots & 0 \\ 0 & -1-i & 2 & \vdots & 0 \end{bmatrix} \quad \text{and}$$

$$(\mathbf{A} - 3\mathbf{I})\mathbf{u} = \mathbf{0} \quad \text{or} \quad \begin{bmatrix} -3 & 0 & 0 & \vdots & 0 \\ 0 & -2 & -1+i & \vdots & 0 \\ 0 & -1-i & -1 & \vdots & 0 \end{bmatrix},$$

and Gauss elimination yields the corresponding solutions $t_1 \begin{bmatrix} 0 & 1 & -i & 1 \end{bmatrix}^T + t_2 \begin{bmatrix} 1 & 0 & 0 \end{bmatrix}^T$ and $t_3 \begin{bmatrix} 0 & -1+i & 2 \end{bmatrix}^T$. As predicted by Theorem 4, the subspace of eigenvectors for the double eigenvalue 0 has dimension 2; the choices $t_1 = 0$, $t_2 = 1$ and $t_1 = 1$, $t_2 = 0$ yield mutually orthogonal eigenvectors in this subspace, each orthogonal to the third eigenvector.[5] ∎

By applying the Gram–Schmidt algorithm to the multidimensional eigenspace in Theorem 4 we arrive at the more economical formulation:

Corollary 2. A Hermitian n-by-n matrix has n orthogonal eigenvectors.

[5]Since these eigenvectors are already orthogonal, the Gram–Schmidt algorithm is unnecessary.

A fairly safe rule of thumb: to complexify (yes, that *is* a word) any statement about real matrices, change all transposes to conjugate transposes. (But see Problem 13.) Also note that for complex scalars c,

$$< c\mathbf{v}|\mathbf{w} > = \bar{c} < \mathbf{v}|\mathbf{w} > \quad \text{but} \quad < \mathbf{v}|c\mathbf{w} > = c < \mathbf{v}|\mathbf{w} > .$$

Exercises 5.3

Find two orthogonal eigenvectors for the symmetric matrices in Problems 1–2.

1. $\begin{bmatrix} -7 & 6 \\ 6 & 2 \end{bmatrix}$
2. $\begin{bmatrix} 4 & 1 \\ 1 & 4 \end{bmatrix}$

Find three orthogonal eigenvectors for the symmetric matrices in Problems 3–6.

3. $\begin{bmatrix} 1 & 4 & 3 \\ 4 & 1 & 0 \\ 3 & 0 & 1 \end{bmatrix}$
4. $\begin{bmatrix} 2 & 0 & -1 \\ 0 & 2 & -1 \\ -1 & -1 & 3 \end{bmatrix}$
5. $\begin{bmatrix} 2 & 1 & 1 \\ 1 & 2 & -1 \\ 1 & -1 & 2 \end{bmatrix}$
6. $\begin{bmatrix} 0 & 1 & 1 \\ 1 & 0 & 1 \\ 1 & 1 & 0 \end{bmatrix}$

Find two orthogonal eigenvectors for the Hermitian matrices in Problems 7–8.

7. $\begin{bmatrix} 0 & i \\ -i & 0 \end{bmatrix}$
8. $\begin{bmatrix} 4 & 2+i \\ 2-i & 0 \end{bmatrix}$

Find three orthogonal eigenvectors for the Hermitian matrices in Problems 9–12.

9. $\begin{bmatrix} -1 & 0 & -1+i \\ 0 & -1 & 0 \\ -1-i & 0 & 0 \end{bmatrix}$
10. $\begin{bmatrix} 2 & 0 & -i \\ 0 & 3 & 0 \\ i & 0 & 2 \end{bmatrix}$
11. $\begin{bmatrix} 0 & 0 & -i \\ 0 & 0 & i \\ i & -i & 0 \end{bmatrix}$

12. $\begin{bmatrix} 0 & -2-i & -1+i \\ -2+i & -3 & 3i \\ -1-i & -3i & 0 \end{bmatrix}$

13. **Violation of the rule of thumb.** Show by example that $\mathbf{A}$ and $\mathbf{A}^H$ do *not* always possess the same eigenvalues. Then prove that the eigenvalues of $\mathbf{A}^H$ are complex conjugates of eigenvalues of $\mathbf{A}$. Show that if $\mathbf{A}\mathbf{v} = r\mathbf{v}$ and $\mathbf{A}^H\mathbf{w} = s\mathbf{w}$, then $\mathbf{w}^H\mathbf{v} = 0$ unless r and s are conjugate.

14. Verify the statement that changing transposes to conjugate transposes in the Gram–Schmidt algorithm results in a set of vectors that are orthogonal in the sense of Definition 2. [*Hint*: Simply modify the analysis of the algorithm as given in Section 4.1.]

15. Perform the Gram–Schmidt algorithm on the vectors $[i \quad 0 \quad 0]$, $[i \quad 1 \quad 1]$, and $[i \quad i \quad 1]$ to extract an orthonormal set spanning $\mathbf{C}^3_{\text{row}}$.

16. Are there any Hermitian matrices, other than multiples of the identity, with only one eigenvalue?

17. You may have caught the slight gap between real symmetric matrices and (complex) Hermitian matrices: namely, matrices that are complex and symmetric. Explore the example $\begin{bmatrix} 6i & -1-5i \\ -1-5i & 4i \end{bmatrix}$ to argue that such matrices have no special properties:

    (a) Are the eigenvalues real?

    (b) Do the eigenvalues occur in complex conjugate pairs?

    (c) Are the eigenvectors corresponding to distinct eigenvalues orthogonal $(\mathbf{v}_1^H \mathbf{v}_2 = 0)$?

    (d) Are the eigenvectors corresponding to distinct eigenvalues orthogonal in the "old" sense $(\mathbf{v}_1^T \mathbf{v}_2 = 0)$? Explain. (Recall Equation (2).)

18. Prove that a skew-symmetric matrix, that is, one that satisfies $\mathbf{A} = -\mathbf{A}^T$, has only pure imaginary eigenvalues, and eigenvectors corresponding to different eigenvalues are orthogonal. [*Hint*: Consider $i\mathbf{A}$.]

5.4 SUMMARY

An eigenvector **u** for a square matrix **A** is a vector whose *direction* is unchanged (or, possibly, reversed) by multiplication by **A**. Thus if $\mathbf{Au} = r\mathbf{u}$ (and $\mathbf{u} \neq \mathbf{0}$), then **u** is an eigenvector and the scalar r is the associated eigenvalue. Sometimes the eigenvectors can be visualized for familiar matrix operators, but they can always be gleaned by examining the matrix $\mathbf{A} - r\mathbf{I}$; the eigenvalues of **A** are the scalars r that make the determinant $|\mathbf{A} - r\mathbf{I}|$ (the characteristic polynomial) equal to zero, and its null vectors for such r are the eigenvectors.

An n-by-n matrix always has n eigenvalues (real and complex) if the multiplicities are counted. Each eigenvalue has at least one eigenvector—more precisely, a one-dimensional subspace of eigenvectors, since only the direction of **u** is significant. And eigenvectors corresponding to distinct eigenvalues are linearly independent. Therefore all matrices possessing n distinct eigenvalues have n linearly independent eigenvectors, but some matrices with multiple eigenvalues may be defective, possessing fewer than n independent eigenvectors.

Symmetric and Hermitian Matrices

Real symmetric matrices have favorable eigenvector properties: their eigenvalues are real, the associated eigenvectors can be taken to be real, and a complete orthogonal basis for the vector space can be assembled out of the eigenvectors.

A handy and highly reliable rule of thumb for dealing with complex matrices is to replace "transpose" in the theorems and definitions for real matrices by "conjugate transpose." With this proviso, the favorable eigenvector properties of real symmetric matrices extend to Hermitian matrices.

6

SIMILARITY

6.1 SIMILARITY TRANSFORMATIONS AND DIAGONALIZABILITY

The *similarity transformation is* an important concept in linear algebra. We have a matrix **A** that describes a certain operation in $\mathbf{R}^n$, but we would like to know the form that the matrix will take if we employ a different basis for $\mathbf{R}^n$; that is, we seek to translate the action of the given matrix into the "language" associated with a different basis.

To be specific, in Example 3 of Section 2.2, we saw that the matrix

$$\mathbf{M}_{\text{ref}} = \begin{bmatrix} 0 & 1 \\ 1 & 0 \end{bmatrix} \tag{1}$$

effects a reflection in the mirror lying on the line $y = x$ (Fig. 6.1a). That is, if $[x\,y]^T$ gives the coordinates of a point before the reflection, then after the reflection its coordinates are $[y\,x]^T = \mathbf{M}_{\text{ref}}[x\,y]^T$.

But suppose we wish to describe points in a different coordinate system; say, one that is rotated counterclockwise $30°$ around the origin as in Figure 6.1b. If $[x',\ y']^T$ designates a point's coordinates in the new system *before* the mirror reflection, then what are the coordinates (in the new system) after the reflection?

One way to answer this question would be to compute the mirror's normal in the new coordinates and carry out the calculation of the revised reflector matrix (by the process described in Example 3 of Section 2.2). However at this point, we prefer to employ an

Fundamentals of Matrix Analysis with Applications,
First Edition. Edward Barry Saff and Arthur David Snider.
© 2016 John Wiley & Sons, Inc. Published 2016 by John Wiley & Sons, Inc.

(a)

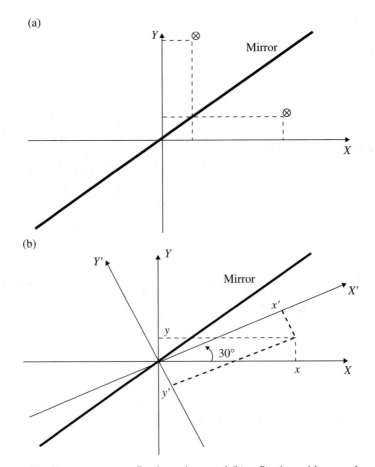

(b)

Fig. 6.1   (a) The line $y = x$ as a reflecting mirror and (b) reflection with rotated coordinates.

operator to translate the new coordinates into old coordinates; then we apply (1) in the old system, and finally translate back. The operator is a rotation matrix, and we use the equations derived in Example 1 of Section 2.2.[1]

To "translate" new coordinates $[x' \ y']^T$ into old $[x \ y]^T$ coordinates we multiply by the rotator matrix $\mathbf{M}_{\text{rot}}$:

$$
\begin{bmatrix} x \\ y \end{bmatrix} = \mathbf{M}_{\text{rot}} \begin{bmatrix} x' \\ y' \end{bmatrix} = \begin{bmatrix} \cos 30° & -\sin 30° \\ \sin 30° & \cos 30° \end{bmatrix} \begin{bmatrix} x' \\ y' \end{bmatrix} = \begin{bmatrix} \sqrt{3}/2 & -1/2 \\ 1/2 & \sqrt{3}/2 \end{bmatrix} \begin{bmatrix} x' \\ y' \end{bmatrix} \tag{2}
$$

So, if we are given the coordinates of a point $[x' \ y']^T$ in the new system and want to determine its image in the mirror, first we translate into old-system coordinates

---

[1]Remember that here we are not rotating the vector, but rather rotating the coordinate system. The effect on the coordinates is the same as rotating the vector by $30°$ in the new system.

$[x \ y]^T = \mathbf{M}_{\text{rot}}[x' \ y']^T$, multiply by the reflector $\mathbf{M}_{\text{ref}}$, and use $\mathbf{M}_{\text{rot}}^{-1}$ to restore to the new coordinate system:

$$\begin{bmatrix} x' \\ y' \end{bmatrix}_{\text{reflected}} = \mathbf{M}_{\text{rot}}^{-1}\mathbf{M}_{\text{ref}}\mathbf{M}_{\text{rot}} \begin{bmatrix} x' \\ y' \end{bmatrix}.$$

We see that the reflector matrix, expressed in the new coordinate system, is $\mathbf{M}'_{\text{ref}} = \mathbf{M}_{\text{rot}}^{-1}\mathbf{M}_{\text{ref}}\mathbf{M}_{\text{rot}}$.

If we denote the unit vectors along the x, y axes as $\mathbf{i}$ and $\mathbf{j}$, and those along the x', y' axes as $\mathbf{i}'$ and $\mathbf{j}'$, then these pairs are bases for $\mathbf{R}_{\text{col}}^2$. Thus the point to be reflected has the equivalent descriptions $x\mathbf{i} + y\mathbf{j} = x'\mathbf{i}' + y'\mathbf{j}'$; the matrix $\mathbf{M}_{\text{rot}}$ effects a *change of basis*. In fact, *whenever we encounter a relation* $\mathbf{A}' = \mathbf{P}^{-1}\mathbf{A}\mathbf{P}$, *we can view* $\mathbf{A}'$ *as a matrix performing the same operation as* $\mathbf{A}$, *but in different coordinates; the matrix* $\mathbf{P}$ *plays the role of an interpreter, mapping the coordinates of a vector in one basis to its coordinates in another.*

Similar Matrices

Definition 1. If two given n-by-n matrices $\mathbf{A}$ and $\mathbf{B}$ satisfy

$$\mathbf{A} = \mathbf{P}^{-1}\mathbf{B}\mathbf{P} \tag{3}$$

for some invertible matrix $\mathbf{P}$, then we say $\mathbf{A}$ is **similar** to $\mathbf{B}$, and $\mathbf{P}$ generates a **similarity transformation** from $\mathbf{B}$ to $\mathbf{A}$.

Obviously if $\mathbf{A}$ is similar to $\mathbf{B}$, then $\mathbf{B}$ is similar to $\mathbf{A}$, with $\mathbf{P}^{-1}$ generating the similarity transformation, since by (3) we have $\mathbf{B} = \mathbf{P}\mathbf{A}\mathbf{P}^{-1}$. And if $\mathbf{B}$ is also similar to $\mathbf{C}$, then $\mathbf{A}$ is similar to $\mathbf{C}$:

$$\mathbf{C} = \mathbf{Q}\mathbf{B}\mathbf{Q}^{-1} = \mathbf{Q}\left(\mathbf{P}\mathbf{A}\mathbf{P}^{-1}\right)\mathbf{Q}^{-1} = (\mathbf{Q}\mathbf{P})\mathbf{A}(\mathbf{Q}\mathbf{P})^{-1}.$$

From what we have seen, we can interpret similar matrices as carrying out identical operations, but in different coordinate systems. This gives rise to a useful rule of thumb:

Essential Feature of Similar Matrices

If a matrix $\mathbf{A}$ has a property that can be described without reference to any specific basis or coordinate system, then every matrix similar to $\mathbf{A}$ shares that property.

A few examples will make this "folk theorem" clear.

Corollary 1. If **A** and **B** are similar matrices, then the following hold:

- **A** and **B** have the same eigenvalues.
- If **A** has k linearly independent eigenvectors corresponding to the eigenvalue r, then **B** also has k linearly independent eigenvectors corresponding to the eigenvalue r.
- **A** and **B** are both singular or both nonsingular.
- **A** and **B** have the same rank.

Folk proof: "**A** has the eigenvalue r" means there is at least a one-dimensional family of solutions to the equation $\mathbf{Au} = r\mathbf{u}$. This statement contains no reference to any coordinate system or basis, so it applies to any matrix similar to **A**.

Similarly, the statement "**A** has a k-dimensional family of solutions to $\mathbf{Au} = r\mathbf{u}$" is basis-free; as is the claim that **A** has a nontrivial null vector. Finally, recall that the rank is the range's dimension, which is basis-free.   ∎

Example 1. Construct a rigorous proof demonstrating that similar matrices have the same eigenvalues.

Solution. If $\mathbf{Au} = r\mathbf{u}$, then $\mathbf{Pu}$ is an eigenvector for $\mathbf{PAP}^{-1}$ with eigenvalue r, since $[(\mathbf{PAP}^{-1})(\mathbf{Pu}) = \mathbf{PAu} = \mathbf{P}(r\mathbf{u}) = r(\mathbf{Pu})]$.   ∎

Indeed, the same logic readily verifies the other claims of Corollary 1. The generalization "similar matrices have the same eigen*vectors*" is *not* justified, because a statement like "**A** has an eigenvector $[1 \; 0 \; 1]^T$" refers specifically to the coordinates of the eigenvector. (In fact Example 1 shows that the eigenvector of $\mathbf{PAP}^{-1}$ that corresponds to **u** is **Pu**.)

Other, more subtle properties shared by similar matrices are described in Corollary 2.

Corollary 2. If **A** and **B** are similar, they have the same characteristic polynomial, determinant, and trace.

Proof. The relation between the characteristic polynomials $p_\mathbf{A}(r)$ and $p_\mathbf{B}(r)$ is revealed by a chain of reasoning exploiting the determinant properties:

$$\begin{aligned}
p_\mathbf{A}(r) &= \det(\mathbf{A} - r\mathbf{I}) = \det\left(\mathbf{P}^{-1}\mathbf{BP} - r\mathbf{I}\right) \\
&= \det\left(\mathbf{P}^{-1}\mathbf{BP} - r\mathbf{P}^{-1}\mathbf{IP}\right) = \det\left(\mathbf{P}^{-1}[\mathbf{B} - r\mathbf{I}]\mathbf{P}\right) \\
&= \det\left(\mathbf{P}^{-1}\right)\det(\mathbf{B} - r\mathbf{I})\det(\mathbf{P}) = \det(\mathbf{B} - r\mathbf{I}) = p_\mathbf{B}(r).
\end{aligned}$$

Problems 25 and 26 of Exercises 5.2 showed that the determinant and the trace (the sum of the diagonal elements) of a matrix appear as coefficients in its characteristic polynomial.   ∎

Since the simplest matrices to compute with are the diagonal matrices, the most important similarity transformations are those that render a matrix diagonal. Such transformations are generated by matrices made up of linearly independent eigenvectors, as shown by the following:

> **Matrix Form of Eigenvector Equations**
>
> **Lemma 1.** If $\{\mathbf{u}_1, \mathbf{u}_2, \ldots, \mathbf{u}_p\}$ is any collection of p eigenvectors of an n-by-n matrix $\mathbf{A}$ and we arrange them into the columns of a n-by-p matrix $\mathbf{P}$, then the set of eigenvector equations $\mathbf{A}\mathbf{u}_j = r_j\mathbf{u}_j (j = 1, 2, \ldots, p)$ is equivalent to the matrix equation
>
> $$\mathbf{AP} = \mathbf{PD}, \tag{4}$$
>
> where $\mathbf{D}$ is a diagonal matrix with eigenvalues on the diagonal.

Proof. It's immediate, because

$$\mathbf{AP} = [\mathbf{A}] \begin{bmatrix} \vdots & \vdots & & \vdots \\ \mathbf{u}_1 & \mathbf{u}_2 & \cdots & \mathbf{u}_p \\ \vdots & \vdots & & \vdots \end{bmatrix} = \begin{bmatrix} \vdots & \vdots & & \vdots \\ \mathbf{A}\mathbf{u}_1 & \mathbf{A}\mathbf{u}_2 & \cdots & \mathbf{A}\mathbf{u}_p \\ \vdots & \vdots & & \vdots \end{bmatrix} = \begin{bmatrix} \vdots & \vdots & & \vdots \\ r_1\mathbf{u}_1 & r_2\mathbf{u}_2 & \cdots & r_p\mathbf{u}_p \\ \vdots & \vdots & & \vdots \end{bmatrix}$$

and

$$\mathbf{PD} = \begin{bmatrix} \vdots & \vdots & & \vdots \\ \mathbf{u}_1 & \mathbf{u}_2 & \cdots & \mathbf{u}_p \\ \vdots & \vdots & & \vdots \end{bmatrix} \begin{bmatrix} r_1 & 0 & \cdots & 0 \\ 0 & r_2 & \cdots & 0 \\ \vdots & & \ddots & \vdots \\ 0 & 0 & \cdots & r_p \end{bmatrix} = \begin{bmatrix} \vdots & \vdots & & \vdots \\ r_1\mathbf{u}_1 & r_2\mathbf{u}_2 & \cdots & r_p\mathbf{u}_p \\ \vdots & \vdots & & \vdots \end{bmatrix}. \quad \blacksquare$$

For instance, the eigenvector equations derived in Example 5, Section 5.2,

$$\mathbf{A} = \begin{bmatrix} 1 & -2 & 2 \\ -2 & 1 & 2 \\ 2 & 2 & 1 \end{bmatrix}; \tag{5}$$

$$\mathbf{A}\begin{bmatrix} 1 \\ 0 \\ 1 \end{bmatrix} = 3\begin{bmatrix} 1 \\ 0 \\ 1 \end{bmatrix}, \quad \mathbf{A}\begin{bmatrix} -1 \\ 1 \\ 0 \end{bmatrix} = 3\begin{bmatrix} -1 \\ 1 \\ 0 \end{bmatrix}, \quad \mathbf{A}\begin{bmatrix} -1 \\ -1 \\ 1 \end{bmatrix} = -3\begin{bmatrix} -1 \\ -1 \\ 1 \end{bmatrix}, \tag{6}$$

can be rewritten in the format (4):

$$\mathbf{A}\begin{bmatrix} 1 & -1 & -1 \\ 0 & 1 & -1 \\ 1 & 0 & 1 \end{bmatrix} = \begin{bmatrix} 1 & -1 & -1 \\ 0 & 1 & -1 \\ 1 & 0 & 1 \end{bmatrix} \begin{bmatrix} 3 & 0 & 0 \\ 0 & 3 & 0 \\ 0 & 0 & -3 \end{bmatrix}. \tag{7}$$

If the eigenvector collection is linearly independent and spans the vector space, then $\mathbf{P}^{-1}$ exists and we can premultiply $\mathbf{AP} = \mathbf{PD}$ by $\mathbf{P}^{-1}$ to get a similarity transformation. We have shown

---

Diagonalization of Matrices

> **Theorem 1.** Every n-by-n matrix $\mathbf{A}$ that possesses n linearly independent eigenvectors is similar to a diagonal matrix $\mathbf{D} = \mathbf{P}^{-1}\mathbf{A}\mathbf{P}$ (and $\mathbf{A} = \mathbf{P}\mathbf{D}\mathbf{P}^{-1}$); the matrix generating the similarity transformation comprises columns of independent eigenvectors, and we say that $\mathbf{P}$ diagonalizes $\mathbf{A}$.

Example 2.   Diagonalize the matrix in (5).

Solution. From Equation (7), we derive

$$\begin{bmatrix} 1 & -1 & -1 \\ 0 & 1 & -1 \\ 1 & 0 & 1 \end{bmatrix}^{-1} \mathbf{A} \begin{bmatrix} 1 & -1 & -1 \\ 0 & 1 & -1 \\ 1 & 0 & 1 \end{bmatrix} = \begin{bmatrix} 1/3 & 1/3 & 2/3 \\ -1/3 & 2/3 & 1/3 \\ -1/3 & -1/3 & 1/3 \end{bmatrix} \mathbf{A} \begin{bmatrix} 1 & -1 & -1 \\ 0 & 1 & -1 \\ 1 & 0 & 1 \end{bmatrix}$$

$$= \begin{bmatrix} 3 & 0 & 0 \\ 0 & 3 & 0 \\ 0 & 0 & -3 \end{bmatrix}, \tag{8}$$

or

$$\mathbf{A} = \begin{bmatrix} 1 & -1 & -1 \\ 0 & 1 & -1 \\ 1 & 0 & 1 \end{bmatrix} \begin{bmatrix} 3 & 0 & 0 \\ 0 & 3 & 0 \\ 0 & 0 & -3 \end{bmatrix} \begin{bmatrix} 1/3 & 1/3 & 2/3 \\ -1/3 & 2/3 & 1/3 \\ -1/3 & -1/3 & 1/3 \end{bmatrix}. \tag{9}$$

∎

By Corollary 1, Section 5.3, the eigenvectors for a *symmetric* matrix can be taken to be orthogonal, and hence orthonormal. The matrix $\mathbf{P}$ then becomes orthogonal [orthonormal columns, transpose equals inverse (Section 4.2)], and we have the refinement

Corollary 3. A real symmetric matrix $\mathbf{A}$ can be diagonalized by an orthogonal matrix $\mathbf{Q}$:

$$\mathbf{D} = \mathbf{Q}^T \mathbf{A} \mathbf{Q}.$$

Example 3.   Use Corollary 3 to diagonalize the matrix $\mathbf{A}$ in (5).

Solution. Even though $\mathbf{A}$ is symmetric, the eigenvectors displayed in (6) and (8) are not orthogonal (due to the multiplicity of $\mathbf{A}$'s eigenvalue 3). However, orthogonal eigenvectors (computed using the Gram–Schmidt algorithm) for $\mathbf{A}$ were exhibited in equation (4) of Section 5.3. Reduced to unit vectors they are $[1 \quad 0 \quad 1]^T/\sqrt{2}$, $[-1/2 \quad 1 \quad 1/2]^T/\sqrt{3/2}$, and $[-1 \quad -1 \quad 1]^T/\sqrt{3}$, and Corollary 3 produces

$$
\mathbf{Q}^T \mathbf{AQ} = \begin{bmatrix} 1/\sqrt{2} & 0 & 1/\sqrt{2} \\ -1/\sqrt{6} & 2/\sqrt{6} & 1/\sqrt{6} \\ -1/\sqrt{3} & -1/\sqrt{3} & 1/\sqrt{3} \end{bmatrix} \begin{bmatrix} 1 & -2 & 2 \\ -2 & 1 & 2 \\ 2 & 2 & 1 \end{bmatrix} \begin{bmatrix} 1/\sqrt{2} & -1/\sqrt{6} & -1/\sqrt{3} \\ 0 & 2/\sqrt{6} & -1/\sqrt{3} \\ 1/\sqrt{2} & 1/\sqrt{6} & 1/\sqrt{3} \end{bmatrix}
$$

$$
= \begin{bmatrix} 3 & 0 & 0 \\ 0 & 3 & 0 \\ 0 & 0 & -3 \end{bmatrix}. \tag{10}
$$

∎

The complex version of Corollary 3 is

Corollary 4. A Hermitian matrix can be diagonalized by a unitary matrix $\mathbf{Q}$:

$$
\mathbf{D} = \mathbf{Q}^H \mathbf{AQ}.
$$

The set of eigenvalues of a Hermitian matrix is sometimes called its *spectrum*, and this corollary is known as the *spectral theorem*; the display $\mathbf{A} = \mathbf{QDQ}^H$ is called the *spectral factorization*. Why? The eigenvalues of the (Hermitian) quantum mechanical energy operator for the bound electrons in an atom are proportional to the frequencies of the light that it emits. Frequencies determine color; hence, we say "spectrum."

The benefits of being able to diagonalize a matrix are enormous:

(i) High powers of $\mathbf{A}$, which would normally be time-consuming to calculate, become trivial:

$$
\mathbf{A}^N = \left(\mathbf{PDP}^{-1}\right)^N = \left(\mathbf{PDP}^{-1}\right)\left(\mathbf{PDP}^{-1}\right)\cdots\left(\mathbf{PDP}^{-1}\right) = \mathbf{PD}^N \mathbf{P}^{-1}
$$

$$
= \mathbf{P} \begin{bmatrix} r_1^N & 0 & \cdots & 0 \\ 0 & r_2^N & \cdots & 0 \\ \vdots & & \ddots & \vdots \\ 0 & 0 & \cdots & r_n^N \end{bmatrix} \mathbf{P}^{-1}. \tag{11}
$$

The inner cancellation of factors in (11) is called "telescoping." For the matrix $\mathbf{A}$ in (9) this reads

$$
\begin{bmatrix} 1 & -2 & 2 \\ -2 & 1 & 2 \\ 2 & 2 & 1 \end{bmatrix}^5 = \begin{bmatrix} 1 & -1 & -1 \\ 0 & 1 & -1 \\ 1 & 0 & 1 \end{bmatrix} \begin{bmatrix} 3^5 & 0 & 0 \\ 0 & 3^5 & 0 \\ 0 & 0 & (-3)^5 \end{bmatrix} \begin{bmatrix} 1/3 & 1/3 & 2/3 \\ -1/3 & 2/3 & 1/3 \\ -1/3 & -1/3 & 1/3 \end{bmatrix}.
$$

(ii) $\mathbf{A}$ is trivial to invert:

$$
\mathbf{A}^{-1} = \left(\mathbf{PDP}^{-1}\right)^{-1} = \mathbf{PD}^{-1}\mathbf{P}^{-1} = \mathbf{P} \begin{bmatrix} r_1^{-1} & 0 & \cdots & 0 \\ 0 & r_2^{-1} & \cdots & 0 \\ \vdots & & \ddots & \vdots \\ 0 & 0 & \cdots & r_n^{-1} \end{bmatrix} \mathbf{P}^{-1}. \tag{12}
$$

Thus

$$
\begin{bmatrix} 1 & -2 & 2 \\ -2 & 1 & 2 \\ 2 & 2 & 1 \end{bmatrix}^{-1} = \begin{bmatrix} 1 & -1 & -1 \\ 0 & 1 & -1 \\ 1 & 0 & 1 \end{bmatrix} \begin{bmatrix} 3^{-1} & 0 & 0 \\ 0 & 3^{-1} & 0 \\ 0 & 0 & (-3)^{-1} \end{bmatrix}
$$
$$
\times \begin{bmatrix} 1/3 & 1/3 & 2/3 \\ -1/3 & 2/3 & 1/3 \\ -1/3 & -1/3 & 1/3 \end{bmatrix}.
$$

(iii) "Square roots" of $\mathbf{A}$ can be calculated:

$$
\mathbf{A}^{1/2} = \mathbf{P}\mathbf{D}^{1/2}\mathbf{P}^{-1} = \mathbf{P} \begin{bmatrix} r_1^{1/2} & 0 & \cdots & 0 \\ 0 & r_2^{1/2} & \cdots & 0 \\ \vdots & & \ddots & \vdots \\ 0 & 0 & \cdots & r_n^{1/2} \end{bmatrix} \mathbf{P}^{-1} \tag{13}
$$

satisfies $\left(\mathbf{A}^{1/2}\right)^2 = \mathbf{P}\mathbf{D}^{1/2}\mathbf{P}^{-1}\mathbf{P}\mathbf{D}^{1/2}\mathbf{P}^{-1} = \mathbf{P}\mathbf{D}\mathbf{P}^{-1} = \mathbf{A}$, so a square root of the matrix in (9) is given by

$$
\begin{bmatrix} 1 & -2 & 2 \\ -2 & 1 & 2 \\ 2 & 2 & 1 \end{bmatrix}^{1/2} = \begin{bmatrix} 1 & -1 & -1 \\ 0 & 1 & -1 \\ 1 & 0 & 1 \end{bmatrix} \begin{bmatrix} \sqrt{3} & 0 & 0 \\ 0 & \sqrt{3} & 0 \\ 0 & 0 & \sqrt{3}i \end{bmatrix} \begin{bmatrix} 1/3 & 1/3 & 2/3 \\ -1/3 & 2/3 & 1/3 \\ -1/3 & -1/3 & 1/3 \end{bmatrix}.
$$

Considering the $\pm$ ambiguities for $r^{1/2}$, (13) spawns 2^n square roots for $\mathbf{A}$. (In general, matrices may have more, or less, roots than this. See Problem 21.)

In fact, for any real number x

$$
\mathbf{A}^x = \mathbf{P}\mathbf{D}^x\mathbf{P}^{-1} = \mathbf{P} \begin{bmatrix} r_1^x & 0 & \cdots & 0 \\ 0 & r_2^x & \cdots & 0 \\ \vdots & & \ddots & \vdots \\ 0 & 0 & \cdots & r_n^x \end{bmatrix} \mathbf{P}^{-1}
$$

is a valid (multiple-valued) definition.

(iv) Because of (11), polynomials in $\mathbf{A}$ are easy to compute:

$$
q(\mathbf{A}) := c_m\mathbf{A}^m + c_{m-1}\mathbf{A}^{m-1} + \cdots + c_1\mathbf{A} + c_0\mathbf{I}
$$
$$
= \mathbf{P} \begin{bmatrix} q(r_1) & 0 & \cdots & 0 \\ 0 & q(r_2) & \cdots & 0 \\ \vdots & & \ddots & \vdots \\ 0 & 0 & \cdots & q(r_n) \end{bmatrix} \mathbf{P}^{-1}. \tag{14}
$$

Example 4. Evaluate $2\mathbf{A}^3 + \mathbf{I}$ for $\mathbf{A}$ as displayed in (9).

Solution.

$$2\mathbf{A}^3 + \mathbf{I} = \begin{bmatrix} 1 & -1 & -1 \\ 0 & 1 & -1 \\ 1 & 0 & 1 \end{bmatrix} \begin{bmatrix} 2 \cdot 3^3 + 1 & 0 & 0 \\ 0 & 2 \cdot 3^3 + 1 & 0 \\ 0 & 0 & 2 \cdot (-3)^3 + 1 \end{bmatrix}$$

$$\times \begin{bmatrix} 1/3 & 1/3 & 2/3 \\ -1/3 & 2/3 & 1/3 \\ -1/3 & -1/3 & 1/3 \end{bmatrix}$$

$$= \begin{bmatrix} 1 & -1 & -1 \\ 0 & 1 & -1 \\ 1 & 0 & 1 \end{bmatrix} \begin{bmatrix} 55 & 0 & 0 \\ 0 & 55 & 0 \\ 0 & 0 & -53 \end{bmatrix} \begin{bmatrix} 1/3 & 1/3 & 2/3 \\ -1/3 & 2/3 & 1/3 \\ -1/3 & -1/3 & 1/3 \end{bmatrix}. \qquad \blacksquare$$

Functions that are limits of polynomials, such as power series, can be extended to matrix arguments if attention is paid to the domains of convergence. For example, the exponential e^x is the limit of polynomials:

$$e^x = \sum_{k=0}^{\infty} \frac{x^k}{k!} = \lim_{m \to \infty} \left[1 + x + \frac{x^2}{2!} + \frac{x^3}{3!} + \cdots + \frac{x^m}{m!} \right],$$

for every value of x. If we insert a diagonalizable matrix $\mathbf{A} = \mathbf{PDP}^{-1}$ into the polynomial in brackets, we find

$$\mathbf{I} + \mathbf{A} + \mathbf{A}^2/2! + \cdots + \mathbf{A}^m/m!$$

$$= \mathbf{P} \begin{bmatrix} 1 + r_1 + r_1^2/2! + \cdots + r_1^m/m! & 0 & \cdots & 0 \\ 0 & 1 + r_2 + r_2^2/2! + \cdots + r_2^m/m! & \cdots & 0 \\ \vdots & \vdots & \ddots & \vdots \\ 0 & 0 & \cdots & 1 + r_n + r_n^2/2! + \cdots + r_n^m/m! \end{bmatrix} \mathbf{P}^{-1}$$

As $m \to \infty$ this matrix converges entrywise to

$$\mathbf{I} + \mathbf{A} + \mathbf{A}^2/2! + \cdots + \mathbf{A}^m/m! + \cdots = \mathbf{P} \begin{bmatrix} e^{r_1} & 0 & \cdots & 0 \\ 0 & e^{r_2} & \cdots & 0 \\ \vdots & & \ddots & \vdots \\ 0 & 0 & \cdots & e^{r_n} \end{bmatrix} \mathbf{P}^{-1} =: e^{\mathbf{A}}, \quad (15)$$

the "matrix exponential."

Example 5. Compute the matrix exponential $e^{\mathbf{A}}$ for $\mathbf{A}$ as displayed in (9).

Solution. From (15), we have

$$e^{\mathbf{A}} = \begin{bmatrix} 1 & -1 & -1 \\ 0 & 1 & -1 \\ 1 & 0 & 1 \end{bmatrix} \begin{bmatrix} e^3 & 0 & 0 \\ 0 & e^3 & 0 \\ 0 & 0 & e^{-3} \end{bmatrix} \begin{bmatrix} 1/3 & 1/3 & 2/3 \\ -1/3 & 2/3 & 1/3 \\ -1/3 & -1/3 & 1/3 \end{bmatrix}. \qquad \blacksquare$$

In Chapter 7, we will see that the matrix exponential is extremely useful in solving certain systems of ordinary differential equations. In fact, this feature will motivate the study of *generalized eigenvectors*, which enable the computation $e^{\mathbf{A}}$ when $\mathbf{A}$ is not diagonalizable.

Exercises 6.1

1. Prove: if $\mathbf{A}$ is similar to $\mathbf{B}$, then $\mathbf{A}^n$ is similar to $\mathbf{B}^n$ for every positive integer n. If $\mathbf{A}$ is invertible, the statement holds for negative integers as well.

2. Suppose $\mathbf{B}$ is obtained from $\mathbf{A}$ by switching its ith and jth columns and its ith and jth rows. Prove that $\mathbf{A}$ and $\mathbf{B}$ have the same eigenvalues. How are their eigenvectors related?

3. Prove that if $\mathbf{A}$ is diagonalizable and has only one eigenvalue, it is a scalar multiple of the identity.

4. Use the last assertion in Corollary 1 to prove that **the rank of a diagonalizable matrix equals the number of its nonzero eigenvalues (counting multiplicities)**.

5. Prove that if two matrices have the same eigenvalues with the same multiplicities and are both diagonalizable, then they are similar.

6. Prove or disprove: the trace of $\mathbf{AB}$ equals the trace of $\mathbf{BA}$, when $\mathbf{A}$ and $\mathbf{B}$ are n-by-n matrices. Does this hold for all square matrices?

7. Show that if $\mathbf{A}$ and $\mathbf{B}$ are both n-by-n matrices and one of them is invertible, then $\mathbf{AB}$ is similar to $\mathbf{BA}$. Construct a counterexample if neither is invertible.

8. Is the trace of $\mathbf{AB}$ equal to the trace of $\mathbf{A}$ times the trace of $\mathbf{B}$?

Diagonalize the matrices in Problems 9–20; that is, factor the matrix into the form PDP^{-1} with an invertible P and a diagonal D. Express the inverse (when it exists) and the exponential of the matrix, in factored form. The eigenvalues and eigenvectors of these matrices were assigned in Exercises 5.2.

9. $\begin{bmatrix} 9 & -3 \\ -2 & 4 \end{bmatrix}$ 10. $\begin{bmatrix} -1 & -3 \\ 4 & 7 \end{bmatrix}$ 11. $\begin{bmatrix} 1 & -2 \\ -2 & 1 \end{bmatrix}$ 12. $\begin{bmatrix} -3 & 2 \\ -5 & 4 \end{bmatrix}$ 13. $\begin{bmatrix} 1 & 1 \\ 1 & 1 \end{bmatrix}$

14. $\begin{bmatrix} 1 & 0 & 2 \\ 3 & 4 & 2 \\ 0 & 0 & 0 \end{bmatrix}$ 15. $\begin{bmatrix} 4 & -1 & -2 \\ 2 & 1 & -2 \\ 1 & -1 & 1 \end{bmatrix}$ 16. $\begin{bmatrix} 2 & -1 & -1 \\ -1 & 3 & 0 \\ -1 & 0 & 3 \end{bmatrix}$

17. $\begin{bmatrix} 0 & -1 \\ 1 & 0 \end{bmatrix}$ 18. $\begin{bmatrix} 0 & 1 \\ -1+i & -1 \end{bmatrix}$ 19. $\begin{bmatrix} 1 & 2i \\ -1 & 3+i \end{bmatrix}$ 20. $\begin{bmatrix} -6 & 2 & 16 \\ -3 & 1 & 11 \\ -3 & 1 & 7 \end{bmatrix}$

21. Show that the following matrix form is a square root of the zero matrix for *every* value of a, yet itself has *no* square root if $a \neq 0$:

$$\begin{bmatrix} 0 & a \\ 0 & 0 \end{bmatrix}.$$

22. The **geometric series** $1 + x + x^2 + x^3 + \cdots$ is the Maclaurin series for $(1 - x)^{-1}$ and converges for $|x| < 1$. Insert the diagonalizable matrix $\mathbf{A} = \mathbf{PDP}^{-1}$ into $\mathbf{I} + \mathbf{A} + \mathbf{A}^2 + \cdots$ and show that it converges to

$$
\mathbf{I} + \mathbf{A} + \mathbf{A}^2 + \cdots = \mathbf{P} \begin{bmatrix} \frac{1}{1-r_1} & 0 & \cdots & 0 \\ 0 & \frac{1}{1-r_2} & \cdots & 0 \\ \vdots & \vdots & \ddots & \vdots \\ 0 & 0 & \cdots & \frac{1}{1-r_n} \end{bmatrix} \mathbf{P}^{-1} =: \mathbf{B}
$$

as long as each eigenvalue of $\mathbf{A}$ has magnitude less than one. Prove that $\mathbf{B} = (\mathbf{I} - \mathbf{A})^{-1}$ by computing $(\mathbf{I} - \mathbf{A})\mathbf{B}\mathbf{u}_i$ for each of the (eigenvector) columns of $\mathbf{P}$.

23. Show that if $\mathbf{A}$ is a symmetric matrix whose diagonalization is expressed by $\mathbf{A} = \mathbf{QDQ}^T$, and if $\mathbf{u}_1, \mathbf{u}_2, \ldots, \mathbf{u}_n$ are the columns of the orthogonal matrix $\mathbf{Q}$, then $\mathbf{A}$ can be expressed as follows:

$$
\mathbf{A} = r_1 \mathbf{u}_1 \mathbf{u}_1^T + r_2 \mathbf{u}_2 \mathbf{u}_2^T + \cdots + r_n \mathbf{u}_n \mathbf{u}_n^T, \tag{16}
$$

where the r_i are the eigenvalues of $\mathbf{A}$. [*Hint*: Test (16) on the basis consisting of the eigenvectors.] What is the form of (16) for a Hermitian matrix? Is (16) valid for a real nonsymmetric matrix? (Many authors cite (16) as an expression of the spectral theorem.)

24. **Every polynomial with leading term** $(-r)^n$ **is the characteristic polynomial for some matrix.** Prove this by showing that the characteristic polynomial of

$$
\mathbf{C}_q = \begin{bmatrix} 0 & 1 & 0 & \cdots & 0 \\ 0 & 0 & 1 & \cdots & 0 \\ & & & \vdots & \\ 0 & 0 & 0 & \cdots & 1 \\ -a_0 & -a_1 & -a_2 & \cdots & -a_{n-1} \end{bmatrix}
$$

is $q(r) = (-1)^n \{ r^n + a_{n-1} r^{n-1} + a_{n-2} r^{n-2} + \cdots + a_1 r + a_0 \}$. In fact $\mathbf{C}_q$ is known as the *companion matrix* for $q(r)$.

25. Suppose the polynomial $q(r)$ has distinct roots $r_1, r_2, \ldots, r_n$. The **Vandermonde** matrix (Problem 23, Exercises 2.5) for these numbers equals

$$
\mathbf{V}_q = \begin{bmatrix} 1 & 1 & \cdots & 1 \\ r_1 & r_2 & \cdots & r_n \\ r_1^2 & r_2^2 & \cdots & r_n^2 \\ \vdots & \vdots & & \vdots \\ r_1^{n-1} & r_2^{n-1} & \cdots & r_n^{n-1} \end{bmatrix}.
$$

Show that each column of $\mathbf{V}_q$ is an eigenvector of the companion matrix $\mathbf{C}_q$ of the preceding problem. What is $\mathbf{V}_q^{-1}\mathbf{C}_q\mathbf{V}_q$?

26. The **Frobenius norm** of a real m-by-n matrix $\mathbf{A}$ is the square root of the sum of the squares of its entries, $||\mathbf{A}||_{\text{Frob}} = \left(\sum_{i=1}^{m}\sum_{j=1}^{n} a_{ij}^2\right)^{1/2}$.

   Recall from Section 4.2 that multiplication of a vector by an orthogonal matrix preserves the Euclidean norm of the vector. Derive an extension of this statement for the Frobenius norm by the following reasoning:

   (a) Verify that $||\mathbf{A}||_{\text{Frob}}^2$ equals the sum, over all the columns, of the squared Euclidean norms of the columns of $\mathbf{A}$. It also equals the sum, over all the rows, of the squared Euclidean norms of the rows of $\mathbf{A}$.

   (b) Verify that if $\mathbf{Q}$ is an orthogonal m-by-m matrix, then the Euclidean norms of the columns of $\mathbf{A}$ are the same as the Euclidean norms of the columns of $\mathbf{QA}$. Similarly, if $\mathbf{Q}$ is an orthogonal n-by-n matrix, then the Euclidean norms of the rows of $\mathbf{A}$ are the same as the Euclidean norms of the rows of $\mathbf{AQ}$.

   (c) From these observations fashion a proof that **if a similarity transformation $\mathbf{B} = \mathbf{QAQ}^T$ for a square matrix A is generated by an orthogonal matrix Q, then A and B have the same Frobenius norms.**

27. Extend the reasoning of the previous problem to show that $||\mathbf{A}||_{\text{Frob}} = ||\mathbf{Q}_1\mathbf{AQ}_2||_{\text{Frob}}$ for any matrix $\mathbf{A}$, if $\mathbf{Q}_1$ and $\mathbf{Q}_2$ are both orthogonal matrices (of appropriate dimensions). What is the complex version of this statement?

28. The order of appearance of the eigenvalues in a diagonal form of a nondefective matrix $\mathbf{A}$ is not fixed, of course; they can be extracted in any order. Find a similarity transform that changes the diagonal matrix with 1, 2, 3, 4 along its diagonal to one with 1,3,2,4 along the diagonal. [*Hint:* You know how to switch the second and third rows; how will you switch the second and third columns?]

6.2 PRINCIPLE AXES AND NORMAL MODES

6.2.1 Quadratic Forms

Quadratic forms (in three variables) are expressions that look like

$$ax^2 + by^2 + cz^2 + dxy + eyz + fxz. \tag{1}$$

It is easy to show (Problems 1 and 2) that the quadratic form is compactly expressed in matrix notation as $\mathbf{x}^T\mathbf{Ax}$, with a symmetric matrix $\mathbf{A}$ and the identifications

$$\mathbf{x} = \begin{bmatrix} x_1 & x_2 & x_3 \end{bmatrix}^T = \begin{bmatrix} x & y & z \end{bmatrix}^T, \quad \mathbf{A} = \begin{bmatrix} a_{11} & a_{12} & a_{13} \\ a_{21} & a_{22} & a_{23} \\ a_{31} & a_{32} & a_{33} \end{bmatrix} = \begin{bmatrix} a & \frac{d}{2} & \frac{f}{2} \\ \frac{d}{2} & b & \frac{e}{2} \\ \frac{f}{2} & \frac{e}{2} & c \end{bmatrix}. \tag{2}$$

The slightly more general *quadratic polynomials* (in three variables) incorporate additional linear terms and a constant:

$$(ax^2 + by^2 + cz^2 + dxy + eyz + fxz) + (gx + hy + jz) + (k) = \mathbf{x}^T \mathbf{A} \mathbf{x} + \mathbf{b}^T \mathbf{x} + (k). \quad (3)$$

These notions generalize to n dimensions, of course. The n-dimensional quadratic polynomial is $\mathbf{x}^T \mathbf{A} \mathbf{x} + \mathbf{b}^T \mathbf{x} + k$ with $\mathbf{x}$ and $\mathbf{b}$ in $\mathbf{R}^n_{\text{col}}$ and $\mathbf{A}$ symmetric in $\mathbf{R}^{n,n}$.

Quadratic polynomials arise in formulas for moments of inertia, potential energy (elastic, electromagnetic), and surface functions (i.e., their level surfaces describe familiar shapes such as spheres, cylinders, cones, and ellipsoids). Before we explore their interaction with similarity transforms, it is convenient to tabulate two elementary properties. The first generalizes the familiar process of "completing the square," and effectively reduces quadratic polynomials to quadratic forms (plus constants).

Lemma 1. If $\mathbf{A}$ is symmetric and invertible and $\mathbf{y} = (\mathbf{x} + \mathbf{A}^{-1}\mathbf{b}/2)$, then

$$\mathbf{x}^T \mathbf{A} \mathbf{x} + \mathbf{b}^T \mathbf{x} + k = \mathbf{y}^T \mathbf{A} \mathbf{y} + \left(k - \mathbf{b}^T \mathbf{A}^{-1}\mathbf{b}/4\right). \quad (4)$$

Proof. Start with the quadratic form in $\mathbf{y}$:

$$\mathbf{y}^T \mathbf{A} \mathbf{y} = \left(\mathbf{x} + \mathbf{A}^{-1}\mathbf{b}/2\right)^T \mathbf{A} \left(\mathbf{x} + \mathbf{A}^{-1}\mathbf{b}/2\right)$$
$$= \mathbf{x}^T \mathbf{A} \mathbf{x} + \frac{1}{2}\mathbf{x}^T \mathbf{A} \mathbf{A}^{-1}\mathbf{b} + \frac{1}{2}\mathbf{b}^T \left(\mathbf{A}^{-1}\right)^T \mathbf{A} \mathbf{x} + \frac{1}{4}\mathbf{b}^T \left(\mathbf{A}^{-1}\right)^T \mathbf{A} \mathbf{A}^{-1}\mathbf{b}.$$

Since $\mathbf{A}$ (and its inverse) is symmetric, this reduces to

$$\mathbf{y}^T \mathbf{A} \mathbf{y} = \mathbf{x}^T \mathbf{A} \mathbf{x} + \mathbf{b}^T \mathbf{x} + \mathbf{b}^T \mathbf{A}^{-1}\mathbf{b}/4;$$

transpose the constant and add k, to derive (4). ∎

For example, Problem 1(b) asks you to confirm that

$$\begin{bmatrix} x_1 & x_2 \end{bmatrix} \begin{bmatrix} 1 & 2 \\ 2 & 3 \end{bmatrix} \begin{bmatrix} x_1 \\ x_2 \end{bmatrix} + \begin{bmatrix} 4 & 6 \end{bmatrix} \begin{bmatrix} x_1 \\ x_2 \end{bmatrix} = \begin{bmatrix} y_1 & y_2 \end{bmatrix} \begin{bmatrix} 1 & 2 \\ 2 & 3 \end{bmatrix} \begin{bmatrix} y_1 \\ y_2 \end{bmatrix} - 3 \quad \text{if} \quad \mathbf{y} = \mathbf{x} + \begin{bmatrix} 0 \\ 1 \end{bmatrix}.$$
$$(5)$$

The second property is the formula for differentiating a quadratic polynomial. (A matrix *function* $\mathbf{A}(t) = [a_{ij}(t)]$ is differentiated entrywise: $\mathbf{A}'(t) = [a'_{ij}(t)]$.)

Lemma 2. If $\mathbf{A}(t)$ and $\mathbf{B}(t)$ are differentiable matrices, then the **product rule** holds:

$$\frac{d}{dt}[\mathbf{A}(t)\mathbf{B}(t)] = \left[\frac{d}{dt}\mathbf{A}(t)\right]\mathbf{B}(t) + \mathbf{A}(t)\frac{d}{dt}[\mathbf{B}(t)]. \quad (6)$$

(Note that we are careful to respect the order of appearance of $\mathbf{A}$ and $\mathbf{B}$, because matrix multiplication is noncommutative.)

Proof. This is simple bookkeeping. Recall that the (i,j)th entry of $\mathbf{AB}$ is given by

$$[\mathbf{AB}]_{ij} = \sum_{k=1}^{p} a_{ik}b_{kj},$$

so its derivative is

$$\left[\frac{d(\mathbf{AB})}{dt}\right]_{ij} = \sum_{k=1}^{p}\left[\frac{da_{ik}}{dt}b_{kj} + a_{ik}\frac{db_{kj}}{dt}\right] = \left[\frac{d\mathbf{A}}{dt}\mathbf{B} + \mathbf{A}\frac{d\mathbf{B}}{dt}\right]_{ij}. \qquad \blacksquare$$

(The derivative of a sum of matrices is simply the sum of their derivatives.)

Now if $\mathbf{x}$ is a vector in $\mathbf{R}^n_{\text{col}}$ and $f(\mathbf{x})$ is a scalar-valued function, the gradient of f is the column vector of its partial derivatives.

$$\mathbf{grad}f = \left[\frac{\partial f}{\partial x_1} \quad \frac{\partial f}{\partial x_2} \quad \cdots \quad \frac{\partial f}{\partial x_n}\right]^T.$$

What is the formula for the gradient of the quadratic polynomial (3)?

Lemma 3. If $\mathbf{A}$ is a constant symmetric n-by-n matrix and b is a constant n-by-1 vector, then

$$\mathbf{grad}\left(\mathbf{x}^T\mathbf{Ax} + \mathbf{b}^T\mathbf{x}\right) = 2\mathbf{Ax} + \mathbf{b}. \qquad (7)$$

Proof. Using the product rule to differentiate with respect to x_2 (for example), we find that

$$\frac{\partial}{\partial x_2}\mathbf{b}^T\mathbf{x} = \mathbf{b}^T\frac{\partial}{\partial x_2}\mathbf{x} = \mathbf{b}^T\frac{\partial}{\partial x_2}\begin{bmatrix} x_1 \\ x_2 \\ x_3 \\ \vdots \\ x_n \end{bmatrix} = \mathbf{b}^T\begin{bmatrix} 0 \\ 1 \\ 0 \\ \vdots \\ 0 \end{bmatrix} = \begin{bmatrix} b_1 & b_2 & \cdots & b_n \end{bmatrix}\begin{bmatrix} 0 \\ 1 \\ 0 \\ \vdots \\ 0 \end{bmatrix},$$

which simply "extracts" the second component of $\mathbf{b}$. Generalizing from this, we see

$$\mathbf{grad}\,\mathbf{b}^T\mathbf{x} = \left[\frac{\partial \mathbf{b}^T\mathbf{x}}{\partial x_1} \quad \frac{\partial \mathbf{b}^T\mathbf{x}}{\partial x_2} \quad \cdots \quad \frac{\partial \mathbf{b}^T\mathbf{x}}{\partial x_n}\right]^T = \mathbf{b}.$$

For the quadratic form we have

$$\frac{\partial}{\partial x_2}\mathbf{x}^T\mathbf{Ax} = \frac{\partial\mathbf{x}^T}{\partial x_2}\mathbf{Ax} + \mathbf{x}^T\mathbf{A}\frac{\partial\mathbf{x}}{\partial x_2} = \begin{bmatrix} 0 & 1 & 0 & 0 & \cdots & 0 \end{bmatrix}\mathbf{Ax} + \mathbf{x}^T\mathbf{A}\begin{bmatrix} 0 \\ 1 \\ \vdots \\ 0 \end{bmatrix},$$

extracting the second component of $\mathbf{Ax}$ plus the second component of $\mathbf{x}^T\mathbf{A}$—which are identical, since $\mathbf{A}$ is symmetric ($[\mathbf{Ax}]^T = \mathbf{x}^T\mathbf{A}^T = \mathbf{x}^T\mathbf{A}$). In general, then,

$$\mathbf{grad}\,\mathbf{x}^T\mathbf{Ax} = \left[\frac{\partial\mathbf{x}^T\mathbf{Ax}}{\partial x_1}\quad\frac{\partial\mathbf{x}^T\mathbf{Ax}}{\partial x_2}\cdots\frac{\partial\mathbf{x}^T\mathbf{Ax}}{\partial x_n}\right]^T = 2\mathbf{Ax}.$$ ∎

For instance, it is easy to confirm that

$$\mathbf{grad}\left\{[x_1\quad x_2]\begin{bmatrix}1 & 2\\2 & 3\end{bmatrix}\begin{bmatrix}\mathbf{x}_1\\\mathbf{x}_2\end{bmatrix} + [4\quad 6]\begin{bmatrix}\mathbf{x}_1\\\mathbf{x}_2\end{bmatrix}\right\} = \mathbf{grad}\,\{x_1^2 + 4x_1x_2 + 3x_2^2 + 4x_1 + 6x_2\}$$

$$= \begin{bmatrix}2x_1 + 4x_2 + 4\\4x_1 + 6x_2 + 6\end{bmatrix}$$

$$= 2\begin{bmatrix}1 & 2\\2 & 3\end{bmatrix}\begin{bmatrix}\mathbf{x}_1\\\mathbf{x}_2\end{bmatrix} + \begin{bmatrix}4\\6\end{bmatrix}.$$

What do similarity transformations have to do with quadratic forms? Suppose we rotate our coordinate system, as we did in the previous section (Fig. 6.1b), so that the new coordinates of a point ($\mathbf{x}'$) are related to its old coordinates ($\mathbf{x}$) through a matrix equation $\mathbf{x} = \mathbf{Qx}'$, where $\mathbf{Q}$ is an orthogonal matrix. Then the quadratic form becomes expressed in the new coordinates as

$$\mathbf{x}^T\mathbf{Ax} = (\mathbf{Qx}')^T\mathbf{A}(\mathbf{Qx}') = \mathbf{x}'^T\mathbf{Q}^T\mathbf{AQx}' = \mathbf{x}'^T(\mathbf{Q}^T\mathbf{AQ})\mathbf{x}' = \mathbf{x}'^T(\mathbf{Q}^{-1}\mathbf{AQ})\mathbf{x}'.$$

The symmetric matrix $\mathbf{A}$ in the quadratic form is replaced by the *similar* matrix $\mathbf{Q}^{-1}\mathbf{AQ}$. But more importantly, because $\mathbf{A}$ *is* symmetric we could use the matrix of its (unit) eigenvectors to define the generator $\mathbf{Q}$—in which case by Corollary 3 of the previous section the matrix $\mathbf{Q}^{-1}\mathbf{AQ}$ would be diagonal, and there would be no cross terms in the quadratic form! Indeed, the coefficients of the surviving terms would be the eigenvalues. The eigenvector directions are the **principal axes** of $\mathbf{A}$.

Principal Axes Theorem

Theorem 2. If $\mathbf{A}$ is a real symmetric matrix, its eigenvectors generate an orthogonal change of coordinates that transforms the quadratic form $\mathbf{x}^T\mathbf{Ax}$ into a quadratic form with no cross terms.

Corollary 5. If $\mathbf{A}$ is a Hermitian matrix, its eigenvectors generate a unitary change of coordinates that transforms the quadratic form $\mathbf{x}^H\mathbf{Ax}$ into a quadratic form with no cross terms.

The following example brings to bear all of the important features of the theory of real quadratic forms.

Example 1. Rewrite $x_1^2 + x_2^2 + x_3^2 - 4x_1x_2 + 4x_1x_3 + 4x_2x_3$ in new coordinates so that it has no cross terms. Determine its maximum and minimum values on the sphere $x_1^2 + x_2^2 + x_3^2 = 1$.

Solution. From (1) and (2), we determine that the symmetric matrix appearing in the quadratic form is the same matrix **A** that we used in Example 5, Section 5.2; that is,

$$x_1^2 + x_2^2 + x_3^2 - 4x_1x_2 + 4x_1x_3 + 4x_2x_3 = \begin{bmatrix} x_1 & x_2 & x_2 \end{bmatrix} \begin{bmatrix} 1 & -2 & 2 \\ -2 & 1 & 2 \\ 2 & 2 & 1 \end{bmatrix} \begin{bmatrix} x_1 \\ x_2 \\ x_3 \end{bmatrix} = \mathbf{x}^T \mathbf{A} \mathbf{x}.$$

Its diagonalization via the (orthogonal) matrix of eigenvectors is displayed in Section 6.1 Equation (10). Implementing the construction described above Theorem 2, we find

$$\mathbf{x}^T \mathbf{A} \mathbf{x} = \mathbf{x}^T \left(\mathbf{Q} \mathbf{Q}^T \right) \mathbf{A} \left(\mathbf{Q} \mathbf{Q}^T \right) \mathbf{x} = \left(\mathbf{Q}^T \mathbf{x} \right)^T \mathbf{Q}^T \mathbf{A} \mathbf{Q} \left(\mathbf{Q}^T \mathbf{x} \right)$$

$$= \left(\mathbf{Q}^T \mathbf{x} \right)^T \begin{bmatrix} 1/\sqrt{2} & 0 & 1/\sqrt{2} \\ -1/\sqrt{6} & 2/\sqrt{6} & 1/\sqrt{6} \\ -1/\sqrt{3} & -1/\sqrt{3} & 1/\sqrt{3} \end{bmatrix} \begin{bmatrix} 1 & -2 & 2 \\ -1 & 1 & 2 \\ 2 & 2 & 1 \end{bmatrix}$$

$$\times \begin{bmatrix} 1/\sqrt{2} & -1/\sqrt{6} & -1/\sqrt{3} \\ 0 & 2/\sqrt{6} & -1/\sqrt{3} \\ 1/\sqrt{2} & 1/\sqrt{6} & 1/\sqrt{3} \end{bmatrix} \left(\mathbf{Q}^T \mathbf{x} \right)$$

$$= \mathbf{y}^T \begin{bmatrix} 3 & 0 & 0 \\ 0 & 3 & 0 \\ 0 & 0 & -3 \end{bmatrix} \mathbf{y} = 3y_1^2 + 3y_2^2 - 3y_3^2,$$

where $\mathbf{y} = \mathbf{Q}^T \mathbf{x}$. The principal axes, in y coordinates, are $[1\,0\,0]^T$, $[0\,1\,0]^T$, and $[0\,0\,1]^T$; in the original x coordinates they are the columns of **Q** (i.e., the eigenvectors), because $\mathbf{x} = \mathbf{Q} \mathbf{y}$.

Since multiplication by orthogonal matrices preserves lengths (Euclidean norms), the set of points where $x_1^2 + x_2^2 + x_3^2 = 1$ is the same as the set where $y_1^2 + y_2^2 + y_3^2 = 1$. The maximum and minimum of $3y_1^2 + 3y_2^2 - 3y_3^2 \, (= 3 - 6y_3^2)$ are obviously 3 and -3, the largest and smallest eigenvalues. ∎

6.2.2 Normal Modes

Figure 6.2a shows an old (1920–1940) microphone used in radio broadcasting. (A mockup of a similar microphone was used in "The King's Speech," the academy award-winning movie produced in 2011.) The mounting is spring loaded to cushion the microphone from environmental vibrations. Example 2 will demonstrate the use of eigenvectors and similarity transformations to analyze the oscillations. To keep our calculations down to a manageable size, we employ the simplified geometry shown in Figure 6.2b.

(a)

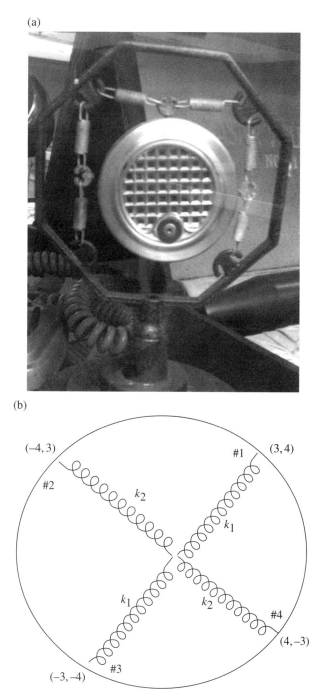

(b)

Fig. 6.2   (a) Spring-Mounted microphone, (b) Simplified model, (c) Displaced microphone, and (d) Multiple spring model.

(c)

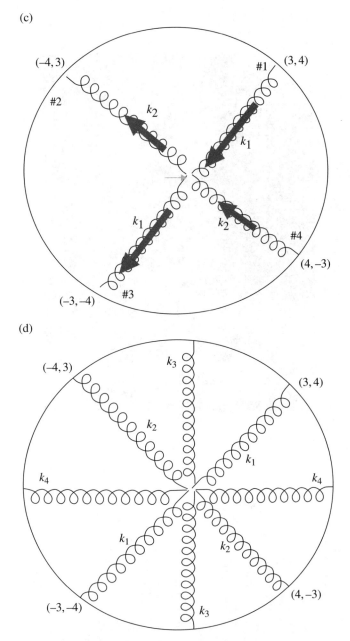

Fig. 6.2 Continued.

Before we start the analysis, let's speculate on the results. Suppose in Figure 6.2b that springs #1 and #3 are much stiffer than #2 and #4, so that $k_1 \gg k_2$. Suppose further that the microphone is displaced horizontally, to the right, as in Figure 6.2c. Then the stronger springs 1 and 3 will force it to the left and down with more intensity than the weaker springs force it right and up. When released from rest, the microphone will execute some complicated curvilinear motion.

But there are displacement directions for which the restoring force is directed *along the displacement*: namely, displacements along the lines of the springs, at $\tan^{-1}(3/4) \approx 36.5°$ to the axes. If we rotate our coordinate system to point along the springs, the x and y equations of motion should "uncouple" and become simpler to solve. Such motions are called the **normal modes** of the system.

Example 2. Determine the restoring force when the microphone in Figure 6.2c is released from rest at the position (x, y) slightly displaced from the origin, and display the transformation that uncouples the equations of motion.

Solution. To hold the microphone tightly, in practice all the springs are stretched to some extent from their "relaxed" lengths before the microphone is loaded—and its weight will distort them even further. However, all the spring forces are presumably at equilibrium when the microphone is centered, and we need only consider each spring's *incremental* force, given by the product of its stiffness times the incremental change in length.

If the microphone is shifted to the position (x, y), the length of spring #1 is changed from $\sqrt{3^2 + 4^2} = 5$ to $\sqrt{(3-x)^2 + (4-y)^2} = \sqrt{9 - 6x + x^2 + 16 - 8y + y^2}$.

This nonlinear function bears no resemblance to the linear combinations that are susceptible to matrix analysis, and we must approximate. Problem 16 describes the Maclaurin series which shows that if the shift in position is very small, then the length of spring #1 is contracted by $5 - \sqrt{25 - 6x - 8y + x^2 + y^2} \approx 0.6x + .8y$.

The restoring force is directed along the spring, which for small displacements will lie in the direction of $-3\mathbf{i} - 4\mathbf{j}$. Inserting the spring constant we find this force to be approximated by

$$\frac{-3\mathbf{i} - 4\mathbf{j}}{\sqrt{3^2 + 4^2}} k_1(0.6x + 0.8y) = \begin{bmatrix} -3/5 \\ -4/5 \end{bmatrix} k_1 \begin{bmatrix} 0.6 & 0.8 \end{bmatrix} \begin{bmatrix} x \\ y \end{bmatrix} = -k_1 \begin{bmatrix} 0.36 & 0.48 \\ 0.48 & 0.64 \end{bmatrix} \begin{bmatrix} x \\ y \end{bmatrix}. \tag{8}$$

Spring #3 is *stretched* by the same amount and thus doubles this restoring force.

Problem 16 also dictates that spring #2 is compressed/stretched by $-0.8x + 0.6y$, and the restoring force, directed along $4\mathbf{i} - 3\mathbf{j}$, is approximately

$$\begin{bmatrix} 4/5 \\ -3/5 \end{bmatrix} k_2 \begin{bmatrix} -0.8 & 0.6 \end{bmatrix} \begin{bmatrix} x \\ y \end{bmatrix} = -k_2 \begin{bmatrix} 0.64 & -0.48 \\ -0.48 & 0.36 \end{bmatrix} \begin{bmatrix} x \\ y \end{bmatrix}, \tag{9}$$

doubled due to the identical force from spring #4.

So the total restoring force equals $-\mathbf{Kx}$, where $\mathbf{x} = \begin{bmatrix} x & y \end{bmatrix}^T$ and the stiffness matrix $\mathbf{K}$ is

$$\mathbf{K} = 2k_1 \begin{bmatrix} 0.36 & 0.48 \\ 0.48 & 0.64 \end{bmatrix} + 2k_2 \begin{bmatrix} 0.64 & -0.48 \\ -0.48 & 0.36 \end{bmatrix} = \begin{bmatrix} 0.72k_1 + 1.28k_2 & 0.96k_1 - 0.96k_2 \\ 0.96k_1 - 0.96k_2 & 1.28k_1 + 0.72k_2 \end{bmatrix}. \tag{10}$$

Note that the stiffness matrix is symmetric.

The displacements for which the restoring force $\mathbf{Kx}$ is aligned with the displacement $\mathbf{x}$ are, of course, eigenvectors of $\mathbf{K}$. If k_1 happens to equal k_2, then

$$\mathbf{K} = k_2 \begin{bmatrix} 2 & 0 \\ 0 & 2 \end{bmatrix}$$

is a multiple of the identity and *every* displacement is, indeed, an eigenvector; the spring support is isotropic.

For $k_1 = 2k_2$ the similarity transformation

$$\mathbf{K} = \begin{bmatrix} 2.72k_2 & 0.96k_2 \\ 0.96\,k_2 & 3.28k_2 \end{bmatrix} = \begin{bmatrix} 0.6 & -0.8 \\ 0.8 & 0.6 \end{bmatrix} \begin{bmatrix} 4k_2 & 0 \\ 0 & 2k_2 \end{bmatrix} \begin{bmatrix} 0.6 & 0.8 \\ -0.8 & 0.6 \end{bmatrix} = \mathbf{QDQ}^{-1}.$$

reveals the eigenvalues $(4k_2, 2k_2)$ and eigenvectors (columns of $\mathbf{Q}$) of the stiffness matrix. As predicted, the eigenvectors $[0.6\ \ 0.8]^T$ and $[-0.8\ \ 0.6]^T$ are parallel to the springs. If a displacement has coordinates $\mathbf{x} = [x\ \ y]^T$ in the original system, its coordinates $\mathbf{z} = [u\ \ v]^T$ in the new system are given by $\mathbf{z} = \mathbf{Q}^{-1}\mathbf{x}$; Newton's law *force equals mass times acceleration* simplifies from

$$m\ddot{\mathbf{x}} = -\mathbf{Kx} \quad \text{or} \quad m\begin{bmatrix} \ddot{x} \\ \ddot{y} \end{bmatrix} = \begin{bmatrix} -2.72k_2 & -0.96k_2 \\ -0.96\,k_2 & -3.28k_2 \end{bmatrix} \begin{bmatrix} x \\ y \end{bmatrix}$$

to

$$m\ddot{\mathbf{z}} = m\mathbf{Q}^{-1}\ddot{\mathbf{x}} = -\mathbf{Q}^{-1}\mathbf{Kx} = -\mathbf{Q}^{-1}\mathbf{KQQ}^{-1}\mathbf{x} = \mathbf{Dz}' \quad \text{or}$$

$$m\ddot{\mathbf{z}} = m\begin{bmatrix} \ddot{u} \\ \ddot{v} \end{bmatrix} = \begin{bmatrix} -4k_2 & 0 \\ 0 & -2k_2 \end{bmatrix} \begin{bmatrix} u \\ v \end{bmatrix} = \begin{bmatrix} -4k_2u \\ -2k_2v \end{bmatrix}. \qquad \blacksquare$$

What about Figure 6.2d? Are there still directions for which the restoring force is aligned along the displacement? The answer is *yes*; it is easy to generalize (8) and (9) and see that every spring adds a *symmetric* contribution to the total stiffness matrix. Thus $\mathbf{K}$ will always be symmetric and possess orthogonal eigenvectors defining, through the associated similarity transformation, a coordinate system in which the equations are uncoupled (see Problem 14).

These observations form the foundations of the engineering discipline known as **modal analysis**.

Exercises 6.2

1. (a) Verify directly that $ax^2 + by^2 + cxy = [x\ \ y] \begin{bmatrix} a & c/2 \\ c/2 & b \end{bmatrix} \begin{bmatrix} x \\ y \end{bmatrix}$

   (b) Verify equation (5).

2. (a) Show that if $\mathbf{v}$ is a column vector with all zeros except for a "1" in its ith entry, and if $\mathbf{w}$ is a column vector with all zeros except for a "1" in its jth entry, then $\mathbf{v}^T\mathbf{A}\mathbf{w}$ equals the isolated entry a_{ij}.

(b) Use (a) to show that in the product $\mathbf{x}^T\mathbf{A}\mathbf{x}$ appearing in (3) the term a_{ij} appears with the factors x_i and x_j. [*Hint*: Observe

$$
\begin{aligned}
\mathbf{x}^T &= \begin{bmatrix} x_1 & x_2 & \cdots & x_n \end{bmatrix} \\
&= \begin{bmatrix} x_1 & 0 & \cdots & 0 \end{bmatrix} + \begin{bmatrix} 0 & x_2 & \cdots & 0 \end{bmatrix} + \cdots + \begin{bmatrix} 0 & 0 & \cdots & x_n \end{bmatrix}
\end{aligned}
$$

and apply the distributive law of matrix multiplication.]

(c) Use (b) to show that formula (1) and $\mathbf{x}^T\mathbf{A}\mathbf{x}$ are identical expressions if $a_{11} = a$, $a_{12} + a_{21} = d$, etc.

Note that we don't *have* to select equal values for a_{12} and a_{21} (namely $d/2$), but doing so gives us the desirable eigenvector properties of a *symmetric* matrix, leading to the principal axes theorem.

Find a change of coordinates that eliminates the cross terms in the quadratic forms in Problems 3–7, and find the maximum and minimum of the forms on the unit circle $(x_1^2 + x_2^2 = 1)$ or sphere $(x_1^2 + x_2^2 + x_3^2 = 1)$, whichever is appropriate. [*Hint*: Check Problems 1–6, Exercises 5.3, to see if you have already diagonalized the matrix.]

3. $-7x_1^2 + 2x_2^2 + 12x_1x_2$ **4.** $4x_1^2 + 4x_2^2 + 2x_1x_2$ **5.** $x_1^2 + x_2^2 + x_3^2 + 8x_1x_2 + 6x_1x_3$

6. $2x_1^2 + 2x_2^2 + 3x_3^2 - 2x_2x_3 - 2x_1x_3$ **7.** $2x_1^2 + 2x_2^2 + 2x_3^2 + 2x_1x_2 + 2x_1x_3 - 2x_2x_3$

8. Find the maximum on the "unit hypersphere" $(x_1^2 + x_2^2 + x_3^2 + x_4^2 = 1)$ of the quadratic form $4x_1^2 + 4x_2^2 - 7x_3^2 + 2x_4^2 + 2x_1x_2 + 12x_3x_4$.

9. **Rayleigh Quotient** Suppose the eigenvalues of $\mathbf{A}$ are distinct and are ordered as $r_1 > r_2 > \cdots > r_n$.

(a) Show that the quadratic form $\mathbf{v}^T\mathbf{A}\mathbf{v}$ achieves its maximum value, over all *unit* vectors $\mathbf{v}$, when $\mathbf{v}$ is an eigenvector $\mathbf{u}_1$ corresponding to the maximum eigenvalue r_1 of $\mathbf{A}$; and that this maximum value is r_1.

(b) Show that $\mathbf{v}^T\mathbf{A}\mathbf{v}$ achieves its maximum value, over all unit vectors $\mathbf{v}$ *orthogonal to* $\mathbf{u}_1$, when $\mathbf{v}$ is an eigenvector $\mathbf{u}_2$ corresponding to the second largest eigenvalue r_2; and that this maximum value is r_2.

(c) Characterize the eigenvalue $r_i (i > 2)$ similarly, in turns of $\mathbf{v}^T\mathbf{A}\mathbf{v}$.

(d) Argue that if we replace $\mathbf{v}^T\mathbf{A}\mathbf{v}$ by the *Rayleigh Quotient*, $\mathbf{v}^T\mathbf{A}\mathbf{v}/\mathbf{v}^T\mathbf{v}$, in (a–c) above, we don't have to restrict the search to *unit* vectors.

10. Combine what you have learned about the Rayleigh Quotient (in the previous problem), quadratic forms, and eigenvalues of circulant matrices (Problem 14, Exercises 5.1) to deduce the maximum of the expression $x_1x_2 + x_2x_3 + x_3x_4 + \cdots + x_{n-1}x_n + x_nx_1$ when $x_1^2 + x_2^2 + \cdots + x_n^2 = 1$.

11. Replace the quadratic form in the previous problem by the sum of $x_i x_j$ over *all* unmatched indices $x_1 x_2 + x_2 x_3 + x_1 x_3 + x_3 x_4 + x_2 x_4 + x_1 x_4 \cdots + x_{n-1} x_n = \sum_{i=1}^{n} \sum_{\substack{j=1 \\ j \neq i}}^{n} x_i x_j$, and rework it.

12. **Linear Invariant Discrete Time Systems** (*homogeneous*). A sequence of vectors $\mathbf{x}_n := \mathbf{x}(n \, \Delta t)$ $(0 \leq n < \infty$, n integer) that describe the state of some system at multiples of a given time step Δt is called a **discrete-time signal**. If $\mathbf{x}_n$ evolves according to the *transition equation*

$$\mathbf{x}_{n+1} = \mathbf{A} \mathbf{x}_n \tag{11}$$

with a constant m-by-m matrix $\mathbf{A}$, then the discrete time system is said to be *linear homogenous time-invariant*. An example is the Markov chain described in Problem 22 of Section 2.1. Note that we easily derive $\mathbf{x}_n = \mathbf{A}^n \mathbf{x}_0$. For simplification, assume that $\mathbf{A}$ is diagonalizable.

(a) Let the initial state $\mathbf{x}_0$ be expressed in terms of m linearly independent eigenvectors $\{\mathbf{u}_i\}$ of $\mathbf{A}$ as $\mathbf{x}_0 = c_1 \mathbf{u}_1 + c_2 \mathbf{u}_2 + \cdots c_m \mathbf{u}_m$. Show that

$$\mathbf{x}_n = c_1 r_1^n \mathbf{u}_1 + c_2 r_2^n \mathbf{u}_2 + \cdots + c_m r_m^n \mathbf{u}_m \tag{12}$$

for any $n > 0$, where r_i is the eigenvalue corresponding to $\mathbf{u}_i$.

(b) Show that if $\mathbf{x}_0$ is an eigenvector of $\mathbf{A}$ whose corresponding eigenvalue equals one, then $\mathbf{x}_0$ is a *stationary state* in the sense that the constant signal $\mathbf{x}_n \equiv \mathbf{x}_0$ is a solution of (11).

(c) Show that if some of the eigenvalues of $\mathbf{A}$ equal one and all of the others are less than one in magnitude, then every solution of (11) approaches either $\mathbf{0}$ or a stationary state as $n \to \infty$. The system is said to be *stable* under such conditions.

(d) What are the solution possibilities if some of the eigenvalues of $\mathbf{A}$ are greater than 1 in magnitude?

13. **Linear Invariant Discrete-Time Systems** *(nonhomogeneous)*. If the transition equation of the system in the previous problem takes the form $\mathbf{x}_{n+1} = \mathbf{A} \mathbf{x}_n + \mathbf{b}_n$ with a constant m-by-m matrix $\mathbf{A}$, the discrete system is called linear *non*homogeneous time-invariant, even if the vectors $\mathbf{b}_n$ vary with n. For simplicity, we address the situation where the $\mathbf{b}_n$ *are* constant, $\mathbf{b}_n \equiv \mathbf{b}$:

$$\mathbf{x}_{n+1} = \mathbf{A} \mathbf{x}_n + \mathbf{b}. \tag{13}$$

As in the previous problem, assume $\mathbf{A}$ is diagonalizable.

(a) Verify that $\mathbf{x}_n \equiv (\mathbf{I} - \mathbf{A})^{-1} \mathbf{b}$ is a stationary solution to (13) if none of the eigenvalues of $\mathbf{A}$ equals one.

(b) Show that all solutions of (13) can be represented as $\mathbf{x}_n = \mathbf{A}^n \mathbf{x}_0 + \left(\mathbf{A}^{n-1} + \mathbf{A}^{n-2} + \cdots + \mathbf{A} + \mathbf{I} \right) \mathbf{b}$.

(c) Apply the geometric series discussed in Problem 22 of Exercises 6.1 to show that the solutions in (b) approach $(\mathbf{I} - \mathbf{A})^{-1}\mathbf{b}$ if all of the eigenvalues of $\mathbf{A}$ have magnitude less than one.

(d) Show that if $\mathbf{x}_n$ satisfies the nonhomogeneous equation (13), then $\mathbf{y}_n$ satisfies the *homogeneous* equation $\mathbf{y}_{n+1} = \mathbf{B}\mathbf{y}_n$ where

$$[\mathbf{y}_n] := \begin{bmatrix} \mathbf{x}_n \\ 1 \end{bmatrix}, \qquad [\mathbf{B}] = \begin{bmatrix} \mathbf{A} & \vdots & \mathbf{b} \\ 0 \cdots 0 & \vdots & 1 \end{bmatrix}. \tag{14}$$

(e) Conversely, show that if $\mathbf{y}_{n+1} = \mathbf{B}\mathbf{y}_n$ then the $(m+1)$st component of $\mathbf{y}_n$ is constant; and if this constant is not zero a solution to (13) can be extracted from a solution to (14) by dividing $\mathbf{y}_n$ by this constant and taking the first m components.

(f) How are the eigenvalues of $\mathbf{A}$ and $\mathbf{B}$ related?

(g) How are the eigen*vectors* of $\mathbf{A}$ and $\mathbf{B}$ related? For simplicity, assume that one is not an eigenvalue of $\mathbf{A}$. [*Hint*: Consider the eigenvectors of $\mathbf{A}$, padded with a final zero, and the vector $(\mathbf{I} - \mathbf{A})^{-1}\mathbf{b}$, padded with a final one.]

(h) Put this all together: given an initial state $\mathbf{x}_0$ for the nonhomogeneous equation (13), explain thoroughly how to combine the reformulation in (d), the eigenvectors uncovered in (g), the technique for solving homogeneous transition equations described in the previous problem, and the extraction procedure in (e) to find an expression generalizing (12) to the nonhomogeneous case.

14. Construct the stiffness matrix for the microphone support in Figure 6.2d and verify that it is symmetric. Find its eigenvectors and sketch the directions in which uncoupled oscillations are possible.

15. The inertia tensor I of a rigid body is a symmetric matrix calculated from the body's mass density function $\rho(x, y, z)$ (grams per cubic centimeter): with the coordinates enumerated as $(x, y, z) = (x_1, x_2, x_3)$, we have

$$I = [I_{ij}] = \left\{ \int \int \int \rho(x_1, x_2, x_3)\, x_i x_j\, dx_1\, dx_2\, dx_3 \right\} \quad (i, j = 1, 2, 3).$$

For any unit vector $\mathbf{n}$ (in $\mathbf{R}_{\text{col}}^3$) the quadratic form $\mathbf{n}^T I \mathbf{n}$ gives the moment of inertia of the body about the axis $\mathbf{n}$. If I is diagonalized by an orthogonal matrix $\mathbf{Q}$, then classical mechanics shows that (due to the absence of the off-diagonal "products of inertia") the body can spin smoothly about its principal axes (columns of $\mathbf{Q}$) without wobbling. (Think of a spiral pass or an end-over-end placement kick of a football. A punt, however, typically imparts a spin about a random axis, and the result is a wobbly motion.)

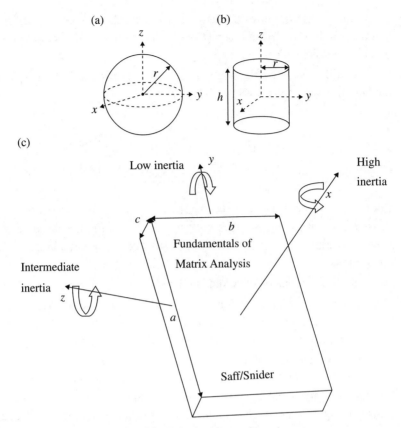

Fig. 6.3   Principal axes of inertia.

(a) Show that for a uniform sphere of mass M and radius r, the axes in Figure 6.3a are principal axes and

$$I = \begin{bmatrix} \frac{2}{5}Mr^2 & 0 & 0 \\ 0 & \frac{2}{5}Mr^2 & 0 \\ 0 & 0 & \frac{2}{5}Mr^2 \end{bmatrix}.$$

(b) Show that for a uniform cylinder of mass M, radius r, and height h, the axes in Figure 6.3b are principal axes and

$$I = \begin{bmatrix} M\left(\frac{r^2}{4} + \frac{h^2}{12}\right) & 0 & 0 \\ 0 & M\left(\frac{r^2}{4} + \frac{h^2}{12}\right) & 0 \\ 0 & 0 & M\frac{r^2}{2} \end{bmatrix}.$$

(c) Show that for a uniform rectangular prism as shown in Figure 6.3c the displayed axes are principal axes and

$$I = \begin{bmatrix} \frac{1}{12}M(a^2+b^2) & 0 & 0 \\ 0 & \frac{1}{12}M(c^2+b^2) & 0 \\ 0 & 0 & \frac{1}{12}M(a^2+c^2) \end{bmatrix}.$$

You can experiment with the moments of inertia of your (hardcopy) textbook by tying it shut with a rubber band and tossing it in the air. You will be able to spin it smoothly around the "high inertia" and "low inertia" axes; but even though the "intermediate inertia" axis is a principal axis, the revolutions are unstable with respect to small perturbations. Try it out.

16. The first-order expansion of the Maclauren series for a suitably smooth function f is given by

$$f(x,y) = f(0,0) + \frac{\partial f(0,0)}{\partial x}x + \frac{\partial f(0,0)}{\partial y}y + O\left(x^2+y^2\right),$$

where $O(x^2+y^2)$ is a term bounded by a constant times (x^2+y^2). Apply this to the length functions in Example 2 to justify the approximations made there.

6.3   SCHUR DECOMPOSITION AND ITS IMPLICATIONS

Section 6.1 demonstrated that the diagonalization of a matrix is extremely useful in facilitating computations. But the goal of attaining a diagonal $\mathbf{D} = \mathbf{P}^{-1}\mathbf{A}\mathbf{P}$, with an invertible $\mathbf{P}$, is not always possible, since Example 5 of Section 5.2 showed that not all matrices possess a full basis of eigenvectors (the columns of $\mathbf{P}$); such matrices are said to be **defective**. If we can't get a diagonal matrix, what is the simplest matrix form that we *can* get, using similarity transformations?[2]

One answer is provided by a result known as the Schur decomposition.[3] The form it achieves, upper triangular, may seem a far cry from diagonal, but (as we shall see) the result is quite powerful, the diagonalizing matrix $\mathbf{P}$ is unitary (and thus easy and accurate to invert), and the derivation is not hard.

---

[2]The upper triangular matrix $\mathbf{U}$ produced by the $\mathbf{A} = \mathbf{L}\mathbf{U}$ factorization using Gauss elimination (Project A, Part I) is not *similar* to $\mathbf{A}$ (there is no right multiplier $\mathbf{L}^{-1}$).
[3]Issai Schur (1875–1941) is most remembered for this decomposition and his Lemma on group representations.

Upper Triangular Form

> **Theorem 3.** (*Schur Decomposition*) Every complex n-by-n matrix $\mathbf{A}$ is similar to an upper triangular matrix $\mathbf{U} = \mathbf{Q}^H \mathbf{A} \mathbf{Q}$, with a similarity transformation generated by a unitary[4] matrix $\mathbf{Q}$.

Before we get to the proof, look at what this implies about Hermitian matrices. If $\mathbf{A}$ is Hermitian, so is the upper triangular matrix $\mathbf{U}$:

$$\mathbf{U} = \mathbf{Q}^H \mathbf{A} \mathbf{Q} = \mathbf{Q}^H \mathbf{A}^H \mathbf{Q} = \left(\mathbf{Q}^H \mathbf{A} \mathbf{Q} \right)^H = \mathbf{U}^H.$$

But an upper triangular Hermitian matrix must be diagonal! Thanks to Schur, we now know that Hermitian (and real symmetric) matrices can *always* be diagonalized. Specifically,

Corollary 6. Any Hermitian matrix can be diagonalized by a unitary matrix. Its multiple eigenvalues have eigenspaces of dimension equal to the multiplicity of the corresponding eigenvalues.

Just to refresh your memory: the eigenvalues are real and eigenvectors corresponding to distinct eigenvalues are orthogonal; real symmetric matrices are Hermitian, and real unitary matrices are orthogonal. If $\mathbf{A}$ is real and symmetric, the Schur decomposition can be carried out in real arithmetic.

Proof of the Schur Decomposition. To maintain consistency in the discussion, we temporarily rename $\mathbf{A}$ as $\mathbf{A}_1$. Roughly, the strategy of the proof is this. First we use (any) unit eigenvector $\mathbf{u}_1$ of $\mathbf{A}_1$ to construct a similarity transformation that "upper-triangulates its first column"—that is, introduces zeros below the diagonal. Then we operate similarly on the lower $(n-1)$-by-$(n-1)$ submatrix $\mathbf{A}_2$ to upper-triangulate *its* first column—*without overwriting the zeros in $\mathbf{A}_1$'s first column*. And so on. Pictorially,

$$\begin{bmatrix} \mathbf{A}_1 \end{bmatrix} \rightarrow \begin{bmatrix} r_1 & \cdots & \\ 0 & & \\ \vdots & & \overline{|\mathbf{A}_2|} \\ 0 & & \end{bmatrix} \rightarrow \begin{bmatrix} r_1 & \cdots & & \\ 0 & r_2 & \cdots & \\ 0 & 0 & & \\ \vdots & \vdots & & |\,\overline{\mathbf{A}_3}| \\ 0 & 0 & & \end{bmatrix}.$$

So we start with a unit eigenvector $\mathbf{A}_1 \mathbf{u}_1 = r_1 \mathbf{u}_1$, $\|\mathbf{u}_1\| = 1$, and (using, say, the Gram–Schmidt algorithm) we construct an orthonormal basis for $\mathbf{C}_{\text{col}}^n$ containing $\mathbf{u}_1$. The matrix $\mathbf{Q}_1$ containing this basis as columns (with $\mathbf{u}_1$ first) is unitary. Then the similarity transformation $\mathbf{Q}_1^H \mathbf{A}_1 \mathbf{Q}_1$ starts us on the way to upper triangular form, as exhibited by the equations

---

[4]Perhaps you see our notation conflict; "$\mathbf{U}$" for upper triangular, or "$\mathbf{U}$" for unitary? We opt for the former, and use "$\mathbf{Q}$" for the unitary matrix which, if real, is orthogonal.

$$\underbrace{\text{columns orthogonal to } \mathbf{u}_1}$$

$$\mathbf{Q}_1 = \begin{bmatrix} \mathbf{u}_1 & \overbrace{\cdots}^{\downarrow} \\ \downarrow & \downarrow \cdots \downarrow \end{bmatrix}, \quad \mathbf{A}_1\mathbf{Q}_1 = \begin{bmatrix} r_1\mathbf{u}_1 & \cdots \\ \downarrow & \downarrow \cdots \downarrow \end{bmatrix},$$

$$\mathbf{Q}_1^H\mathbf{A}_1\mathbf{Q}_1 = \begin{bmatrix} \rightarrow \mathbf{u}_1^H \rightarrow \\ \rightarrow \\ \left(\text{rows orthogonal to } \mathbf{u}_1{}^H\right) \\ \rightarrow \end{bmatrix} \begin{bmatrix} r_1\mathbf{u}_1 & \cdots \\ \downarrow & \downarrow \cdots \downarrow \end{bmatrix}$$

$$= \begin{bmatrix} r_1 & \cdots \\ 0 & \\ \vdots & |\underline{\mathbf{A}_2}| \\ 0 & \end{bmatrix}. \tag{1}$$

We don't care about the details of the horizontal $\cdots$ entries until the last stage in Equation (1), when they define $\mathbf{A}_2$, the lower right corner of $\mathbf{Q}_1^H\mathbf{A}_1\mathbf{Q}_1$.

$\mathbf{A}_2$ is an $(n-1)$-by-$(n-1)$ matrix, and we pick one of its unit eigenvectors $\mathbf{u}_2$, say $\mathbf{A}_2\mathbf{u}_2 = r_2\mathbf{u}_2$; then we use the Gram–Schmidt algorithm to flesh out the rest of an orthonormal basis of $\mathbf{C}_{\text{col}}^{n-1}$. These columns are padded with a row and column of zeros and a "one" to construct the n-by-n unitary matrix $\mathbf{Q}_2$ as shown:

$$\mathbf{Q}_2 = \begin{bmatrix} 1 & 0 & 0 \cdots 0 \\ 0 & \mathbf{u}_2 & \cdots \\ \downarrow & \downarrow & \downarrow \cdots \downarrow \end{bmatrix}.$$

$$\underbrace{\hspace{3cm}}_{\text{columns orthogonal to } \mathbf{u}_2}$$

The product $\mathbf{Q}_2^H\mathbf{Q}_1^H\mathbf{A}_1\mathbf{Q}_1\mathbf{Q}_2$ then displays how we have introduced zeros below the diagonal in the second column, while leaving the first column intact:

$$(\mathbf{Q}_1^H\mathbf{A}_1\mathbf{Q}_1)\mathbf{Q}_2 = \begin{bmatrix} r_1 & \cdots \\ 0 & \\ \vdots & |\underline{\mathbf{A}_2}| \\ 0 & \end{bmatrix} \begin{bmatrix} 1 & 0 & 0 & \cdots & 0 \\ 0 & \mathbf{u}_2 & & \cdots & \\ \downarrow & \downarrow & \downarrow & \cdots & \downarrow \end{bmatrix}$$

$$= \begin{bmatrix} r_1 & \cdots & \cdots & \cdots & \cdots \\ 0 & \mathbf{A}_2\mathbf{u}_2 & & \cdots & \\ \downarrow & \downarrow & \downarrow & \cdots & \downarrow \end{bmatrix} = \begin{bmatrix} r_1 & \cdots & \cdots \\ 0 & r_2\mathbf{u}_2 & \cdots \\ \downarrow & \downarrow & \cdots \end{bmatrix},$$

$$Q_2^H \left(Q_1^H A_1 Q_1 Q_2 \right) = \begin{bmatrix} 1 & 0 & \cdots & & 0 \\ 0 & \mathbf{u}_2^H & \rightarrow & & \\ & & \rightarrow & & \\ \downarrow & \begin{pmatrix} \text{rows orthogonal} \\ \text{to} \quad \mathbf{u}_2^H \end{pmatrix} & & \\ & & \rightarrow & & \end{bmatrix} \begin{bmatrix} r_1 & \cdots & \cdots \\ 0 & r_2 \mathbf{u}_2 & \cdots \\ \downarrow & \downarrow & \cdots \end{bmatrix}$$

$$= \begin{bmatrix} r_1 & \cdots & \\ 0 & r_2 & \cdots \\ \downarrow & \mathbf{0} & \\ & \downarrow & \boxed{\mathbf{A}_3} \end{bmatrix}.$$

In a similar fashion, a unit eigenvector $\mathbf{v}_3$ of $\mathbf{A}_3$ spawns the unitary matrix

$$Q_3 = \begin{bmatrix} 1 & 0 & 0 & \cdots & 0 \\ 0 & 1 & 0 & \cdots & 0 \\ \downarrow & \mathbf{0} & \mathbf{u}_3 & \cdots & \\ & \downarrow & \downarrow & \cdots & \end{bmatrix}$$

and so on. After $n - 1$ of these maneuvers, we achieve the desired upper triangular form. The complete similarity transformation, generated by the unitary matrix product $(Q_1 Q_2 \cdots Q_{n-1})$, is

$$(Q_1 Q_2 \cdots Q_{n-1})^H A (Q_1 Q_2 \cdots Q_{n-1}) = \begin{bmatrix} r_1 & \cdots & \cdots & \cdots & \cdots \\ 0 & r_2 & \cdots & \cdots & \cdots \\ \downarrow & \mathbf{0} & r_3 & \cdots & \cdots \\ & \downarrow & \mathbf{0} & \cdots & \cdots \\ & & \downarrow & \vdots & \vdots \\ & & & \cdots & r_n \end{bmatrix} =: \mathbf{U}. \quad \blacksquare$$

Corollary 7. The upper triangular matrix in Theorem 3 has the eigenvalues of $\mathbf{A}$ on its diagonal.

Proof. U is similar to **A** and thus shares its eigenvalues, which for a triangular matrix are displayed on its diagonal (Section 5.2). $\quad \blacksquare$

We now know that unitary matrices can be used to *diagonalize*, and not merely triangularize, *Hermitian* matrices (through similarity transformations). What is the most general class that can be so diagonalized?

Normal Matrices

Definition 7. Any matrix that commutes with its conjugate transpose is said to be **normal**:

$$AA^H = A^H A. \tag{2}$$

Diagonal matrices are clearly normal, as are Hermitian matrices (since $\mathbf{A} = \mathbf{A}^H$) and unitary matrices (since $\mathbf{Q}^H = \mathbf{Q}^{-1}$). Note the following:

Remark 1. *If a unitary matrix* $\mathbf{Q}$ *effects a similarity transformation between* $\mathbf{A}$ *and* $\mathbf{B} = \mathbf{Q}^H \mathbf{A} \mathbf{Q}$, *and* $\mathbf{A}$ *is normal, then so is* $\mathbf{B}$. This is verified by contrived juggling:

$$
\begin{aligned}
\mathbf{BB}^H &= \left(\mathbf{Q}^H \mathbf{A} \mathbf{Q}\right)\left(\mathbf{Q}^H \mathbf{A} \mathbf{Q}\right)^H = \mathbf{Q}^H \mathbf{A} \mathbf{Q} \mathbf{Q}^H \mathbf{A}^H \mathbf{Q} \quad \left(\text{because } [\mathbf{CDE}]^H = \mathbf{E}^H \mathbf{D}^H \mathbf{C}^H\right) \\
&= \mathbf{Q}^H \mathbf{A} \mathbf{A}^H \mathbf{Q} = \mathbf{Q}^H \mathbf{A}^H \mathbf{A} \mathbf{Q} \qquad \left(\text{because } \mathbf{QQ}^H = \mathbf{I} \text{ and } \mathbf{AA}^H = \mathbf{A}^H \mathbf{A}\right) \\
&= \mathbf{Q}^H \mathbf{A}^H \mathbf{Q} \mathbf{Q}^H \mathbf{A} \mathbf{Q} \qquad\qquad\qquad\quad \left(\text{because } \mathbf{QQ}^H = \mathbf{I}\right) \\
&= \mathbf{B}^H \mathbf{B}.
\end{aligned}
$$

In particular, $\mathbf{A}$ is normal if and only if its upper triangular form $\mathbf{U}$, generated by Schur's decomposition, is normal.

Remark 2. *Triangular matrices are normal only if they are diagonal.* To see why, look at the 4-by-4 case:

$$
\mathbf{UU}^H = \begin{bmatrix} u_{11} & u_{12} & u_{13} & u_{14} \\ 0 & u_{22} & u_{23} & u_{24} \\ 0 & 0 & u_{33} & u_{34} \\ 0 & 0 & 0 & u_{44} \end{bmatrix} \begin{bmatrix} u_{11}^* & 0 & 0 & 0 \\ u_{12}^* & u_{22}^* & 0 & 0 \\ u_{13}^* & u_{23}^* & u_{33}^* & 0 \\ u_{14}^* & u_{24}^* & u_{34}^* & u_{44}^* \end{bmatrix},
$$

$$
\mathbf{U}^H \mathbf{U} = \begin{bmatrix} u_{11}^* & 0 & 0 & 0 \\ u_{12}^* & u_{22}^* & 0 & 0 \\ u_{13}^* & u_{23}^* & u_{33}^* & 0 \\ u_{14}^* & u_{24}^* & u_{34}^* & u_{44}^* \end{bmatrix} \begin{bmatrix} u_{11} & u_{12} & u_{13} & u_{14} \\ 0 & u_{22} & u_{23} & u_{24} \\ 0 & 0 & u_{33} & u_{34} \\ 0 & 0 & 0 & u_{44} \end{bmatrix}.
$$

The jth diagonal entry of $\mathbf{UU}^H$ is the squared norm of the jth row of $\mathbf{U}$, but the jth diagonal entry of $\mathbf{U}^H \mathbf{U}$ is the squared norm of the jth column of $\mathbf{U}$. If they are equal we conclude, in turn, that the first row of $\mathbf{U}$ is diagonal, then the second, and so on.

Merging these two observations, we have Theorem 4.

Diagonalization of Normal Matrices

Theorem 4. A matrix can be diagonalized by a similarity transformation generated by a unitary matrix if, and only if, it is normal.

Corollary 8. Any n-by-n normal matrix has n linearly independent eigenvectors (the columns of the unitary matrix in Theorem 4).

We close this section by noting a widely-known corollary to the Schur decomposition: namely, every square matrix satisfies its characteristic equation.

Cayley–Hamilton Theorem

Corollary 9. Let $p(r)$ be the characteristic polynomial of the square matrix $\mathbf{A}$,

$$p(r) \equiv \det(\mathbf{A} - r\mathbf{I}) = (-r)^n + c_{n-1}r^{n-1} + \cdots + c_1 r + c_0 r^0.$$

Then if r^j is replaced by $\mathbf{A}^j$ (and r^0 by $\mathbf{I}$) in $p(r)$, the result is the zero matrix. Briefly,

$$p(\mathbf{A}) = (-\mathbf{A})^n + c_{n-1}\mathbf{A}^{n-1} + \cdots + c_1\mathbf{A} + c_0\mathbf{I} = \mathbf{0}. \tag{3}$$

Note that the Cayley–Hamilton theorem does *not* make the (obvious) assertion that the determinant $|\mathbf{A} - \mathbf{A}\mathbf{I}|$ equals zero; it is a matrix equality, equivalent to n^2 scalar equalities.

Proof of Corollary 9. If the list $\{r_1, r_2, \ldots, r_n\}$ contains the eigenvalues of $\mathbf{A}$ including multiplicities, then we know that multiplying out

$$(-1)^n(r - r_1)(r - r_2) \cdots (r - r_n) \tag{4}$$

results in the characteristic polynomial

$$p(r) = (-r)^n + c_{n-1}r^{n-1} + \cdots + c_1 r + c_0.$$

But collecting powers of $\mathbf{A}$ in

$$(-1)^n(\mathbf{A} - r_1\mathbf{I})(\mathbf{A} - r_2\mathbf{I}) \cdots (\mathbf{A} - r_n\mathbf{I})$$

involves the same algebra as collecting powers of r in (4). So we also know that

$$p(\mathbf{A}) = (-1)^n(\mathbf{A} - r_1\mathbf{I})(\mathbf{A} - r_2\mathbf{I}) \cdots (\mathbf{A} - r_n\mathbf{I}).$$

Now thanks to the telescoping effect (recall Equation (11), Section 6.1), if $\mathbf{A}$ has the Schur decomposition $\mathbf{A} = \mathbf{Q}\mathbf{U}\mathbf{Q}^H$, then $\mathbf{A}^j = \mathbf{Q}\mathbf{U}^j\mathbf{Q}^H$ and $p(\mathbf{A}) = \mathbf{Q}p(\mathbf{U})\mathbf{Q}^H$. But $p(\mathbf{U}) = (-1)^n(\mathbf{U} - r_1\mathbf{I})(\mathbf{U} - r_2\mathbf{I}) \cdots (\mathbf{U} - r_n\mathbf{I})$ turns out to be the zero matrix; observe how the zeros bulldoze their way through, as we carry out the products in this 4-by-4 example. (Remember that the diagonal entries of $\mathbf{U}$ are the eigenvalues r_j.)

$$p(\mathbf{U}) = (-1)^n (\mathbf{U} - r_1\mathbf{I}) \times (\mathbf{U} - r_2\mathbf{I}) \times (\mathbf{U} - r_3\mathbf{I}) \times (\mathbf{U} - r_4\mathbf{I})$$

$$= (-1)^n \underbrace{\begin{bmatrix} 0 & \# & \# & \# \\ 0 & \# & \# & \# \\ 0 & 0 & \# & \# \\ 0 & 0 & 0 & \# \end{bmatrix} \times \begin{bmatrix} \# & \# & \# & \# \\ 0 & 0 & \# & \# \\ 0 & 0 & \# & \# \\ 0 & 0 & 0 & \# \end{bmatrix}}$$

$$\times \begin{bmatrix} \# & \# & \# & \# \\ 0 & \# & \# & \# \\ 0 & 0 & 0 & \# \\ 0 & 0 & 0 & \# \end{bmatrix} \times \begin{bmatrix} \# & \# & \# & \# \\ 0 & \# & \# & \# \\ 0 & 0 & \# & \# \\ 0 & 0 & 0 & 0 \end{bmatrix}$$

$$= (-1)^n \underbrace{\begin{bmatrix} \mathbf{0} & \mathbf{0} & \# & \# \\ \mathbf{0} & \mathbf{0} & \# & \# \\ \mathbf{0} & \mathbf{0} & \# & \# \\ 0 & 0 & 0 & \# \end{bmatrix} \times \begin{bmatrix} \# & \# & \# & \# \\ 0 & \# & \# & \# \\ 0 & 0 & 0 & \# \\ 0 & 0 & 0 & \# \end{bmatrix}} \times \begin{bmatrix} \# & \# & \# & \# \\ 0 & \# & \# & \# \\ 0 & 0 & \# & \# \\ 0 & 0 & 0 & 0 \end{bmatrix}$$

$$= (-1)^n \begin{bmatrix} \mathbf{0} & \mathbf{0} & \mathbf{0} & \# \\ \mathbf{0} & \mathbf{0} & \mathbf{0} & \# \\ \mathbf{0} & \mathbf{0} & \mathbf{0} & \# \\ 0 & 0 & 0 & \# \end{bmatrix} \times \begin{bmatrix} \# & \# & \# & \# \\ 0 & \# & \# & \# \\ 0 & 0 & \# & \# \\ 0 & 0 & 0 & 0 \end{bmatrix} = \begin{bmatrix} \mathbf{0} & \mathbf{0} & \mathbf{0} & \mathbf{0} \\ \mathbf{0} & \mathbf{0} & \mathbf{0} & \mathbf{0} \\ \mathbf{0} & \mathbf{0} & \mathbf{0} & \mathbf{0} \\ \mathbf{0} & \mathbf{0} & \mathbf{0} & \mathbf{0} \end{bmatrix}$$

The n-by-n generalization is clear and (3) is proved.   ∎

Problems 20 and 21 demonstrate how the Cayley–Hamilton theorem can be exploited to simplify many algebraic expressions involving $\mathbf{A}$.

Exercises 6.3

Express the Schur decomposition $\mathbf{A} = \mathbf{Q}\mathbf{U}\mathbf{Q}^H$ *for the matrices in Problems 1–5.*

1. $\begin{bmatrix} 7 & 6 \\ -9 & -8 \end{bmatrix}$ 2. $\begin{bmatrix} 3 & 1 \\ -1 & 1 \end{bmatrix}$ 3. $\begin{bmatrix} 2 & 10 \\ -4 & -2 \end{bmatrix}$

4. $\begin{bmatrix} 6 & -3 & 4 \\ 4 & -2 & 4 \\ -4 & 2 & -2 \end{bmatrix}$ 5. $\begin{bmatrix} 0 & -1 & 2 \\ -1 & 0 & 0 \\ -1 & 1 & -1 \end{bmatrix}$

6. Show that the upper triangular matrix in the Schur decomposition can be chosen so that the eigenvalues appear on the diagonal in any specified order.

7. Find a diagonalizable matrix that is not normal. (*Hint*: there are several examples in Section 5.2.)

8. Show that if $\mathbf{A}$ is normal, then so is any polynomial in $\mathbf{A}$.

9. Give examples of two normal matrices $\mathbf{A}$ and $\mathbf{B}$ such that $\mathbf{A} + \mathbf{B}$ is *not* normal; similarly for $\mathbf{AB}$.

10. Show that any circulant matrix (Problem 12, Exercises 5.1) is normal.

11. Prove that the norm of $\mathbf{Av}$ equals the norm of $\mathbf{A}^H\mathbf{v}$ if $\mathbf{A}$ is normal, for any $\mathbf{v}$.

12. Show that any eigenvector of a normal matrix is also an eigenvector of its conjugate transpose.

   Show that the nonhermitian matrices in Problems 13–16 are normal, and find an orthonormal basis of eigenvectors.

13. $\begin{bmatrix} 1 & i \\ i & 1 \end{bmatrix}$ 14. $\begin{bmatrix} 1-i & -1 & 1 \\ -1 & 2-i & 0 \\ 1 & 0 & 2-i \end{bmatrix}$

15. $\begin{bmatrix} 0 & -2 & -2 \\ 2 & 0 & 1 \\ 2 & -1 & 0 \end{bmatrix}$ 16. $\begin{bmatrix} 1 & 1 & -1 & -1 \\ 1 & 1 & 1 & 1 \\ 1 & -1 & -1 & 1 \\ 1 & -1 & 1 & -1 \end{bmatrix}$

17. Show that if a matrix is normal, the norm of its ith row equals the norm of its ith column.

18. Two normal matrices $\mathbf{A}$ and $\mathbf{B}$ are said to be **simultaneously diagonalizable** if and only if there is a single unitary matrix $\mathbf{Q}$ that renders both $\mathbf{Q}^H\mathbf{AQ}$ and $\mathbf{Q}^H\mathbf{BQ}$ diagonal. Show that this can happen if, and only if, $\mathbf{AB} = \mathbf{BA}$.

19. Find a unitary matrix that simultaneously diagonalizes

$$\begin{bmatrix} 0 & -1 & 1 \\ 1 & 0 & -1 \\ -1 & 1 & 0 \end{bmatrix} \quad \text{and} \quad \begin{bmatrix} 0 & 1 & 1 \\ 1 & 0 & 1 \\ 1 & 1 & 0 \end{bmatrix}.$$

20. Show that if $\mathbf{A}$ is n-by-n and $p_m(\mathbf{A})$ is a polynomial in $\mathbf{A}$ of degree $m \geq n$, then there is a polynomial $p_s(\mathbf{A})$ of degree $s < n$ such that $p_m(\mathbf{A}) = p_s(\mathbf{A})$.

21. Use the Cayley–Hamilton Theorem to construct a polynomial formula for $\mathbf{A}^{-1}$ if $\mathbf{A}$ is nonsingular.

6.4   THE SINGULAR VALUE DECOMPOSITION[5]

The bad news about matrix multiplication is that it distorts space. The good news is— it does it in a regular way. If $\mathbf{A}$ is a real m-by-n matrix with rank r, multiplication by $\mathbf{A}$ takes vectors in $\mathbf{R}_{\text{col}}^n$ into vectors in $\mathbf{R}_{\text{col}}^m$. If we focus on all the vectors in $\mathbf{R}_{\text{col}}^n$ with unit norm—the n-*dimensional unit hypersphere*—multiplication by $\mathbf{A}$ maps this hypersphere into an r-dimensional "hyperellipsoid," distorting the sphere in the manner of Figure 6.4.

   However we will see more, as suggested by the figure. Each hyperellipsoid contains a vector $\mathbf{u}_1$ of maximal length (more than one, really; pick any one). It is the image of some hypersphere vector $\mathbf{v}_1$. Now restrict multiplication by $\mathbf{A}$ to the vectors in the

[5]The authors express their gratitude to Prof. Michael Lachance (University of Michigan-Dearborn) for valuable tutelage and contributions to this section.

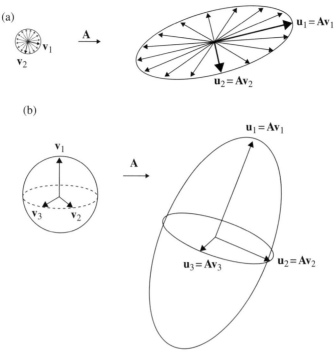

Fig. 6.4 Ellipsoidal image of unit sphere under multiplication by **A**. (a) Two dimensions and (b) Three dimensions.

lower-dimensional hypersphere *orthogonal* to $\mathbf{v}_1$ (the equatorial circle in Fig. 6.4b), and pick a vector $\mathbf{v}_2$ whose image $\mathbf{u}_2 := \mathbf{A}\mathbf{v}_2$ has maximal length among these. Continuing, pick a unit vector $\mathbf{v}_3$ orthogonal to $\mathbf{v}_1$ and $\mathbf{v}_2$ with $\|\mathbf{u}_3\| := \mathbf{A}\mathbf{v}_3$ maximal: then $\mathbf{u}_4 = \mathbf{A}\mathbf{v}_4$, $\mathbf{u}_5 = \mathbf{A}\mathbf{v}_5$, until we exhaust the rank of **A** with $\mathbf{u}_r = \mathbf{A}\mathbf{v}_r$. Let us express this situation in matrix jargon. Despite the clutter, it will be useful to occasionally post a reminder of the various matrix dimensions. Columnwise, we have

$$\mathbf{A}_{m-\text{by}-n} \begin{bmatrix} \mathbf{v}_1 & \mathbf{v}_2 & \cdots & \mathbf{v}_r \end{bmatrix}_{n-\text{by}-r} = \begin{bmatrix} \mathbf{u}_1 & \mathbf{u}_2 & \cdots & \mathbf{u}_r \end{bmatrix}_{m-\text{by}-r}. \tag{1}$$

Of course the unit vectors $\mathbf{v}_1, \mathbf{v}_2, \ldots, \mathbf{v}_r$ are orthogonal, by construction. Surprisingly, however, the columns $\mathbf{u}_1, \mathbf{u}_2, \ldots, \mathbf{u}_r$ are also orthogonal, as indicated in the figure. The proof of this statement involves some ideas outside the realm of matrix theory. We will guide you through them in Problems 25–27. For now let's go on, to see the benefits.

It is traditional to tidy up (1) using orthogonal matrices. First we render the columns of the **u**-matrix so that they are unit vectors (like the **v**-matrix):

$$\mathbf{A} \begin{bmatrix} \mathbf{v}_1 & \mathbf{v}_2 & \cdots & \mathbf{v}_r \end{bmatrix}$$

$$= \begin{bmatrix} \dfrac{\mathbf{u}_1}{\|\mathbf{u}_1\|} & \dfrac{\mathbf{u}_2}{\|\mathbf{u}_2\|} & \cdots & \dfrac{\mathbf{u}_r}{\|\mathbf{u}_r\|} \end{bmatrix}_{m-\text{by}-r} \begin{bmatrix} \|\mathbf{u}_1\| & 0 & 0 & \\ 0 & \|\mathbf{u}_2\| & 0 & 0 \\ 0 & 0 & \ddots & 0 \\ 0 & 0 & 0 & \|\mathbf{u}_r\| \end{bmatrix}_{r-\text{by}-r}. \tag{2}$$

Now there is a string of vectors $\mathbf{v}_{r+1}, \mathbf{v}_{r+2}, \ldots, \mathbf{v}_n$ in the hypersphere, each orthogonal to all of its predecessors and each mapped to the *zero* vector in $\mathbf{R}_{col}^m$. Append them to the **v**-matrix, and compensate by appending $(n - r)$ zero column vectors to the right-hand matrix:

$$\mathbf{A}_{m-by-n}\begin{bmatrix} \mathbf{v}_1 & \mathbf{v}_2 & \cdots & \mathbf{v}_r & \mathbf{v}_{r+1} & \cdots & \mathbf{v}_n \end{bmatrix}_{n-by-n}$$

$$= \begin{bmatrix} \dfrac{\mathbf{u}_1}{||\mathbf{u}_1||} & \dfrac{\mathbf{u}_2}{||\mathbf{u}_2||} & \cdots & \dfrac{\mathbf{u}_r}{||\mathbf{u}_r||} \end{bmatrix}_{m-by-r} \begin{bmatrix} ||\mathbf{u}_1|| & 0 & 0 & 0 & 0 & \cdots & 0 \\ 0 & ||\mathbf{u}_2|| & 0 & 0 & 0 & \cdots & 0 \\ \vdots & \vdots & \ddots & \vdots & \vdots & \cdots & \vdots \\ 0 & 0 & 0 & ||\mathbf{u}_r|| & 0 & \cdots & 0 \end{bmatrix}_{r-by-n}.$$

The **v**-matrix is now orthogonal; convention dictates that we call it **V** (not **Q**).

Similarly the **u**-matrix can be supplemented with $m - r$ unit column vectors $\mathbf{u}_{r+1}, \mathbf{u}_{r+2}, \ldots, \mathbf{u}_m$ to create an orthogonal matrix, **U** (convention again); this is compensated by adding $(m - r)$ zero *row* vectors to form the right-hand side diagonal matrix ... diagonal matrix **bold capital Sigma**:

$$\mathbf{AV} = \begin{bmatrix} \dfrac{\mathbf{u}_1}{||\mathbf{u}_1||} & \dfrac{\mathbf{u}_2}{||\mathbf{u}_2||} & \cdots & \dfrac{\mathbf{u}_r}{||\mathbf{u}_r||} & \mathbf{u}_{r+1} & \cdots & \mathbf{u}_m \end{bmatrix}_{m-by-m}$$

$$\times \begin{bmatrix} ||\mathbf{u}_1|| & 0 & 0 & 0 & 0 & \cdots & 0 \\ 0 & ||\mathbf{u}_2|| & 0 & 0 & 0 & \cdots & 0 \\ 0 & 0 & \ddots & 0 & 0 & \cdots & 0 \\ 0 & 0 & 0 & ||\mathbf{u}_r|| & 0 & \cdots & 0 \\ 0 & 0 & 0 & 0 & 0 & \cdots & 0 \\ \vdots & \vdots & \vdots & \vdots & \vdots & \ddots & \vdots \\ 0 & 0 & 0 & 0 & 0 & \cdots & 0 \end{bmatrix}_{m-by-n}$$

$$=: \mathbf{U\Sigma}.$$

Multiplying on the right by the inverse $\mathbf{V}^{-1} = \mathbf{V}^T$, we obtain the following result.

Singular Value Decomposition Theorem

Theorem 5. Any real m-by-n matrix **A** can be decomposed (factored) into the product of an m-by-m orthogonal matrix **U**, times an m-by-n diagonal matrix $\mathbf{\Sigma}$ with nondecreasing, nonnegative numbers down its diagonal, times an n-by-n orthogonal matrix[6] $\mathbf{V}^T$:

---

[6]The rightmost factor is traditionally written as $\mathbf{V}^T$, in keeping with the protocol of expressing the diagonalization of a symmetric matrix as $\mathbf{A} = \mathbf{QDQ}^T$. The singular value decomposition subroutine in many software packages returns the matrix **V**, and it must be transposed to verify the factorization (3). Of course equation (3) does not express a similarity transform, unless $\mathbf{U} = \mathbf{V}$.

$$\mathbf{A}_{m-\text{by}-n} = \mathbf{U\Sigma V}^T$$

$$= \begin{bmatrix} \mathbf{U} \end{bmatrix}_{m-\text{by}-m} \begin{bmatrix} \sigma_1 & 0 & 0 & 0 & 0 & \cdots & 0 \\ 0 & \sigma_2 & 0 & 0 & 0 & \cdots & 0 \\ 0 & 0 & \ddots & 0 & 0 & \cdots & 0 \\ 0 & 0 & 0 & \sigma_r & 0 & \cdots & 0 \\ 0 & 0 & 0 & 0 & 0 & \cdots & 0 \\ \vdots & \vdots & \vdots & \vdots & \vdots & \ddots & \vdots \\ 0 & 0 & 0 & 0 & 0 & \cdots & 0 \end{bmatrix}_{m-\text{by}-n} \begin{bmatrix} \mathbf{V}^T \end{bmatrix}_{n-\text{by}-n}$$

(3)

The entries on the diagonal of Σ are called the *singular values* of $\mathbf{A}$; there are $\min(m, n)$ of them. The number of nonzero singular values is $r = \text{rank}(\mathbf{A})$, and their values are the norms of the vectors $\mathbf{u}_1, \mathbf{u}_2, \ldots, \mathbf{u}_r$. (Some authors restrict the term "singular values" to the nonzero set.)

Recall that every real nondefective square matrix $\mathbf{A}$ can be diagonalized: $\mathbf{A} = \mathbf{PDP}^{-1}$ (Theorem 1, Section 6.1). And, in fact, $\mathbf{P}$ can be taken to be orthogonal when $\mathbf{A}$ is symmetric. How does the singular value decomposition differ from this?

- The SVD holds for *all* real matrices—including defective and rectangular.
- The diagonalizing matrices $\mathbf{U}$ and $\mathbf{V}$ are *always* orthogonal. They are not always equal.
- The singular values are always real and nonnegative.

Some examples of singular value decompositions appear below. They were obtained using software. The matrix in display (4) was the coefficient matrix for an underdetermined system solved in Example 4, Section 1.3:

$$\begin{bmatrix} 1 & 2 & 1 & 1 & -1 & 0 \\ 1 & 2 & 0 & -1 & 1 & 0 \\ 0 & 0 & 1 & 2 & -2 & 1 \end{bmatrix}$$

$$= \begin{bmatrix} -0.533 & -0.607 & -0.589 \\ 0.225 & -0.773 & 0.593 \\ -0.815 & 0.184 & 0.549 \end{bmatrix} \begin{bmatrix} 3.793 & 0 & 0 & 0 & 0 & 0 \\ 0 & 3.210 & 0 & 0 & 0 & 0 \\ 0 & 0 & 0.566 & 0 & 0 & 0 \end{bmatrix}$$

$$\times \begin{bmatrix} -0.081 & -0.430 & 0.007 & -0.560 & 0.560 & 0.426 \\ -0.163 & -0.860 & 0.015 & 0.342 & -0.342 & 0.000 \\ -0.356 & -0.132 & -0.071 & -0.248 & 0.248 & -0.853 \\ -0.630 & 0.166 & -0.149 & 0.562 & 0.438 & 0.213 \\ 0.630 & -0.166 & 0.149 & 0.438 & 0.562 & -0.213 \\ -0.215 & 0.057 & 0.975 & 0 & 0 & 0 \end{bmatrix} \tag{4}$$

The matrix **A** in (5) clearly has two independent rows; accordingly, its singular value decomposition displays two nonzero singular values.

$$\begin{bmatrix} 1 & -1 & 1 \\ 1 & 1 & 3 \\ 1 & 1 & 3 \\ 1 & 1 & 3 \end{bmatrix} = \begin{bmatrix} -0.166 & 0.986 & 0 & 0 \\ -0.569 & -0.096 & -0.577 & -0.577 \\ -0.569 & -0.096 & 0.789 & -0.211 \\ -0.569 & -0.096 & -0.211 & 0.789 \end{bmatrix} \begin{bmatrix} 5.820 & 0 & 0 \\ 0 & 1.458 & 0 \\ 0 & 0 & 0 \\ 0 & 0 & 0 \end{bmatrix}$$

$$\times \begin{bmatrix} -0.322 & 0.479 & 0.817 \\ -0.265 & -0.874 & 0.408 \\ -0.909 & 0.085 & -0.408 \end{bmatrix}^T. \tag{5}$$

The eigenvalues for the (square) matrix in (6) were determined to be 1, 2, and 3 in Example 2, Section 5.2. Note the singular values are quite different.

$$\begin{bmatrix} 1 & 2 & -1 \\ 1 & 0 & 1 \\ 4 & -4 & 5 \end{bmatrix} = \begin{bmatrix} 0.163 & 0.940 & 0.299 \\ -0.152 & 0.323 & -0.934 \\ -0.975 & 0.107 & 0.195 \end{bmatrix} \begin{bmatrix} 7.740 & 0 & 0 \\ 0 & 2.231 & 0 \\ 0 & 0 & 0.348 \end{bmatrix}$$

$$\times \begin{bmatrix} -0.502 & 0.758 & 0.416 \\ 0.546 & 0.652 & -0.527 \\ -0.671 & -0.037 & -0.741 \end{bmatrix}^T. \tag{6}$$

A short calculation shows that the eigenvalues of the matrix in (7) are complex: $1 \pm i$. Nonetheless, the singular values are real and positive:

$$\begin{bmatrix} 1 & 1 \\ -1 & 1 \end{bmatrix} = \begin{bmatrix} -1/\sqrt{2} & 1/\sqrt{2} \\ 1/\sqrt{2} & 1/\sqrt{2} \end{bmatrix} \begin{bmatrix} \sqrt{2} & 0 \\ 0 & \sqrt{2} \end{bmatrix} \begin{bmatrix} -1 & 0 \\ 0 & 1 \end{bmatrix}^T. \tag{7}$$

The 4-by-4 Hilbert matrix (Problem 6, Exercises 1.5) is a symmetric matrix with positive eigenvalues; note that in its singular value decomposition in (8), we see $\mathbf{U} = \mathbf{V}$, so (8) merely expresses its usual diagonalization.

$$
\begin{bmatrix} 1 & 1/2 & 1/3 & 1/4 \\ 1/2 & 1/3 & 1/4 & 1/5 \\ 1/3 & 1/4 & 1/5 & 1/6 \\ 1/4 & 1/5 & 1/6 & 1/7 \end{bmatrix} = \begin{bmatrix} -0.793 & 0.582 & -0.179 & -0.029 \\ -0.452 & -0.371 & 0.742 & 0.329 \\ -0.322 & -0.510 & -0.100 & -0.791 \\ -0.255 & -0.514 & -0.638 & 0.515 \end{bmatrix}
$$

$$
\times \begin{bmatrix} 1.5002 & 0 & 0 & 0 \\ 0 & 0.1691 & 0 & 0 \\ 0 & 0 & 0.0067 & 0 \\ 0 & 0 & 0 & 0.0001 \end{bmatrix}
$$

$$
\times \begin{bmatrix} -0.793 & 0.582 & -0.179 & -0.029 \\ -0.452 & -0.371 & 0.742 & 0.329 \\ -0.322 & -0.510 & -0.100 & -0.791 \\ -0.255 & -0.514 & -0.638 & 0.515 \end{bmatrix}^T \tag{8}
$$

However, the symmetric matrix in (9) has a negative eigenvalue (3 twice and -3; Example 5, Section 5.2). The singular values are all $+3$'s, and $\mathbf{U} \neq \mathbf{V}$.

$$
\begin{bmatrix} 1 & -2 & 2 \\ -2 & 1 & 2 \\ 2 & 2 & 1 \end{bmatrix} = \begin{bmatrix} -1/3 & 2/3 & -2/3 \\ 2/3 & -1/3 & -2/3 \\ -2/3 & -2/3 & -1/3 \end{bmatrix} \begin{bmatrix} 3 & 0 & 0 \\ 0 & 3 & 0 \\ 0 & 0 & 3 \end{bmatrix} \begin{bmatrix} -1 & 0 & 0 \\ 0 & -1 & 0 \\ 0 & 0 & -1 \end{bmatrix}^T. \tag{9}
$$

The matrix in (10) is defective; the nullity of $\mathbf{A}\text{-}3\mathbf{I}$ is only 1. A study of the zeros in $\mathbf{U}$ and $\mathbf{V}$ reveals that the singular value decomposition maintains the block structure, and the eigenvalue -2 in the 1-by-1 block shows up, in magnitude, among the singular values.

$$
\begin{bmatrix} 3 & 1 & 0 \\ 0 & 3 & 0 \\ 0 & 0 & -2 \end{bmatrix} = \begin{bmatrix} 0.763 & -0.646 & 0 \\ 0.646 & 0.763 & 0 \\ 0 & 0 & 1 \end{bmatrix} \begin{bmatrix} 3.541 & 0 & 0 \\ 0 & 2.541 & 0 \\ 0 & 0 & 2 \end{bmatrix} \begin{bmatrix} 0.646 & -0.763 & 0 \\ 0.763 & 0.646 & 0 \\ 0 & 0 & -1 \end{bmatrix}. \tag{10}
$$

The singular value decomposition is a factorization of the matrix $\mathbf{A}$ that identifies the semi-axis lengths (the nonzero singular values in $\boldsymbol{\Sigma}$) and axes of symmetry (the columns of $\mathbf{U}$) of the image ellipsoid, together with the pre-images of those axes of symmetry (the columns of $\mathbf{V}$).

The computation of the $\mathbf{U}$ and $\mathbf{V}$ factors in the decomposition requires some fussiness, and we have relegated it to Group Project D at the end of Part III; however, the computation of $\boldsymbol{\Sigma}$ is easily enabled by the following theorem.

Computation of Singular Values

Theorem 6. Let $\mathbf{A}$ be a real m-by-n matrix. Then the eigenvalues of $\mathbf{A}^T\mathbf{A}$ are nonnegative, and the positive square roots of the nonzero eigenvalues are identical (including multiplicities) with the nonzero singular values of $\mathbf{A}$.

Proof. Since $A^T A$ is symmetric, it can be diagonalized: $A^T A = QDQ^T$. But from $A = U\Sigma V^T$ we derive $A^T A = V\Sigma^T U^T U\Sigma V^T = V\Sigma^T \Sigma V^T$. Therefore $QDQ^T = V\Sigma^T \Sigma V^T$, implying D and $\Sigma^T \Sigma$ are similar and hence possess the same eigenvalues and multiplicities (Corollary 1, Section 6.1). Since $\Sigma^T \Sigma$ is the m-by-m matrix

$$
\Sigma^T \Sigma = \begin{bmatrix} \sigma_1 & 0 & 0 & 0 & \cdots & 0 \\ 0 & \ddots & 0 & 0 & \cdots & 0 \\ 0 & 0 & \sigma_r & 0 & \cdots & 0 \\ 0 & 0 & 0 & 0 & \cdots & 0 \\ \vdots & \vdots & \vdots & \vdots & \ddots & \vdots \\ 0 & 0 & 0 & 0 & \cdots & 0 \end{bmatrix}_{n-\text{by}-m} \begin{bmatrix} \sigma_1 & 0 & 0 & 0 & \cdots & 0 \\ 0 & \ddots & 0 & 0 & \cdots & 0 \\ 0 & 0 & \sigma_r & 0 & \cdots & 0 \\ 0 & 0 & 0 & 0 & \cdots & 0 \\ \vdots & \vdots & \vdots & \vdots & \ddots & \vdots \\ 0 & 0 & 0 & 0 & \cdots & 0 \end{bmatrix}_{m-\text{by}-n}
$$

$$
= \begin{bmatrix} \sigma_1^2 & 0 & 0 & 0 & \cdots & 0 \\ 0 & \ddots & 0 & 0 & \cdots & 0 \\ 0 & 0 & \sigma_r^2 & 0 & \cdots & 0 \\ 0 & 0 & 0 & 0 & \cdots & 0 \\ \vdots & \vdots & \vdots & \vdots & \ddots & \vdots \\ 0 & 0 & 0 & 0 & \cdots & 0 \end{bmatrix}_{n-\text{by}-n} , \tag{11}
$$

we can get the singular values by taking nonnegative square roots of the diagonal entries of D (and reordering if necessary).   ∎

We have introduced the singular value decomposition $A = U\Sigma V^T$ through the interpretation of Figure 6.4, and we know that the factors are not unique. For example, there is flexibility in selecting the hypersphere vectors comprising the columns of V (we could replace any v_i by $-v_i$, e.g.). So what if we happened upon *another* orthogonal/diagonal/orthogonal factorization $\widetilde{U}\widehat{\Sigma}\widetilde{V}^T$ of A? What can we say about the factors?

- The *absolute values* of the diagonal entries of $\widehat{\Sigma}$ must match up with the singular values of A, including multiplicities, in some order.
- *If* the diagonal entries of $\widehat{\Sigma}$ are nonnegative and occur in nonincreasing order, the columns of $\widetilde{U}$ and $\widetilde{V}$ are simply alternative choices to the columns of U and V, as regards Figure 6.4.

These properties are proved in Problems 25–27.

Note that if A is a real, symmetric matrix, it admits the orthogonal/diagonal/orthogonal factorization $A = QDQ^T$. Therefore, we have shown the following.

Corollary 10. The singular values of a real, symmetric matrix equal the absolute values of its eigenvalues.

Example 1. Calculate the singular values of the matrix

$$\mathbf{A} = \begin{bmatrix} 1 & 1 \\ -1 & 1 \\ 1 & 0 \end{bmatrix}.$$

Solution. We find

$$\mathbf{A}^T \mathbf{A} = \begin{bmatrix} 1 & -1 & 1 \\ 1 & 1 & 0 \end{bmatrix} \begin{bmatrix} 1 & 1 \\ -1 & 1 \\ 1 & 0 \end{bmatrix} = \begin{bmatrix} 3 & 0 \\ 0 & 2 \end{bmatrix}.$$

whose eigenvalues are, obviously, 3 and 2. Thus the singular values of $\mathbf{A}$ are $\sqrt{3}$ and $\sqrt{2}$. ∎

6.4.1   Application 1. The Matrix Norm

The generalization of the concept of the length of a vector is its (Euclidean) norm, introduced in Section 4.1: for $\mathbf{v}$ in $\mathbf{R}_{col}^n$, $||\mathbf{v}||^2 := \mathbf{v}^T \mathbf{v}$. It is convenient to extend this notion to matrices, so that $||\mathbf{A}||$ measures the extent to which multiplication by the matrix $\mathbf{A}$ can *stretch* a vector. If the stretch is quantified as the ratio $||\mathbf{Av}||/||\mathbf{v}||$, we arrive at

---

Norm

Definition 8. The (Euclidean) **norm** of a matrix $\mathbf{A}$ equals

$$||\mathbf{A}|| = \max_{\mathbf{v} \neq \mathbf{0}} \frac{||\mathbf{Av}||}{||\mathbf{v}||}. \tag{12}$$

Note that this is identical with

$$||\mathbf{A}|| = \max_{||\mathbf{v}||=1} ||\mathbf{Av}||. \tag{13}$$

Indeed at the beginning of this section, we identified $\mathbf{v} = \mathbf{v}_1$ as the unit vector achieving this maximum, and so the corresponding norm is the largest singular value:

$$||\mathbf{A}|| = \sigma_1 = ||\mathbf{u}_1|| = ||\mathbf{Av}_1||. \tag{14}$$

Observe also that (12) validates the extremely useful inequality

$$||\mathbf{Av}|| \leq ||\mathbf{A}|| \, ||\mathbf{v}||. \tag{15}$$

Since orthogonal matrices preserve length, we have immediately

Corollary 11. The Euclidean norm of an orthogonal matrix equals one.

If A is n-by-n (square), then so are the matrices $U, \Sigma,$ and V^T. If A is invertible, A^{-1} can be expressed $A^{-1} = V\Sigma^{-1}U^T$ which, strictly speaking, isn't the singular value decomposition of A (since the diagonal entries $\sigma_1^{-1}, \sigma_2^{-1}, \ldots, \sigma_n^{-1}$ are in *increasing* order). Nonetheless, the display demonstrates the following.

Corollary 12.   If A is invertible, then $||A^{-1}|| = \sigma_{\min}^{-1}$.

6.4.2   Application 2. The Condition Number

In Problem 6, Exercises 1.5, we stated that the relative accuracy of a solution to $Ax = b$ was related to the relative accuracy in the data b through the condition number μ of the matrix A. Now we can be more specific about this relationship.

How do we assess the accuracy of an approximation x_{app} to a *vector* x—an n-component entity? A reasonable answer (not the only one) would be to take, as the *error*, the norm of the difference $||x - x_{app}|| = ||\Delta x||$; and the relative error would be the ratio $||\Delta x||/||x||$. Now if the vector x is the solution to the system $Ax = b$, and there is an error Δx in the solution due to an error Δb in the measurement of b (that is to say $A(x + \Delta x) = b + \Delta b$), then two easy applications of inequality (15) give us a relation between the relative errors:

$$\Delta x = A^{-1}\Delta b \quad \Rightarrow ||\Delta x|| \leq ||A^{-1}||\,||\Delta b||,$$
$$b = Ax \quad\quad\quad \Rightarrow ||b|| \leq ||A||\,||x|| \Rightarrow ||x|| \geq ||b||/||A||.$$

Dividing inequalities (in the right direction) yields

$$\frac{||\Delta x||}{||x||} \leq ||A^{-1}||\,||A||\frac{||\Delta b||}{||b||}.$$

We summarize with a formal definition and a corollary.

Condition Number

Definition 9.   The **condition number** $\mu(A)$ of an invertible matrix A is the product $||A^{-1}||\,||A||$. If A is square and noninvertible, its condition number is taken to be infinity.

Corollary 13.   For the n-by-n invertible system $Ax = b$, the ratio of the relative error in the solution x to the relative error in the data b is upper-bounded by the condition number:

$$\frac{||\Delta x||/||x||}{||\Delta b||/||b||} \leq \mu(A) = ||A^{-1}||\,||A|| = \sigma_{\max}/\sigma_{\min}. \tag{16}$$

As is evident from (16), all condition numbers are greater than or equal to one. We cannot expect to *gain* accuracy when we solve a linear system. But if the coefficient matrix is orthogonal, we won't lose accuracy either. That is why Householder reflectors are used in sensitive algorithms.[7]

Example 2.   Verify the statement made in Problem 6, Exercises 1.5, that the condition number for the 4-by-4 Hilbert matrix (8) is approximately 15,000.

Solution. Display (8) reveals $\sigma_{\max} = 1.5002$ and $\sigma_{\min} = 0.0001$. From (16), then, we have $\mu(\mathbf{A}) = 1.5002/0.0001 \approx 15,000$.   ∎

One might be tempted to think of $\det \mathbf{A}$ as a measure of how sensitive a linear system is to errors in the data, due to its appearance in the denominator of Cramer's rule and the adjoint formula for $\mathbf{A}^{-1}$ (Section 2.5). This is false, as can be seen by the example $\mathbf{A} = (0.1)\mathbf{I}_{100}$, whose $\det \mathbf{A} = 10^{-100}$, a very small number; yet $\mathbf{Ax} = \mathbf{b}$ can be solved perfectly with ease, since $\mathbf{A}^{-1} = (10)\mathbf{I}_{100}$. Clearly $\mu(\mathbf{A})$ (= 1 for this $\mathbf{A}$) is a much better measure of sensitivity.

6.4.3   Application 3. The Pseudoinverse Matrix

If $\mathbf{A}$ has the singular value decomposition (3),

$$
\mathbf{A}_{m\text{-by-}n} = \mathbf{U}\boldsymbol{\Sigma}\mathbf{V}^T = \begin{bmatrix} \mathbf{U} \end{bmatrix}_{m\text{-by-}m} \begin{bmatrix} \sigma_1 & & 0 & 0 & & 0 \\ & \ddots & & & & \\ 0 & & \sigma_r & 0 & & 0 \\ 0 & & 0 & 0 & & 0 \\ & & & & \ddots & \\ 0 & & 0 & 0 & & 0 \end{bmatrix}_{m\text{-by-}n} \begin{bmatrix} \mathbf{V}^T \end{bmatrix}_{n\text{-by-}n},
$$

it is constructive to try to create an inverse for $\mathbf{A}$ by transposing and reverse-ordering the matrix factors and inverting the nonzero singular values:

$$
\begin{aligned}
\mathbf{A}^{\psi}_{n\text{-by-}m} &:= \mathbf{V}\boldsymbol{\Sigma}^{\psi}\mathbf{U}^T \\[2mm]
&:= \begin{bmatrix} \mathbf{V} \end{bmatrix}_{n\text{-by-}n} \begin{bmatrix} \sigma_1^{-1} & & 0 & 0 & & 0 \\ & \ddots & & & & \\ 0 & & \sigma_r^{-1} & 0 & & 0 \\ 0 & & 0 & 0 & & 0 \\ & & & & \ddots & \\ 0 & & 0 & 0 & & 0 \end{bmatrix}_{n\text{-by-}m} \begin{bmatrix} \mathbf{U}^T \end{bmatrix}_{m\text{-by-}m}.
\end{aligned} \tag{17}
$$

[7]See Section 4.2 and Group Project B of Part II.

$\mathbf{A}^{\psi}$ is "very nearly" an inverse; observe:

$$\mathbf{A}^{\psi}\mathbf{A} = \mathbf{V}\boldsymbol{\Sigma}^{\Psi}\mathbf{U}^{T}\mathbf{U}\boldsymbol{\Sigma}\mathbf{V}^{T} = \mathbf{V}\boldsymbol{\Sigma}^{\Psi}\boldsymbol{\Sigma}\mathbf{V}^{T} = \mathbf{V}\begin{bmatrix} 1 & & 0 & 0 & & 0 \\ & \ddots & & & & \\ 0 & & 1 & 0 & & 0 \\ 0 & & 0 & 0 & & 0 \\ & & & & \ddots & \\ 0 & & 0 & 0 & & 0 \end{bmatrix}_{n-\text{by}-n} \mathbf{V}^{T}.$$

$$\mathbf{A}\mathbf{A}^{\psi} = \mathbf{U}\boldsymbol{\Sigma}\mathbf{V}^{T}\mathbf{V}\boldsymbol{\Sigma}^{\psi}\mathbf{U}^{T} = \mathbf{U}\boldsymbol{\Sigma}\boldsymbol{\Sigma}^{\psi}\mathbf{U}^{T} = \mathbf{U}\begin{bmatrix} 1 & & 0 & 0 & & 0 \\ & \ddots & & & & \\ 0 & & 1 & 0 & & 0 \\ 0 & & 0 & 0 & & 0 \\ & & & & \ddots & \\ 0 & & 0 & 0 & & 0 \end{bmatrix}_{m-\text{by}-m} \mathbf{U}^{T}.$$

If $\mathbf{A}$ is invertible, then $\mathbf{A}^{\psi} = \mathbf{A}^{-1}$, but in general neither of the above products is the identity, although one can easily verify the identity-like formulas $\mathbf{A}(\mathbf{A}^{\psi}\mathbf{A}) = \mathbf{A} = (\mathbf{A}\mathbf{A}^{\psi})\mathbf{A}$.

Pseudoinverse

Definition 10. If $\mathbf{A}$ has the singular value decomposition $\mathbf{A} = \mathbf{U}\boldsymbol{\Sigma}\mathbf{V}^{T}$, the **pseudoinverse** of $\mathbf{A}$ is given by $\mathbf{A}^{\psi} := \mathbf{V}\boldsymbol{\Sigma}^{\psi}\mathbf{U}^{T}$, where $\boldsymbol{\Sigma}^{\psi}$ is formed by inverting the nonzero singular values in $\boldsymbol{\Sigma}$ and transposing.

The pseudoinverse is more than a mathematical plaything; it fills a key role in the least-squares theory described in Section 4.3.

Minimal Least Squares Solution

Theorem 7. For any m-by-n matrix $\mathbf{A}$, the vector $\mathbf{x} = \mathbf{A}^{\psi}\mathbf{b}$ provides a least-squares solution to the (possibly) inconsistent system $\mathbf{A}\mathbf{x} = \mathbf{b}$, in the sense that the norm of $\mathbf{b} - \mathbf{A}\mathbf{x}$ will be less than or equal to the norm of $\mathbf{b} - \mathbf{A}\mathbf{y}$ for any vector $\mathbf{y}$ in $\mathbf{R}_{\text{col}}^{n}$. *Moreover*, $\mathbf{x}$ itself has the least norm of all such least-square solutions. ("$\mathbf{x}$ is the least-squares solution of minimal norm.")

Proof. Let y be a generic vector in $\mathbf{R}_{\text{col}}^{n}$. Since $\mathbf{U}$ in the decomposition is orthogonal,

$$||\mathbf{b} - \mathbf{A}\mathbf{y}|| = ||\mathbf{U}^{T}(\mathbf{b} - \mathbf{A}\mathbf{y})|| = ||\mathbf{U}^{T}\mathbf{b} - \mathbf{U}^{T}\mathbf{U}\boldsymbol{\Sigma}\mathbf{V}^{T}\mathbf{y}|| = ||\mathbf{U}^{T}\mathbf{b} - \boldsymbol{\Sigma}\mathbf{V}^{T}\mathbf{y}||. \quad (18)$$

The theorem will become clear when we scrutinize this expression componentwise. Write $\mathbf{U}^{T}\mathbf{b}$ as $[\beta_1 \, \beta_2 \, \cdots \, \beta_n]^{T}$ and $\mathbf{V}^{T}\mathbf{y}$ as $[\nu_1 \, \nu_2 \, \cdots \, \nu_n]^{T}$. Then, since only r of the singular values are nonzero, $\boldsymbol{\Sigma}\mathbf{V}^{T}\mathbf{y}$ equals the m-component vector $[\sigma_1\nu_1 \, \sigma_2\nu_2 \, \ldots \, \sigma_r\nu_r \, 0 \, 0 \, \cdots \, 0]^{T}$. From (17), then,

$$||\mathbf{b} - \mathbf{Ay}||^2 = ||\mathbf{U}^T\mathbf{b} - \mathbf{\Sigma}\mathbf{V}^T\mathbf{y}||^2 = (\beta_1 - \sigma_1\nu_1)^2 + (\beta_2 - \sigma_2\nu_2)^2$$
$$+ \cdots + (\beta_r - \sigma_r\nu_r)^2 + (\beta_{r+1} - 0)^2 + \cdots + (\beta_m - 0)^2.$$

How can we choose $\mathbf{y}$, or equivalently $\mathbf{V}^T\mathbf{y}$, to minimize $||\mathbf{b} - \mathbf{Ay}||^2$? Clearly we take $\nu_i = \beta_i/\sigma_i$ for $i = 1, 2, \ldots, r$ to make the first r terms zero; and there's nothing we can do about the remaining terms. In terms of $\mathbf{y}$, then, a least-squares solution is obtained by any vector of the form

$$\mathbf{y} = \mathbf{V} \begin{bmatrix} \beta_1/\sigma_1 \\ \vdots \\ \beta_r/\sigma_r \\ \nu_{r+1} \\ \vdots \\ \nu_n \end{bmatrix} \quad (\nu_{r+1}, \ldots, \nu_n \text{ arbitary}),$$

and (since $\mathbf{V}$ is orthogonal) the shortest of these is

$$\mathbf{x} = \mathbf{V} \begin{bmatrix} \beta_1/\sigma_1 \\ \vdots \\ \beta_r/\sigma_r \\ 0 \\ \vdots \\ 0 \end{bmatrix}.$$

Remembering that $[\beta_1 \ \beta_2 \ \ldots \ \beta_m]^T = \mathbf{U}^T\mathbf{b}$, compare this with $\mathbf{A}^{\psi}\mathbf{b}$ (Equation (17)):

$$\mathbf{x} = \mathbf{A}^{\psi}\mathbf{b} = \mathbf{V}_{n-\text{by}-n} \begin{bmatrix} \sigma_1^{-1} & 0 & 0 & 0 \\ 0 & \sigma_2^{-1} & 0 & 0 \\ 0 & 0 & \sigma_3^{-1} & 0 \\ 0 & 0 & 0 & 0 \\ 0 & 0 & 0 & 0 \end{bmatrix}_{n-\text{by}-n} \mathbf{U}^T_{m-\text{by}-m}\mathbf{b};$$

they are the same.   ∎

Example 3.   Find the least-squares solution of minimal norm to $\mathbf{Ax} = \mathbf{b}$, where $\mathbf{A}$ is the matrix in display (5) and $\mathbf{b} = [1\ 0\ 1\ 2]^T$.

Solution.   Using the singular value decomposition factors (5) we calculate the minimal least-squares solution as

$$\mathbf{x} = \mathbf{A}^{\psi}\mathbf{b} = \mathbf{V}\boldsymbol{\Sigma}^{\psi}\mathbf{U}^{T}\mathbf{b} = \mathbf{V} \begin{bmatrix} 0.172 & 0 & 0 & 0 \\ 0 & 0.686 & 0 & 0 \\ 0 & 0 & 0 & 0 \end{bmatrix} \mathbf{U}^{T}\mathbf{b} = \begin{bmatrix} 0 \\ -0.5000 \\ 0.5000 \end{bmatrix}. \quad (19)$$

In fact, in Example 1 of Section 4.3 we used the normal equations to show that all least-square solutions took the form

$$\begin{bmatrix} 1 \\ 0 \\ 0 \end{bmatrix} + t \begin{bmatrix} -2 \\ -1 \\ 1 \end{bmatrix}. \quad (20)$$

The minimal-norm vector of this family occurs when $t = 1/2$, and coincides with (19) (Problem 9). ∎

6.4.4   Application 4. Rank Reduction

Suppose an engineer is trying to characterize a certain machine, for computer simulation. She knows that the machine's action can be described theoretically as converting an input vector **x** into an output vector **y** via a matrix product $\mathbf{y} = \mathbf{Ax}$, and that the matrix **A** has rank one. By making test measurements (with imperfect instruments), she estimates the matrix **A** to be

$$\mathbf{A} = \begin{bmatrix} 0.956 & 1.912 & 2.868 \\ 1.912 & 3.871 & 5.726 \\ 3.824 & 7.647 & 11.518 \end{bmatrix}. \quad (21)$$

How should she choose an approximation to this **A** that has rank one, as required by, say, her simulation software?

We need to choose a rank-one matrix **B** that is as close as possible to **A**. We'll interpret this to mean that the Frobenius norm of $(\mathbf{A} - \mathbf{B})$, that is,

$$\|\mathbf{A} - \mathbf{B}\|_{\text{Frob}} := \sqrt{\sum_{i=1}^{3}\sum_{j=1}^{3}(a_{ij} - b_{ij})^2},$$

is as small as possible. Now Problem 27 of Exercises 6.1 demonstrates that left or right multiplication of a matrix by an orthogonal matrix does not change its Frobenius norm. So, consider the singular value decomposition of **A**:

$$\mathbf{A} = \mathbf{U}\boldsymbol{\Sigma}\mathbf{V}^{T} = \begin{bmatrix} -0.218 & 0.015 & 0.976 \\ -0.436 & 0.893 & -0.111 \\ -0.873 & -0.450 & -0.188 \end{bmatrix} \begin{bmatrix} 16.431 & 0 & 0 \\ 0 & 0.052 & 0 \\ 0 & 0 & 0.003 \end{bmatrix} \times$$

$$\begin{bmatrix} -0.267 & 0.007 & 0.964 \\ -0.534 & 0.831 & -0.154 \\ -0.802 & -0.556 & -0.218 \end{bmatrix}.$$

Now if **B** is any other matrix (of the same dimensions),

$$||\mathbf{A} - \mathbf{B}||_{\text{Frob}} = ||\mathbf{U\Sigma V}^T - \mathbf{B}||_{\text{Frob}} = ||\mathbf{U}^T(\mathbf{U\Sigma V}^T - \mathbf{B})\mathbf{V}||_{\text{Frob}}$$
$$= ||\mathbf{\Sigma} - \mathbf{U}^T\mathbf{BV}||_{\text{Frob}}.$$

Thus we can find a closest rank-one approximation to **A** by finding a closest rank one approximation to $\mathbf{\Sigma}$—call it $\mathbf{\Sigma}_1$, and forming $\mathbf{U\Sigma}_1\mathbf{V}^T$ (which has the same rank). The obvious candidate for $\mathbf{\Sigma}_1$ is

$$\mathbf{\Sigma}_1 = \begin{bmatrix} 16.431 & 0 & 0 \\ 0 & 0 & 0 \\ 0 & 0 & 0 \end{bmatrix}, \tag{22}$$

and it is certainly the closest *diagonal* rank one approximation to $\mathbf{\Sigma}$. But for the proof that $\mathbf{\Sigma}_1$ is indeed the closest overall rank-one approximation, we must refer the reader to the Eckart–Young Theorem described in *Matrix Computations* 4th ed., G. H. Golub and Ch. F. van Loan, Johns Hopkins Press, 2013.

The rank one approximation corresponding to (22) is

$$\begin{aligned} \mathbf{A}_1 &= \mathbf{U\Sigma}_1\mathbf{V}^T \\ &= \begin{bmatrix} -0.218 & 0.015 & 0.976 \\ -0.436 & 0.893 & -0.111 \\ -0.873 & -0.450 & -0.188 \end{bmatrix} \begin{bmatrix} 16.431 & 0 & 0 \\ 0 & 0 & 0 \\ 0 & 0 & 0 \end{bmatrix} \\ &\quad \times \begin{bmatrix} -0.267 & 0.007 & 0.964 \\ -0.534 & 0.831 & -0.154 \\ -0.802 & -0.556 & -0.218 \end{bmatrix} \\ &= 16.431 \begin{bmatrix} -0.218 \\ -0.436 \\ -0.873 \end{bmatrix} \begin{bmatrix} -0.267 & 0.007 & 0.964 \end{bmatrix} \\ &= \begin{bmatrix} 0.954 & 1.912 & 2.869 \\ 1.912 & 3.833 & 5.752 \\ 3.824 & 7.667 & 11.505 \end{bmatrix}. \end{aligned} \qquad \blacksquare$$

It is easy to generalize from this example.

Theorem 7. If **A** has the singular value decomposition

$$
\mathbf{A} = \mathbf{U}
\begin{bmatrix}
\sigma_1 & 0 & 0 & 0 & 0 & \dots & 0 \\
0 & \ddots & 0 & 0 & 0 & \dots & 0 \\
0 & 0 & \ddots & 0 & 0 & \dots & 0 \\
0 & 0 & 0 & \sigma_r & 0 & \dots & 0 \\
0 & 0 & 0 & 0 & 0 & \dots & 0 \\
& & \vdots & & & \ddots & \\
0 & 0 & 0 & 0 & 0 & \dots & 0
\end{bmatrix}_{m-\text{by}-n}
\mathbf{V}^T,
$$

then the closest matrix of lower rank $s < r$ to **A** is given by

$$
\mathbf{A}_s = \mathbf{U}
\begin{bmatrix}
\sigma_1 & 0 & 0 & 0 & 0 & \dots & 0 \\
0 & \ddots & 0 & 0 & 0 & \dots & 0 \\
0 & 0 & \sigma_s & 0 & 0 & \dots & 0 \\
0 & 0 & 0 & 0 & 0 & \dots & 0 \\
0 & 0 & 0 & 0 & 0 & \dots & 0 \\
& & \vdots & & & \ddots & \\
0 & 0 & 0 & 0 & 0 & \dots & 0
\end{bmatrix}_{m-\text{by}-n}
\mathbf{V}^T.
\tag{23}
$$

We can use the singular value decomposition to find *reduced rank approximations* to any matrix. This practice has had tremendous payoffs in signal processing and statistics in recent years, underlying the disciplines known as *image compression* and *principal component analysis*.

Other aspects of the singular value decomposition are discussed in Cleve Moler's article at http://www.mathworks.com/company/newsletters/articles/professor-svd.html.

Exercises 6.4

1. Find (by hand) the norm and condition number of $\begin{bmatrix} 1 & 0 \\ 1 & 1 \end{bmatrix}$.

2. Find (by hand) the norm and condition number of $\begin{bmatrix} 1 & 0 & 1 \\ 0 & 1 & 0 \\ 0 & 0 & 1 \end{bmatrix}$.

Use software (if necessary) to determine the norms and condition numbers of the matrices in Problems 3–6.

3. $\begin{bmatrix} 1 & 2 & 3 \\ 4 & 5 & 6 \\ 7 & 8 & 9 \end{bmatrix}$ 4. $\begin{bmatrix} 1 & 1 & 0 \\ 0 & 1 & 1 \\ 0 & 0 & 1 \end{bmatrix}$ 5. $\begin{bmatrix} 1 & 0 & 0 & 0 \\ 0 & -2 & 0 & 0 \\ 0 & 0 & 3 & 0 \\ 0 & 0 & 0 & -4 \end{bmatrix}$

6. The Hilbert matrices of order 5, 8, and 10.

7. Find nonzero vectors $\mathbf{w}_1$, $\mathbf{w}_2$ rendering the inequalities

$$||\mathbf{A}\mathbf{w}_1|| \leq ||\mathbf{A}|| \, ||\mathbf{w}_1||, \quad ||\mathbf{A}^{-1}\mathbf{w}_2|| \leq ||\mathbf{A}^{-1}|| \, ||\mathbf{w}_2||$$

as equalities. (*Hint*: Search among the columns of $\mathbf{V}$ in (1).)

8. From among the columns of $\mathbf{V}$ in (1), find $\mathbf{b}$ and $\Delta\mathbf{b}$ producing solutions to $\mathbf{A}\mathbf{x} = \mathbf{b}$ and $\mathbf{A}(\mathbf{x} + \Delta\mathbf{x}) = \mathbf{b} + \Delta\mathbf{b}$ that render the inequality (16) as an equality (for invertible $\mathbf{A}$). (*Hint*: Trace the derivation of the inequality.)

9. Show that the minimum-norm vector in the family (20) is given by (19).

10. If the rank of the m-by-n matrix $\mathbf{A}$ is either m or n, prove that $\mathbf{A}^{\psi}$ is a one-sided inverse of $\mathbf{A}$.

11. Prove that if $\mathbf{A}$ is m-by-n and $\mathbf{B}$ is n-by-p, then $||\mathbf{A}\mathbf{B}|| \leq ||\mathbf{A}|| \, ||\mathbf{B}||$.

12. Prove that if $\mathbf{A}$ is square, $||\mathbf{A}^n|| \leq ||\mathbf{A}||^n$ for every $n = 1, 2, \ldots$.

13. Prove that if $\mathbf{A}$ is symmetric then $||\mathbf{A}^n|| = ||\mathbf{A}||^n$ for every $n = 1, 2, \ldots$.

14. Prove that if $\mathbf{A}_s$ is the closest rank-s matrix to $\mathbf{A}$, then $||\mathbf{A} - \mathbf{A}_s||_{\text{Frob}} = \sigma_{s+1}^2 + \cdots + \sigma_r^2$ where $\sigma_1, \sigma_2, \ldots, \sigma_r$ are the nonzero singular values of $\mathbf{A}$.

15. Show that Theorem 6 also holds with $\mathbf{A}^T\mathbf{A}$ replaced by $\mathbf{A}\mathbf{A}^T$.

16. Find the closest rank-one approximation to $\begin{bmatrix} 1 & 2 \\ 2 & 4 \end{bmatrix}$.

17. Find the closest rank-one approximation to $\begin{bmatrix} 1 & 1 \\ 1 & -1 \end{bmatrix}$.

18. Find the closest rank-one approximation to $\begin{bmatrix} 1 & 0 & 0 & 0 \\ 0 & -2 & 0 & 0 \\ 0 & 0 & 3 & 0 \\ 0 & 0 & 0 & -4 \end{bmatrix}$.

19. Find the closest rank-two approximation to the first matrix given in the following. Compare the approximation with the second matrix. Compare the quality of the approximations, in the Frobenius norm.

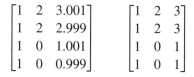

$$\begin{bmatrix} 1 & 2 & 3.001 \\ 1 & 2 & 2.999 \\ 1 & 0 & 1.001 \\ 1 & 0 & 0.999 \end{bmatrix} \qquad \begin{bmatrix} 1 & 2 & 3 \\ 1 & 2 & 3 \\ 1 & 0 & 1 \\ 1 & 0 & 1 \end{bmatrix}$$

20. Show that the singular value decomposition $\mathbf{A} = \mathbf{U}\Sigma\mathbf{V}^T$ in Theorem 5 can be replaced by an "economical" decomposition $\mathbf{A} = \mathbf{U}_r\Sigma_r\mathbf{V}_r^T$, using only the first r columns of $\mathbf{U}$ and $\mathbf{V}$ and the first r columns and rows of Σ (where r is the rank of $\mathbf{A}$).

21. For the data given in the following text, it is desired to express the y values as a linear combination of the x values.

x_1	0	0.1	0.2	0.3	0.4	$(\mathbf{x}_1^T)$
x_2	0	0.01	0.04	0.09	0.16	$(\mathbf{x}_2^T)$
x_3	0	0.21	0.44	0.69	0.95	$(\mathbf{x}_3^T)$
y	0	0.40	0.96	1.69	2.57	$(\mathbf{y}^T)$

   (a) Use the pseudoinverse to find coefficients r, s, t providing a least squares fit of the form $y = rx_1 + sx_2 + tx_3$.

   (b) Use the pseudoinverse to find coefficients r, s providing a least squares fit of the form $y = rx_1 + sx_2$.

   (c) Replace the matrix $\mathbf{X} = [\mathbf{x}_1 \ \mathbf{x}_2 \ \mathbf{x}_3]$ by its best rank two approximation $\mathbf{X}_2$, and find the least squares solution of minimum norm to $\mathbf{X}_2[r \ s]^T = \mathbf{y}$.

   (d) Tabulate the sum of squares of errors for (a), (b), and (c). Compare the performance of the approximations.

22. When a matrix $\mathbf{A}$ is replaced by a lower-rank approximation $\mathbf{A}_s = \mathbf{U}_s\Sigma_s\mathbf{V}_s^T$ using the singular value decomposition as in (23), each column of $\mathbf{A}$ is replaced by a linear combination of the first s columns of $\mathbf{U}$.

   (a) Where, in the display $\mathbf{A}_s = \mathbf{U}_s\Sigma_s\mathbf{V}_s^T$, would you find the *coefficients* in the linear combination of the first s columns of $\mathbf{U}_s$ that approximates the first column of $\mathbf{A}$? The jth column?

   (b) In fact, $\mathbf{A}_s$ replaces each column of $\mathbf{A}$ with a linear combination of *its own columns*. Where, in the display $\mathbf{A}_s = \mathbf{U}_s\Sigma_s\mathbf{V}_s^T$, would you find coefficients for a linear combination of the columns of $\mathbf{A}$ that replaces the first column of $\mathbf{A}$? The jth column? (*Hint*: Derive the identity $\mathbf{A}_s = \mathbf{A}\mathbf{V}_s\mathbf{V}_s^T$.)

23. If $\mathbf{A}$ has a singular value decomposition $\mathbf{A} = \mathbf{U}\Sigma\mathbf{V}^T$, prove that $\det \mathbf{A} = \pm \det \Sigma$. Give an examples of each case.

24. (a) If $\mathbf{A} = \mathbf{U}\Sigma\mathbf{V}^T$ is a singular value decomposition of $\mathbf{A}$, and $\mathbf{A} = \hat{\mathbf{U}}\hat{\Sigma}\hat{\mathbf{V}}^T$ is *any* orthogonal/diagonal/orthogonal factorization of $\mathbf{A}$, show that the absolute values of the nonzero entries of $\hat{\Sigma}$ must equal $\mathbf{A}$'s singular values, and their multiplicities, in some order. (*Hint*: First show that $\hat{\Sigma}^T\hat{\Sigma}$ is similar to $\Sigma^T\Sigma$ by expressing $\mathbf{A}^T\mathbf{A}$ in terms of both factorizations.)

   (b) If the diagonal entries of $\hat{\Sigma}$ in (a) are nonnegative and occur in nonincreasing order (implying $\hat{\Sigma} = \Sigma$), show that $\mathbf{A}$ can be written in terms of the columns of $\hat{\mathbf{U}}$ and $\hat{\mathbf{V}}$ as

$$\mathbf{A} = \sigma_1\hat{\mathbf{u}}_1\hat{\mathbf{v}}_1^T + \sigma_2\hat{\mathbf{u}}_2\hat{\mathbf{v}}_2^T + \cdots + \sigma_r\hat{\mathbf{u}}_r\hat{\mathbf{v}}_r^T, \tag{24}$$

   where $\sigma_1, \sigma_2, \ldots, \sigma_r$ are the nonzero singular values of $\mathbf{A}$. (*Hint*: Recall Problem 23, Section 6.1.)

(c) Use (24) to demonstrate that alternative orthogonal/diagonal/orthogonal factorizations of **A** can be created by changing the signs of certain columns in $\hat{\mathbf{U}}$ and $\hat{\mathbf{V}}$.

(d) Use (24) to describe how alternative orthogonal/diagonal/orthogonal factorizations of **A** can be generated when some of its singular values are equal.

(e) Use (24) to show that $\hat{\mathbf{v}}_1$ is one of the unit vectors maximizing $||\mathbf{Av}||$ over all unit vectors **v**, and that $\mathbf{A}\hat{\mathbf{v}}_1 = \sigma_1 \hat{\mathbf{u}}_1$.

(f) Use (24) to show that $\hat{\mathbf{v}}_2$ is one of the unit vectors maximizing $||\mathbf{Av}||$ over all unit vectors **v** orthogonal to $\hat{\mathbf{v}}_2$, and that $\mathbf{A}\hat{\mathbf{v}}_2 = \sigma_2 \hat{\mathbf{v}}_2$. (And so on.)

Problems 25–27 guide you through the missing step in the proof of the singular value decomposition: that is, the proof that the vectors $\mathbf{u}_i$ in Equation (1) are orthogonal. The analysis hinges on the following lemma.

Lemma 4. Let S be the set of unit vectors in a vector subspace **W** of $\mathbf{R}^n_{\text{col}}$:

$$S = \{\mathbf{v} \text{ in } \mathbf{W} \text{ such that } ||\mathbf{v}|| = 1\},$$

and let $\mathbf{v}_1$ be a vector in S such that $\mathbf{u}_1 := \mathbf{A}\mathbf{v}_1$ has maximal norm among all vectors in S. Then for every vector **v** in S and orthogonal to $\mathbf{v}_1$, the vectors **Av** and $\mathbf{A}\mathbf{v}_1$ are orthogonal.

We shall guide you through two derivations of Lemma 4. The first, Problem 25, is geometric, based on a diagram similar to Figure 6.4; so it can be visualized, but it is not rigorous. The second, Problem 26, is incontrovertible but lacking in motivation.

25. (*Geometric proof of Lemma 4*) (a) Show that proving **Av** is orthogonal to $\mathbf{A}\mathbf{v}_1$ is equivalent to proving that **v** is orthogonal to $\mathbf{A}^T\mathbf{A}\mathbf{v}_1$.

(b) In Figure 6.5, we depict the (hyper)sphere S. The vector $\mathbf{v}_1$ that maximizes $||\mathbf{Av}||$ on the sphere is displayed as the North Pole, and the set of vectors in S that are orthogonal to $\mathbf{v}_1$ lie in the equatorial plane. Now the function $f(\mathbf{x}) := ||\mathbf{Ax}||^2 = \mathbf{x}^T\mathbf{A}^T\mathbf{Ax}$ is maximal, over the sphere, when $\mathbf{x} = \mathbf{v}_1$; starting from the North Pole there is no direction we can move on the sphere that will increase this function. Use vector calculus to show that this implies that all the tangent vectors to the sphere, at the North Pole, must be orthogonal to the gradient of this function.

(c) Use Lemma 3 in Section 6.2 to compute the gradient of $\mathbf{x}^T\mathbf{A}^T\mathbf{Ax}$ at the North Pole, where $\mathbf{x} = \mathbf{v}_1$.

(d) The tangent vectors at the North Pole are displacements of the vectors in the equatorial plane. So every vector **v** in the equatorial plane is orthogonal to the gradient. Use this fact and (a) to complete the proof. ∎

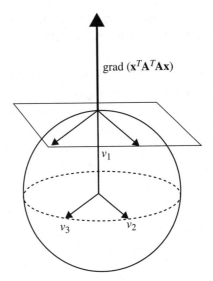

grad $(\mathbf{x}^T\mathbf{A}^T\mathbf{A}\mathbf{x})$

v_1

v_3 v_2

Fig. 6.5 Gradient vector configuration.

26. (*Analytic proof of Lemma 4*) (a) Show that all vectors in the family

$$\mathbf{v}_1 \cos\theta + \mathbf{v}\sin\theta, \quad -\pi < \theta < \pi$$

($\mathbf{v}$ orthogonal to $\mathbf{v}_0$) have unit norm.

(b) Explain why the function $f(\theta) = \|\mathbf{A}(\mathbf{v}_1\cos\theta + \mathbf{v}\sin\theta)\|^2$ is maximal when $\theta = 0$.

(c) Expand $f(\theta)$ using the dot product.

(d) Set $f'(\theta) = 0$ at $\theta = 0$ and conclude that $\mathbf{A}\mathbf{v}$ is orthogonal to $\mathbf{A}\mathbf{v}_1$.

27. (a) For what choice of subspace $\mathbf{W}$ does Lemma 4 imply that $\mathbf{u}_2, \mathbf{u}_3, \ldots$, and $\mathbf{u}_r$ are each orthogonal to $\mathbf{u}_1$?

(b) For what choice of subspace $\mathbf{W}$ does Lemma 4 imply that $\mathbf{u}_3, \mathbf{u}_4, \ldots$, and $\mathbf{u}_r$ are each orthogonal to $\mathbf{u}_2$?

(c) For what choice of subspace $\mathbf{W}$ does Lemma 4 imply that $\mathbf{u}_{k+1}, \mathbf{u}_{k+2}, \ldots$, and $\mathbf{u}_r$ are each orthogonal to $\mathbf{u}_k$?

28. (*For students who have studied real analysis.*) Why, in the proof of the singular value decomposition, are we justified in stating that there are vectors like $\mathbf{v}_1$ in the hypersphere such that $\|\mathbf{A}\mathbf{v}_1\|$ is maximal?

6.5 THE POWER METHOD AND THE QR ALGORITHM

In all our examples, we have employed a straightforward method of calculating eigenvalues for a matrix: namely, working out the characteristic polynomial $\det(\mathbf{A} - r\mathbf{I})$ and finding its zeros. In most applications, this is hopelessly impractical for two reasons:

(i) Since $\mathbf{A} - r\mathbf{I}$ is a *symbolic* matrix, its determinant cannot be evaluated by Gauss elimination. It must be expanded using the original Definition 3 or Theorem 5 (cofactor expansions) of Section 2.4. And at the end of that section, we estimated that the evaluation of a modest 25-by-25 determinant would take a modern computer hundreds of years.

(ii) Even if we could work out the characteristic polynomial, we only have formulas for its roots when the degree is less than 5. In fact, Galois and Abel proved in the early 1800s that no such formula can possibly be stated for the general fifth-degree polynomial.

So in practice, eigenvalues have to be computed numerically. We will briefly describe two algorithms: the power method, which is classic, well-understood, but often slow and unreliable; and the QR method, which is contemporary (or so it seems to the elder author) and fast, but not completely understood.

An example will elucidate the power method. For simplicity, suppose that $\mathbf{A}$ is 3-by-3 and its eigenvalues are 0.5, 2, and 4, corresponding to eigenvectors $\mathbf{u}_1, \mathbf{u}_2$, and $\mathbf{u}_3$ respectively. If $\mathbf{v}$ is any vector expressed in terms of the eigenvectors as $\mathbf{v} = c_1\mathbf{u}_1 + c_2\mathbf{u}_2 + c_3\mathbf{u}_3$, and we repeatedly multiply $\mathbf{v}$ by $\mathbf{A}$ the results will look like

$$\mathbf{A}\mathbf{v} = \frac{c_1}{2}\mathbf{u}_1 + 2c_2\mathbf{u}_2 + 4c_3\mathbf{u}_3, \quad \mathbf{A}^2\mathbf{v} = \frac{c_1}{4}\mathbf{u}_1 + 4c_2\mathbf{u}_2 + 16c_3\mathbf{u}_3, \ldots,$$
$$\mathbf{A}^{10}\mathbf{v} = \frac{c_1}{1,024}\mathbf{u}_1 + 1,024c_2\mathbf{u}_2 + 1,048,576c_3\mathbf{u}_3, \ldots \tag{1}$$

and the *direction* of $\mathbf{A}^p\mathbf{v}$ would be rotated toward the eigenvector with the dominant eigenvalue, $\mathbf{u}_3$ (or $-\mathbf{u}_3$ if $c_3 < 0$). Computationally, then, we could form the powers $\mathbf{A}^p\mathbf{v}$ until the direction stabilized, giving us an eigenvector (recall that only the *direction* of an eigenvector matters); then we could apply $\mathbf{A}$ once more and extract the value of the dominant eigenvalue r (namely, 4) from the equation $\mathbf{A}^{p+1}\mathbf{v} \approx r\mathbf{A}^p\mathbf{v}$. The behavior is depicted in Figure 6.6.

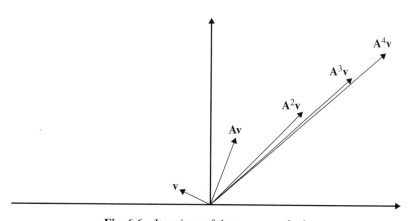

Fig. 6.6   Iterations of the power method.

Example 1.   Estimate the speed of convergence of the power method for an n-by-n matrix $\mathbf{A}$ possessing n linearly independent eigenvectors.

Solution.   Let $\mathbf{u}_1, \mathbf{u}_2, \ldots, \mathbf{u}_{n-1}, \mathbf{u}_n$ be an independent set of eigenvectors *of unit norm*, whose eigenvalues are sorted by nondecreasing magnitude: $|r_1| \le |r_2| \le \cdots \le |r_{n-1}| < r_n$. Represent the starting vector as

$\mathbf{v} = c_1\mathbf{u}_1 + \cdots + c_{n-1}\mathbf{u}_{n-1} + c_n\mathbf{u}_n$ with $c_n > 0$. (We'll discuss these restrictions later.) Then the vector $\mathbf{A}^p\mathbf{v}$ equals $c_1 r_1^p \mathbf{u}_1 + \cdots + c_{n-1} r_{n-1}^p \mathbf{u}_{n-1} + c_n r_n^p \mathbf{u}_n$. By a simple extension of the triangular inequality (Problem 24, Exercises 4.1),

$$
\begin{aligned}
\left\| \frac{\mathbf{A}^p\mathbf{v}}{c_n r_n^p} - \mathbf{u}_n \right\| &= \left\| \frac{c_1}{c_n}\left(\frac{r_1}{r_n}\right)^p \mathbf{u}_1 + \cdots + \frac{c_{n-1}}{c_n}\left(\frac{r_{n-1}}{r_n}\right)^p \mathbf{u}_{n-1} \right\| \\
&\le \left| \frac{c_1}{c_n}\left(\frac{r_1}{r_n}\right)^p \right| \|\mathbf{u}_1\| + \cdots + \left| \frac{c_{n-1}}{c_n}\left(\frac{r_{n-1}}{r_n}\right)^p \right| \|\mathbf{u}_{n-1}\| \\
&\le \left| \frac{c_1}{c_n}\left(\frac{r_{n-1}}{r_n}\right)^p \right| + \cdots + \left| \frac{c_{n-1}}{c_n}\left(\frac{r_{n-1}}{r_n}\right)^p \right| \\
&\le \left(\frac{|r_{n-1}|}{|r_n|}\right)^p \left\{ \left|\frac{c_1}{c_n}\right| + \cdots + \left|\frac{c_{n-1}}{c_n}\right| \right\} \\
&=: \left(\frac{|r_{n-1}|}{|r_n|}\right)^p C.
\end{aligned}
$$

Therefore, if $|r_{n-1}| < r_n$, the vectors $\mathbf{A}^p\mathbf{v}$, suitably scaled, converge to the eigenvector $\mathbf{u}_n$; their tips lie inside a set of shrinking circles, of radii $r_p = \left(\frac{|r_{n-1}|}{|r_n|}\right)^p C$, centered at the tip of $\mathbf{u}_n$; see Figure 6.7.   ∎

Of course in practice, we cannot rescale by the factor $c_n r_n^p$, since we know neither c_n nor r_n. But we can compute the *unit* vector $\mathbf{A}^p\mathbf{v}/\|\mathbf{A}^p\mathbf{v}\|$; and since (see Fig. 6.7)

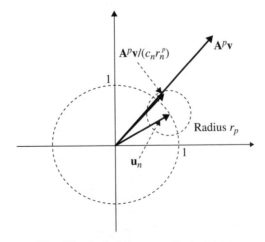

Fig. 6.7   Scaled power method vectors.

the latter is closer to $\frac{A^p v}{c_n r_n^p}$ than any other unit vector - in particular, $\mathbf{u}_n$ - we have the estimate

$$
\left\| \frac{\mathbf{A}^p \mathbf{v}}{\|\mathbf{A}^p \mathbf{v}\|} - \mathbf{u}_n \right\| = \left\| \frac{\mathbf{A}^p \mathbf{v}}{\|\mathbf{A}^p \mathbf{v}\|} - \frac{\mathbf{A}^p \mathbf{v}}{c_n r_n^p} + \frac{\mathbf{A}^p \mathbf{v}}{c_n r_n^p} - \mathbf{u}_n \right\|
$$

$$
\leq \left\| \frac{\mathbf{A}^p \mathbf{v}}{\|\mathbf{A}^p \mathbf{v}\|} - \frac{\mathbf{A}^p \mathbf{v}}{c_n r_n^p} \right\| + \left\| \frac{\mathbf{A}^p \mathbf{v}}{c_n r_n^p} - \mathbf{u}_n \right\|
$$

$$
\leq \left\| \mathbf{u}_n - \frac{\mathbf{A}^p \mathbf{v}}{c_n r_n^p} \right\| + \left\| \frac{\mathbf{A}^p \mathbf{v}}{c_n r_n^p} - \mathbf{u}_n \right\|
$$

$$
\leq 2 \left(\frac{|r_{n-1}|}{|r_n|} \right)^p C.
$$

Therefore, the distances from the normalized power iterates to the eigenvector $\mathbf{u}_n$ diminish by, at least, the factor $|r_{n-1}/r_n|$ on each iteration.

The power method is easy to understand. But it is subject to several shortcomings:

(i) The rescaling can be deferred, but the computer may overflow if we fail to perform *some* rescaling.

(ii) Only the dominant eigenvalue is calculated.

(iii) If the dominant eigenvalue is negative or multiple, or if there are other eigenvalues with the same magnitude (allowing $|r_{n-1}/r_n| = 1$), the directional convergence could be thwarted.

(iv) Even if the dominant eigenvalue is unique, the speed of directional convergence ($|r_{n-1}/r_n|$) is slowed if there are other eigenvalues with magnitude close to that of the dominant. (If the second eigenvalue were 3.9 instead of 2 in (1), the displayed dominance of $\mathbf{v}_3$ would not occur until $\mathbf{A}^{270}\mathbf{v}$.)

(v) If the computist has really bad luck, he/she might choose a starting vector with no component along $\mathbf{u}_n$—that is, c_n might be zero. Then the powers $\mathbf{A}^n \mathbf{v}$ would converge to the direction of the second most dominant eigenvector $\mathbf{u}_{n-1}$(if $c_{n-1} > 0$)—or, very likely, a small component along $\mathbf{u}_n$ might creep into the calculations due to roundoff error, and very slowly steer the powers $\mathbf{A}^p \mathbf{v}$ back to the direction of $\mathbf{u}_n$. The performance of the algorithm would appear to be bizarre indeed.

The element of "luck" arising in (v) is alien to rigorous mathematics, of course. But we shall see that it pervades the eigenvalue problem in other ways.

There are workarounds for all of these difficulties (i–v) with the power method (see Problems 8–11.) But the best technique for calculating eigenvalues is known as the QR method, originally published by H. Rutishauser in 1958. It relies first of all on the fact that if we use Gauss elimination to factor a square matrix $\mathbf{A}$ into lower triangular

(**L**) and upper triangular (**U**) factors (as described in Project A of Chapter 1), and then construct a new matrix **A′** by multiplying **L** and **U** in the reverse order,

$$\mathbf{A} = \mathbf{LU}, \quad \mathbf{A'} = \mathbf{UL},$$

we get a matrix that is similar to **A**, and thus has the same eigenvalues:

$$\mathbf{A'} = \mathbf{UL} = (\mathbf{L}^{-1}\mathbf{L})\mathbf{UL} = \mathbf{L}^{-1}(\mathbf{LU})\mathbf{L} = \mathbf{L}^{-1}\mathbf{AL}.$$

But the marvelous thing about the algorithm is that if it is performed repeatedly (factoring **A′** and reversing the factors, etc.), the "reconstructed" products usually converge to a triangular form, from which we can read off the eigenvalues! For example, a typical sequence of 1, 2, and 10 iterations of the **LU, UL** factorization/multiplication process produces the following *similar* matrices

$$\begin{bmatrix} 3 & -3 & 3 \\ 2 & -2 & 3 \\ 2 & 4 & 3 \end{bmatrix}, \begin{bmatrix} 3 & 3 & -3 \\ 4.667 & 1 & 6 \\ 0.667 & 1 & 0 \end{bmatrix}, \begin{bmatrix} 4.857 & 6.849 & 6 \\ 1.378 & -2.494 & -6.857 \\ 0.234 & 0.595 & 1.636 \end{bmatrix},$$

$$\dots, \begin{bmatrix} 5.993 & 6.667 & 6 \\ 0.009 & -2.993 & -7.494 \\ 0.000 & 0.000 & 1.000 \end{bmatrix}$$

Observing the diminishing values of the entries below the diagonal, we see that the diagonal estimates are close to the true eigenvalues: 6, −3, and 1.

Rutishauser further noted that a factorization into a *unitary* matrix **Q** and an upper triangular matrix **R**, followed by reverse multiplication $(\mathbf{A} = \mathbf{QR}, \mathbf{A'} = \mathbf{RQ} = \mathbf{RQRR}^{-1} = \mathbf{RAR}^{-1})$, would also converge, and would be less sensitive to rounding errors. The Gram–Schmidt algorithm provides one way of achieving this factorization (see Problem 18). (In practice, Householder reflectors are used; see Section 4.2.) The timeline of the 1st, 2nd, and 10th iterations of **QR, RQ** for our example reads as follows:

$$\begin{bmatrix} 3 & -3 & 3 \\ 2 & -2 & 3 \\ 2 & 4 & 3 \end{bmatrix}, \begin{bmatrix} 4.882 & 3.116 & -3.296 \\ 2.839 & -0.882 & 4.366 \\ 0.404 & 0.728 & 0 \end{bmatrix}, \begin{bmatrix} 5.987 & 0.581 & -1.255 \\ 0.826 & -3.272 & -4.700 \\ 0.095 & 0.312 & 1.286 \end{bmatrix},$$

$$\dots, \begin{bmatrix} 6.000 & 0.004 & -1.730 \\ 0.004 & -3.000 & -4.900 \\ 0.000 & 0.000 & 1.000 \end{bmatrix}.$$

However, these schemes don't always converge. For instance, if **A** itself happened to be unitary, then the iterations would go nowhere: $\mathbf{A} = \mathbf{Q} = \mathbf{QI}, \mathbf{A'} = \mathbf{IQ} = \mathbf{A}$. This is another instance of bad luck on the part of the computist!

The "cure" for a nonconvergent **QR** episode is to introduce a random shift into the iterations. (Another affront to our standards of mathematical rigor!) In fact, judiciously chosen shifts will speed up the convergence in general. And sophisticated logic can render the coding of the QR algorithm *in real arithmetic* (for a real matrix **A**), even when the eigenvalues are complex.

Amazingly, Rutishauser's single 1958 paper covered the **LU** *and* **QR** forms of the algorithm, the shifts, and the real arithmetic coding, in one publication![8]

All Rusthishauser left for the rest of us to do was to explain why it works, and this has occupied numerical analysts ever since. It has been shown that the rationale of the power method dwells inside the workings of the QR algorithm, but the necessity of the random shift chaffs us. Yet the authors are aware of no documented instance of the algorithm's failure to converge (when the shifts are truly random).

There is one last point we would like to make about the role of luck in eigenvalue problems. If a matrix has distinct eigenvalues nice things happen; it has a full set of linearly independent eigenvectors and it is diagonalizable. Troubles are caused by multiple eigenvalues. So what are the odds that a random matrix has multiple eigenvalues?

Here is a formula for the eigenvalues of a 2-by-2 matrix

$$\mathbf{A} = \begin{bmatrix} a_{11} & a_{12} \\ a_{21} & a_{22} \end{bmatrix}:$$

$$r_{1,2} = \frac{a_{11} + a_{22} \pm \sqrt{(a_{11} + a_{22})^2 - 4(a_{11}a_{22} - a_{12}a_{21})}}{2} \tag{2}$$

(Problem 17). If they are equal, then the entries of **A** must satisfy the equation

$$(a_{11} + a_{22})^2 - 4(a_{11}a_{22} - a_{12}a_{21}) = 0 \tag{3}$$

or, equivalently,

$$a_{12} = \frac{-(a_{11} + a_{22})^2 + 4a_{11}a_{22}}{4a_{21}} \tag{4}$$

if a_{21} is not zero. If the entries are random numbers, what is the probability that (4) occurs? In other words, what is the probability that the random number a_{12} takes the exact value indicated? If there is a continuum of possible values for a_{12} and they are all equally likely, this probability is zero. In technical jargon, the probability that a matrix with random data will have multiple eigenvalues is the probability that a collection of n^2 random numbers satisfy an algebraic relation, that is they fall on an algebraic "hypersurface of measure zero"; the probability is zero![9]

In fact, even if the random datum a_{12} in formula (2) happened to satisfy Equation (4), it is unlikely that your computer would detect it; practically all computations are

---

[8] Actually, Ruthishauser's designation for the L$_{ower}$ U$_{pper}$ factorization was "LR"—possibly referring to Left ("Links") and Right ("Richtig"). Heinz Rutishauser (1918–1970) was a Swiss mathematician who worked with Eduard Stiefel on the development of the first Swiss computer ERMETH.

[9] For the same reasons, the probability that a random matrix is singular is zero. Zero probability does not imply impossibility.

performed in floating point mode, so even as simple a number as 1/3 is rounded and stored incorrectly. Unless your computer were programmed in rational arithmetic (which is slow and typically impractical), it would report that **A** had distinct, very close eigenvalues and would try to diagonalize it, with ugly results.

On the other hand, we don't always pick matrices at random. In university, your teachers go to great pains to construct homework and test problems with small integers that lead to answers with small-denominator fractions. In practice, engineers are going to design control systems with coefficients that are rounded to one or two digits. Indeed, we shall see in Chapter 7 that Newton's Law $F = ma$, when expressed in "matrix normal form," leads to a double eigenvalue of zero for the simplest case, when there is no force!

So there is a place in applied mathematics for the zero-probability events of multiple eigenvalues and singular systems. But they are seldom evaluated correctly on computers, and rational arithmetic or hand calculations have to be made.

Many interesting aspects of the QR algorithm are discussed in Cleve Moler's article at http://www.mathworks.com/tagteam/72899_92026v00Cleve_QR_Algorithm_Sum_1995.pdf .

Exercises 6.5

A MATLAB® code suitable for experimenting with the power method is given below. Similar codes for many other softwares can be designed easily.

(First define the matrix **M** *and the initial vector* **v**, *and initialize the* IterationCounter *to 0.)*

$$v = M * v; \ v = v/\text{norm}(v); \ \text{IterationCounter} = \text{IterationCounter} + 1;$$
$$\text{disp}(\text{IterationCounter}); \ \text{disp}([v \ M * v]);$$

Copy this sequence of commands; then paste it into MATLAB™ *and execute it repeatedly. You will be able to see the succession of power-method iterates, and to compare* **v** *with* **Mv** *for each.*

Execute the power-method iterations with **M** = **A** *to find the largest-in-magnitude eigenvalue and its associated eigenvector for the matrices displayed in Problems 1 and 2. Iterate until the first three decimals in the eigenvector entries stabilize. Repeat with a different, linearly independent, initial vector.*

1. $\begin{bmatrix} 1 & 0 & 2 \\ 3 & 4 & 2 \\ 0 & 0 & 0 \end{bmatrix}$ 2. $\begin{bmatrix} 2 & -1 & -1 \\ -1 & 3 & 0 \\ -1 & 0 & 3 \end{bmatrix}$

3. From Equation (1), Figure 6.6, and Example 1 we see that the speed of convergence of the power method is roughly governed by the ratio of the largest-in-magnitude

eigenvalue to the second largest. The *inverse power method* looks at the iterates $\mathbf{v}, \mathbf{A}^{-1}\mathbf{v}, \mathbf{A}^{-2}\mathbf{v}, \ldots$. What ratio governs its speed of convergence? Why is this variation so effective on nearly singular matrices?

Execute the power-method iterations with $\mathbf{M} = \mathbf{A}$ and with $\mathbf{M} = \mathbf{A}^{-1}$ to find the largest-in-magnitude and smallest-in-magnitude eigenvalues and their associated eigenvectors for the matrices $\mathbf{A}$ displayed in Problems 4 through 7. Iterate until the first three decimals in the eigenvector entries stabilize. Repeat with a different, linearly independent, initial vector.

4. $\begin{bmatrix} 9 & -3 \\ -2 & 4 \end{bmatrix}$ **5.** $\begin{bmatrix} -1 & -3 \\ 4 & 7 \end{bmatrix}$ **6.** $\begin{bmatrix} 1 & -2 \\ -2 & 1 \end{bmatrix}$ **7.** $\begin{bmatrix} -3 & 2 \\ -5 & 4 \end{bmatrix}$

8. Let $\mathbf{A} = \begin{bmatrix} 2 & 1 \\ -3 & -2 \end{bmatrix}$, $\mathbf{v} = \begin{bmatrix} 0 \\ -4 \end{bmatrix}$.

(a) Apply your power-method code to $\mathbf{A}$. You will observe that the iterates eventually oscillate between two vectors. Note the calculations of the eigenvalues and eigenvectors of $\mathbf{A}$ in Example 1, Section 5.2; compare with Equation (2), and explain what you see.

(b) Let $\mathbf{M} = \mathbf{A} + 0.5\mathbf{I}$, and apply your power-method code to $\mathbf{M}$. Observe the convergence, and deduce one of the eigenvalue–eigenvector pairs for $\mathbf{A}$.

(c) Let $\mathbf{M} = \mathbf{A} - 0.5\mathbf{I}$, and apply your power-method code to $\mathbf{M}$. Observe the convergence, and deduce the other eigenvalue–eigenvector pair for $\mathbf{A}$.

9. If the dominant eigenvalues of $\mathbf{A}$ are known to be negatives of each other, how can the eigenvectors be extracted from the iterates of the power method?

10. The *shifted inverse power method* looks at the iterates $\mathbf{v}, (\mathbf{A} - m\mathbf{I})^{-1}\mathbf{v}$, $(\mathbf{A} - m\mathbf{I})^{-2}\mathbf{v}, \ldots$, where m is an approximation to any "interior" eigenvalue of $\mathbf{A}$. What ratio governs its speed of convergence?

11. Let $\mathbf{A} = \begin{bmatrix} 1 & 2 & -1 \\ 1 & 0 & 1 \\ 4 & -4 & 5 \end{bmatrix}$, $\mathbf{v} = \begin{bmatrix} 1 \\ 1 \\ 1 \end{bmatrix}$.

(a) Apply your power-method code with $\mathbf{M} = \mathbf{A}$. How many iterations are required until the eigenvector entries stabilize to three decimals? Deduce the largest-in-magnitude eigenvalue of $\mathbf{A}$.

(b) Apply your power-method code with $\mathbf{M} = \mathbf{A}^{-1}$. How many iterations are required until the eigenvector entries stabilize to three decimals? Deduce the smallest-in-magnitude eigenvalue of $\mathbf{A}$.

(c) Apply your power-method code with $\mathbf{M} = (\mathbf{A} - 1.9\mathbf{I})^{-1}$. How many iterations are required until the eigenvector entries stabilize to three decimals? Deduce remaining eigenvalue of $\mathbf{A}$.

(d) The eigenvalues of $\mathbf{A}$ were calculated in Example 2, Section 5.2. Referring to Example 1, explain the rapid convergence in (c), as compared to (a) and (b).

Execute the power-method iterations with $\mathbf{M} = \mathbf{A}$ for the matrices $\mathbf{A}$ displayed in Problems 12 through 14. Refer to the text cited to see the true eigenvalues, and explain the failure of the algorithm to converge.

12. $\mathbf{A} = \begin{bmatrix} 1 & 2 & 0 \\ 0 & 1 & -2 \\ 2 & 2 & -1 \end{bmatrix}$, $\quad \mathbf{v} = \begin{bmatrix} 1 \\ 1 \\ 1 \end{bmatrix}$ (Example 3, Section 5.2)

13. $\mathbf{A} = \begin{bmatrix} 1 & -2 & 2 \\ -2 & 1 & 2 \\ 2 & 2 & 1 \end{bmatrix}$, $\quad \mathbf{v} = \begin{bmatrix} 1 \\ 1 \\ 1 \end{bmatrix}$ (Example 5, Section 5.2)

14. $\mathbf{A} = \begin{bmatrix} 0.8 & 0.6 \\ 0.6 & -0.8 \end{bmatrix}$, $\quad \mathbf{v} = \begin{bmatrix} 1 \\ 1 \end{bmatrix}$ (Part III, Introduction)

15. Suppose we have computed an approximation $\mathbf{v}$ to an eigenvector of $\mathbf{A}$. If $\mathbf{v}$ is not exact, then the relation $\mathbf{A}\mathbf{v} = r\mathbf{v}$ will be false, for every value of r. Apply least-squares theory (Section 4.3) to $\mathbf{A}\mathbf{v} = r\mathbf{v}$, *regarded as an inconsistent linear system of n equations in the single unknown r.* Show that the normal equations dictate that the best approximation for the eigenvalue is the *Rayleigh quotient* (recall Problem 9, Exercises 6.2) $r \approx \mathbf{v}^T \mathbf{A}\mathbf{v}/\mathbf{v}^T\mathbf{v}$.

16. Problem 22 of Exercises 2.1 directed the reader to experiment with the limit (as $k \to \infty$) of $\mathbf{A}^k \mathbf{x}_0$, where $\mathbf{A}$ is a stochastic matrix and $\mathbf{x}_0$ is a probability vector. You observed the convergence to a steady-state probability vector.

    (a) Explain how the definition of "stochastic matrix" ensures that it will have an eigenvalue equal to one.

    (b) It can be shown that all of the other eigenvalues of a stochastic matrix are less than one in magnitude. Use the logic behind the power method to explain the observed behavior in Problem 22. How is the steady-state probability vector related to the eigenvectors of $\mathbf{A}$?

17. Derive Equation (2).

18. (**QR Factorization**) If $\{\mathbf{v}_1, \mathbf{v}_2, \ldots, \mathbf{v}_m\}$ are the columns of an *m*-by-*m* matrix $\mathbf{A}$, show how the Equations (9–11), Section 4.1, that define the Gram–Schmidt algorithm can be interpreted as a factorization of $\mathbf{A}$ into a unitary matrix times an upper triangular matrix.

6.6   SUMMARY

Similarity Transformations and Diagonalization

When square matrices are related by an equality $\mathbf{A} = \mathbf{P}\mathbf{B}\mathbf{P}^{-1}$, we say that $\mathbf{A}$ is similar to $\mathbf{B}$ and that $\mathbf{P}$ generates the similarity transformation. Similar matrices can be viewed as representatives of the same linear operation, expressed in different bases; they share the same eigenvalues, rank, determinant, and trace.

The most important similarity transformations are those that produce a diagonal matrix $\mathbf{D}$, and the relation $\mathbf{AP} = \mathbf{PD}$ explicitly displays the fact that the columns of a matrix $\mathbf{P}$ diagonalizing $\mathbf{A}$ are eigenvectors of $\mathbf{A}$. Thus diagonalization is only possible if $\mathbf{A}$ is not defective. The telescoping of the adjacent factors of $\mathbf{P}$ and $\mathbf{P}^{-1}$ in powers $\mathbf{A}^n = (\mathbf{PDP}^{-1})^n$ reduces the computation of matrix polynomials and power series to mere evaluation of the corresponding scalar functions, for each of the eigenvalues.

Besides their role in facilitating the computation of the important matrix exponential function, similarity transformations are used in uncoupling the degrees of freedom of vibrating systems and eliminating cross terms in bilinear forms.

Schur Decomposition

The *full* diagonalization of a matrix via a similarity transformation generated by a unitary matrix is only possible if the matrix is normal, that is, commutes with its conjugate transpose. But in fact, *any* matrix can be upper-triangularized by a unitary matrix. This theorem of Schur can be used to show that every matrix satisfies its own characteristic equation—the Cayley–Hamilton theorem. As a result, any integer power $\mathbf{A}^m$ of an n-by-n matrix $\mathbf{A}$ can be expressed as a combination of lower powers, if m equals or exceeds n.

The Singular Value Decomposition

Any m-by-n matrix $\mathbf{A}$ can be decomposed (factored) into the product of an m-by-m orthogonal matrix $\mathbf{U}$, times an m-by-n diagonal matrix $\mathbf{\Sigma}$ with nondecreasing, nonnegative numbers down its diagonal (the singular values), times an n-by-n orthogonal matrix $\mathbf{V}^T$.

From the singular values one can deduce the matrix's norm and its condition number. Further, the decomposition enables one to devise optimal lower-rank approximations to the matrix, and to express minimal least squares solutions using the pseudoinverse.

Numerical Computations

The exact calculation of the eigenvalues of a matrix of dimension larger than 4 is, in general, not possible, according to the insolvability theory of Abel and Galois. The power method for estimating eigenvalues is based on the observation that successive multiplications of a vector $\mathbf{v}$ by higher and higher powers of the matrix $\mathbf{A}$ tend to steer $\mathbf{v}$ into the direction of the eigenvector of $\mathbf{A}$ of dominant eigenvalue. And adjustments of this strategy have been devised to accommodate most of the exceptional cases.

However, the prevailing procedure for approximating the eigenvalues is the QR algorithm, which is based on two observations:

(i) if **A** is factored into the product of a unitary and an upper triangular matrix $\mathbf{A} = \mathbf{QR}$, the matrix $\mathbf{A}' = \mathbf{RQ}$ is similar to **A**;

(ii) repetition of step (i) typically results in a near-upper-triangular matrix $\mathbf{A}'$, from which the eigenvalues can easily be estimated;

Failure of (ii) to converge (and sluggishness in its convergence rate) can be overcome by suitable shifting strategies applied to **A**.

Nonetheless, the presence of multiple eigenvalues - which, indeed, may be a symptom of nondiagonalizability - can thwart the algorithm. Although such coincidences can result from human choices, they rarely occur in random matrices.

7

LINEAR SYSTEMS OF DIFFERENTIAL EQUATIONS

7.1 FIRST-ORDER LINEAR SYSTEMS

Although our textbook does not presume familiarity with differential equations, there is one type of equation whose resolution is so simple that mere calculus suffices to solve it completely. And fortunately, the matrix version of that equation pervades many of the advanced applications of matrix analysis. The scalar version is illustrated in the following radioactive decay example.

Example 1. The plutonium isotope with the shortest half-life, 20 min, is Pu-233. Formulate a differential equation for the time evolution of a sample of PU-233 of mass $x(t)$.

Solution. The mass of a substance with a half-life of 20 min decays to one-half of its value every 20 min, so it seems reasonable to propose that the decay rate is proportional to the amount of substance currently available:

$$\frac{dx}{dt} = ax, \tag{1}$$

with $a < 0$. We can solve Equation (1) using a simple calculus trick. Notice that by the product rule for differentiation,

$$\frac{d}{dt}[e^{-at}x] = e^{-at}[-ax] + e^{-at}\frac{dx}{dt} = e^{-at}\left[\frac{dx}{dt} - ax\right], \tag{2}$$

Fundamentals of Matrix Analysis with Applications,
First Edition. Edward Barry Saff and Arthur David Snider.
© 2016 John Wiley & Sons, Inc. Published 2016 by John Wiley & Sons, Inc.

which is zero according to (1). Therefore, for any two times t_1 and t_2,

$$e^{-at_2}x(t_2) - e^{-at_1}x(t_1) = \int_{t_1}^{t_2} \frac{d}{dt}[e^{-at}x]\, dt = \int_{t_1}^{t_2} [0]\, dt = 0,$$

or

$$x(t_2) = e^{a(t_2-t_1)}x(t_1). \tag{3}$$

Thus $x(t_2)$ will be one-half of $x(t_1)$ whenever $e^{a(t_2-t_1)} = 1/2$; that is, whenever the time interval $a(t_2 - t_1)$ equals $\log_e(1/2)$. Equation (3) does indeed imply a half-life for $x(t)$ of 20 min if

$$a = \frac{\log_e(1/2)}{20} \approx -0.0347 \text{ min}^{-1}. \qquad\blacksquare$$

Equation (1) describes a host of physical phenomena besides radioactive decay. Any process whose growth or decay rate is proportional to its size will exhibit the exponential behavior predicted by (3). This includes the Malthusian model for population growth, Newtonian cooling, compound interest, resistive electromagnetic attenuation, and dielectric relaxation.

In the parlance of differential equations, (1) is known as a **linear homogeneous differential equation of first order with constant coefficient** (a). Its **nonhomogeneous** form is

$$\frac{dx}{dt} = ax + f(t), \tag{4}$$

where $f(t)$ is a known function; one can interpret (4) physically as describing a material decaying radioactively while being replenished at the rate $f(t)$. The calculus trick for analyzing (4) is almost as simple as that for the homogeneous equations:

$$\frac{d}{dt}[e^{-at}x] = e^{-at}[-ax] + e^{-at}\frac{dx}{dt} = e^{-at}\left[\frac{dx}{dt} - ax\right] = e^{-at}f(t),$$

so

$$e^{-at_2}x(t_2) - e^{-at_1}x(t_1) = \int_{t_1}^{t_2} \frac{d}{dt}[e^{-at}x]\, dt = \int_{t_1}^{t_2} [e^{-at}f(t)]\, dt$$

or

$$x(t_2) = e^{a(t_2-t_1)}x(t_1) + e^{at_2}\int_{t_1}^{t_2} [e^{-at}f(t)]\, dt. \tag{5}$$

Thus $x(t_2)$ can be expressed in terms of $x(t_1)$ for any interval $[t_1,\ t_2]$ in which $f(t)$ is known (and integrable).

Since a differential equation is an equation for the derivative, or rate of change, of a function, it is not surprising that it would yield an expression for the new value, $x(t_2)$, *in terms of the original value* $x(t_1)$. Commonly t_1 is taken to be zero and $x(t_1)$ is called the **initial value**.

Now a *linear nonhomogeneous* **system** *of differential equations of first order with constant coefficients* is formally identical to (4), with a, x, and f replaced by (appropriately sized) matrices:

$$\frac{dx}{dt} = \mathbf{A}\mathbf{x} + \mathbf{f}(t); \tag{6}$$

here the coefficient matrix $\mathbf{A}$ is constant, and the system is homogeneous if $\mathbf{f}(t)$ is identically zero. Let's be sure the interpretation of everything in (6) is clear. The symbols $\mathbf{x}$ and $\mathbf{f}$ are matrix column functions—each entry is a function of t. And the entries in $d\mathbf{x}/dt$ are the derivatives of the entries of $\mathbf{x}$:

$$\mathbf{x} = \begin{bmatrix} x_1(t) \\ x_2(t) \\ \vdots \\ x_n(t) \end{bmatrix}; \quad \mathbf{f} = \begin{bmatrix} f_1(t) \\ f_2(t) \\ \vdots \\ f_n(t) \end{bmatrix}; \quad \frac{d\mathbf{x}}{dt} = \begin{bmatrix} dx_1(t)/dt \\ dx_2(t)/dt \\ \vdots \\ dx_n(t)/dt \end{bmatrix}$$

Similarly, integrals of matrix functions are evaluated entrywise:

$$\int \mathbf{A}(t)\,dt = \int \begin{bmatrix} a_{11}(t) & a_{12}(t) & \cdots & a_{1n}(t) \\ \vdots & \vdots & \vdots & \vdots \\ a_{m1}(t) & a_{m2}(t) & \cdots & a_{mn}(t) \end{bmatrix} dt$$

$$= \begin{bmatrix} \int a_{11}(t)\,dt & \int a_{12}(t)\,dt & \cdots & \int a_{1n}(t)\,dt \\ \vdots & \vdots & & \vdots \\ \int a_{m1}(t)\,dt & \int a_{m2}(t)\,dt & \cdots & \int a_{mn}(t)\,dt \end{bmatrix}.$$

Example 2. Show that $\mathbf{x}(t) = \begin{bmatrix} \cos \omega t \\ -\omega \sin \omega t \end{bmatrix}$ is a solution of the matrix differential equation $d\mathbf{x}/dt = \mathbf{A}\mathbf{x}$, where $\mathbf{A} = \begin{bmatrix} 0 & 1 \\ -\omega^2 & 0 \end{bmatrix}$.

Solution. We simply check that $d\mathbf{x}/dt$ and $\mathbf{A}\mathbf{x}(t)$ are the same vector function:

$$\frac{d\mathbf{x}(t)}{dt} = \begin{bmatrix} -\omega \sin \omega t \\ -\omega^2 \cos \omega t \end{bmatrix};$$

$$\mathbf{A}\mathbf{x} = \begin{bmatrix} 0 & 1 \\ -\omega^2 & 0 \end{bmatrix} \begin{bmatrix} \cos \omega t \\ -\omega \sin \omega t \end{bmatrix} = \begin{bmatrix} -\omega \sin \omega t \\ -\omega^2 \cos \omega t \end{bmatrix}. \qquad \blacksquare$$

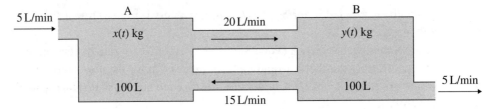

Fig. 7.1 Mixing problem for interconnected tanks.

Example 3. Two large tanks, each holding 100 L of a brine solution, are interconnected by pipes as shown in Figure 7.1. Fresh water flows into tank A at a rate of 5 L/min, and fluid is drained out of tank B at the same rate; also 20 L/min of fluid are pumped from tank A to tank B, and 15 L/min from tank B to tank A. The liquids inside each tank are kept well stirred so that each mixture is homogeneous. If tank A contains $x(t)$ kg of salt and tank B contains $y(t)$ kg, write a linear homogeneous system of differential equations of first order with constant coefficients describing the rates of change of $x(t)$ and $y(t)$.

Solution. Note that the *volume* of liquid in each tank remains constant at 100 L because of the balance between the inflow and outflow volume rates. To formulate the equations for this system, we equate the rate of change of salt in each tank with the *net* rate at which salt is transferred to that tank.

The salt *concentration* in tank A is $x/100$ kg/L, so the upper interconnecting pipe carries salt out of tank A at a rate of $20x/100$ kg/min; similarly, the lower interconnecting pipe brings salt into tank A at the rate $15y/100$ kg/min, since the concentration of salt in tank B is $y/100$ kg/L. The fresh water inlet, of course, transfers no salt (it simply maintains the volume in tank A at 100 L). Thus for tank A,

$$\frac{dx}{dt} = \text{input rate} - \text{output rate} = \frac{15y}{100} - \frac{20x}{100};$$

and for tank B,

$$\frac{dy}{dt} = \frac{20x}{100} - \frac{15y}{100} - \frac{5y}{100}.$$

Using matrices, we write the system in the prescribed form (6):

$$\frac{d}{dt}\begin{bmatrix} x \\ y \end{bmatrix} = \begin{bmatrix} -0.20 & 0.15 \\ 0.20 & -0.20 \end{bmatrix}\begin{bmatrix} x \\ y \end{bmatrix}. \tag{7}$$

■

There are two remarkable observations to be made about the system (6):

(i) It is far more general than it seems, because differential equations of order higher than one can be recast in this form;

(ii) Employing the matrix exponential function, we can execute the solution for (6) almost as easily as for (4).

We shall elaborate on (i) in the remainder of this section. The rest of Chapter 7 will describe the matrix exponential solution for linear systems.

Example 4. The mass-spring oscillator depicted in Figure I.1 in the introduction to Part I obeys Newton's law, force equals mass time acceleration, or $m\, d^2x/dt^2$. The forces on the mass include the spring force $-kx$ (Hooke's law, with $k > 0$ denoting the spring constant) and damping $-b\, dx/dt$ with damping constant $b \geq 0$. Therefore Newton's law becomes

$$m\frac{d^2x}{dt^2} = -kx - b\frac{dx}{dt}. \tag{8}$$

Express (8) as a linear homogeneous system of differential equations of *first order* with constant coefficients.

Solution. Observe that Equation (8) is of *second order*; it involves the second derivative of $x(t)$. To recast it in terms of first derivatives, we are going to utilize the formula

$$v = dx/dt \tag{9}$$

in two ways:

(a) It defines an "artificial" new variable v whose first derivative can replace the second derivative of x.

(b) It creates a new differential equation which we append to (8) to form an equivalent linear first-order system.

$$\begin{aligned}\frac{dx}{dt} &= v \\ m\frac{dv}{dt} &= -kx - bv\end{aligned} \quad \text{or} \quad \frac{d}{dt}\begin{bmatrix} x \\ v \end{bmatrix} = \begin{bmatrix} 0 & 1 \\ -k/m & -b/m \end{bmatrix}\begin{bmatrix} x \\ v \end{bmatrix}. \tag{10}$$

A solution to (8) generates a solution to (10) through the substitution (9), and the x component of a solution to (10) is a solution to (8). They are equivalent.   ■

The same rule can be applied to a third order equation, or a system of coupled second-order equations.

Example 5. Write the following as linear systems of differential equations of first order:

$$\frac{d^3x}{dt^3} + 4\frac{d^2x}{dt^2} + 6\frac{dx}{dt} + 8x = f(t);$$

(11)

$$\begin{cases} \frac{d^2x}{dt^2} = -2x + y, \\ \frac{d^2y}{dt^2} = -2y + x \end{cases}$$

(12)

Solution. For (11) the identifications

$$v = \dot{x}, \quad a = \ddot{x}$$

(motivated by "velocity" and "acceleration") spawn the system

$$\frac{dx}{dt} = v$$
$$\frac{dv}{dt} = a \qquad \text{or} \qquad \frac{d}{dt}\begin{bmatrix} x \\ v \\ a \end{bmatrix} = \begin{bmatrix} 0 & 1 & 0 \\ 0 & 0 & 1 \\ -8 & -6 & -4 \end{bmatrix}\begin{bmatrix} x \\ v \\ a \end{bmatrix} + \begin{bmatrix} 0 \\ 0 \\ f(t) \end{bmatrix}.$$
$$\frac{da}{dt} = -4a - 6v - 8x + f(t)$$

For (12) the expedient tabulation

$$x_1 = x, \quad x_2 = \frac{dx}{dt}, \quad x_3 = y, \quad x_4 = \frac{dy}{dt}$$

results in the system

$$\frac{dx_1}{dt} = x_2$$
$$\frac{dx_2}{dt} = -2x_1 + x_3 \qquad \text{or} \qquad \frac{d}{dt}\begin{bmatrix} x_1 \\ x_2 \\ x_3 \\ x_4 \end{bmatrix} = \begin{bmatrix} 0 & 1 & 0 & 0 \\ -2 & 0 & 1 & 0 \\ 0 & 0 & 0 & 1 \\ 1 & 0 & -2 & 0 \end{bmatrix}\begin{bmatrix} x_1 \\ x_2 \\ x_3 \\ x_4 \end{bmatrix}. \qquad ■$$
$$\frac{dx_3}{dt} = x_4$$
$$\frac{dx_4}{dt} = -2x_3 + x_1$$

Equations (12) can be interpreted as describing the motion of a pair of coupled mass-spring oscillators.

As the differential equation textbooks show, matrix reformulations can be helpful in consolidating the theory and computer algorithms for this subject. But the full power of the matrix approach is unleashed in the case of linear first-order systems with constant coefficients. We devote the remainder of our book to this topic.

Exercises 7.1

In Problems 1 and 2, find $d\mathbf{X}/dt$ for the given matrix functions.

1. $\mathbf{X}(t) = \begin{bmatrix} e^{5t} & 3e^{5t} \\ -2e^{5t} & -e^{5t} \end{bmatrix}$ **2.** $\mathbf{X}(t) = \begin{bmatrix} \sin 2t & \cos 2t & e^{-2t} \\ -\sin 2t & 2\cos 2t & 3e^{-2t} \\ 3\sin 2t & \cos 2t & e^{-2t} \end{bmatrix}$

In Problems 3 and 4, the matrices $\mathbf{A}(t)$ and $\mathbf{B}(t)$ are given. Find

(a) $\int \mathbf{A}(t)\, dt$.   (b) $\int_0^1 \mathbf{B}(t)\, dt$.   (c) $\dfrac{d}{dt}[\mathbf{A}(t)\mathbf{B}(t)]$.

3. $\mathbf{A}(t) = \begin{bmatrix} t & e^t \\ 1 & e^t \end{bmatrix}$, $\mathbf{B}(t) = \begin{bmatrix} \cos t & -\sin t \\ \sin t & \cos t \end{bmatrix}$.

4. $\mathbf{A}(t) = \begin{bmatrix} 1 & e^{-2t} \\ 3 & e^{-2t} \end{bmatrix}$, $\mathbf{B}(t) = \begin{bmatrix} e^{-t} & e^{-t} \\ -e^{-t} & 3e^{-t} \end{bmatrix}$.

In Problems 5 and 6, verify that the given vector function satisfies the given system.

5. $\mathbf{x}' = \begin{bmatrix} 1 & 1 \\ -2 & 4 \end{bmatrix}\mathbf{x}$, $\mathbf{x}(t) = \begin{bmatrix} e^{3t} \\ 2e^{3t} \end{bmatrix}$.

6. $\mathbf{x}' = \begin{bmatrix} 0 & 0 & 0 \\ 0 & 1 & 0 \\ 1 & 0 & 1 \end{bmatrix}\mathbf{x}$, $\mathbf{x}(t) = \begin{bmatrix} 0 \\ e^t \\ -3e^t \end{bmatrix}$.

In Problems 7 and 8, verify that the given matrix function satisfies the given matrix differential equation.

7. $\mathbf{X}' = \begin{bmatrix} 1 & -1 \\ 2 & 4 \end{bmatrix}\mathbf{X}$, $\mathbf{X}(t) = \begin{bmatrix} e^{2t} & e^{3t} \\ -e^{2t} & -2e^{3t} \end{bmatrix}$.

8. $\mathbf{X}' = \begin{bmatrix} 1 & 0 & 0 \\ 0 & 3 & -2 \\ 0 & -2 & 3 \end{bmatrix}\mathbf{X}$, $\mathbf{X}(t) = \begin{bmatrix} e^t & 0 & 0 \\ 0 & e^t & e^{5t} \\ 0 & e^t & -e^{5t} \end{bmatrix}$.

9. Verify that the vector functions

$$\mathbf{x}_1 = \begin{bmatrix} e^t & e^t \end{bmatrix}^T \qquad \text{and} \qquad \mathbf{x}_2 = \begin{bmatrix} e^{-t} & 3e^{-t} \end{bmatrix}^T$$

are solutions to the homogeneous system

$$\mathbf{x}' = \mathbf{A}\mathbf{x} = \begin{bmatrix} 2 & -1 \\ 3 & -2 \end{bmatrix}\mathbf{x}$$

on $(-\infty, \infty)$ and that

$$\mathbf{x}_p = \frac{3}{2}\begin{bmatrix} te^t \\ te^t \end{bmatrix} - \frac{1}{4}\begin{bmatrix} e^t \\ 3e^t \end{bmatrix} + \begin{bmatrix} t \\ 2t \end{bmatrix} - \begin{bmatrix} 0 \\ 1 \end{bmatrix}$$

is a solution to the nonhomogeneous system $\mathbf{x}' = \mathbf{A}\mathbf{x} + \mathbf{f}(t)$, where $\mathbf{f}(t) = \begin{bmatrix} e^t & t \end{bmatrix}^T$.

10. Verify that the vector functions

$$\mathbf{x}_1 = \begin{bmatrix} e^{3t} \\ 0 \\ e^{3t} \end{bmatrix}, \quad \mathbf{x}_2 = \begin{bmatrix} -e^{3t} \\ e^{3t} \\ 0 \end{bmatrix}, \quad \mathbf{x}_3 = \begin{bmatrix} -e^{-3t} \\ -e^{-3t} \\ e^{-3t} \end{bmatrix}$$

are solutions to the homogeneous system

$$\mathbf{x}' = \mathbf{A}\mathbf{x} = \begin{bmatrix} 1 & -2 & 2 \\ -2 & 1 & 2 \\ 2 & 2 & 1 \end{bmatrix} \mathbf{x}$$

on $(-\infty, \infty)$ and that

$$\mathbf{x}_p = \begin{bmatrix} 5t + 1 \\ 2t \\ 4t + 2 \end{bmatrix}$$

is a solution to $\mathbf{x}' = \mathbf{A}\mathbf{x} + \mathbf{f}(t)$, where $\mathbf{f}(t) = [-9t \ \ 0 \ -18t]^T$.

In Problems 11–14, express the given system of differential equations in the matrix form $\mathbf{x}' = \mathbf{A}\mathbf{x}$.

11. $x' = 7x + 2y,$
$\quad y' = 3x - 2y.$

12. $x' = y,$
$\quad y' = -x.$

13. $x' = x + y + z,$
$\quad y' = 2z - x,$
$\quad z' = 4y.$

14. $x_1' = x_1 - x_2 + x_3 - x_4,$
$\quad x_2' = x_1 + x_4,$
$\quad x_3' = \sqrt{\pi}x_1 - x_3,$
$\quad x_4' = 0.$

In Problems 15–17, write the given system in the matrix form $\mathbf{x}' = \mathbf{A}\mathbf{x} + \mathbf{f}$.

15. $x'(t) = 3x(t) - y(t) + t^2,$
$\quad y'(t) = -x(t) + 2y(t) + e^t.$

16. $r'(t) = 2r(t) + \sin t,$
$\quad \theta'(t) = r(t) - \theta(t) + 1.$

17. $x' = x + y + z + 3,$
$\quad y' = 2x - y + 3z + \sin t,$
$\quad z' = x + 5z.$

In Problems 18 and 19, express the given system of higher-order differential equations as a first-order matrix system.

18. $x'' + 3x + 2y = 0,$
$\quad y'' - 2x = 0.$

19. $x'' + 3x' - y' + 2y = 0,$
$\quad y'' + x' + 3y' + y = 0.$

In Problems 20–23, rewrite the given scalar equation as a first-order linear system. Express the system in the matrix form $\mathbf{x}' = \mathbf{Ax} + \mathbf{f}$.

20. $y''(t) - 3y'(t) - 10y(t) = \sin t$.

21. $x''(t) + x(t) = t^2$.

22. $\dfrac{d^4 w}{dt^4} + w = t^2$.

23. $\dfrac{d^3 y}{dt^3} - \dfrac{dy}{dt} + y = \cos t$.

24. Two large tanks, each holding 50 L of liquid, are interconnected by pipes, with the liquid flowing from tank A into tank B at a rate of 3 L/min and from B into A at a rate of 1 L/min (see Fig. 7.2). The liquid inside each tank is kept well stirred. A brine solution with a concentration of 0.2 kg/L of salt flows into tank A at a rate of 5 L/min. The (diluted) solution flows out of the system from tank A at 3 L/min and from tank B at 2 L/min. Write a system of differential equations in the form $\mathbf{x}' = \mathbf{Ax} + \mathbf{f}$ for the mass of salt in each tank.

25. Two tanks A and B, each holding 75 L of liquid, are interconnected by pipes. The liquid flows from tank A into tank B at a rate of 4 L/min and from B into A at a rate of 1 L/min (see Fig. 7.3). The liquid inside each tank is kept well stirred. A brine solution that has a concentration of 0.2 kg/L of salt flows into tank A at a rate of 4 L/min. A brine solution that has a concentration of 0.1 kg/L of salt flows into tank B at a rate of 1 L/min. The solutions flow out of the system from both tanks—from tank A at 1 L/min and from tank B at 5 L/min. (So the volume in tank B decreases.) Write a system of differential equations in the form $\mathbf{x}' = \mathbf{Ax} + \mathbf{f}$ for the mass of salt in each tank (until tank B is depleted).

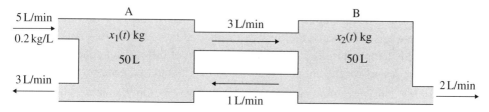

Fig. 7.2   Mixing problem for interconnected tanks.

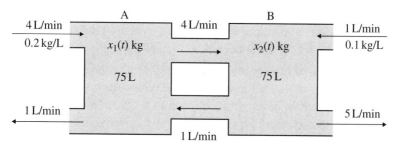

Fig. 7.3   Mixing problem for interconnected tanks.

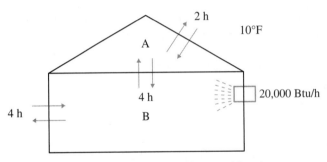

Fig. 7.4   Air-conditioned house with attic.

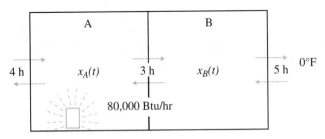

Fig. 7.5   Two-zone building with one zone heated.

Problems 26 through 32 are directed to mechanical engineering students.

26. A house, for cooling purposes, consists of two zones: the attic area zone A and the living area zone B (see Fig. 7.4). The living area is cooled by a 2-ton air conditioning unit that removes 20,000 Btu/h. The heat capacity of zone B is $1/2°F$ per thousand Btu. The time constant for heat transfer between zone A and the outside is 2 h, between zone B and the outside is 4 h, and between the two zones is 4 h. The outside temperature is 100°F. Write a system of differential equations in the form $\mathbf{x}' = \mathbf{Ax} + \mathbf{f}$ for the temperatures, x_A and x_B, in the two zones.

27. A building consists of two zones A and B (see Fig. 7.5). Only zone A is heated by a furnace, which generates 80,000 Btu/h. The heat capacity of zone A is $1/4°F$ per thousand Btu. The time constant for heat transfer between zone A and the outside is 4 h, between the unheated zone B and the outside is 5 h, and between the two zones is 3 h. The outside temperature is 0°F. Write a system of differential equations in the form $\mathbf{x}' = \mathbf{Ax} + \mathbf{f}$ for the temperatures, x_A and x_B, in the two zones.

28. Two springs and two masses are attached in a straight line on a horizontal frictionless surface as illustrated in Figure 7.6. Express Newton's laws $F = ma$ for the system as a matrix differential equation in the form $\mathbf{x}' = \mathbf{Ax}$. [*Hint*: Write Newton's law for each mass separately by applying Hooke's law.]

29. Four springs with the same spring constant and three equal masses are attached in a straight line on a horizontal frictionless surface as illustrated in Figure 7.7.

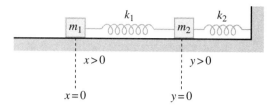

Fig. 7.6 Coupled mass–spring system with one end free.

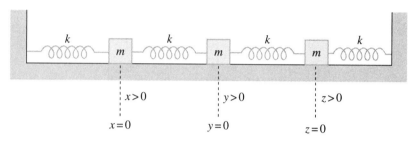

Fig. 7.7 Coupled mass–spring system with three degrees of freedom.

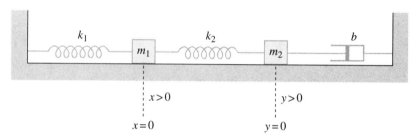

Fig. 7.8 Coupled mass–spring system with one end damped.

Express Newton's law for the system as a matrix differential equation in the form $\mathbf{x}' = \mathbf{A}\mathbf{x}$.

30. Two springs, two masses, and a dashpot are attached in a straight line on a horizontal frictionless surface as shown in Figure 7.8. The dashpot provides a damping force on mass m_2, given by $F = -by'$. Derive the matrix system of differential equations for the displacements x and y.

31. Two springs, two masses, and a dashpot are attached in a straight line on a horizontal frictionless surface as shown in Figure 7.9. The dashpot damps both m_1 and m_2 with a force whose magnitude equals $b|y' - x'|$. Derive the matrix system of differential equations for the displacements x and y.

32. Referring to the coupled mass–spring system discussed in Problem 28, suppose an external force $E(t) = 3\cos 3t$ is applied to the mass m_1. Express Newton's law for the system as a matrix differential equation in the form $\mathbf{x}' = \mathbf{A}\mathbf{x} + \mathbf{f}$.

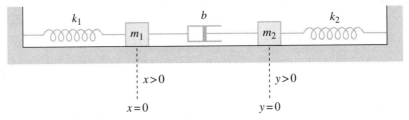

Fig. 7.9   Coupled mass–spring system with damping between the masses.

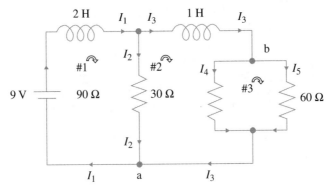

Fig. 7.10   *RL* network for Problem 33.

Problems 33–39 are directed to electrical engineering students.

33. **RL Network.** The currents I_1 through $I5$ in the *RL* network given by the schematic diagram in Figure 7.10 are governed by the following equations:

Kirchhoff's voltage law around loop #1: $2I_1' + 90I_2 = 9,$
Kirchhoff's voltage law around loop #2: $I_3' + 30I_4 - 90I_2 = 0,$
Kirchhoff's voltage law around loop #3: $60I_5 - 30I_4 = 0,$
Kirchhoff's current law at node a : $I_1 = I_2 + I_3,$
Kirchhoff's current law at node b : $I_3 = I_4 + I_5.$

Use Gauss elimination on the *algebraic* equations (the last three) to express I_2, I_4, and I_5 in terms of I_1 and I_3. Then write a matrix differential equation for I_1 and I_3.

34. Use the strategy of Problem 33 to write a matrix differential equation for the currents I_1 and I_2 in the *RL* network given by the schematic diagram in Figure 7.11.

35. Use the strategy of Problem 33 to write a matrix differential equation for the currents I_2 and I_3 in the *RL* network given by the schematic diagram in Figure 7.12.

36. **RLC Network.** The currents and charges in the *RLC* network given by the schematic diagram in Figure 7.13 are governed by the following equations:

Kirchhoff's voltage law for loop #1: $4I_2' + 52q_1 = 10,$
Kirchhoff's voltage law for loop #2: $13I_3 + 52q_1 = 10,$
Kirchhoff's current law for node a : $I_1 = I_2 + I_3,$

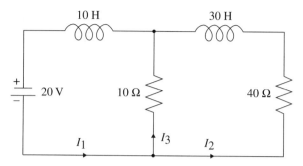

Fig. 7.11 *RL* network for Problem 34.

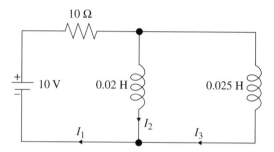

Fig. 7.12 *RL* network for Problem 35.

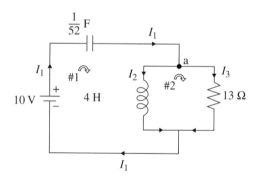

Fig. 7.13 *RLC* network for Problem 36.

where q_1 is the charge on the capacitor and $I_1 = q_1'$. Eliminate the nuisance charge variable q_1 by differentiating the first two equations and replacing q_1' by I_1; then replace I_1 using the third equation. Express the result as a *first*-order matrix differential equation system for I_2 and I_3.

37. Use the strategy of Problem 36 to write a first-order matrix differential equation system for the currents I_1 and I_2 in the *RLC* network given by the schematic diagram in Figure 7.14.

38. Use the strategy of Problem 36 to write a first-order matrix differential equation system for the currents I_1 and I_2 in the *RLC* network given by the schematic diagram in Figure 7.15.

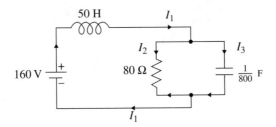

Fig. 7.14 *RLC* network for Problem 37.

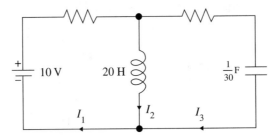

Fig. 7.15 *RLC* network for Problem 38.

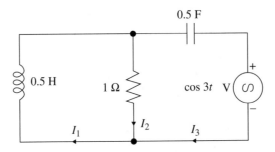

Fig. 7.16 *RLC* network for Problem 39.

39. Use the strategy of Problem 36 to write a first-order matrix differential equation system for the currents I_1 and I_2 in the *RLC* network given by the schematic diagram in Figure 7.16.

7.2 THE MATRIX EXPONENTIAL FUNCTION

In the preceding section, we promised that the "quick and dirty" formula for a solution to the (scalar) differential equation

$$\frac{dx}{dt} = ax + f(t) \tag{1}$$

namely,

$$x(t_2) = e^{a(t_2-t_1)}x(t_1) + e^{at_2} \int_{t_1}^{t_2} [e^{-as}f(s)] \, ds \tag{2}$$

would have a matrix analog. This is given in the following result.

Solution Formula for Linear Systems

Theorem 1. Any solution of the linear system of differential equations of first order with constant coefficients

$$\frac{d\mathbf{x}}{dt} = \mathbf{A}\mathbf{x} + \mathbf{f}(t) \tag{3}$$

(with $\mathbf{f}(t)$ continuous) satisfies

$$\mathbf{x}(t) = e^{\mathbf{A}(t-t_1)}\mathbf{x}(t) + e^{\mathbf{A}t} \int_{t_1}^{t} [e^{-\mathbf{A}s}\mathbf{f}(s)] \, ds. \tag{4}$$

In order to verify that (4) is a solution, we have to examine the matrix exponential function. Section 6.1 demonstrated that the formula

$$e^{\mathbf{A}} = \mathbf{I} + \mathbf{A} + \mathbf{A}^2/2! + \cdots + \mathbf{A}^m/m! + \cdots \tag{5}$$

converges and defines a legitimate function for every *diagonalizable* matrix $\mathbf{A}$:

$$\mathbf{A} = \mathbf{P} \begin{bmatrix} r_1 & 0 & \cdots & 0 \\ 0 & r_2 & \cdots & 0 \\ \vdots & & \ddots & \vdots \\ 0 & 0 & \cdots & r_p \end{bmatrix} \mathbf{P}^{-1} \Rightarrow e^{\mathbf{A}} = \mathbf{P} \begin{bmatrix} e^{r_1} & 0 & \cdots & 0 \\ 0 & e^{r_2} & \cdots & 0 \\ \vdots & & \ddots & \vdots \\ 0 & 0 & \cdots & e^{r_p} \end{bmatrix} \mathbf{P}^{-1}. \tag{6}$$

And of course if $\mathbf{A}$ is diagonalizable, so is $\mathbf{A}t$, so for such $\mathbf{A}$ we can define the matrix exponential via

$$e^{\mathbf{A}t} = \mathbf{I} + \mathbf{A}t + \mathbf{A}^2 t^2/2! + \cdots + \mathbf{A}^m t^m/m! + \cdots \tag{7}$$

$$= \mathbf{P} \begin{bmatrix} e^{r_1 t} & 0 & \cdots & 0 \\ 0 & e^{r_2 t} & \cdots & 0 \\ \vdots & & \ddots & \vdots \\ 0 & 0 & \cdots & e^{r_p t} \end{bmatrix} \mathbf{P}^{-1}$$

But as a matter of fact, the power series in (5) and (7) converge for *every* matrix **A,** diagonalizable or not. The proof is outlined in Problem 28. Indeed the limit function $e^{\mathbf{A}t}$ is differentiable and its derivative can be obtained by differentiating (5) term-by-term,[1] yielding the expected formula

$$
\begin{aligned}
\frac{d}{dt}e^{\mathbf{A}t} &= \frac{d}{dt}[\mathbf{I} + \mathbf{A}t + \mathbf{A}^2 t^2/2! + \cdots + \mathbf{A}^m t^m/m! + \cdots] \\
&= \mathbf{0} + \mathbf{A} + 2\mathbf{A}^2 t/2! + \cdots + m\mathbf{A}^m t^{m-1}/m! + \cdots \\
&= \mathbf{A}[\mathbf{I} + \mathbf{A}t + \mathbf{A}^2 t^2/2! + \cdots \mathbf{A}^{m-1} t^{m-1}/(m-1)! + \cdots] \\
&= \mathbf{A}e^{\mathbf{A}t}.
\end{aligned}
\tag{8}
$$

Equation (8) expresses the most important feature of the matrix exponential function. Other properties that the matrix exponential $e^{\mathbf{A}t}$ shares with the scalar function e^{at} are listed in the following theorem.

Properties of the Matrix Exponential Function

Theorem 2.   Let **A** and **B** be n-by-n constant matrices and $r, s,$ and t be real (or complex) numbers. Then

(i) $e^{\mathbf{A}t}$ satisfies the matrix differential equation $\mathbf{X}' = \mathbf{A}\mathbf{X}$.

(ii) $e^{\mathbf{A}0} = e^{\mathbf{0}} = \mathbf{I}$.

(iii) $e^{\mathbf{I}t} = e^t \mathbf{I}$.

(iv) $e^{\mathbf{A}(t+s)} = e^{\mathbf{A}t} e^{\mathbf{A}s}$.

(v) $(e^{\mathbf{A}t})^{-1} = e^{-\mathbf{A}t}$.

(vi) $e^{(\mathbf{A}+\mathbf{B})t} = e^{\mathbf{A}t} e^{\mathbf{B}t}$ provided $\mathbf{A}\mathbf{B} = \mathbf{B}\mathbf{A}$.

Proof.   Property (i) simply restates (8). Property (ii) is obvious from the series (5), and (iii) follows from (7). (**I** is diagonalizable, of course.). We'll confirm (iv) by examining a few terms of the Taylor series for each side, expanded around $t = 0$ for fixed s.

$$
e^{\mathbf{A}(t+s)} = e^{\mathbf{A}(0+s)} + \left[\frac{d}{dt}e^{\mathbf{A}(t+s)}\right]_{t=0} t + \left[\frac{d^2}{dt^2}e^{\mathbf{A}(t+s)}\right]_{t=0} t^2/2! + \cdots .
\tag{9}
$$

---

[1]For rigorous proofs of these and other properties of the matrix exponential function, see *Matrix Computations,* 4th ed., by Gene H. Golub and Charles F. van Loan (Johns Hopkins University Press, Baltimore, 2013), Chapter 11. See also the amusing articles "Nineteen Dubious Ways to Compute the Exponential of a Matrix," by Cleve Moler and Charles van Loan, *SIAM Review*, Vol. 20, No. 4 (Oct. 1978), and "Nineteen Dubious Ways to Compute the Exponential of a Matrix, Twenty-Five Years Later," ibid., Vol 45, No. 1 (Jan. 2003).

By the chain rule and formula (7),

$$\left[\frac{d}{dt}e^{\mathbf{A}(t+s)}\right]_{t=0} = \left[\left(\frac{d}{d(t+s)}e^{\mathbf{A}(t+s)}\right)\frac{d(t+s)}{dt}\right]_{t=0}$$

$$= \left[(\mathbf{A}e^{\mathbf{A}(t+s)})(1)\right]_{t=0} = \mathbf{A}e^{\mathbf{A}s},$$

$$\left[\frac{d^2}{dt^2}e^{\mathbf{A}(t+s)}\right]_{t=0} = \left[\frac{d}{d(t+s)}(\mathbf{A}e^{\mathbf{A}(t+s)})\frac{d(t+s)}{dt}\right]_{t=0}$$

$$= [(\mathbf{A}^2 e^{\mathbf{A}(t+s)})(1)]_{t=0} = \mathbf{A}^2 e^{\mathbf{A}s}$$

and so on, leading to the series

$$e^{\mathbf{A}(t+s)} = e^{\mathbf{A}s} + \mathbf{A}e^{\mathbf{A}s}t + \mathbf{A}^2 e^{\mathbf{A}s}t^2/2! + \cdots. \tag{10}$$

However from (7) we have

$$e^{\mathbf{A}t}e^{\mathbf{A}s} = \{\mathbf{I} + \mathbf{A}t + \mathbf{A}^2 t^2/2! + \cdots\}e^{\mathbf{A}s}, \tag{11}$$

which matches up with (10).

We confirm (v) by letting $s = -t$ in (iv) and invoking (ii).

Finally, the first three terms of the Taylor series expansions of the members in (vi) demonstrate the need for the proviso:

$$e^{(\mathbf{A}+\mathbf{B})t} = \mathbf{I} + (\mathbf{A}+\mathbf{B})t + (\mathbf{A}+\mathbf{B})^2 t^2/2! + \cdots \tag{12}$$

by (7); on the other hand,

$$e^{\mathbf{A}t}e^{\mathbf{B}t} = e^{\mathbf{A}0}e^{\mathbf{B}0} + \left[\frac{d}{dt}(e^{\mathbf{A}t}e^{\mathbf{B}t})\right]_{t=0}t + \left[\frac{d^2}{dt^2}(e^{\mathbf{A}t}e^{\mathbf{B}t})\right]_{t=0}t^2/2! + \cdots$$

$$= \mathbf{I} + [\mathbf{A}e^{\mathbf{A}t}e^{\mathbf{B}t} + e^{\mathbf{A}t}\mathbf{B}e^{\mathbf{B}t}]_{t=0}t \quad (\text{product rule, Section 6.2})$$

$$+ [\mathbf{A}^2 e^{\mathbf{A}t}e^{\mathbf{B}t} + \mathbf{A}e^{\mathbf{A}t}\mathbf{B}e^{\mathbf{B}t} + \mathbf{A}e^{\mathbf{A}t}\mathbf{B}e^{\mathbf{B}t} + e^{\mathbf{A}t}\mathbf{B}^2 e^{\mathbf{B}t}]_{t=0} t^2/2! + \cdots \tag{13}$$

$$= \mathbf{I} + (\mathbf{A}+\mathbf{B})t + (\mathbf{A}^2 + 2\mathbf{A}\mathbf{B} + \mathbf{B}^2) t^2/2! + \cdots.$$

Note that $(\mathbf{A}+\mathbf{B})^2 = \mathbf{A}^2 + \mathbf{A}\mathbf{B} + \mathbf{B}\mathbf{A} + \mathbf{B}^2$ in (12), and this equals $\mathbf{A}^2 + 2\mathbf{A}\mathbf{B} + \mathbf{B}^2$ in (13) only if $\mathbf{A}\mathbf{B} = \mathbf{B}\mathbf{A}$. ∎

Indeed, any algebraic proof of (vi) for the scalar case can be directly translated into the matrix case as long as $\mathbf{A}\mathbf{B} = \mathbf{B}\mathbf{A}$. A slick proof that exploits differential equation theory is given in Problem 30.

The serendipity of properties (i) through (v) tempts us to try to derive the promised formula (4) by reinterpreting the scalar derivation:

$$\frac{d\mathbf{x}}{dt} = \mathbf{A}\mathbf{x} + \mathbf{f}(t) \Rightarrow \mathbf{x}(t_2) = e^{\mathbf{A}(t_2-t_1)}\mathbf{x}(t_1) + e^{\mathbf{A}t_2}\int_{t_1}^{t_2}[e^{-\mathbf{A}s}\mathbf{f}(s)]\,ds. \tag{14}$$

Proof of (14).   The differential equation is equivalent to

$$\frac{d}{dt}\left[e^{-\mathbf{A}t}\mathbf{x}\right] = -\mathbf{A}e^{-\mathbf{A}t}\mathbf{x} + e^{-\mathbf{A}t}\frac{d\mathbf{x}}{dt} = e^{-\mathbf{A}t}\left[\frac{d\mathbf{x}}{dt} - \mathbf{A}\mathbf{x}\right] = e^{-\mathbf{A}t}\mathbf{f} \tag{15}$$

(since $\mathbf{A}e^{-\mathbf{A}t} = e^{-\mathbf{A}t}\mathbf{A}$; look at the power series), and hence

$$e^{-\mathbf{A}t_2}\mathbf{x}(t_2) - e^{-\mathbf{A}t_1}\mathbf{x}(t_1) = \int_{t_1}^{t_2}[e^{-\mathbf{A}t}\mathbf{f}(t)]\,dt,$$

implying (14) and hence, Theorem 1.   ∎

We have seen how to compute $e^{\mathbf{A}t}$ when $\mathbf{A}$ is diagonalizable:

$$e^{\mathbf{A}t} = \mathbf{P}\begin{bmatrix} e^{r_1t} & 0 & \cdots & 0 \\ 0 & e^{r_2t} & \cdots & 0 \\ \vdots & & \ddots & \vdots \\ 0 & 0 & \cdots & e^{r_pt} \end{bmatrix}\mathbf{P}^{-1}.$$

Thus we can see that the corresponding solutions to a homogeneous ($\mathbf{f} = \mathbf{0}$) linear constant coefficient system are combinations of constant vectors times exponentials (e^{r_it}) for diagonalizable $\mathbf{A}$. Of course, these will be oscillatory if the eigenvalues r_i are complex.

Example 1.   Example 2 of Section 7.1 demonstrated that

$$\mathbf{x}(t) = \begin{bmatrix} \cos\omega t \\ -\omega\sin\omega t \end{bmatrix} \tag{16}$$

is a solution of the matrix differential equation $d\mathbf{x}/dt = \mathbf{A}\mathbf{x}$, where

$$\mathbf{A} = \begin{bmatrix} 0 & 1 \\ -\omega^2 & 0 \end{bmatrix}.$$

Verify this solution using Equation (4) of Theorem 1.

Solution. Taking $t_1 = 0$ (and $\mathbf{f}(t) = \mathbf{0}$) in (4) produces

$$\mathbf{x}(t) = e^{\mathbf{A}t}\mathbf{x}(0) = e^{\mathbf{A}t}\begin{bmatrix} 1 \\ 0 \end{bmatrix}$$

for the proposed $\mathbf{x}(0)$ (specified by equation 16). Since

$$\det(\mathbf{A} - r\mathbf{I}) = \det\begin{bmatrix} -r & 1 \\ -\omega^2 & -r \end{bmatrix} = r^2 + \omega^2,$$

the eigenvalues of $\mathbf{A}$ are $\pm i\omega$. The corresponding eigenvectors are found to be $[1 \ \pm i\omega]^T$, and the diagonal form of $\mathbf{A}$ is

$$\mathbf{A} = \mathbf{P}\begin{bmatrix} r_1 & 0 \\ 0 & r_2 \end{bmatrix}\mathbf{P}^{-1} = \begin{bmatrix} 1 & 1 \\ i\omega & -i\omega \end{bmatrix}\begin{bmatrix} i\omega & 0 \\ 0 & -i\omega \end{bmatrix}\begin{bmatrix} \dfrac{1}{2} & -\dfrac{i}{2\omega} \\ \dfrac{1}{2} & \dfrac{i}{2\omega} \end{bmatrix},$$

so that

$$e^{\mathbf{A}t} = \begin{bmatrix} 1 & 1 \\ i\omega & -i\omega \end{bmatrix}\begin{bmatrix} e^{i\omega t} & 0 \\ 0 & e^{-i\omega t} \end{bmatrix}\begin{bmatrix} \dfrac{1}{2} & -\dfrac{i}{2\omega} \\ \dfrac{1}{2} & \dfrac{i}{2\omega} \end{bmatrix}.$$

Hence,

$$e^{\mathbf{A}t}\mathbf{x}(0) = \begin{bmatrix} 1 & 1 \\ i\omega & -i\omega \end{bmatrix}\begin{bmatrix} e^{i\omega t} & 0 \\ 0 & e^{-i\omega t} \end{bmatrix}\begin{bmatrix} \dfrac{1}{2} & -\dfrac{i}{2\omega} \\ \dfrac{1}{2} & \dfrac{i}{2\omega} \end{bmatrix}\begin{bmatrix} 1 \\ 0 \end{bmatrix}$$

$$= \begin{bmatrix} 1 & 1 \\ i\omega & -i\omega \end{bmatrix}\begin{bmatrix} e^{i\omega t} & 0 \\ 0 & e^{-i\omega t} \end{bmatrix}\begin{bmatrix} 1/2 \\ 1/2 \end{bmatrix}$$

$$= \begin{bmatrix} 1 & 1 \\ i\omega & -i\omega \end{bmatrix}\begin{bmatrix} e^{i\omega t}/2 \\ e^{-i\omega t}/2 \end{bmatrix}$$

$$= \begin{bmatrix} e^{i\omega t}/2 + e^{-i\omega t}/2 \\ i\omega\left(e^{i\omega t}/2 - e^{-i\omega t}/2\right) \end{bmatrix} = \begin{bmatrix} \cos\omega t \\ -\omega\sin\omega t \end{bmatrix},$$

the oscillatory solution promised in (16). ■

What if $\mathbf{A}$ cannot be diagonalized? There are other methods for *numerically* computing $e^{\mathbf{A}t}$; indeed, with patience we can sum the power series (7) (but see Problem 31). However it would be nice to know the nature of the solutions for defective (Section 5.2) coefficient matrices $\mathbf{A}$; if they're not constants times exponentials, what are they? The

Jordan normal form that we will describe in the next section answers this question completely. (They are constants times exponentials times polynomials.)

Final Remarks for Readers with Differential Equations Experience

1. Formula (4) settles the existence and uniqueness issues for linear first-order systems with constant coefficients. The explicit formula is a solution, and its derivation demonstrates that it is the only one.

2. The individual columns of $e^{\mathbf{A}t}$ are solutions to $d\mathbf{x}/dt = \mathbf{A}\mathbf{x}$. They are linearly independent matrix functions; Theorem 2 states that $e^{\mathbf{A}t}$ is invertible and therefore nonsingular (for every t). Therefore if $\mathbf{c}$ is a generic constant column vector, $e^{\mathbf{A}t}\mathbf{c}$ is a *general solution* and $e^{\mathbf{A}t}$ is a *fundamental matrix*.

3. The matrix function $e^{-\mathbf{A}t}$ plays the role of an *integrating factor* in (15). The display (4) can be interpreted as an instance of the *variation of parameters* formula.

Exercises 7.2

Compute $e^{\mathbf{A}t}$ for the matrices in Problems 1–12 *using diagonalizability.* [*Hint:* Their eigenvalues and eigenvectors were calculated in Exercises 6.1.]

1. $\begin{bmatrix} 9 & -3 \\ -2 & 4 \end{bmatrix}$ 2. $\begin{bmatrix} -1 & -3 \\ 4 & 7 \end{bmatrix}$ 3. $\begin{bmatrix} 1 & -2 \\ -2 & 1 \end{bmatrix}$ 4. $\begin{bmatrix} -3 & 2 \\ -5 & 4 \end{bmatrix}$ 5. $\begin{bmatrix} 1 & 1 \\ 1 & 1 \end{bmatrix}$

6. $\begin{bmatrix} 1 & 0 & 2 \\ 3 & 4 & 2 \\ 0 & 0 & 0 \end{bmatrix}$ 7. $\begin{bmatrix} 4 & -1 & -2 \\ 2 & 1 & -2 \\ 1 & -1 & 1 \end{bmatrix}$ 8. $\begin{bmatrix} 2 & -1 & -1 \\ -1 & 3 & 0 \\ -1 & 0 & 3 \end{bmatrix}$

9. $\begin{bmatrix} 0 & -1 \\ 1 & 0 \end{bmatrix}$ 10. $\begin{bmatrix} 0 & 1 \\ -1+i & -1 \end{bmatrix}$ 11. $\begin{bmatrix} 1 & 2i \\ -1 & 3+i \end{bmatrix}$ 12. $\begin{bmatrix} -6 & 2 & 16 \\ -3 & 1 & 11 \\ -3 & 1 & 7 \end{bmatrix}$

If a matrix has only one eigenvalue (i.e., all of its eigenvalues are equal), then there is a slick trick enabling the easy computation of its exponential:

(i) *its characteristic polynomial (Section 5.2) must take the form $p(r) = (r_1 - r)^n$;*

(ii) *by the Cayley–Hamilton theorem (Corollary 9, Section 6.3), $(\mathbf{A} - r_1\mathbf{I})^n = \mathbf{0}$;*

(iii) *the power series for $e^{(\mathbf{A}-r_1\mathbf{I})}$ terminates:*

$$e^{(\mathbf{A}-r_1\mathbf{I})} = \mathbf{I} + (\mathbf{A} - r_1\mathbf{I}) + (\mathbf{A} - r_1\mathbf{I})^2/2! + \cdots$$
$$+ (\mathbf{A} - r_1\mathbf{I})^{n-1}/(n-1)! + \mathbf{0} + \mathbf{0} + \cdots$$

(iv) $e^{\mathbf{A}} = e^{r_1 \mathbf{I}} e^{(\mathbf{A} - r_1 \mathbf{I})} = e^{r_1} e^{(\mathbf{A} - r_1 \mathbf{I})}$ *provides a terminating power series for* $e^{\mathbf{A}}$.

Use this trick to find the matrix exponential $e^{\mathbf{A}t}$ *for the matrices in Problems* 13 *through* 18

13. $\begin{bmatrix} 1 & -1 \\ 1 & 3 \end{bmatrix}$　　14. $\begin{bmatrix} 3 & -2 \\ 0 & 3 \end{bmatrix}$　　15. $\begin{bmatrix} 2 & 1 & -1 \\ -3 & -1 & 1 \\ 9 & 3 & -4 \end{bmatrix}$　　16. $\begin{bmatrix} 2 & 1 & 3 \\ 0 & 2 & -1 \\ 0 & 0 & 2 \end{bmatrix}$

17. $\begin{bmatrix} 0 & 1 & 0 \\ 0 & 0 & 1 \\ -1 & -3 & -3 \end{bmatrix}$　　18. $\begin{bmatrix} -2 & 0 & 0 \\ 4 & -2 & 0 \\ 1 & 0 & -2 \end{bmatrix}$.

19. Use the results of Problem 15 and formula (4) to find the solution to the initial-value problem

$$\mathbf{x}'(t) = \begin{bmatrix} 2 & 1 & -1 \\ -3 & -1 & 1 \\ 9 & 3 & -4 \end{bmatrix} \mathbf{x}(t) + \begin{bmatrix} 0 \\ t \\ 0 \end{bmatrix}, \quad \mathbf{x}(0) = \begin{bmatrix} 0 \\ 3 \\ 0 \end{bmatrix}.$$

20. Use the result of Problem 18 to find the solution to the initial-value problem

$$\mathbf{x}'(t) = \begin{bmatrix} -2 & 0 & 0 \\ 4 & -2 & 0 \\ 1 & 0 & -2 \end{bmatrix} \mathbf{x}(t), \quad \mathbf{x}(0) = \begin{bmatrix} 1 \\ 1 \\ -1 \end{bmatrix}.$$

21. Use the result of Problem 17 to find the solution to the initial-value problem

$$\mathbf{x}'(t) = \begin{bmatrix} 0 & 1 & 0 \\ 0 & 0 & 1 \\ -1 & -3 & -3 \end{bmatrix} \mathbf{x}(t) + t \begin{bmatrix} 1 \\ -1 \\ 2 \end{bmatrix} + \begin{bmatrix} 3 \\ 1 \\ 0 \end{bmatrix}, \quad \mathbf{x}(0) = \mathbf{0}.$$

In Problems 22–27, use formula (4) to determine a solution of the system $\mathbf{x}'(t) = \mathbf{A}\mathbf{x}(t) + \mathbf{f}(t)$, *where* $\mathbf{A}$ *and* $\mathbf{f}(t)$ *are given. Take* $\mathbf{x}(0) = \mathbf{0}$.

22. $\mathbf{A} = \begin{bmatrix} 0 & 1 \\ 2 & 0 \end{bmatrix}, \quad \mathbf{f}(t) = \begin{bmatrix} \sin 3t \\ t \end{bmatrix}.$

23. $\mathbf{A} = \begin{bmatrix} -1 & 0 \\ 2 & 2 \end{bmatrix}, \quad \mathbf{f}(t) = \begin{bmatrix} t^2 \\ t+1 \end{bmatrix}.$

24. $\mathbf{A} = \begin{bmatrix} 0 & -1 & 0 \\ -1 & 0 & 1 \\ 0 & 0 & 1 \end{bmatrix}, \quad \mathbf{f}(t) = \begin{bmatrix} 0 \\ 0 \\ 1 \end{bmatrix}.$

25. $\mathbf{A} = \begin{bmatrix} 0 & 1 \\ -1 & 0 \end{bmatrix}, \quad \mathbf{f}(t) = \begin{bmatrix} 1 \\ 0 \end{bmatrix}.$

26. Let

$$\mathbf{A} = \begin{bmatrix} 1 & -2 & 2 \\ -2 & 1 & 2 \\ 2 & 2 & 1 \end{bmatrix}.$$

For which initial conditions $\mathbf{x}(0)$ do the solutions to $\mathbf{x}' = \mathbf{A}\mathbf{x}$ remain bounded for all $t > 0$?

27. For the matrix $\mathbf{A}$ in the previous problem, determine the solution to the initial value problem

$$\mathbf{x}' = \mathbf{A}\mathbf{x} + \sin(2t) \begin{bmatrix} 2 \\ 0 \\ 4 \end{bmatrix}, \quad \mathbf{x}(0) = \begin{bmatrix} 0 \\ 1 \\ -1 \end{bmatrix}.$$

28. Convergence of the matrix exponential series We shall guide you through a proof that the series (7) converges by invoking the Comparison Test, cited in most calculus texts:

Comparison Test. Suppose that $\sum_{k=0}^{\infty} M_k$ is a convergent series with each $M_k \geq 0$, and suppose that $|\alpha_k| \leq M_k$ for all subscripts k sufficiently large. Then $\sum_{k=0}^{\infty} \alpha_k$ also converges.

(a) For any n-by-n matrix $\mathbf{A}$, define $\nu(\mathbf{A})$ to be the largest-in-magnitude entry in $\mathbf{A}$. Prove that $\nu(\mathbf{A}^2) \leq n\nu(\mathbf{A})^2$.

(b) Prove that for all nonnegative integers k, $\nu(\mathbf{A}^k) \leq n^{k-1}\nu(\mathbf{A})^k$.

(c) Argue that the matrix entries in the series (7) are dominated in magnitude by the corresponding terms in the series $\sum_{k=0}^{\infty} n^{k-1}\nu(\mathbf{A}t)^k/k!$, and that the latter converges to $e^{n\nu(\mathbf{A}t)}/n$.

(d) Apply the comparison test to conclude that (7) converges.

Although the matrix exponential series (7) converges in theory, its numerical implementation is vexing. Problem 29 is drawn from the exposition in "Nineteen Dubious Ways to Compute the Exponential of a Matrix," by Cleve Moler and Charles van Loan, SIAM Review, Vol. 20, No. 4 (Oct. 1978).

29. The matrix $\mathbf{A}$ has a diagonalization expressed as follows:

$$\mathbf{A} := \begin{bmatrix} -49 & 24 \\ -64 & 31 \end{bmatrix} = \begin{bmatrix} 1 & 3 \\ 2 & 4 \end{bmatrix} \begin{bmatrix} -1 & 0 \\ 0 & -17 \end{bmatrix} \begin{bmatrix} 1 & 3 \\ 2 & 4 \end{bmatrix}^{-1}$$

(a) Use formula (6) to calculate

$$e^{\mathbf{A}} = \begin{bmatrix} 1 & 3 \\ 2 & 4 \end{bmatrix} \begin{bmatrix} e^{-1} & 0 \\ 0 & e^{-17} \end{bmatrix} \begin{bmatrix} 1 & 3 \\ 2 & 4 \end{bmatrix}^{-1} \approx \begin{bmatrix} -0.7357589 & 0.5518191 \\ -1.4715176 & 1.1036382 \end{bmatrix} \quad (17)$$

(b) Use a computer to show that, in the power series

$$e^{\mathbf{A}} = \mathbf{I} + \mathbf{A} + \mathbf{A}^2/2! + \cdots + \mathbf{A}^m/m! + \cdots ,$$

the 16th and 17th terms are approximately given by

$$\frac{\mathbf{A}^{16}}{16!} \approx \begin{bmatrix} 6,977,251.906101090 & -3,488,625.953050545 \\ 9,303,002.541468118 & -4,651,501.270734060 \end{bmatrix}$$

$$\frac{\mathbf{A}^{17}}{17!} \approx \begin{bmatrix} -6,977,251.906101089 & 3,488,625.953050546 \\ -9,303,002.541468109 & 4,651,501.270734056 \end{bmatrix}$$

Hence in the summation $(\mathbf{A}^{16}/16! + \mathbf{A}^{17}/17!)$, 15 *significant (decimal) digits are lost through cancellation*. A 16-digit computer has no chance of performing this summation correctly.

(c) Moler and van Loan reported that on their 1978 computer, the first 59 terms of the series

$$e^{\mathbf{A}} = \mathbf{I} + \mathbf{A} + \mathbf{A}^2/2! + \cdots + \mathbf{A}^m/m! + \cdots$$

resulted in the *horrible* approximation

$$e^{\mathbf{A}} \approx \begin{bmatrix} -22.25880 & -1.432766 \\ -61.49931 & -3.474280 \end{bmatrix}$$

(compare with (17)). Today's computers are much more accurate. Write a program to calculate

$$\mathbf{I} + \mathbf{A} + \mathbf{A}^2/2! + \cdots + \mathbf{A}^m/m!$$

for any input m. How many terms must be summed to obtain an answer close to the true value (17) on your computer?

(d) Change $\mathbf{A}$ to

$$\mathbf{A} := \begin{bmatrix} -100 & 49.5 \\ -132 & 65 \end{bmatrix} = \begin{bmatrix} 1 & 3 \\ 2 & 4 \end{bmatrix} \begin{bmatrix} -1 & 0 \\ 0 & -34 \end{bmatrix} \begin{bmatrix} 1 & 3 \\ 2 & 4 \end{bmatrix}^{-1}$$

and use the diagonal form to show that

$$e^{\mathbf{A}} \approx \begin{bmatrix} -0.7357589 & 0.5518192 \\ -1.4715178 & 1.1036383 \end{bmatrix} \tag{18}$$

(e) Now how many terms of

$$e^{\mathbf{A}} = \mathbf{I} + \mathbf{A} + \mathbf{A}^2/2! + \cdots + \mathbf{A}^m/m! + \cdots$$

must be summed to obtain an answer close to the true value on your computer?

30. Theorem 1 implies that the initial value problem

$$\mathbf{x}' = \mathbf{Mx}, \quad \mathbf{x}(0) = \mathbf{c}$$

has one, and only one, solution, given by equation (4) in the theorem (with $\mathbf{f}$ and t_1 each zero), when $\mathbf{M}$ and $\mathbf{c}$ are any constant matrices of the appropriate dimensions. Use this to prove statement (vi) in Theorem 2 by demonstrating that $e^{(\mathbf{A}+\mathbf{B})t}\mathbf{c}$ and $e^{\mathbf{A}t}e^{\mathbf{B}t}\mathbf{c}$ both satisfy $\mathbf{x}' = (\mathbf{A} + \mathbf{B})\mathbf{x}$, $\mathbf{x}(0) = \mathbf{c}$. [*Hint:* argue that

$$\frac{d}{dt}e^{\mathbf{A}t}e^{\mathbf{B}t} = \mathbf{A}e^{\mathbf{A}t}e^{\mathbf{B}t} + e^{\mathbf{A}t}\mathbf{B}e^{\mathbf{B}t}$$

and also that $e^{\mathbf{A}t}\mathbf{B}e^{\mathbf{B}t} = \mathbf{B}e^{\mathbf{A}t}e^{\mathbf{B}t}$ if $\mathbf{AB} = \mathbf{BA}$.]

31. Let $\mathbf{A} = \begin{bmatrix} 1 & 2 \\ -1 & 3 \end{bmatrix}$ and $\mathbf{B} = \begin{bmatrix} 2 & 1 \\ 0 & 1 \end{bmatrix}$.

(a) Show that $\mathbf{AB} \neq \mathbf{BA}$.

(b) Show that property (*vi*) in Theorem 2 does not hold for these matrices. That is, show that $e^{(\mathbf{A}+\mathbf{B})t} \neq e^{\mathbf{A}t}e^{\mathbf{B}t}$.

In Problems 32–35, use a linear algebra software package for help in determining $e^{\mathbf{A}t}$

32. $\mathbf{A} = \begin{bmatrix} 0 & 1 & 0 & 0 & 0 \\ 0 & 0 & 1 & 0 & 0 \\ 1 & -3 & 3 & 0 & 0 \\ 0 & 0 & 0 & 0 & 1 \\ 0 & 0 & 0 & -1 & 0 \end{bmatrix}$.

33. $\mathbf{A} = \begin{bmatrix} 1 & 0 & 0 & 0 & 0 \\ 0 & 0 & 1 & 0 & 0 \\ 0 & -1 & -2 & 0 & 0 \\ 0 & 0 & 0 & 0 & 1 \\ 0 & 0 & 0 & -1 & 0 \end{bmatrix}$.

34. $\mathbf{A} = \begin{bmatrix} 0 & 1 & 0 & 0 & 0 \\ 0 & 0 & 1 & 0 & 0 \\ -1 & -3 & -3 & 0 & 0 \\ 0 & 0 & 0 & 0 & 1 \\ 0 & 0 & 0 & -4 & -4 \end{bmatrix}$.

35. $\mathbf{A} = \begin{bmatrix} -1 & 0 & 0 & 0 & 0 \\ 0 & 0 & 1 & 0 & 0 \\ 0 & -1 & -2 & 0 & 0 \\ 0 & 0 & 0 & 0 & 1 \\ 0 & 0 & 0 & -4 & -4 \end{bmatrix}$.

7.3   THE JORDAN NORMAL FORM

We have seen three possibilities for the set of eigenvalues of an arbitrary n-by-n matrix $\mathbf{A}$:

(i) $\mathbf{A}$ has n distinct eigenvalues. Then $\mathbf{A}$ must have n linearly independent eigenvectors, and it can be diagonalized by a similarity transformation generated by a matrix $\mathbf{P}$ of eigenvectors: $\mathbf{A} = \mathbf{PDP}^{-1}$.

(ii) **A** has multiple eigenvalues, but it still has n linearly independent eigenvectors. It can, therefore, still be diagonalized by a similarity transformation generated by a matrix of eigenvectors: $\mathbf{A} = \mathbf{PDP}^{-1}$.

(iii) **A** has multiple eigenvalues, but it has less than n linearly independent eigenvectors. Then it cannot be diagonalized. It is *defective*.

Practically, all random matrices fall into class (i), but "home-grown" matrices frequently fall into classes (ii) and (iii) (Section 6.5).

In all three cases, **A** can be upper-triangularized by a similarity transformation generated by a unitary matrix (thanks to the Schur decomposition theorem of Section 6.3). But the advantages of full diagonalization are enormous: powers, and power series, can be evaluated readily, and linear differential equation systems can be solved. So it is worthwhile to seek a form "as close to diagonal as possible" that is valid for *all* matrices. The *Jordan form* is difficult to derive and devilish to calculate, but it's the best that mathematics has come up with.

To start with, it is *block*-diagonal, a concept which is much easier to diagram (Fig. 7.17) than to define:

Remark. If two matrices are block diagonal with compatible dimensions, their sums and products can be calculated "blockwise," as illustrated in Figure 7.18.[2] In particular, any power of a block-diagonal matrix is calculated by raising the *blocks* to that power.

$$
A = \begin{bmatrix} \# & \# & 0 & 0 & 0 & 0 \\ \# & \# & 0 & 0 & 0 & 0 \\ 0 & 0 & \# & \# & \# & 0 \\ 0 & 0 & \# & \# & \# & 0 \\ 0 & 0 & \# & \# & \# & 0 \\ 0 & 0 & 0 & 0 & 0 & \# \end{bmatrix} = \begin{bmatrix} \boxed{\mathbf{A}_1} & \boxed{\mathbf{0}} & \boxed{\mathbf{0}} \\ \boxed{\mathbf{0}} & \boxed{\mathbf{A}_2} & \boxed{\mathbf{0}} \\ \boxed{\mathbf{0}} & \boxed{\mathbf{0}} & \boxed{\mathbf{A}_3} \end{bmatrix}
$$

Fig. 7.17 Block-diagonal matrix.

The individual diagonal blocks in the *Jordan* form have the structure illustrated in Figures 7.19 and 7.20. (For convenience we have written $\mathbf{J}_{1\times 1}$ for $\mathbf{J}_{1\text{-by-}1}$.)

Jordan Form

Definition 1. A square matrix **J** is said to be a **Jordan block** if its diagonal entries are all equal, its superdiagonal entries are all ones, and all other entries

[2] The best "proof" is simply to try these rules out.

are zero. A square matrix is said to be in **Jordan form** if it is block diagonal and each of its blocks is a Jordan block.

Theorem 3. Every square matrix **A** is similar to a matrix **J** in Jordan form: $\mathbf{A} = \mathbf{PJP}^{-1}$. The dimensions of the Jordan blocks in **J** are uniquely determined by **A**.

$$
\begin{bmatrix} [A] & [0] \\ [0] & [B] \end{bmatrix} + \begin{bmatrix} [C] & [0] \\ [0] & [D] \end{bmatrix}
$$

$$
= \begin{bmatrix} [A+C] & [0] \\ [0] & [B+D] \end{bmatrix}
$$

$$
\begin{bmatrix} [A] & [0] \\ [0] & [B] \end{bmatrix} \times \begin{bmatrix} [C] & [0] \\ [0] & [D] \end{bmatrix}
$$

$$
= \begin{bmatrix} [AC] & [0] \\ [0] & [BD] \end{bmatrix}
$$

Fig. 7.18 Blockwise addition and multiplication.

$$
\mathbf{J}_{1\times1} = \begin{bmatrix} r \end{bmatrix}, \quad \mathbf{J}_{2\times2} = \begin{bmatrix} r & 1 \\ 0 & r \end{bmatrix}, \quad \mathbf{J}_{3\times3} = \begin{bmatrix} r & 1 & 0 \\ 0 & r & 1 \\ 0 & 0 & r \end{bmatrix}, \quad \mathbf{J}_{4\times4} = \begin{bmatrix} r & 1 & 0 & 0 \\ 0 & r & 1 & 0 \\ 0 & 0 & r & 1 \\ 0 & 0 & 0 & r \end{bmatrix}
$$

Fig. 7.19 Jordan blocks.

$$
\begin{bmatrix} r_1 & 1 & 0 & 0 & 0 & 0 \\ 0 & r_1 & 0 & 0 & 0 & 0 \\ 0 & 0 & r_2 & 1 & 0 & 0 \\ 0 & 0 & 0 & r_2 & 1 & 0 \\ 0 & 0 & 0 & 0 & r_2 & 0 \\ 0 & 0 & 0 & 0 & 0 & r_3 \end{bmatrix} = \begin{bmatrix} |\mathbf{J}_{2\times2}| & |0| & |0| \\ |0| & |\mathbf{J}_{3\times3}| & |0| \\ |0| & |0| & |\mathbf{J}_{1\times1}| \end{bmatrix}
$$

Fig. 7.20 Jordan form.

Hordes of authors have sought to publish simple proofs of Theorem 3. We shall not add to their number; suffice it to say we prefer the derivation of Filippov as expostulated by Strang.[3]

As discussed in Section 6.5, even the *detection* of multiple eigenvalues is impractical with finite-precision computation; calculations have to be performed in rational arithmetic.[4] Nonetheless knowledge of the mere *existence* of the Jordan form is of immense value in some applications, so we explore it further.

The eigenvalues (and their multiplicities) of a matrix in Jordan form are displayed along its diagonal (since the Jordan form is upper triangular). These are also the eigenvalues (and multiplicities) of the original matrix $\mathbf{A}$, because similarity transformations do not change the characteristic polynomial of a matrix (Corollary 2, Section 6.1). Note, however, that the same eigenvalue may appear in more than one Jordan block. Thus, **the multiplicity of a particular eigenvalue equals the sum of the dimensions of the Jordan blocks in which it appears**. If r_1 is distinct from r_2 and r_3 in Figure 7.20, its multiplicity is two; if $r_1 = r_2 \neq r_3$, the multiplicity of r_1 is 5.

Of course the statement $\mathbf{A} = \mathbf{PJP}^{-1}$ is equivalent to $\mathbf{AP} = \mathbf{PJ}$. For the matrix in Figure 7.20, if $\mathbf{P}$'s columns are displayed as

$$
\mathbf{P} = \begin{bmatrix} \vdots & \vdots & & \vdots \\ \mathbf{u}_1 & \mathbf{u}_2 & \cdots & \mathbf{u}_6 \\ \vdots & \vdots & & \vdots \end{bmatrix},
$$

then $\mathbf{AP} = \mathbf{PJ}$ takes the form

$$
\mathbf{AP} = \mathbf{A} \begin{bmatrix} \vdots & \vdots & & \vdots \\ \mathbf{u}_1 & \mathbf{u}_2 & \cdots & \mathbf{u}_6 \\ \vdots & \vdots & & \vdots \end{bmatrix} = \begin{bmatrix} \vdots & \vdots & & \vdots \\ \mathbf{Au}_1 & \mathbf{Au}_2 & \cdots & \mathbf{Au}_6 \\ \vdots & \vdots & & \end{bmatrix}
$$

$$
= \mathbf{PJ} = \begin{bmatrix} \vdots & \vdots & & \vdots \\ \mathbf{u}_1 & \mathbf{u}_2 & \cdots & \mathbf{u}_6 \\ \vdots & \vdots & & \vdots \end{bmatrix} \begin{bmatrix} r_1 & 1 & 0 & 0 & 0 & 0 \\ 0 & r_1 & 0 & 0 & 0 & 0 \\ 0 & 0 & r_2 & 1 & 0 & 0 \\ 0 & 0 & 0 & r_2 & 1 & 0 \\ 0 & 0 & 0 & 0 & r_2 & 0 \\ 0 & 0 & 0 & 0 & 0 & r_3 \end{bmatrix} =
$$

---

[3]See "A short proof of the theorem on reduction of a matrix to Jordan form (Russian)", Filippov, A. F., *Vestnik Moskov. Univ. Ser. I Mat. Meh.* 26 1971 no. 2, 18–19; and *Linear Algebra and Its Applications* 4th ed., Strang, G., 2005, Cengage Learning (Appendix B. The Jordan Form). The web page http://math.fullerton.edu/mathews/n2003/jordanform/JordanFormBib/Links/JordanFormBib_lnk_3.html lists 64 proofs, published between 1962 and 2003.

[4]Some software codes perform rational arithmetic using symbol manipulation in their Jordan subroutines. The resulting slowdown is apparent to the user.

$$
= \begin{bmatrix} \vdots & \vdots & \vdots & \vdots & \vdots & \vdots \\ r_1\mathbf{u}_1 & [r_1\mathbf{u}_2 + \mathbf{u}_1] & r_2\mathbf{u}_3 & [r_2\mathbf{u}_4 + \mathbf{u}_3] & [r_2\mathbf{u}_5 + \mathbf{u}_4] & r_6\mathbf{u}_6 \\ \vdots & \vdots & \vdots & \vdots & \vdots & \vdots \end{bmatrix} \tag{1}
$$

Therefore column $\mathbf{u}_i$ of $\mathbf{P}$ satisfies either

$$
\mathbf{A}\mathbf{u}_i = r\mathbf{u}_i \tag{2}
$$

or

$$
\mathbf{A}\mathbf{u}_i = r\mathbf{u}_i + \mathbf{u}_{i-1} \tag{3}
$$

for some eigenvalue r. Of course, in case (2), $\mathbf{u}_i$ is an eigenvector corresponding to the eigenvalue r on the ith diagonal. Thus in (1), the eigenvectors are $\mathbf{u}_1$, $\mathbf{u}_3$, and $\mathbf{u}_6$.

Example 1.   What are the possible Jordan forms for a 5-by-5 matrix $\mathbf{A}$ with a single eigenvalue r (of multiplicity 5)?

Solution. $\mathbf{A}$ can have 5, 4, 3, 2, or 1 linearly independent eigenvectors. They are manifested in the Jordan form $\mathbf{J}$ by the columns with zeros in the (first) superdiagonal. From Figure 7.20 or Equation (1), we can see that, informally speaking, every time we "lose" an eigenvector, we gain a "1" on the superdiagonal. Keep in mind that only the *dimensions*, not the *locations*, of the Jordan blocks are fixed.

5 (linearly independent) eigenvectors: $\mathbf{J}$ is diagonal:

$$
\begin{bmatrix} r & 0 & 0 & 0 & 0 \\ 0 & r & 0 & 0 & 0 \\ 0 & 0 & r & 0 & 0 \\ 0 & 0 & 0 & r & 0 \\ 0 & 0 & 0 & 0 & r \end{bmatrix}.
$$

4 eigenvectors: Adding a 1 on the superdiagonal will create three 1-by-1 Jordan blocks and one 2-by-2. For example, if the 1 is placed at the $(4, 5)$ location,

$$
\begin{bmatrix} r & 0 & 0 & 0 & 0 \\ 0 & r & 0 & 0 & 0 \\ 0 & 0 & r & 0 & 0 \\ 0 & 0 & 0 & r & 1 \\ 0 & 0 & 0 & 0 & r \end{bmatrix}.
$$

3 eigenvectors: J contains two 1's on the superdiagonal. This will create either a 3-by-3 block or two 2-by-2 blocks. For example,

$$
\begin{bmatrix}
r & 0 & 0 & 0 & 0 \\
0 & r & 0 & 0 & 0 \\
0 & 0 & r & 1 & 0 \\
0 & 0 & 0 & r & 1 \\
0 & 0 & 0 & 0 & r
\end{bmatrix}
\quad \text{or} \quad
\begin{bmatrix}
r & 0 & 0 & 0 & 0 \\
0 & r & 1 & 0 & 0 \\
0 & 0 & r & 0 & 0 \\
0 & 0 & 0 & r & 1 \\
0 & 0 & 0 & 0 & r
\end{bmatrix}.
$$

2 eigenvectors: J contains three 1's on the superdiagonal. Since there are four superdiagonal entries, the only issue is where to put the 0. As we see in the following display,

$$
\begin{bmatrix}
r & 0 & 0 & 0 & 0 \\
0 & r & 1 & 0 & 0 \\
0 & 0 & r & 1 & 0 \\
0 & 0 & 0 & r & 1 \\
0 & 0 & 0 & 0 & r
\end{bmatrix}
\quad \text{or} \quad
\begin{bmatrix}
r & 1 & 0 & 0 & 0 \\
0 & r & 0 & 0 & 0 \\
0 & 0 & r & 1 & 0 \\
0 & 0 & 0 & r & 1 \\
0 & 0 & 0 & 0 & r
\end{bmatrix},
$$

either there will be a 4-by-4 block, or a 2-by-2 together with a 3-by-3 block.

1 eigenvector: the superdiagonal is "full" and **J** is a single 5-by-5 Jordan block:

$$
\begin{bmatrix}
r & 1 & 0 & 0 & 0 \\
0 & r & 1 & 0 & 0 \\
0 & 0 & r & 1 & 0 \\
0 & 0 & 0 & r & 1 \\
0 & 0 & 0 & 0 & r
\end{bmatrix}. \qquad\blacksquare
$$

Shortly, we shall see how these considerations can be used to compute the matrix **P**. But first we list some pertinent observations that follow from Theorem 3 and the display (1).

(i) Since **P** is invertible, its columns form a basis for the underlying vector space $\mathbf{R}^n_{\text{col}}$ (or $\mathbf{C}^n_{\text{col}}$).

(ii) Thanks to (2) and (3), the columns of **P** cluster into groups $\{\mathbf{u}_1, \mathbf{u}_2\}$, $\{\mathbf{u}_3, \mathbf{u}_4, \mathbf{u}_5\}$, and $\{\mathbf{u}_6\}$ with the property that whenever **v** lies in the span of one of the groups, **Av** lies in its span also. For example, if $\mathbf{v} = c_1\mathbf{u}_1 + c_2\mathbf{u}_2$, then

$$
\mathbf{Av} = c_1 r_1 \mathbf{u}_1 + c_2(r_2\mathbf{u}_2 + \mathbf{u}_1) = (c_1 r_1 + c_2)\mathbf{u}_1 + c_2 r_2 \mathbf{u}_2.
$$

(In the jargon of algebra, the span of each group is an *invariant subspace*.)

(iii) Each of the columns $\mathbf{u}_i$ that satisfies (2) or (3) for the eigenvalue r also satisfies

$$
(\mathbf{A} - r\mathbf{I})^p \mathbf{u}_i = 0 \qquad\qquad (4)
$$

for high enough p; if $\mathbf{u}_i$ satisfies (2), then (4) is true for $p = 1$, and if $\mathbf{u}_i$ satisfies (3), then $(\mathbf{A} - r\mathbf{I})^p\,\mathbf{u}_i = (\mathbf{A} - r\mathbf{I})^{p-1}\mathbf{u}_{i-1} = (\mathbf{A} - r\mathbf{I})^{p-2}\mathbf{u}_{i-2}\ldots$ and so on, until we "get down" to a true eigenvector $\mathbf{u}$ for r, whence $(\mathbf{A} - r\mathbf{I})\mathbf{u} = 0$. Thus p can be taken to be the dimension of the largest Jordan block with r on its diagonal.

Generalized Eigenvectors

Definition 2. A nonzero solution $\mathbf{u}$ of $(\mathbf{A} - r\mathbf{I})^p\mathbf{u} = \mathbf{0}$ (for any r and any integer $p > 0$) is called a **generalized eigenvector of A**.

Thus all the columns of $\mathbf{P}$ are (linearly independent) generalized eigenvectors.

(iv) (*An intriguing theoretical note*) Because $\mathbf{A}$ and $\mathbf{J}$ are similar, they have the same characteristic polynomial (Corollary 7.2, Section 6.1). From the (upper triangular) Jordan matrix display in Figure 7.20, we deduce that this polynomial can be expressed in terms of the Jordan parameters as

$$p(r) = \det\,(\mathbf{A} - r\mathbf{I}) = \det\,(\mathbf{J} - r\mathbf{I}) = (r_1 - r)^{p_1}(r_2 - r)^{p_2} \cdots (r_J - r)^{p_J}, \tag{5}$$

when $\mathbf{A}$ has J Jordan blocks of size $p_1, p_2, \ldots, p_J$ containing corresponding eigenvalues $r_1, r_2, \ldots, r_J$. Since some of the eigenvalues might be repeated in the display (5), for convenience let's assume that $\{r_1, r_2, \ldots, r_K\}$ are *distinct* eigenvalues and $\{r_{K+1}, r_{K+2}, \ldots, r_J\}$ replicate these. Then the characteristic polynomial has the alternative display

$$p(r) = (r_1 - r)^{q_1}(r_2 - r)^{q_2} \cdots (r_K - r)^{q_K} \tag{6}$$

where the exponent in $(r_i - r)^{q_i}$ is the *sum* of the exponents of factors in (5) containing the eigenvalue r_i.

Now let $\mathbf{u}$ be any column of $\mathbf{P}$ satisfying either (2) or (3) with $r = r_i$, $i \leq K$; by our count, there are q_i such columns in all. Then $\mathbf{u}$ satisfies (4) with p equal to the dimension of the largest Jordan block with r_i on its diagonal. Hence $\mathbf{u}$ satisfies $(\mathbf{A} - r_i\mathbf{I})^{q_i}\mathbf{u} = \mathbf{0}$ since q_i is at least as big as the dimension of any such Jordan block. The columns of $\mathbf{P}$ attest to the fact that for each eigenvalue r_i the generalized eigenvector equation $(\mathbf{A} - r_i\mathbf{I})^{q_i}\mathbf{u} = \mathbf{0}$ has (at least) q_i solutions that are linearly independent of each other, and linearly independent of those for the other eigenvalues.

Could there be more than q_i linearly independent solutions? No; $q_1 + q_2 + \cdots + q_K$ equals the degree n of the characteristic polynomial (where $\mathbf{A}$ is n-by-n), and there can be no more than n linearly independent vectors. We have derived a theorem that will be of crucial importance in Section 7.4.

Primary Decomposition Theorem

Theorem 4. If the characteristic polynomial of the n-by-n square matrix $\mathbf{A}$ is

$$p(r) = \det(\mathbf{A} - r\mathbf{I}) = (r_1 - r)^{q_1}(r_2 - r)^{q_2} \cdots (r_K - r)^{q_K}, \qquad (7)$$

where the eigenvalues $\{r_i\}$ are distinct, then for each i there are q_i linearly independent generalized eigenvectors $\mathbf{u}$ vectors satisfying

$$(\mathbf{A} - r_i\mathbf{I})^{q_i}\mathbf{u} = \mathbf{0}; \qquad (8)$$

further, the totality of these n generalized eigenvectors is a linearly independent set.

(Indeed, we have shown that q_i in (8) can be replaced by the dimension of the largest Jordan block corresponding to the eigenvalue r_i.)

(v) One important application of the Jordan form is the *direct* calculation of the matrix exponential for defective matrices. Although a defective matrix is not diagonalizable, the powers of its Jordan form are transparent, and this facilitates the summing of the power series. The powers of an m-by-m Jordan block have a pattern that suggests the binomial formula (see Problem 28):

$$\mathbf{J} = \begin{bmatrix} r & 1 & 0 & 0 \\ 0 & r & 1 & 0 \\ 0 & 0 & r & 1 \\ 0 & 0 & 0 & r \end{bmatrix}, \; \mathbf{J}^2 = \begin{bmatrix} r^2 & 2r & 1 & 0 \\ 0 & r^2 & 2r & 1 \\ 0 & 0 & r^2 & 2r \\ 0 & 0 & 0 & r^2 \end{bmatrix}, \; \mathbf{J}^3 = \begin{bmatrix} r^3 & 3r^2 & 3r & 1 \\ 0 & r^3 & 3r^2 & 3r \\ 0 & 0 & r^3 & 3r^2 \\ 0 & 0 & 0 & r^3 \end{bmatrix},$$

$$\mathbf{J}^4 = \begin{bmatrix} r^4 & 4r^3 & 6r^2 & 4r \\ 0 & r^4 & 4r^3 & 6r^2 \\ 0 & 0 & r^4 & 4r^3 \\ 0 & 0 & 0 & r^4 \end{bmatrix} \qquad (9)$$

Hence, the Taylor series for the matrix exponential $e^{\mathbf{J}t}$ looks like

$$e^{\mathbf{J}t} = \mathbf{I} + t\mathbf{J} + t^2\mathbf{J}^2/2! + t^3\mathbf{J}^3/3! + t^4\mathbf{J}^4/4! + t^5\mathbf{J}^5/5! + \cdots$$

$$= \begin{bmatrix} 1 & 0 & 0 & 0 \\ 0 & 1 & 0 & 0 \\ 0 & 0 & 1 & 0 \\ 0 & 0 & 0 & 1 \end{bmatrix} + t\begin{bmatrix} r & 1 & 0 & 0 \\ 0 & r & 1 & 0 \\ 0 & 0 & r & 1 \\ 0 & 0 & 0 & r \end{bmatrix} + \frac{t^2}{2!}\begin{bmatrix} r^2 & 2r & 1 & 0 \\ 0 & r^2 & 2r & 1 \\ 0 & 0 & r^2 & 2r \\ 0 & 0 & 0 & r^2 \end{bmatrix} +$$

$$
+ \frac{t^3}{3!}
\begin{bmatrix}
r^3 & 3r^2 & 3r & 1 \\
0 & r^3 & 3r^2 & 3r \\
0 & 0 & r^3 & 3r^2 \\
0 & 0 & 0 & r^3
\end{bmatrix}
+ \frac{t^4}{4!}
\begin{bmatrix}
r^4 & 4r^3 & 6r^2 & 4r \\
0 & r^4 & 4r^3 & 6r^2 \\
0 & 0 & r^4 & 4r^3 \\
0 & 0 & 0 & r^4
\end{bmatrix}
$$

$$
+ \frac{t^5}{5!}
\begin{bmatrix}
r^5 & 5r^4 & 10r^3 & 10r^2 \\
0 & r^5 & 5r^4 & 10r^3 \\
0 & 0 & r^5 & 5r^4 \\
0 & 0 & 0 & r^4
\end{bmatrix}
+ \cdots . \tag{10}
$$

With a little cleaning up, we see the exponential power series recurring in every entry. Along the diagonal, we get

$$
1 + (tr) + \frac{(tr)^2}{2!} + \frac{(tr)^3}{3!} + \cdots = e^{rt};
$$

on the first superdiagonal, we get

$$
t \left[1 + (tr) + \frac{(tr)^2}{2!} + \frac{(tr)^3}{3!} + \cdots \right] = te^{rt};
$$

and for the second and third superdiagonals, we have

$$
\frac{t^2}{2} \left[1 + (tr) + \frac{(tr)^2}{2!} + \frac{(tr)^3}{3!} + \cdots \right] = \frac{t^2}{2} e^{rt}
$$

and

$$
\frac{t^3}{3!} \left[1 + (tr) + \frac{(tr)^2}{2!} + \cdots \right] = \frac{t^3}{3!} e^{rt}.
$$

So the infinite matrix sum converges to the simple upper triangular form

$$
e^{\mathbf{J}t} =
\begin{bmatrix}
\dfrac{e^{rt}}{0!} & \dfrac{te^{rt}}{1!} & \dfrac{t^2 e^{rt}}{2!} & \dfrac{t^3 e^{rt}}{3!} \\[2ex]
0 & \dfrac{e^{rt}}{0!} & \dfrac{te^{rt}}{1!} & \dfrac{t^2 e^{rt}}{2!} \\[2ex]
0 & 0 & \dfrac{e^{rt}}{0!} & \dfrac{te^{rt}}{1!} \\[2ex]
0 & 0 & 0 & \dfrac{e^{rt}}{0!}
\end{bmatrix} . \tag{11}
$$

Thus for a matrix whose Jordan form is given by the pattern in Figure 7.20, we have

$$\mathbf{A} = \mathbf{P} \begin{bmatrix} r_1 & 1 & 0 & 0 & 0 & 0 \\ 0 & r_1 & 0 & 0 & 0 & 0 \\ 0 & 0 & r_2 & 1 & 0 & 0 \\ 0 & 0 & 0 & r_2 & 1 & 0 \\ 0 & 0 & 0 & 0 & r_2 & 0 \\ 0 & 0 & 0 & 0 & 0 & r_3 \end{bmatrix} \mathbf{P}^{-1},$$

$$e^{\mathbf{A}t} = \mathbf{P} \begin{bmatrix} e^{r_1 t} & te^{r_1 t} & 0 & 0 & 0 & 0 \\ 0 & e^{r_1 t} & 0 & 0 & 0 & 0 \\ 0 & 0 & e^{r_2 t} & te^{r_2 t} & \frac{t^2}{2}e^{r_2 t} & 0 \\ 0 & 0 & 0 & e^{r_2 t} & te^{r_2 t} & 0 \\ 0 & 0 & 0 & 0 & e^{r_2 t} & 0 \\ 0 & 0 & 0 & 0 & 0 & e^{r_3 t} \end{bmatrix} \mathbf{P}^{-1}.$$

(12)

For the reasons we have mentioned, it is virtually impossibly to find Jordan forms without using exact arithmetic on exact data; the Jordan form is numerically unstable and impractical. To satisfy the reader's curiosity; however, we shall describe a pencil-and-paper method for finding a set of generalized eigenvectors which, inserted into the columns of $\mathbf{P}$, produce a Jordan form $\mathbf{P}^{-1}\mathbf{A}\mathbf{P} = \mathbf{J}$. It has the singular merit of being very straightforward; it does not strive for efficiency. We presume that the eigenvalues of the matrix $\mathbf{A}$ are known—*exactly, of course*.

Example 2. Find a Jordan form and $e^{\mathbf{A}t}$ for $\mathbf{A} = \begin{bmatrix} 1 & 1 & 2 \\ 0 & 1 & 3 \\ 0 & 0 & 2 \end{bmatrix}$.

Solution. The eigenvalues of $\mathbf{A}$ are obviously 2 and 1, the latter having multiplicity two. So possible Jordan forms for $\mathbf{A}$ are

$$\begin{bmatrix} 2 & 0 & 0 \\ 0 & 1 & 0 \\ 0 & 0 & 1 \end{bmatrix} \quad \text{and} \quad \begin{bmatrix} 2 & 0 & 0 \\ 0 & 1 & 1 \\ 0 & 0 & 1 \end{bmatrix},$$

(13)

according to whether the multiple eigenvalue 1 has two linearly independent eigenvectors

$$\mathbf{A}\mathbf{u}^{(1)} = (2)\mathbf{u}^{(1)}, \quad \mathbf{A}\mathbf{u}^{(2)} = (1)\mathbf{u}^{(2)}, \quad \mathbf{A}\mathbf{u}^{(3)} = (1)\mathbf{u}^{(3)},$$

(14)

or one eigenvector and one generalized eigenvector

$$\mathbf{A}\mathbf{u}^{(1)} = (2)\mathbf{u}^{(1)}, \quad \mathbf{A}\mathbf{u}^{(2)} = (1)\mathbf{u}^{(2)}, \quad \mathbf{A}\mathbf{u}^{(3)} = (1)\mathbf{u}^{(3)} + \mathbf{u}^{(2)}.$$

(15)

Now here's the trick. Equations (14) and (15) can each be interpreted as single linear systems, *for the unknowns* $\mathbf{u} = [u_1^{(1)}, u_2^{(1)}, u_3^{(1)}, u_1^{(2)}, u_2^{(2)}, u_3^{(2)}, u_1^{(3)}, u_2^{(3)}, u_3^{(3)}]^T$. That is, (14) can be expressed as

$$
\begin{bmatrix} \mathbf{0} \\ \mathbf{0} \\ \mathbf{0} \end{bmatrix} = \begin{bmatrix} A\mathbf{u}^{(1)} - 2\mathbf{u}^{(1)} \\ A\mathbf{u}^{(2)} - \mathbf{u}^{(2)} \\ A\mathbf{u}^{(3)} - \mathbf{u}^{(3)} \end{bmatrix} = \begin{bmatrix} A - 2I & \mathbf{0} & \mathbf{0} \\ \mathbf{0} & A - I & \mathbf{0} \\ \mathbf{0} & \mathbf{0} & A - I \end{bmatrix} \begin{bmatrix} \mathbf{u}^{(1)} \\ \mathbf{u}^{(2)} \\ \mathbf{u}^{(3)} \end{bmatrix}, \qquad (16)
$$

or equivalently, in terms of the components of $\mathbf{u}$,

$$
\begin{bmatrix}
1-2 & 1 & 2 & 0 & 0 & 0 & 0 & 0 & 0 & : & 0 \\
0 & 1-2 & 3 & 0 & 0 & 0 & 0 & 0 & 0 & : & 0 \\
0 & 0 & 2-2 & 0 & 0 & 0 & 0 & 0 & 0 & : & 0 \\
0 & 0 & 0 & 1-1 & 1 & 2 & 0 & 0 & 0 & : & 0 \\
0 & 0 & 0 & 0 & 1-1 & 3 & 0 & 0 & 0 & : & 0 \\
0 & 0 & 0 & 0 & 0 & 2-1 & 0 & 0 & 0 & : & 0 \\
0 & 0 & 0 & 0 & 0 & 0 & 1-1 & 1 & 2 & : & 0 \\
0 & 0 & 0 & 0 & 0 & 0 & 0 & 1-1 & 3 & : & 0 \\
0 & 0 & 0 & 0 & 0 & 0 & 0 & 0 & 2-1 & : & 0
\end{bmatrix}. \qquad (17)
$$

And (15) becomes

$$
\begin{bmatrix} \mathbf{0} \\ \mathbf{0} \\ \mathbf{0} \end{bmatrix} = \begin{bmatrix} A\mathbf{u}^{(1)} - 2\mathbf{u}^{(1)} \\ A\mathbf{u}^{(2)} - \mathbf{u}^{(2)} \\ A\mathbf{u}^{(3)} - \mathbf{u}^{(3)} - \mathbf{u}^{(2)} \end{bmatrix} = \begin{bmatrix} A - 2I & \mathbf{0} & \mathbf{0} \\ \mathbf{0} & A - I & \mathbf{0} \\ \mathbf{0} & -I & A - I \end{bmatrix} \begin{bmatrix} \mathbf{u}^{(1)} \\ \mathbf{u}^{(2)} \\ \mathbf{u}^{(3)} \end{bmatrix}. \qquad (18)
$$

Now according to Theorem 3 the *dimensions* of the Jordan blocks are uniquely determined by A, so only one of the matrices in (13) is similar to A. Therefore, only one of the systems (16), (18) generates solutions that separate into *linearly independent* vectors $\mathbf{u}_1, \mathbf{u}_2, \mathbf{u}_3$, suitable as columns of the invertible matrix P in the equation $P^{-1}AP = J$. Thus our procedure is to

- solve (16) and (18) by Gauss (or Gauss-Jordan) elimination (with exact arithmetic),
- repartition the solutions into column vectors, and
- test for independence.

Only one system will pass the test.

Direct calculation shows that a general solution of (16) is

$$\mathbf{u} = \begin{bmatrix} 5t_3 & 3t_3 & t_3 & t_2 & 0 & 0 & t_1 & 0 & 0 \end{bmatrix}^T \quad \text{or}$$

$$\mathbf{u}_1 = t_3 \begin{bmatrix} 5 \\ 3 \\ 1 \end{bmatrix}, \quad \mathbf{u}_2 = t_2 \begin{bmatrix} 1 \\ 0 \\ 0 \end{bmatrix}, \quad \mathbf{u}_3 = t_1 \begin{bmatrix} 1 \\ 0 \\ 0 \end{bmatrix} \tag{19}$$

and that of (18) is

$$\mathbf{u} = \begin{bmatrix} 5t_3 & 3t_3 & t_3 & t_1 & 0 & 0 & t_2 & t_1 & 0 \end{bmatrix}^T \quad \text{or}$$

$$\mathbf{u}_1 = t_3 \begin{bmatrix} 5 \\ 3 \\ 1 \end{bmatrix}, \quad \mathbf{u}_2 = t_1 \begin{bmatrix} 1 \\ 0 \\ 0 \end{bmatrix}, \quad \mathbf{u}_3 = \begin{bmatrix} t_2 \\ t_1 \\ 0 \end{bmatrix}. \tag{20}$$

The columns in (19) are obviously linearly dependent (compare $\mathbf{u}_2$ and $\mathbf{u}_3$). Alternatively, we can test for independence by computing the determinants of the matrix $\begin{bmatrix} \mathbf{u}_1 & \mathbf{u}_2 & \mathbf{u}_3 \end{bmatrix}$ for each case;

$$\text{system (19) determinant} = \begin{vmatrix} 5t_3 & t_2 & t_1 \\ 3t_3 & 0 & 0 \\ t_3 & 0 & 0 \end{vmatrix} = 0;$$

$$\text{system (20) determinant} = \begin{vmatrix} 5t_3 & t_1 & t_2 \\ 3t_3 & 0 & t_1 \\ t_3 & 0 & 0 \end{vmatrix} = t_3 t_1^2.$$

So the simple choice $t_1 = 1$, $t_2 = 0$, $t_3 = 1$ produces a nonzero determinant, and independent $\mathbf{u}_1, \mathbf{u}_2, \mathbf{u}_3$, for system (18); the correct Jordan form is the *second* matrix in (13). The generalized eigenvectors thus created are

$$\mathbf{u}_1 = \begin{bmatrix} 5 \\ 3 \\ 1 \end{bmatrix}, \quad \mathbf{u}_2 = \begin{bmatrix} 1 \\ 0 \\ 0 \end{bmatrix}, \quad \mathbf{u}_3 = \begin{bmatrix} 0 \\ 1 \\ 0 \end{bmatrix},$$

the similarity transform to Jordan form is

$$\mathbf{A} = \begin{bmatrix} 5 & 1 & 0 \\ 3 & 0 & 1 \\ 1 & 0 & 0 \end{bmatrix} \begin{bmatrix} 2 & 0 & 0 \\ 0 & 1 & 1 \\ 0 & 0 & 1 \end{bmatrix} \begin{bmatrix} 5 & 1 & 0 \\ 3 & 0 & 1 \\ 1 & 0 & 0 \end{bmatrix}^{-1} \quad \text{or}$$

$$\begin{bmatrix} 5 & 1 & 0 \\ 3 & 0 & 1 \\ 1 & 0 & 0 \end{bmatrix}^{-1} \mathbf{A} \begin{bmatrix} 5 & 1 & 0 \\ 3 & 0 & 1 \\ 1 & 0 & 0 \end{bmatrix} = \begin{bmatrix} 2 & 0 & 0 \\ 0 & 1 & 1 \\ 0 & 0 & 1 \end{bmatrix},$$

and the matrix exponential function is

$$e^{\mathbf{A}t} = \begin{bmatrix} 5 & 1 & 0 \\ 3 & 0 & 1 \\ 1 & 0 & 0 \end{bmatrix} \begin{bmatrix} e^{2t} & 0 & 0 \\ 0 & e^t & te^t \\ 0 & 0 & e^t \end{bmatrix} \begin{bmatrix} 5 & 1 & 0 \\ 3 & 0 & 1 \\ 1 & 0 & 0 \end{bmatrix}^{-1} . \qquad \blacksquare$$

Example 3.   Find a Jordan form and $e^{\mathbf{B}t}$ for

$$\mathbf{B} = \begin{bmatrix} 1 & 0 & -1 & -2 \\ -2 & 1 & 3 & -4 \\ 0 & 0 & 1 & 4 \\ 0 & 0 & 0 & -3 \end{bmatrix} .$$

Solution.   We'll just state that the eigenvalues of $\mathbf{B}$ turn out to be 1 (of multiplicity 3) and -3. So the candidates for Jordan forms are

$$\begin{bmatrix} 1 & 0 & 0 & 0 \\ 0 & 1 & 0 & 0 \\ 0 & 0 & 1 & 0 \\ 0 & 0 & 0 & -3 \end{bmatrix}, \quad \begin{bmatrix} 1 & 0 & 0 & 0 \\ 0 & 1 & 1 & 0 \\ 0 & 0 & 1 & 0 \\ 0 & 0 & 0 & -3 \end{bmatrix}, \quad \text{and} \quad \begin{bmatrix} 1 & 1 & 0 & 0 \\ 0 & 1 & 1 & 0 \\ 0 & 0 & 1 & 0 \\ 0 & 0 & 0 & -3 \end{bmatrix}, \qquad (21)$$

corresponding to generalized eigenvector equations given by

(4 eigenvectors) $\mathbf{B}\mathbf{u}^{(1)} = \mathbf{u}^{(1)}$, $\mathbf{B}\mathbf{u}^{(2)} = \mathbf{u}^{(2)}$, $\mathbf{B}\mathbf{u}^{(3)} = \mathbf{u}^{(3)}$, and $\mathbf{B}\mathbf{u}^{(4)} = -3\mathbf{u}^{(4)}$.

(3 eigenvectors) $\mathbf{B}\mathbf{u}^{(1)} = \mathbf{u}^{(1)}$, $\mathbf{B}\mathbf{u}^{(2)} = \mathbf{u}^{(2)}$, $\mathbf{B}\mathbf{u}^{(3)} = \mathbf{u}^{(3)} + \mathbf{u}^{(2)}$, and
$\mathbf{B}\mathbf{u}^{(4)} = -3\mathbf{u}^{(4)}$.

(2 eigenvectors) $\mathbf{B}\mathbf{u}^{(1)} = \mathbf{u}^{(1)}$, $\mathbf{B}\mathbf{u}^{(2)} = \mathbf{u}^{(2)} + \mathbf{u}^{(1)}$, $\mathbf{B}\mathbf{u}^{(3)} = \mathbf{u}^{(3)} + \mathbf{u}^{(2)}$, and
$\mathbf{B}\mathbf{u}^{(4)} = -3\mathbf{u}^{(4)}$.

The linear systems describing these three possibilities are

$$\left[\begin{array}{cccc:c} \mathbf{B}-\mathbf{I} & 0 & 0 & 0 & 0 \\ 0 & \mathbf{B}-\mathbf{I} & 0 & 0 & 0 \\ 0 & 0 & \mathbf{B}-\mathbf{I} & 0 & 0 \\ 0 & 0 & 0 & \mathbf{B}+3\mathbf{I} & 0 \end{array} \right], \quad \left[\begin{array}{cccc:c} \mathbf{B}-\mathbf{I} & 0 & 0 & 0 & 0 \\ 0 & \mathbf{B}-\mathbf{I} & 0 & 0 & 0 \\ 0 & -\mathbf{I} & \mathbf{B}-\mathbf{I} & 0 & 0 \\ 0 & 0 & 0 & \mathbf{B}+3\mathbf{I} & 0 \end{array} \right],$$

$$\left[\begin{array}{cccc:c} \mathbf{B}-\mathbf{I} & 0 & 0 & 0 & 0 \\ -\mathbf{I} & \mathbf{B}-\mathbf{I} & 0 & 0 & 0 \\ 0 & -\mathbf{I} & \mathbf{B}-\mathbf{I} & 0 & 0 \\ 0 & 0 & 0 & \mathbf{B}+3\mathbf{I} & 0 \end{array} \right]. \qquad (22)$$

The solutions of (22) partition into the following sets of vectors:

$$\left[\mathbf{u}^{(1)} \ \mathbf{u}^{(2)} \ \mathbf{u}^{(3)} \ \mathbf{u}^{(4)} \right] = \begin{bmatrix} 0 & 0 & 0 & \frac{1}{4}t_1 \\ t_4 & t_3 & t_2 & \frac{15}{6}t_1 \\ 0 & 0 & 0 & -t_1 \\ 0 & 0 & 0 & t_1 \end{bmatrix}, \quad \begin{bmatrix} 0 & 0 & t_3 & \frac{1}{4}t_1 \\ t_4 & -2t_3 & t_2 & \frac{15}{6}t_1 \\ 0 & 0 & 0 & -t_1 \\ 0 & 0 & 0 & t_1 \end{bmatrix},$$

$$\begin{bmatrix} 0 & -t_2 & t_4 & \frac{1}{4}t_1 \\ 2t_2 & (-2t_4 + 3t_2) & t_3 & \frac{15}{8}t_1 \\ 0 & 0 & t_2 & -t_1 \\ 0 & 0 & 0 & t_1 \end{bmatrix} \tag{23}$$

The determinants are respectively 0, 0, and $2t_1 t_2^3$. The simple choice $t_1 = 1$, $t_2 = 1$, $t_3 = t_4 = 0$ makes the latter nonzero; so the Jordan form is the third matrix in (21), there are two linearly independent eigenvectors, and by assembling the corresponding generalized eigenvectors in (23) into a matrix we derive the similarity transform

$$\mathbf{B} = \begin{bmatrix} 1 & 0 & -1 & -2 \\ -2 & 1 & 3 & -4 \\ 0 & 0 & 1 & 4 \\ 0 & 0 & 0 & -3 \end{bmatrix} = \begin{bmatrix} 0 & -1 & 0 & \frac{1}{4} \\ 2 & 3 & 0 & \frac{15}{8} \\ 0 & 0 & 1 & -1 \\ 0 & 0 & 0 & 1 \end{bmatrix} \begin{bmatrix} 1 & 1 & 0 & 0 \\ 0 & 1 & 1 & 0 \\ 0 & 0 & 1 & 0 \\ 0 & 0 & 0 & -3 \end{bmatrix} \begin{bmatrix} 0 & -1 & 0 & \frac{1}{4} \\ 2 & 3 & 0 & \frac{15}{8} \\ 0 & 0 & 1 & -1 \\ 0 & 0 & 0 & 1 \end{bmatrix}^{-1}$$

and the matrix exponential function

$$e^{\mathbf{B}t} = \begin{bmatrix} 0 & -1 & 0 & \frac{1}{4} \\ 2 & 3 & 0 & \frac{15}{8} \\ 0 & 0 & 1 & -1 \\ 0 & 0 & 0 & 1 \end{bmatrix} \begin{bmatrix} e^t & te^t & t^2 e^t/2 & 0 \\ 0 & e^t & te^t & 0 \\ 0 & 0 & e^t & 0 \\ 0 & 0 & 0 & e^{-3t} \end{bmatrix} \begin{bmatrix} 0 & -1 & 0 & \frac{1}{4} \\ 2 & 3 & 0 & \frac{15}{8} \\ 0 & 0 & 1 & -1 \\ 0 & 0 & 0 & 1 \end{bmatrix}^{-1}. \quad \blacksquare$$

For the final example, we compute the solution of a single fourth-order equation using the Jordan form.

Example 4. Use the matrix exponential function to find all solutions to

$$y^{(iv)} - 3y''' + 3y'' - y' = 0. \tag{24}$$

Solution. The substitutions

$$x_1 := y, \quad x_2 := \frac{dy}{dt}, \quad x_3 := \frac{d^2y}{dt^2}, \quad x_4 := \frac{d^3y}{dt^3}$$

result in the system

$$\frac{dx}{dt} = \frac{d}{dt}\begin{bmatrix} x_1 \\ x_2 \\ x_3 \\ x_4 \end{bmatrix} = \begin{bmatrix} 0 & 1 & 0 & 0 \\ 0 & 0 & 1 & 0 \\ 0 & 0 & 0 & 1 \\ 0 & 1 & -3 & 3 \end{bmatrix} \begin{bmatrix} x_1 \\ x_2 \\ x_3 \\ x_4 \end{bmatrix} = Ax.$$

Matrices J, P expressing the similarity transformation of the coefficient matrix A to Jordan form are found by the method described earlier to be

$$A = \begin{bmatrix} 0 & 1 & 0 & 0 \\ 0 & 0 & 1 & 0 \\ 0 & 0 & 0 & 1 \\ 0 & 1 & -3 & 3 \end{bmatrix} = \begin{bmatrix} 1 & -2 & 3 & -3 \\ 1 & -1 & 1 & 0 \\ 1 & 0 & 0 & 0 \\ 1 & 1 & 0 & 0 \end{bmatrix} \begin{bmatrix} 1 & 1 & 0 & 0 \\ 0 & 1 & 1 & 0 \\ 0 & 0 & 1 & 0 \\ 0 & 0 & 0 & 0 \end{bmatrix} \begin{bmatrix} 1 & -2 & 3 & -3 \\ 1 & -1 & 1 & 0 \\ 1 & 0 & 0 & 0 \\ 1 & 1 & 0 & 0 \end{bmatrix}^{-1}$$
$$= PJP^{-1},$$

resulting in the fundamental matrix

$$e^{At} = Pe^{Jt}P^{-1} = P \begin{bmatrix} e^t & te^t & \frac{t^2}{2}e^t & 0 \\ 0 & e^t & te^t & 0 \\ 0 & 0 & e^t & 0 \\ 0 & 0 & 0 & 1 \end{bmatrix} P^{-1}.$$

The solution to $x' = Ax$ is given by $e^{At}x(0)$. The solution to (24) is its first component, or (after some computations)

$$\begin{aligned} y(t) = {}& y(0) - 3y'(0) + 3y''(0) - y'''(0) \\ & + y'(0)e^t(3 - 2t + t^2/2) \\ & + y''(0)e^t(-4 + 4t - t^2) \\ & + y'''(0)e^t(1 - t + t^2/2). \end{aligned}$$
∎

In Section 7.2, we pointed out that the solutions to a homogeneous linear constant coefficient system are combinations of constant vectors times exponentials $(e^{r_i t})$ for diagonalizable A. Now we see that a nondiagonalizable A spawns solutions that are constants time exponentials times polynomials. The corresponding statement, for a single higher-order linear constant coefficient differential equation, is well known in differential theory.

1. Describe the eigenvectors of a matrix in Jordan form.

2. Suppose $\mathbf{A}$ is similar to the matrix $\begin{bmatrix} r & a \\ 0 & r \end{bmatrix}$. Show that $\mathbf{A}$ is diagonalizable if and only if $a = 0$.

3. By experimentation, derive the formula for the inverse of a 3-by-3 nonsingular Jordan block.

4. Suppose $\mathbf{A}$ is similar to the matrix $\begin{bmatrix} r & a & b \\ 0 & r & c \\ 0 & 0 & r \end{bmatrix}$. Show that

   (a) if $a = b = c = 0$, $\mathbf{A}$ has 3 linearly independent eigenvectors;

   (b) if $ac \neq 0$, $\mathbf{A}$ has 1 linearly independent eigenvector;

   (c) otherwise, $\mathbf{A}$ has 2 linearly independent eigenvectors.

   (d) What are the Jordan forms of $\mathbf{A}$ in each of the above cases?

5. What are the possible Jordan forms for a 4-by-4 matrix with a single eigenvalue r (of multiplicity 4)?

6. What are the possible Jordan forms for a 6-by-6 matrix that has 2 distinct eigenvalues r_1, r_2 each of multiplicity 3?

   Find Jordan forms for the matrices in Problems 7–18.

7. $\begin{bmatrix} 4 & -2 \\ 0 & 4 \end{bmatrix}$ 8. $\begin{bmatrix} -5 & 4 \\ -1 & -1 \end{bmatrix}$ 9. $\begin{bmatrix} 1 & 0 & 0 \\ 1 & 1 & 0 \\ 1 & 1 & 1 \end{bmatrix}$ 10. $\begin{bmatrix} 5 & -1 & 0 \\ -1 & 5 & 1 \\ 0 & -1 & 5 \end{bmatrix}$

11. $\begin{bmatrix} 4 & -1 & -1 \\ 2 & 5 & 2 \\ -1 & 1 & 4 \end{bmatrix}$

12. $\begin{bmatrix} 2 & 1 & 1 \\ -1 & -2 & 0 \\ 2 & -2 & 1 \end{bmatrix}$ 13. $\begin{bmatrix} 3 & 0 & 1 & 0 \\ 1 & -1 & 0 & 1 \\ -4 & -6 & -1 & 3 \\ 2 & -6 & 0 & 4 \end{bmatrix}$ [*Hint*: 2 and 1 are eigenvalues]

14. $\begin{bmatrix} 2 & -2 & -2 & 4 \\ 1 & -1 & -1 & 2 \\ -3 & 3 & 3 & -6 \\ -2 & 2 & 2 & -4 \end{bmatrix}$ [*Hint*: 0 is an eigenvalue]

15. $\begin{bmatrix} -1 & 1 & 0 & 1 \\ -2 & 3 & -1 & 3 \\ -1 & 1 & 0 & 1 \\ 1 & -2 & 1 & -2 \end{bmatrix}$ [*Hint*: 0 is an eigenvalue]

16. $\begin{bmatrix} 0 & -6 & -3 & 7 & 3 & 0 \\ 1 & -2 & -1 & 2 & 1 & 1 \\ 0 & -10 & -5 & 10 & 5 & 1 \\ 0 & -2 & -1 & 2 & 1 & 0 \\ 2 & -6 & -3 & 6 & 3 & 3 \\ 0 & 2 & 1 & -2 & -1 & 0 \end{bmatrix}$ [*Hint*: 0 and -2 are eigenvalues]

17. $\begin{bmatrix} 3 & -1 & 1 & 1 & 0 & 0 \\ 1 & 1 & -1 & -1 & 0 & 0 \\ 0 & 0 & 2 & 0 & 1 & 1 \\ 0 & 0 & 0 & 2 & -1 & -1 \\ 0 & 0 & 0 & 0 & 1 & 1 \\ 0 & 0 & 0 & 0 & 1 & 1 \end{bmatrix}$ **18.** $\begin{bmatrix} -2 & -1 & 1 & -1 & 1 & 2 \\ -7 & -5 & 5 & -10 & 5 & 8 \\ -7 & -4 & 4 & -8 & 4 & 7 \\ -3 & -1 & 1 & -2 & 1 & 3 \\ -6 & -2 & 2 & -4 & 2 & 7 \\ -1 & -2 & 2 & -3 & 2 & 1 \end{bmatrix}$

In Problems 19–25, find a Jordan form and $e^{\mathbf{A}t}$ for the given matrix $\mathbf{A}$.

19. $\mathbf{A} = \begin{bmatrix} 3 & -2 \\ 0 & 3 \end{bmatrix}$ **20.** $\mathbf{A} = \begin{bmatrix} 1 & -1 \\ 1 & 3 \end{bmatrix}$ **21.** $\mathbf{A} = \begin{bmatrix} 1 & 0 & 0 \\ 1 & 3 & 0 \\ 0 & 1 & 1 \end{bmatrix}$

22. $\mathbf{A} = \begin{bmatrix} 0 & 1 & 0 \\ 0 & 0 & 1 \\ 1 & -1 & 1 \end{bmatrix}$ **23.** $\mathbf{A} = \begin{bmatrix} 1 & 0 & 1 & 2 \\ 1 & 1 & 2 & 1 \\ 0 & 0 & 2 & 0 \\ 0 & 0 & 1 & 1 \end{bmatrix}$

24. $\mathbf{A} = \begin{bmatrix} 2 & 1 & 0 & 0 & 0 & 0 \\ 0 & 2 & 0 & 0 & 0 & 0 \\ 0 & 0 & 1 & 1 & 0 & 0 \\ 0 & 0 & 0 & 1 & 1 & 0 \\ 0 & 0 & 0 & 0 & 1 & 1 \\ 0 & 0 & 0 & 0 & 0 & 1 \end{bmatrix}$ **25.** $\mathbf{A} = \begin{bmatrix} 2 & 0 & 0 & 0 & 0 \\ 1 & 3 & 0 & 0 & 0 \\ 0 & 1 & 3 & 0 & 0 \\ 0 & 0 & 1 & 3 & 0 \\ 0 & 0 & 0 & 1 & 3 \end{bmatrix}$

26. In elastic deformation theory, a matrix in the (Jordan) form $\mathbf{J} = \begin{bmatrix} r & 1 \\ 0 & r \end{bmatrix}$ is called a *shear* matrix. To see why, sketch the result when $\mathbf{J}$ is applied to the position vectors $[x\ y]^T$ of all the points in the square $\{0 \le x,\ y \le 1\}$.

27. Solve systems (16) and (18).

28. Use the following steps to derive the formula for the powers of a Jordan block:

$$[\mathbf{J}^p]_{1j} = \binom{p}{j-1} r^{p-j+1} = \frac{p!}{(j-1)!(p-j+1)!} r^{p-j+1} \quad \text{for} \quad 1 \le j \le n,$$

$$[\mathbf{J}^p]_{i1} = 0 \quad \text{for} \quad i > 1, \quad [\mathbf{J}^p]_{ij} = [\mathbf{J}^p]_{i-1,\,j-1} \quad \text{for} \quad n \ge i \ge 2,\ n \ge j \ge 2 \quad (25)$$

(a) First show that all powers of $\mathbf{J}$ are upper triangular.

(b) Next show that all powers of $\mathbf{J}$ are constant along any diagonal. This means you only have to find the formula for the first row.

(c) Referring to Equation (9), compare the computation of the first row of $\mathbf{J}^n$ (via $\mathbf{J} \times \mathbf{J}^{n-1}$) with the computation of the powers $(r+1)^n = (r+1) \times (r+1)^{n-1}$; show that the patterns of calculation are identical.

(iv) Cite the binomial formula to derive the first equation in (25). Cite (b) to derive the second formula.

29. Solve systems (22).

30. Prove that if there is a positive integer k, a scalar r, and a nonzero vector $\mathbf{u}$ such that $(\mathbf{A} - r\mathbf{I})^k\mathbf{u} = \mathbf{0}$, then r must be an eigenvalue of $\mathbf{A}$.

31. If the square matrix $\mathbf{A}$ has zeros everywhere but its diagonal and superdiagonal, and its diagonal entries are all equal, then show that a Jordan form of $\mathbf{A}$ is obtained by replacing every nonzero superdiagonal entry by one.

32. The ones on the superdiagonal of the Jordan form can be replaced by any other nonzero number. More precisely, *every matrix in Jordan form is similar to the matrix obtained by replacing the ones on the superdiagonal by any other nonzero number.* Prove this by observing the identity

$$\begin{bmatrix} \varepsilon^3 & 0 & 0 & 0 \\ 0 & \varepsilon^2 & 0 & 0 \\ 0 & 0 & \varepsilon & 0 \\ 0 & 0 & 0 & 1 \end{bmatrix} \begin{bmatrix} r & 1 & 0 & 0 \\ 0 & r & 1 & 0 \\ 0 & 0 & r & 1 \\ 0 & 0 & 0 & r \end{bmatrix} \begin{bmatrix} \varepsilon^{-3} & 0 & 0 & 0 \\ 0 & \varepsilon^{-2} & 0 & 0 \\ 0 & 0 & \varepsilon^{-1} & 0 \\ 0 & 0 & 0 & 1 \end{bmatrix} = \begin{bmatrix} r & \varepsilon & 0 & 0 \\ 0 & r & \varepsilon & 0 \\ 0 & 0 & r & \varepsilon \\ 0 & 0 & 0 & r \end{bmatrix}$$

and using it to devise a suitable similarity transformation. Explain how this justifies the statement there is a diagonalizable matrix arbitrarily close to any nondiagonalizable matrix. This further illustrates the fragility of the Jordan form.

33. The differential equation $y'' = 0$, whose solution is obviously $y(t) = C_1 + C_2t$, can be interpreted as describing the simplest mass–spring oscillator, with no spring and no damping. (Recall Example 4 of Section 7.1.) Recast it as a first-order system in the manner of that example (Equation (10), Section 7.1). Observe that the coefficient matrix is defective, and show that the Jordan form methodology reproduces this obvious solution.

34. Find the solution to the initial value problem $\mathbf{x}' = \mathbf{A}\mathbf{x}$, $\mathbf{x}(0) = [-1 \ 2]^T$, where $\mathbf{A}$ is the matrix in Problem 19.

35. Find the solution to the initial value problem $\mathbf{x}' = \mathbf{A}\mathbf{x}$, $\mathbf{x}(0) = [1 \ 0 \ 1]^T$, where $\mathbf{A}$ is the matrix in Problem 21.

36. Find the solution to the initial value problem $\mathbf{x}' = \mathbf{A}\mathbf{x}$, $\mathbf{x}(0) = [1 \ 0 \ 0 \ 0]^T$, where $\mathbf{A}$ is the matrix in Problem 23.

7.4 MATRIX EXPONENTIATION VIA GENERALIZED EIGENVECTORS

In this section, we will be showing how the primary decomposition theorem provides an alternative, simpler method for constructing the matrix exponential $e^{\mathbf{A}t}$ (as compared to the construction via the Jordan form).

Recall from Section 7.3 that for any n-by-n square matrix $\mathbf{A}$ with characteristic polynomial

$$p(r) = \det(\mathbf{A} - r\mathbf{I}) = (r_1 - r)^{q_1}(r_2 - r)^{q_2} \cdots (r_K - r)^{q_K} \quad (r_i \text{ distinct}),$$

the Primary Decomposition Theorem guarantees the existence of n linearly independent generalized eignvectors $\{\mathbf{u}_1, \mathbf{u}_2, \ldots, \mathbf{u}_n\}$ of $\mathbf{A}$; the first q_1 of them are solutions of the homogeneous linear system $(\mathbf{A} - r_1\mathbf{I})^{q_1}\mathbf{u} = \mathbf{0}$, the next q_2 of them are solutions of the system $(\mathbf{A} - r_2\mathbf{I})^{q_2}\mathbf{u} = \mathbf{0}$, and so on.

Our scheme for calculating the matrix exponential is based on first calculating the product $e^{\mathbf{A}t}\mathbf{U}$, where $\mathbf{U}$ is the n-by-n matrix with the generalized eigenvectors as columns:

$$e^{\mathbf{A}t}\mathbf{P} = e^{\mathbf{A}t}\begin{bmatrix} \mathbf{u}_1 & \mathbf{u}_2 & \cdots & \mathbf{u}_n \end{bmatrix}. \tag{1}$$

This, we shall demonstrate, is quite straightforward. Then we extract $e^{\mathbf{A}t}$ by multiplication by the inverse:

$$e^{\mathbf{A}t} = [e^{\mathbf{A}t}\mathbf{P}]\mathbf{P}^{-1}. \tag{2}$$

We'll calculate $e^{\mathbf{A}t}\mathbf{P}$ columnwise:

$$e^{\mathbf{A}t}\mathbf{P} = \begin{bmatrix} e^{\mathbf{A}t}\mathbf{u}_1 & e^{\mathbf{A}t}\mathbf{u}_2 & \cdots & e^{\mathbf{A}t}\mathbf{u}_n \end{bmatrix}.$$

To get the first column, recall from Section 7.2 (Theorem 2, part vi) that $e^{(\mathbf{B}+\mathbf{C})t} = e^{\mathbf{B}t}e^{\mathbf{C}t}$ when $\mathbf{B}$ and $\mathbf{C}$ commute. Therefore, we can write $e^{\mathbf{A}t} = e^{r_1\mathbf{I}t}e^{(\mathbf{A}-r_1\mathbf{I})t}$, and

$$\begin{aligned} e^{\mathbf{A}t}\mathbf{u}_1 &= e^{r_1\mathbf{I}t}e^{(\mathbf{A}-r_1\mathbf{I})t}\mathbf{u}_1 \\ &= e^{r_1t}\left[\mathbf{I} + t(\mathbf{A}-r_1\mathbf{I}) + \frac{t^2}{2!}(\mathbf{A}-r_1\mathbf{I})^2 + \cdots + \frac{t^{q_1}}{q_1!}(\mathbf{A}-r_1\mathbf{I})^{q_1} + \cdots\right]\mathbf{u}_1. \end{aligned} \tag{3}$$

But because $\mathbf{u}_1$ is a generalized eigenvector, $(\mathbf{A}-r_1\mathbf{I})^{q_1}\mathbf{u}_1 = \mathbf{0}$ and the series terminates! In fact, the series in (3) terminates at (or before) q_1 terms for every generalized eigenvector in the family $\mathbf{u}_1, \mathbf{u}_2, \ldots, \mathbf{u}_{q_1}$. For the next family, $\mathbf{u}_{q_1+1}, \mathbf{u}_{q_1+2}, \ldots, \mathbf{u}_{q_2}$, the series in

$$\begin{aligned} e^{\mathbf{A}t}\mathbf{u}_i &= e^{r_2\mathbf{I}t}e^{(\mathbf{A}-r_2\mathbf{I})t}\mathbf{u}_i \\ &= e^{r_2t}\left[\mathbf{I} + t(\mathbf{A}-r_2\mathbf{I}) + \frac{t^2}{2!}(\mathbf{A}-r_2\mathbf{I})^2 + \cdots + \frac{t^{q_2}}{q_2!}(\mathbf{A}-r_2\mathbf{I})^{q_2} + \cdots\right]\mathbf{u}_i \end{aligned}$$

terminates after q_2 terms (or sooner). And so on.

So we compute $e^{\mathbf{A}t}\mathbf{P}$ (in finite time) and postmultiply by $\mathbf{P}^{-1}$ to extract $e^{\mathbf{A}t}$.

Example 1. Evaluate $e^{\mathbf{A}t}$ for the matrix $\mathbf{A} = \begin{bmatrix} 1 & 1 & 2 \\ 0 & 1 & 3 \\ 0 & 0 & 2 \end{bmatrix}$

Solution. We solved this problem using the Jordan form as Example 2 in Section 7.3. The eigenvalues of $\mathbf{A}$ are obviously 2 and 1, the latter having multiplicity two, and the characteristic polynomial is $p(r) = (2 - r)(1 - r)^2$. Let's compute the generalized eigenvectors.

$$(\mathbf{A} - 2\mathbf{I})\mathbf{u} = \mathbf{0} \iff \begin{bmatrix} -1 & 1 & 2 & : & 0 \\ 0 & -1 & 3 & : & 0 \\ 0 & 0 & 0 & : & 0 \end{bmatrix} \iff \mathbf{u} = \begin{bmatrix} 5t \\ 3t \\ t \end{bmatrix} = t \begin{bmatrix} 5 \\ 3 \\ 1 \end{bmatrix};$$

$$(\mathbf{A} - 1\mathbf{I})^2\mathbf{u} = \begin{bmatrix} 0 & 1 & 2 \\ 0 & 0 & 3 \\ 0 & 0 & 1 \end{bmatrix}^2 \mathbf{u} = \mathbf{0} \iff \begin{bmatrix} 0 & 0 & 5 & : & 0 \\ 0 & 0 & 3 & : & 0 \\ 0 & 0 & 1 & : & 0 \end{bmatrix}$$

$$\iff \mathbf{u} = \begin{bmatrix} s \\ t \\ 0 \end{bmatrix} = s \begin{bmatrix} 1 \\ 0 \\ 0 \end{bmatrix} + t \begin{bmatrix} 0 \\ 1 \\ 0 \end{bmatrix}.$$

As linearly independent generalized eigenvectors, let's choose

$$\mathbf{u}_1 = \begin{bmatrix} 5 \\ 3 \\ 1 \end{bmatrix}, \quad \mathbf{u}_2 = \begin{bmatrix} 1 \\ 1 \\ 0 \end{bmatrix} \ (s = t = 1), \quad \mathbf{u}_3 = \begin{bmatrix} 1 \\ -1 \\ 0 \end{bmatrix} \ (s = 1, \ t = -1).$$

Note that we do not have to choose vectors that form "Jordan chains" (Equations (2) and (3) of Section 7.3); that is the advantage of this method. We have

$$e^{\mathbf{A}t}\mathbf{u}_1 = e^{2t} \left[\mathbf{I}\mathbf{u}_1 + t(\mathbf{A} - 2\mathbf{I})\mathbf{u}_1 \right] = \begin{bmatrix} 5e^{2t} \\ 3e^{2t} \\ e^{2t} \end{bmatrix},$$

$$e^{\mathbf{A}t}\mathbf{u}_2 = e^{t} \left[\mathbf{I}\mathbf{u}_2 + t(\mathbf{A} - \mathbf{I})\mathbf{u}_2 + \frac{t^2}{2!}(\mathbf{A} - \mathbf{I})^2\mathbf{u}_2 \right] = e^{t} \begin{bmatrix} 1 \\ 1 \\ 0 \end{bmatrix} + te^{t} \begin{bmatrix} 0 & 1 & 2 \\ 0 & 0 & 3 \\ 0 & 0 & 1 \end{bmatrix} \begin{bmatrix} 1 \\ 1 \\ 0 \end{bmatrix}$$

$$= \begin{bmatrix} e^{t} + te^{t} \\ e^{t} \\ 0 \end{bmatrix},$$

$$e^{\mathbf{A}t}\mathbf{u}_3 = e^{t} \left[\mathbf{I}\mathbf{u}_3 + t(\mathbf{A} - \mathbf{I})\mathbf{u}_3 + \frac{t^2}{2!}(\mathbf{A} - \mathbf{I})^2\mathbf{u}_3 \right] = e^{t} \begin{bmatrix} 1 \\ -1 \\ 0 \end{bmatrix} + te^{t} \begin{bmatrix} 0 & 1 & 2 \\ 0 & 0 & 3 \\ 0 & 0 & 1 \end{bmatrix} \begin{bmatrix} 1 \\ -1 \\ 0 \end{bmatrix}$$

$$= \begin{bmatrix} e^{t} - te^{t} \\ -e^{t} \\ 0 \end{bmatrix}.$$

Therefore,

$$e^{\mathbf{A}t} = [e^{\mathbf{A}t}\mathbf{P}]\mathbf{P}^{-1} = \begin{bmatrix} e^{\mathbf{A}t}\mathbf{u}_1 & e^{\mathbf{A}t}\mathbf{u}_2 & e^{\mathbf{A}t}\mathbf{u}_3 \end{bmatrix} \begin{bmatrix} \mathbf{u}_1 & \mathbf{u}_2 & \mathbf{u}_3 \end{bmatrix}^{-1}$$

$$= \begin{bmatrix} 5e^{2t} & e^t + te^t & e^t - te^t \\ 3e^{2t} & e^t & e^t \\ e^{2t} & 0 & 0 \end{bmatrix} \begin{bmatrix} 5 & 1 & 1 \\ 3 & 1 & -1 \\ 1 & 0 & 0 \end{bmatrix}^{-1}, \tag{4}$$

and a little arithmetic will confirm that (4) agrees with the answer obtained in Example 2, Section 7.3. ∎

Example 2. Solve the system $\mathbf{x}' = \mathbf{A}\mathbf{x} + \mathbf{f}(t)$, with

$$\mathbf{A} = \begin{bmatrix} 12 & 9 \\ -16 & -12 \end{bmatrix}, \quad \mathbf{f}(t) = \begin{bmatrix} t \\ 0 \end{bmatrix}, \quad \mathbf{x}(0) = \begin{bmatrix} 1 \\ 1 \end{bmatrix}. \tag{5}$$

Solution. The characteristic polynomial of $\mathbf{A}$ is $p(r) = (12-r)(-12-r) - 9(-16) = r^2$, so $\mathbf{A}$ has $r_1 = 0$ as a double eigenvalue. It may be defective, but we only need its generalized eigenvectors. Accordingly, we address the homogeneous system

$$(\mathbf{A} - r_1\mathbf{I})^{q_1}\mathbf{u} = (\mathbf{A} - 0\mathbf{I})^2\mathbf{u} = \mathbf{0} \Rightarrow \begin{bmatrix} 12 & 9 \\ -16 & -12 \end{bmatrix}^2 \mathbf{u} = \begin{bmatrix} 0 & 0 \\ 0 & 0 \end{bmatrix} \mathbf{u} = \mathbf{0},$$

which has the obvious linearly independent solutions $\mathbf{u}_1 = [1\ 0]^T$, $\mathbf{u}_2 = [0\ 1]^T$.

Applying (3) is particularly easy in this case:

$$e^{\mathbf{A}t}\mathbf{u}_1 = e^{0\mathbf{I}t}e^{(\mathbf{A}-0\mathbf{I})t}\mathbf{u}_1$$

$$= \left[\mathbf{I} + t(\mathbf{A} - 0\mathbf{I}) + \frac{t^2}{2!}(\mathbf{A} - 0\mathbf{I})^2 \right]\begin{bmatrix} 1 \\ 0 \end{bmatrix} = \begin{bmatrix} 1 + 12t & 9t \\ -16t & 1 - 12t \end{bmatrix}\begin{bmatrix} 1 \\ 0 \end{bmatrix} = \begin{bmatrix} 1 + 12t \\ -16t \end{bmatrix}.$$

Similarly,

$$e^{\mathbf{A}t}\mathbf{u}_2 = \begin{bmatrix} 1 + 12t & 9t \\ -16t & 1 - 12t \end{bmatrix}\begin{bmatrix} 0 \\ 1 \end{bmatrix} = \begin{bmatrix} 9t \\ 1 - 12t \end{bmatrix}.$$

Therefore, since $[\mathbf{u}_1\ \mathbf{u}_2] = \mathbf{I}$,

$$e^{\mathbf{A}t} = (e^{\mathbf{A}t}[\mathbf{u}_1\ \mathbf{u}_2])[\mathbf{u}_1\ \mathbf{u}_2]^{-1} = \begin{bmatrix} 1 + 12t & 9t \\ -16t & 1 - 12t \end{bmatrix}. \tag{6}$$

The solution formula from Theorem 1, Section 7.2 gives us

$$\mathbf{x}(t) = e^{\mathbf{A}t}\mathbf{x}(0) + e^{\mathbf{A}t} \int_{s=0}^{t} e^{-\mathbf{A}s}\mathbf{f}(s)\, ds$$

$$= \begin{bmatrix} 1 + 12t & 9t \\ -16t & 1 - 12t \end{bmatrix} \begin{bmatrix} 1 \\ 1 \end{bmatrix}$$

$$+ \begin{bmatrix} 1 + 12t & 9t \\ -16t & 1 - 12t \end{bmatrix} \int_{s=0}^{t} \begin{bmatrix} 1 + 12(-s) & 9(-s) \\ -16(-s) & 1 - 12(-s) \end{bmatrix} \begin{bmatrix} s \\ 0 \end{bmatrix} ds$$

$$= \begin{bmatrix} 1 + 21t + t^2/2 + 2t^3 \\ 1 - 28t - 8t^3/3 \end{bmatrix}. \qquad \blacksquare$$

We close with a note on yet another way to calculate the matrix exponential; it is addressed to readers who have experience with differential equations.

Matrix Exponentiation via Fundamental Solutions

Suppose that, by whatever means, you have compiled n vector functions $\{\mathbf{x}_1(t), \mathbf{x}_2(t), \ldots, \mathbf{x}_n(t)\}$ that are solutions of the differential equation $\mathbf{x}' = \mathbf{A}\mathbf{x}$, for an n-by-n matrix $\mathbf{A}$. Suppose also that at $t = 0$ the vectors $\{\mathbf{x}_1(0), \mathbf{x}_2(0), \ldots, \mathbf{x}_n(0)\}$ are linearly independent. Then the assembly $\mathbf{X}(t) = [\mathbf{x}_1(t)\, \mathbf{x}_2(t) \ldots \mathbf{x}_n(t)]$ is known as a *fundamental matrix* for the equation.

Theorem 1 from Section 7.2 states that $\mathbf{x}_i(t) = e^{\mathbf{A}t}\mathbf{x}_i(0)$ for each solution, so columnwise we have $\mathbf{X}(t) = e^{\mathbf{A}t}\mathbf{X}(0)$. By forming the inverse, we derive the formula

$$e^{\mathbf{A}t} = \mathbf{X}(t)\mathbf{X}(0)^{-1}. \qquad (7)$$

So any differential equation solver is a tool for matrix exponentiation. (Of course, this is a moot procedure if the objective for calculating the exponential was to solve the differential equation.)

Exercises 7.4

Use generalized eigenvectors to find $e^{\mathbf{A}t}$ for the matrices in Problems 1–12. (They replicate Problems 7–18 in Section 7.3, where the Jordan form was prescribed for the solution.)

1. $\begin{bmatrix} 4 & -2 \\ 0 & 4 \end{bmatrix}$ **2.** $\begin{bmatrix} -5 & 4 \\ -1 & -1 \end{bmatrix}$ **3.** $\begin{bmatrix} 1 & 0 & 0 \\ 1 & 1 & 0 \\ 1 & 1 & 1 \end{bmatrix}$ **4.** $\begin{bmatrix} 5 & -1 & 0 \\ -1 & 5 & 1 \\ 0 & -1 & 5 \end{bmatrix}$

5. $\begin{bmatrix} 4 & -1 & -1 \\ 2 & 5 & 2 \\ -1 & 1 & 4 \end{bmatrix}$

6. $\begin{bmatrix} 2 & 1 & 1 \\ -1 & -2 & 0 \\ 2 & -2 & 1 \end{bmatrix}$ **7.** $\begin{bmatrix} 3 & 0 & 1 & 0 \\ 1 & -1 & 0 & 1 \\ -4 & -6 & -1 & 3 \\ 2 & -6 & 0 & 4 \end{bmatrix}$ [*Hint*: 2 and 1 are eigenvalues]

8. $\begin{bmatrix} 2 & -2 & -2 & 4 \\ 1 & -1 & -1 & 2 \\ -3 & 3 & 3 & -6 \\ -2 & 2 & 2 & -4 \end{bmatrix}$ [*Hint*: 0 is an eigenvalue]

9. $\begin{bmatrix} -1 & 1 & 0 & 1 \\ -2 & 3 & -1 & 3 \\ -1 & 1 & 0 & 1 \\ 1 & -2 & 1 & -2 \end{bmatrix}$ [*Hint*: 0 is an eigenvalue]

10. $\begin{bmatrix} 0 & -6 & -3 & 7 & 3 & 0 \\ 1 & -2 & -1 & 2 & 1 & 1 \\ 0 & -10 & -5 & 10 & 5 & 1 \\ 0 & -2 & -1 & 2 & 1 & 0 \\ 2 & -6 & -3 & 6 & 3 & 3 \\ 0 & 2 & 1 & -2 & -1 & 0 \end{bmatrix}$ [*Hint*: 0 and −2 are eigenvalues]

11. $\begin{bmatrix} 3 & -1 & 1 & 1 & 0 & 0 \\ 1 & 1 & -1 & -1 & 0 & 0 \\ 0 & 0 & 2 & 0 & 1 & 1 \\ 0 & 0 & 0 & 2 & -1 & -1 \\ 0 & 0 & 0 & 0 & 1 & 1 \\ 0 & 0 & 0 & 0 & 1 & 1 \end{bmatrix}$ **12.** $\begin{bmatrix} -2 & -1 & 1 & -1 & 1 & 2 \\ -7 & -5 & 5 & -10 & 5 & 8 \\ -7 & -4 & 4 & -8 & 4 & 7 \\ -3 & -1 & 1 & -2 & 1 & 3 \\ -6 & -2 & 2 & -4 & 2 & 7 \\ -1 & -2 & 2 & -3 & 2 & 1 \end{bmatrix}$

13. The simplicity of the calculations in Example 2 occurs for any matrix **A** satisfying $\mathbf{A}^p = \mathbf{0}$ for some power p. Such a matrix is called *nilpotent*. Explain why a nilpotent matrix has only 0 as an eigenvalue. Explain why the columns of the identity matrix are generalized eigenvectors for **A**.

14. Show that the matrix in Example 2 is, indeed, defective; find its Jordan form and exponential, and verify that its exponential is given by (6).

15. Solve Problem 34, Exercises 7.3, using the method of generalized eigenvectors.

16. Solve Problem 35, Exercises 7.3, using the method of generalized eigenvectors.

17. Solve Problem 36, Exercises 7.3, using the method of generalized eigenvectors.

18. Prove that every fundamental matrix for $\mathbf{x}' = \mathbf{Ax}$, where $\mathbf{A}$ is n-by-n, must be of the form $e^{\mathbf{A}t}\mathbf{C}$, where $\mathbf{C}$ is an n-by-n constant invertible matrix.

7.5 SUMMARY

First-Order Linear Systems of Differential Equations

The formula $dx/dt = \mathbf{Ax} + \mathbf{f}(t)$ depicts a **linear system of nonhomogeneous differential equations of first order**; the matrix $\mathbf{x}(t)$ is unknown, and the matrices $\mathbf{A}$ and $\mathbf{f}(t)$ are known. The formula is quite general, because single or coupled higher-order linear differential equations can be expressed in this format with suitable renaming of variables.

We have restricted our attention to cases where the coefficient matrix $\mathbf{A}$ is constant; this assumption allows us to express the solution using the matrix exponential:

$$\mathbf{x}(t) = e^{\mathbf{A}(t-t_1)}\mathbf{x}(t_1) + e^{\mathbf{A}t} \int_{t_1}^{t} \left[e^{-\mathbf{A}s}\mathbf{f}(s) \right] ds \tag{1}$$

The success in our utilizing formula (1) rests on our ability to calculate the matrix exponential $e^{\mathbf{A}t}$.

The Matrix Exponential

The matrix exponential function $e^{\mathbf{A}t}$ satisfies most of the properties of the scalar exponential e^{at}, with the exception of the multiplicative identity $e^{\mathbf{A}t}e^{\mathbf{B}t} = e^{(\mathbf{A}+\mathbf{B})t}$, which is assured only when $\mathbf{AB} = \mathbf{BA}$. It can be computed using Taylor series, but a far more reliable procedure is to exploit the diagonal form (when it is available):

$$\mathbf{A} = \mathbf{P} \begin{bmatrix} r_1 & 0 & \cdots & 0 \\ 0 & r_2 & \cdots & 0 \\ \vdots & & \ddots & \vdots \\ 0 & 0 & \cdots & r_p \end{bmatrix} \mathbf{P}^{-1} \Rightarrow e^{\mathbf{A}} = \mathbf{P} \begin{bmatrix} e^{r_1} & 0 & \cdots & 0 \\ 0 & e^{r_2} & \cdots & 0 \\ \vdots & & \ddots & \vdots \\ 0 & 0 & \cdots & e^{r_p} \end{bmatrix} \mathbf{P}^{-1}. \tag{2}$$

The Jordan Normal Form

A defective matrix, by definition, cannot be diagonalized; however, *any* matrix is similar $(A = PJP^{-1})$ to a matrix J in Jordan form:

$$\begin{bmatrix} r_1 & 1 & 0 & 0 & 0 & 0 \\ 0 & r_1 & 0 & 0 & 0 & 0 \\ 0 & 0 & r_2 & 1 & 0 & 0 \\ 0 & 0 & 0 & r_2 & 1 & 0 \\ 0 & 0 & 0 & 0 & r_2 & 0 \\ 0 & 0 & 0 & 0 & 0 & r_3 \end{bmatrix} = \begin{bmatrix} \overline{|J_{2\times 2}|} & \overline{|0|} & \overline{|0|} \\ \overline{|0|} & \overline{|J_{3\times 3}|} & \overline{|0|} \\ \overline{|0|} & \overline{|0|} & \overline{|J_{1\times 1}|} \end{bmatrix}. \tag{3}$$

This renders the expression for the matrix exponential as

$$e^{At} = P \begin{bmatrix} e^{r_1 t} & te^{r_1 t} & 0 & 0 & 0 & 0 \\ 0 & e^{r_1 t} & 0 & 0 & 0 & 0 \\ 0 & 0 & e^{r_2 t} & te^{r_2 t} & \frac{t^2}{2}e^{r_2 t} & 0 \\ 0 & 0 & 0 & e^{r_2 t} & te^{r_2 t} & 0 \\ 0 & 0 & 0 & 0 & e^{r_2 t} & 0 \\ 0 & 0 & 0 & 0 & 0 & e^{r_3 t} \end{bmatrix} P^{-1}. \tag{4}$$

From the forms (2), (4) of the matrix exponential, we see that the solutions to *homogeneous* linear constant-coefficient systems are sums of constants times exponentials for diagonalizable A, and sums of constants times exponentials times polynomials for defective A (interpret (1) when $f = 0$).

The dimensions of the Jordan blocks in formula (3) are uniquely determined by A. This enables a Gauss-elimination-based scheme for deducing the Jordan form, given the eigenvalues of the matrix.

However, the Jordan form is numerically fragile; it is premised on the equality of eigenvalues, and as such must be carried out in exact arithmetic.

Matrix Exponentiation via Generalized Eigenvectors

If u is a generalized eigenvector of A corresponding to the eigenvalue r, $(A - rI)^q u = 0$, then the expression

$$e^{At}u = e^{rIt}e^{(A-rI)t}u$$

$$= e^{rt}\left[I + t(A - rI) + \frac{t^2}{2!}(A - rI)^2 + \cdots + \frac{t^q}{q!}(A - rI)^q + \cdots \right]u$$

terminates. Since the primary decomposition theorem guarantees the existence of n linearly independent generalized eigenvectors $\{u_1, u_2, \ldots, u_n\}$ for any n-by-n matrix,

$$e^{At}P = \begin{bmatrix} e^{At}u_1 & e^{At}u_2 & \cdots & e^{At}u_n \end{bmatrix}$$

can be computed in finite time, and postmultiplication by P^{-1} yields e^{At}.

REVIEW PROBLEMS FOR PART III

1. Describe all matrices that are similar to the zero matrix, and all that are similar to the identity matrix.

2. Show that $r\mathbf{I}$ is the only n-by-n diagonalizable upper triangular matrix with all eigenvalues equal. [*Hint*: Unless all the coefficients in $\mathbf{A} - r\mathbf{I}$ are zeros, back substitution in $(\mathbf{A} - r\mathbf{I})\mathbf{u} = \mathbf{0}$ will yield fewer than n linearly independent solutions.]

3. Each of the basic operations of Gauss elimination, listed in Theorem 1 of Section 1.1, can be represented as left multiplication by an appropriate matrix. What are the eigenvectors of these matrices? Which of them are defective?

4. Prove or disprove: switching 2 rows of a matrix preserves its eigenvalues.

Find the eigenvectors and eigenvalues of the matrices in Problems 5–8.

5. $\begin{bmatrix} -3 & 2 \\ -1 & -6 \end{bmatrix}$ 6. $\begin{bmatrix} 1 & -1 \\ 2 & 1 \end{bmatrix}$ 7. $\begin{bmatrix} 1 & -2 & 1 \\ -1 & 2 & -1 \\ 0 & 1 & 1 \end{bmatrix}$ 8. $\begin{bmatrix} 0 & 1 & -1 \\ 0 & 0 & 3 \\ 1 & 0 & 3 \end{bmatrix}$

9. What are the eigenvalues of a 2-by-2 matrix with trace $= 4$ and determinant $= 6$?

10. Deflate the following matrix using the technique described in Problem 19 of Exercises 5.1, replacing its highest eigenvalue by zero: $\begin{bmatrix} 1 & 2 & -1 \\ 1 & 0 & 1 \\ 4 & -4 & 5 \end{bmatrix}$.

11. Find four real, orthogonal eigenvectors for the symmetric matrix $\begin{bmatrix} 0 & 1 & 0 & 1 \\ 1 & 0 & 1 & 0 \\ 0 & 1 & 0 & 1 \\ 1 & 0 & 1 & 0 \end{bmatrix}$.

12. Find three orthogonal eigenvectors for the Hermitian matrix $\begin{bmatrix} 2 & i & 0 \\ -i & 2 & 0 \\ 0 & 0 & 1 \end{bmatrix}$.

Find similarity transformations that diagonalize the matrices in Problems 13–19. (See Problems 5–8.)

13. $\begin{bmatrix} -3 & 2 \\ -1 & -6 \end{bmatrix}$ 14. $\begin{bmatrix} 1 & -1 \\ 2 & 1 \end{bmatrix}$ 15. $\begin{bmatrix} 1 & -2 & 1 \\ -1 & 2 & -1 \\ 0 & 1 & 1 \end{bmatrix}$ 16. $\begin{bmatrix} 0 & 1 & -1 \\ 0 & 0 & 3 \\ 1 & 0 & 3 \end{bmatrix}$

17. $\begin{bmatrix} 1 & 2 & -1 \\ 1 & 0 & 1 \\ 4 & -4 & 5 \end{bmatrix}$ 18. $\begin{bmatrix} 0 & 1 & 0 & 1 \\ 1 & 0 & 1 & 0 \\ 0 & 1 & 0 & 1 \\ 1 & 0 & 1 & 0 \end{bmatrix}$ 19. $\begin{bmatrix} 2 & i & 0 \\ -i & 2 & 0 \\ 0 & 0 & 1 \end{bmatrix}$

Find a change of coordinates that eliminates the cross terms in each of the quadratic forms appearing in Problems 20–25

20. $7x_1^2 - 12x_1x_2 - 2x_2^2$ 21. $4x_1^2 + 2x_1x_2 + 4x_2^2$ 22. x_1x_2

23. $2x_1^2 + 2x_1x_2 + 2x_2^2 + 2x_1x_3 - 2x_2x_3 + 2x_3^2$

24. $2x_1^2 + 2x_2^2 - 2x_1x_3 - 2x_2x_3 + 3x_3^2$ **25.** $x_1x_2 + x_1x_3 + x_2x_3$

26. Express the Schur decomposition $\mathbf{A} = \mathbf{Q}\mathbf{U}\mathbf{Q}^H$ for the matrix $\begin{bmatrix} 2 & 2 & -1 \\ -6 & -5 & 6 \\ -2 & -1 & 3 \end{bmatrix}$.

27. For what value of a is the matrix $\begin{bmatrix} 0 & -1 & 2 \\ a & 0 & 2 \\ -2 & -2 & 0 \end{bmatrix}$ normal? Find an orthogonal basis of eigenvectors for the resulting matrix.

TECHNICAL WRITING EXERCISES FOR PART III

1. The matrix $\mathbf{A} = \begin{bmatrix} 1 & 0 \\ 1 & 1 \end{bmatrix}$, as we saw in Example 5 of Section 5.2, has only one independent eigenvector for its double eigenvalue $r = 1$. Graph pairs of independent eigenvectors for the matrices $\begin{bmatrix} 1 & 0 \\ 1 & 0.9 \end{bmatrix}$, $\begin{bmatrix} 1 & 0 \\ 1 & 0.99 \end{bmatrix}$, $\begin{bmatrix} 1 & 0 \\ 1 & 0.999 \end{bmatrix}$ approximating $\mathbf{A}$. Do the same for the matrices $\begin{bmatrix} 1 & 0 \\ 0 & 0.9 \end{bmatrix}$, $\begin{bmatrix} 1 & 0 \\ 0 & 0.99 \end{bmatrix}$, $\begin{bmatrix} 1 & 0 \\ 0 & 0.999 \end{bmatrix}$ approximating the identity matrix $\mathbf{I} = \begin{bmatrix} 1 & 0 \\ 0 & 1 \end{bmatrix}$. Write a short paper describing how one of the eigenvectors "disappears" for the matrix $\mathbf{A}$, but not for $\mathbf{I}$.

2. Suppose that the "second" hand on a stopwatch represents the vector $\mathbf{v}$, and that another hand points in the direction of $\mathbf{A}\mathbf{v}$, for some 2-by-2 matrix $\mathbf{A}$. Write a paragraph describing the motions of these hands during the course of one minute, describing in particular what happens as $\mathbf{v}$ approaches an eigenvector of $\mathbf{A}$. Consider the possibilities that $\mathbf{A}$ is singular, defective, symmetric, or antisymmetric. (The MATLAB$^{\circledR}$ function "eigshow.m" demonstrates this for a variety of matrices $\mathbf{A}$.)

GROUP PROJECTS FOR PART III

A.   Positive Definite Matrices

A **positive definite matrix** is any real n-by-n symmetric matrix $\mathbf{A}$ with the property that

$$\mathbf{v}^T\mathbf{A}\mathbf{v} > 0, \text{ for every vector } \mathbf{v} \text{ in } \mathbf{R}_{\text{col}}^n \text{ other than zero.} \tag{1}$$

If every such $\mathbf{v}^T\mathbf{A}\mathbf{v}$ is nonnegative, $\mathbf{A}$ is said to be **positive semidefinite**; and $\mathbf{A}$ is **negative definite** if $-\mathbf{A}$ is positive definite. The concept of positive definiteness extends readily to Hermitian matrices, via the criterion $\mathbf{v}^H\mathbf{A}\mathbf{v} > 0$.

Which of the following properties are necessary and/or sufficient for $\mathbf{A}$ to be positive definite?

(a) The eigenvalues of $\mathbf{A}$ are positive. [*Hint*: Apply (1) to the equation $\mathbf{Av} = \lambda\mathbf{v}$, and to the similarity transform equation $\mathbf{A} = \mathbf{QDQ}^T$ in Corollary 3, Section 6.1.]

(b) The determinant of $\mathbf{A}$ is positive. [*Hint*: How is the determinant related to the eigenvalues?]

(c) All the entries of $\mathbf{A}$ are positive (nonnegative?). [*Hint*: Experiment.]

(d) All of the square submatrices centered on the diagonal of $\mathbf{A}$ are positive definite. [*Hint*: Apply (1) with vectors having appropriate zeros, like $\mathbf{v} = [0\ 0\ x_3\ x_4\ x_5\ 0\ 0]$.]

(e) $\mathbf{A}$ is a *Gram matrix*: that is, the (i, j)th entry of $\mathbf{A}$ is $\mathbf{u}_i^T\mathbf{u}_j$, where $\{\mathbf{u}_1, \mathbf{u}_2, \ldots, \mathbf{u}_n\}$ is any basis for $\mathbf{R}^n_{col}$. (*Hint*: Convince yourself that $\mathbf{A} = \mathbf{P}^T\mathbf{P}$, where the columns of $\mathbf{P}$ are the basis vectors; then apply (1). Why must the columns be linearly independent?)

(f) All of the square submatrices of $\mathbf{A}$ "anchored at the upper left corner" have positive determinants. [*Hint*: This is known as **Sylvester's Criterion**, and the determinants are called **principal minors**. You may find this problem difficult; get help from the Internet.]

(g) (Recall Project A, Part I) $\mathbf{A}$ admits an LU factorization where $\mathbf{U} = \mathbf{L}^T$ (that is, $\mathbf{A} = \mathbf{LL}^T$), with $\mathbf{L}$ real and nonsingular. [*Hint*: This is called the **Cholesky factorization**. It roughly halves the computer time and memory to solve systems $\mathbf{Ax} = \mathbf{b}$. Use the Internet if necessary.]

The most common application of positive definiteness occurs in optimization problems, when the "cost" function to be optimized can be expressed (or approximated) by a quadratic polynomial $f(\mathbf{x}) = \mathbf{x}^T\mathbf{Ax} + \mathbf{b}^T\mathbf{x} + k$. Use Lemma 1, Section 6.2, to show that if $\mathbf{A}$ is positive definite, $f(\mathbf{x})$ has a unique minimum point at $\mathbf{x} = -\mathbf{A}^{-1}\mathbf{b}/2$.

B.   Hessenberg Form

As mentioned at the beginning of Section 6.3, although Gauss elimination reduces a square matrix $\mathbf{A}$ to upper triangular form, the resulting matrix is not similar to $\mathbf{A}$ (so the eigenvalues are altered in the process). Now recall (Section 2.1, Example 2) that each operation in Gauss elimination is equivalent to left-multiplication by a matrix; so if we *right*-multiplied by the *inverse* of that matrix, we would complete a similarity transformation (and preserve the eigenvalues).

(a) Show that if left multiplication of the matrix $\mathbf{A}$ by the matrix $\mathbf{E}$ eliminates the $(2,1)$ entry of $\mathbf{A}$ (by adding an appropriate multiple of the first row to the second), right multiplication by $\mathbf{E}^{-1}$ adds a multiple of the second column to the first, and thus "undoes" the elimination.

(a)   (b)

$$
\begin{bmatrix} a & a & a & a & a \\ a & a & a & a & a \\ 0 & a & a & a & a \\ 0 & 0 & a & a & a \\ 0 & 0 & 0 & a & a \end{bmatrix}
\begin{bmatrix} x & x & x & x & x \\ x & x & x & x & x \\ x & x & x & x & x \\ x & x & x & x & x \\ x & x & x & x & x \end{bmatrix}
\rightarrow
\begin{bmatrix} x & x & x & x & x \\ x & x & x & x & x \\ 0 & y & y & y & y \\ x & x & x & x & x \\ x & x & x & x & x \end{bmatrix}
\rightarrow
\begin{bmatrix} x & z & x & x & x \\ x & z & x & x & x \\ 0 & z & y & y & y \\ x & z & x & x & x \\ x & z & x & x & x \end{bmatrix}
$$

Fig. III.B.1   (a) Hessenberg form and (b) Steps in reduction to Hessenberg form.

(b) Explain why *no* (finite-time) scheme using simple Gauss elimination could serve as an eigenvalue calculator. [*Hint*: Argue that otherwise, we could attain a violation of Galois's unsolvability theorem (Section 6.5) by applying the scheme to the companion matrix (Exercises 6.1, Problem 24) of the polynomial in question.]

Nonetheless, we can use Gauss elimination to find a matrix similar to **A** in *Hessenberg form*, which is "nearly triangular" in that all entries below the first *sub*diagonal are zero. See Figure III.B.1a.

(c) Explain how the strategy depicted in Figure III.B.1b, which eliminates the (3,1) entry (below the *sub*diagonal) by adding an appropriate multiple of the *second* row to the *third*, then completes the similarity transformation by adding a multiple of the third column to the second, can be extended to reduce **A** to Hessenberg form while preserving its eigenvalues.

(d) Show that if **A** is in Hessenberg form and is factored into a unitary matrix **Q** times an upper triangular matrix **R** (as in Exercises 6.5 Problem 18), then **Q**, as well as **RQ**, will also be in Hessenberg form. This means that all the matrices occurring in the **QR** iterations for the estimation of the eigenvalues of **A**, described in Section 6.4, will be in Hessenberg form.

Because of the savings in memory and calculating time, a preliminary reduction to Hessenberg form is a crucial first step in all respectable scientific software featuring the QR algorithm.

C. Discrete Fourier Transform

The response of many engineering systems, particularly electrical circuits and vibrating structures, to an input function $f(t)$ is often easier to characterize when the function is a simple sinusoid,

$$
e^{i\omega t} = \cos \omega t + i \sin \omega t
$$

(complex, but simple!). So an important tool in engineering analysis is the representation of a measured function as a linear combination of sinusoids:

$$
f(t) = \sum_n c_n e^{i\omega_n t}.
$$

The evaluation of the coefficients c_n is particularly convenient when the following conditions are met:

- the function $f(t)$ has been measured ("sampled") at N discrete times $t_0, t_1, \ldots, t_{N-1}$;
- the sampling times are uniform, $t_j = j\Delta t : j = 0, 1, \ldots, N-1$;
- the angular frequencies ω_n are integer multiples ("harmonics") of the fundamental frequency $\omega_a := 2\pi/(N\Delta t)$, $\omega_n = n\omega_a$, $n = 0, 1, \ldots, N-1$.

In such a case, the set of coefficients c_n is called the *discrete Fourier transform* of the data $f(t_j)$.

(a) Formulate the conditions

$$f(t_j) = \sum_n c_n e^{i\omega_n t_j}$$

as a matrix equation for the unknown numbers c_n.

(b) Identify the rows of the coefficient matrix for the system in (a) as rescaled (and transposed) eigenvectors of the N-by-N circulant matrices described in Problem 12 of Exercises 5.1.

(c) Use the geometric sum formula

$$\sum_{j=0}^{m} r^j = (1 - r^{m+1})/(1 - r)$$

to prove that the rows of the coefficient matrix are orthogonal. What are their norms?

(d) Construct $\mathbf{F}$, the Hermitian circulant matrix whose first row comprises the sampled values $f(t_j)$. Show that the eigenvalues of $\mathbf{F}$ are the elements c_n of the discrete Fourier transform of the data $f(t_j)$. Use this observation to construct another proof of the orthogonality of the rows of the coefficient matrix.

(e) Using the orthogonality, construct the inverse of the coefficient matrix, and thereby formulate an explicit formula for the discrete Fourier transform of the sampled values $f(t_j)$.

Note that the resulting formula involves taking inner products of the sampled values of $f(t)$ with sampled values of $e^{i\omega_n t}$. By taking advantage of the redundancies exhibited in the following table, one can considerably accelerate the speed of the calculation; this is the essence of the **Fast Fourier Transform**.

$$\begin{bmatrix} t & \cos t & \sin t \\ 0 & 1.000 & 0 \\ \pi/6 & 0.866 & 0.500 \\ 2\pi/6 & 0.500 & 0.866 \\ 3\pi/6 & 0 & 1.000 \\ 4\pi/6 & -0.500 & 0.866 \\ 5\pi/6 & -0.866 & 0.500 \\ \pi & -1.000 & 0 \\ 7\pi/6 & -0.866 & -0.500 \\ 8\pi/6 & -0.500 & -0.866 \\ 9\pi/6 & 0 & -1.000 \\ 10\pi/6 & 0.500 & -0.866 \\ 11\pi/6 & 0.866 & -0.500 \end{bmatrix}$$

D. Construction of the SVD

In Section 6.4, we saw that in a singular value decomposition of an m-by-n matrix $\mathbf{A} = \mathbf{U\Sigma V}^T$, the nonzero singular values in $\mathbf{\Sigma}$ were the squares roots of the eigenvalues of $\mathbf{A}^T\mathbf{A}$. Here you will show how a SVD factorization can be computed, by justifying the steps in the following algorithm:

(a) Diagonalize the (symmetric) matrix $\mathbf{A}^T\mathbf{A} = \mathbf{QDQ}^T$, with $\mathbf{Q}$ orthogonal and the diagonal entries of $\mathbf{D}$ in nonincreasing order. Construct the m-by-n matrix $\mathbf{\Sigma}$ from $\mathbf{D}$ by taking square roots of the nonzero eigenvalues. Also construct the pseudoinverse $\mathbf{\Sigma}^\psi$ by transposing and inverting:

$$\mathbf{\Sigma} = \begin{bmatrix} \sigma_1 & 0 & 0 & 0 & 0 & 0 \\ 0 & \ddots & 0 & 0 & 0 & 0 \\ 0 & 0 & \sigma_r & 0 & 0 & 0 \\ 0 & 0 & 0 & 0 & 0 & 0 \\ 0 & 0 & 0 & 0 & \ddots & 0 \\ 0 & 0 & 0 & 0 & 0 & 0 \end{bmatrix}_{m\text{-by-}n} := \begin{bmatrix} \mathbf{\Sigma}_r & \mathbf{0} \\ \mathbf{0} & \mathbf{0} \end{bmatrix}_{m\text{-by-}n},$$

$$\mathbf{\Sigma}^T = \begin{bmatrix} \mathbf{\Sigma}_r & \mathbf{0} \\ \mathbf{0} & \mathbf{0} \end{bmatrix}_{n\text{-by-}m}, \qquad \mathbf{\Sigma}^\psi = \begin{bmatrix} \mathbf{\Sigma}_r^{-1} & \mathbf{0} \\ \mathbf{0} & \mathbf{0} \end{bmatrix}_{n\text{-by-}m}. \tag{1}$$

(b) Set $\mathbf{V}$ equal to the orthogonal matrix $\mathbf{Q}$. Thus $\mathbf{V}^T\mathbf{V} = \mathbf{I}_n$ (the n-by-n identity).

(c) Establish that the eigenvector relationship for $\mathbf{A}^T\mathbf{A}$ implies $\mathbf{A}^T\mathbf{A}\mathbf{V} = \mathbf{V}\mathbf{D} = \mathbf{V}\mathbf{\Sigma}^T\mathbf{\Sigma}$.

(d) Calculate the m-by-m matrix $\mathbf{A}\mathbf{V}\mathbf{\Sigma}^\psi$. Justify the following:

$$(\mathbf{A}\mathbf{V}\mathbf{\Sigma}^\psi)^T(\mathbf{A}\mathbf{V}\mathbf{\Sigma}^\psi) = \begin{bmatrix} \mathbf{\Sigma}_r^{-1} & \mathbf{0} \\ \mathbf{0} & \mathbf{0} \end{bmatrix}_{m-\text{by}-n} \mathbf{V}^T\mathbf{A}^T\mathbf{A}\mathbf{V} \begin{bmatrix} \mathbf{\Sigma}_r^{-1} & \mathbf{0} \\ \mathbf{0} & \mathbf{0} \end{bmatrix}_{n-\text{by}-m}$$

$$= \begin{bmatrix} \mathbf{\Sigma}_r^{-1} & \mathbf{0} \\ \mathbf{0} & \mathbf{0} \end{bmatrix}_{m-\text{by}-n} \mathbf{V}^T\mathbf{V}\mathbf{D} \begin{bmatrix} \mathbf{\Sigma}_r^{-1} & \mathbf{0} \\ \mathbf{0} & \mathbf{0} \end{bmatrix}_{n-\text{by}-m}$$

$$= \begin{bmatrix} \mathbf{\Sigma}_r^{-1} & \mathbf{0} \\ \mathbf{0} & \mathbf{0} \end{bmatrix}_{m-\text{by}-n} \mathbf{\Sigma}^T\mathbf{\Sigma} \begin{bmatrix} \mathbf{\Sigma}_r^{-1} & \mathbf{0} \\ \mathbf{0} & \mathbf{0} \end{bmatrix}_{n-\text{by}-m} = \begin{bmatrix} \mathbf{I}_r & \mathbf{0} \\ \mathbf{0} & \mathbf{0} \end{bmatrix}_{m-\text{by}-m}, \quad (2)$$

and conclude that the first r columns of $\mathbf{A}\mathbf{V}\mathbf{\Sigma}^\psi$ are orthogonal unit vectors and its final $m - r$ columns are all zero vectors:

$$\mathbf{A}\mathbf{V}\mathbf{\Sigma}^\psi = \begin{bmatrix} \mathbf{u}_1 & \cdots & \mathbf{u}_r & \mathbf{0} & \cdots & \mathbf{0} \end{bmatrix}_{m-\text{by}-m}. \quad (3)$$

(e) Argue that

$$\mathbf{A}\mathbf{V}\mathbf{\Sigma}^\psi\mathbf{\Sigma}\mathbf{V}^T = \mathbf{A}\mathbf{V}\begin{bmatrix} \mathbf{I}_r & \mathbf{0} \\ \mathbf{0} & \mathbf{0} \end{bmatrix}_{n-\text{by}-n}\mathbf{V}^T = \mathbf{A} - \mathbf{A}\mathbf{V}\begin{bmatrix} \mathbf{0} & \mathbf{0} \\ \mathbf{0} & \mathbf{I}_{n-r} \end{bmatrix}_{n-\text{by}-n}\mathbf{V}^T. \quad (4)$$

(f) Continuing from (4), show that

$$\mathbf{A}\mathbf{V}\begin{bmatrix} \mathbf{0} & \mathbf{0} \\ \mathbf{0} & \mathbf{I}_{n-r} \end{bmatrix}_{n-\text{by}-n} = \mathbf{A}\begin{bmatrix} \mathbf{0} & \cdots & \mathbf{0} & \mathbf{v}_{r+1} & \cdots & \mathbf{v}_n \end{bmatrix}. \quad (5)$$

(g) Prove that because $\mathbf{v}_{r+1}, ..., \mathbf{v}_n$ are eigenvectors of $\mathbf{A}^T\mathbf{A}$ with eigenvalue zero, they are null vectors of $\mathbf{A}$:

$$\mathbf{A}^T\mathbf{A}\mathbf{v}_k = 0\mathbf{v}_k \Rightarrow ||\mathbf{A}\mathbf{v}_k|| = 0, \ k = r + 1, \ldots, n. \quad (6)$$

(h) Equations (4) – (6) demonstrate that $(\mathbf{A}\mathbf{V}\mathbf{\Sigma}^\psi)\mathbf{\Sigma}\mathbf{V}^T$ equals $\mathbf{A}$. The Gram–Schmidt algorithm shows that we can replace the final $n - r$ columns of $(\mathbf{A}\mathbf{V}\mathbf{\Sigma}^\psi)$ in (3) with orthonormal vectors so that the resulting matrix $\mathbf{U}$ is orthogonal:

$$\mathbf{U} = \begin{bmatrix} \mathbf{u}_1 & \cdots & \mathbf{u}_r & \mathbf{u}_{r+1} & \cdots & \mathbf{u}_m \end{bmatrix}_{m-\text{by}-m}.$$

And the form of $\mathbf{\Sigma}$ in (1) shows that the final $m - r$ columns of $(\mathbf{A}\mathbf{V}\mathbf{\Sigma}^\psi)$ don't matter in the product $(\mathbf{A}\mathbf{V}\mathbf{\Sigma}^\psi)\mathbf{\Sigma}\mathbf{V}^T = \mathbf{A}$. Therefore, $\mathbf{A} = \mathbf{U}\mathbf{\Sigma}\mathbf{V}^T$. ∎

Work out singular values decompositions for the following matrices.

(i) $\begin{bmatrix} 3 & 5 \\ 4 & 0 \end{bmatrix}$ (j) $\begin{bmatrix} 1 & -2 \\ -3 & 6 \end{bmatrix}$ (k) $\begin{bmatrix} 1 & 1 \\ -1 & 1 \\ 1 & 0 \end{bmatrix}$ (l) $\begin{bmatrix} 0 & 1 & 0 \\ 0 & 0 & 1 \end{bmatrix}$ (m) $\begin{bmatrix} 2 & 0 \\ 0 & -1 \\ 0 & 0 \end{bmatrix}$

E.   Total Least Squares

(In collaboration with Prof. Michael Lachance, University of Michigan–Dearborn)

In Example 2, Section 4.3 we described how to fit a straight line to a scatter plot of points $\{(x_i, y_i)\}_{i=1}^{N}$ by choosing m and b so as to minimize the sum of squares of errors in y, $\sum_{i=1}^{N}[y_i - (mx_i + b)]^2$. Now we shall see how the singular value decomposition enables us to find the line that minimizes the deviations of *both* x and y from the line.

To clarify this, consider Figure III.E.1a, a scatter plot like Figure 4.6 in Section 4.3. To simplify the derivation initially, we choose a straight line *through the origin*, and tabulate the distances from the data points to this line.

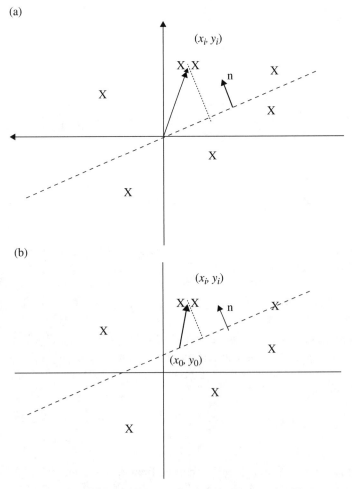

(a)

(b)

Fig. III.E.1   (a) Scatter plot #1. (b) Scatter plot #2.

(a) Argue that this distance is the projection of the vector from the origin to the data point, $\mathbf{r}_i = x_i\mathbf{i} + y_i\mathbf{j}$, onto the direction normal to the line, and is given by the magnitude of the dot product $\mathbf{r}_i \cdot \mathbf{n} = [x_i \; y_i][n_1 \; n_2]^T$ when $\mathbf{n} = n_1\mathbf{i} + n_2\mathbf{j}$ is a unit normal to the line.

(b) Therefore the sums of squares of these distances is given by the norm-squared of the vector

$$
\mathbf{An} := \begin{bmatrix} x_1 & y_1 \\ x_2 & y_1 \\ \vdots & \vdots \\ x_N & y_N \end{bmatrix} \begin{bmatrix} n_1 \\ n_2 \end{bmatrix}
$$

Our goal is to choose n so that $\|\mathbf{An}\|^2$ is minimal. For this purpose, replace $\mathbf{A}$ by its singular value decomposition,

$$
\mathbf{An} = \mathbf{U\Sigma V}^T\mathbf{n} = \mathbf{U} \begin{bmatrix} \sigma_1 & 0 \\ 0 & \sigma_2 \\ 0 & 0 \\ \vdots & \vdots \\ 0 & 0 \end{bmatrix} \mathbf{V}^T\mathbf{n},
$$

and show that $\mathbf{n}$ should be taken to be the second column of $\mathbf{V}$ (to minimize $\|\mathbf{An}\|^2$).

(c) Show that total least squares achievable by a straight line through the origin is the square of $\mathbf{A}$'s second singular value, σ_2^2.

(d) To find the total least squares line passing through a point (x_0, y_0), argue from Figure III.E.1b that the distances are given by the projections of the vectors from the point (x_0, y_0) to the points (x_i, y_i) onto the normal.

(e) Prove the following corollary.

Corollary E.1.   The straight line passing through (x_0, y_0) and achieving the total least squares for the data $\{(x_i, y_i)\}_{i=1}^N$ is normal to the second column of $\mathbf{V}$ in the singular value decomposition $\mathbf{U\Sigma V}^T$ of the matrix

$$
\mathbf{A} = \begin{bmatrix} x_1 - x_0 & y_1 - y_0 \\ x_2 - x_0 & y_2 - y_0 \\ \vdots & \vdots \\ x_N - x_0 & y_N - y_0 \end{bmatrix}.
$$

Typically the point (x_0, y_0) is taken to be the centroid of the data,

$$
\left(\sum_{i=1}^N x_i/N, \sum_{i=1}^N y_i/N \right).
$$

It is rather more complicated to incorporate the choice of (x_0, y_0) into the minimization.

F. Fibonacci Numbers

The Fibonacci numbers F_n form a sequence of integers defined by the recurrence formula

$$F_n = F_{n-1} + F_{n-2}$$

and the starting values $F_0 := 0, F_1 := 1$. Thus $F_2 = F_1 + F_0 = 1$, and $F_3 = F_2 + F_1 = 2$, $F_4 = 3$, etc. In this project we study the behavior of the Fibonacci sequence and its relation to the famous Golden Ratio:

$$\varphi = \frac{1 + \sqrt{5}}{2}.$$

The connection is via properties of the matrix

$$\mathbf{A} := \begin{bmatrix} 1 & 1 \\ 1 & 0 \end{bmatrix}.$$

(a) Show that the eigenvalues of $\mathbf{A}$ are φ and $-1/\varphi$.

(b) Derive an orthogonal matrix $\mathbf{Q}$ that diagonalizes $\mathbf{A}$ via similarity.

(c) Use the result of (b) to prove that for $k = 0, 1, 2, \ldots$

$$\mathbf{A}^k = \frac{1}{1 + \varphi^2} \begin{bmatrix} \varphi^{k+2} + \left(\frac{-1}{\varphi}\right)^k & \varphi^{k+1} + \left(\frac{-1}{\varphi}\right)^{k-1} \\ \varphi^{k+1} + \left(\frac{-1}{\varphi}\right)^{k-1} & \varphi^k + \left(\frac{-1}{\varphi}\right)^{k-2} \end{bmatrix}.$$

(d) Writing

$$\mathbf{A}^k = \begin{bmatrix} a_{11}^{(k)} & a_{12}^{(k)} \\ a_{21}^{(k)} & a_{22}^{(k)} \end{bmatrix}, \quad \mathbf{A}^0 = \mathbf{I},$$

show that $a_{11}^{(k)} = a_{11}^{(k-1)} + a_{11}^{(k-2)}$, $a_{11}^{(0)} = 1$, $a_{11}^{(1)} = 1$. Thereby conclude that

$$a_{11}^{(k)} = F_{k+1} \quad \text{for all } k.$$

How are the three other sequences of entries of $\mathbf{A}^k$ related to the Fibonacci sequence?

Finally show that the ratio F_{n+1}/F_n converges to the golden ratio as n goes to infinity.

(e) The Padovan sequence

$$1, 1, 1, 2, 2, 3, 4, 5, 7, 9, 12, 16, \ldots$$

is named after architect Richard Padovan. These numbers satisfy the recurrence relation $P_n = P_{n-2} + P_{n-3}$ with initial values $P_0 = P_1 = P_2 = 1$. The ratio P_{n+1}/P_n of two consecutive Padovan numbers approximates the **plastic constant** $P = 1.3247\ldots$. Develop a theory analogous to the Fibonacci analysis for the Padovan sequence, starting with a suitable 3-by-3 matrix $\mathbf{A}$.

ANSWERS TO ODD NUMBERED EXERCISES

Exercises 1.1

1. (a) $x_1 = -5/3$, $x_2 = 7/3$, $x_3 = -3$, and $x_4 = 1$.

   (b) $x_1 = 3$, $x_2 = 2$, and $x_3 = 0$.

   (c) $x_1 = -1$, $x_2 = 0$, $x_3 = 1$, and $x_4 = 1$.

   (d) $x_1 = x_2 = x_3 = 1$.

3. $x_1 = -1$, $x_2 = 2$, and $x_3 = -2$.

5. $x_1 = 6$, $x_2 = 1/3$, $x_3 = 2/3$, and $x_4 = -6$.

7. $x_1 = -1.00$, $x_2 = 2.000$. Requires 3 divisions, 3 multiplications, and 3 additions.

9. $x = -1.00$, $y = 2.00$. Requires 2 divisions, 6 multiplications, and 3 additions.

11. $x_1 = 1.63$, $x_2 = -2.41$, $x_3 = 1.28$. Requires 6 divisions, 11 multiplications, and 11 additions.

13. (a) To two decimal places,

$$x_1 = 14.21, \ x_2 = -1.38, \ x_3 = 3.63.$$

Requires 3 divisions, 3 multiplications, and 3 additions.

Fundamentals of Matrix Analysis with Applications,
First Edition. Edward Barry Saff and Arthur David Snider.
© 2016 John Wiley & Sons, Inc. Published 2016 by John Wiley & Sons, Inc.

(b) To three decimal places,

$$x_1 = 5.399, \; x_2 = -6.122, \; x_3 = 4.557, \; x_4 = -4.000.$$

Requires 4 divisions, 6 multiplications, and 6 additions.

15. No. One must subtract the *current*, not the original, first equation from the third.

17. $x^2 = (y-1)^2$ implies $x = \pm(y-1)$.

19.

$$ax_1 + by_1 = c$$
$$ax_2 + by_2 = c.$$

There is always a line passing through two given points, and any multiple of a solution (a, b, c) is also a solution.

21.

$$
\begin{array}{rccccl}
a_0 & +a_1 x_0 & +\cdots+ & a_n x_0^n & = & y_0 \\
a_0 & +a_1 x_1 & +\cdots+ & a_n x_1^n & = & y_1 \\
& & \vdots & & & \\
a_0 & +a_1 x_n & +\cdots+ & a_n x_n^n & = & y_n
\end{array}
$$

23.

$$x_1 = 1 + i, \; x_2 = 0, \; x_3 = 0.$$

25. One cannot perform division in the integer system. $6x - 4y$ is *even* but 9 is odd.

27.

$$x \mod 7 = 6, \quad y \mod 7 = 3.$$

The system is *inconsistent* in integers modulo 6.

Model	n	Typical PC	Tianhe-2
Thermal Stress	10^4	133.333	1.97×10^{-5}
American Sign Language	10^5	1.33×10^5	0.020
Chemical Plant Modeling	3×10^5	3.59×10^6	0.532
Mechanics of Composite Materials	10^6	1.33×10^8	19.723
Electromagnetic Modeling	10^8	1.33×10^{14}	1.97×10^7
Computation Fluid Dynamics	10^9	1.33×10^{17}	1.97×10^{10}

29. (table above)

Exercises 1.2

1. $x_1 = 3, x_2 = 2.$

3. For the first system,

$$x_3 = 10, \ x_2 = 2, \ x_1 = -4.$$

For the second system,

$$x_3 = -2, \ x_2 = -2, \ x_1 = 2.$$

5. (a)

$$\# \, (a_{1,7}) = 43; \quad \# \, (a_{7,1}) = 7;$$
$$\# \, (a_{5,5}) = 33; \quad \# \, (a_{7,8}) = 56.$$

(b)

$$\#4 : a_{4,1}; \quad \#20 : a_{5,4}; \quad \#50 : a_{5,10}.$$

(c)

$$\# \text{ of } a_{ij} = m(j-1) + i.$$

(d) $j = (p/m)$ rounded up, $i = p - m(j-1)$.

7. (a) $x_1 = -1, x_2 = 2,$ and $x_3 = -2.$

(b) The solutions to the given systems, respectively, are

$$x_1 = \frac{3}{2}, \quad x_2 = \frac{1}{2}, \quad x_3 = -\frac{3}{2};$$
$$x_1 = -1, \quad x_2 = 0, \quad x_3 = 1;$$
$$x_1 = \frac{1}{2}, \quad x_2 = -\frac{1}{2}, \quad x_3 = \frac{1}{2}.$$

9. To eliminate (say) a_{12} at the *outset* entails changing a_{13} and a_{14} as well as the right-hand side, but eliminating it at the end entails changing only the right-hand side.

Exercises 1.3

1.

$$x_1 = 0, \; x_2 = \frac{1}{3}, \; x_3 = \frac{2}{3}, \; x_4 = 0.$$

3. Inconsistent.

5. Inconsistent.

7. First system: $x_3 = s, \; x_4 = t, \; x_2 = \frac{5t-3s}{3}, \; x_1 = \frac{3-4t+3s}{3}$.
Second system: $x_3 = s, \; x_4 = t, \; x_2 = \frac{5t-3s}{3}, \; x_1 = \frac{3s-4t}{3}$.

Third system is inconsistent.

9.

$$x_1 = 1 + s - 2t, \; x_2 = s, \; x_3 = t, \; x_4 = 3, \; x_5 = -2.$$

11. First system:

$$x_1 = -\frac{2}{3} + 2s, \; x_2 = s, \; x_3 = \frac{1}{3} - 2s, \; x_4 = -\frac{1}{3}, \; x_5 = \frac{1}{3}.$$

Second system is inconsistent.
Third system:

$$\begin{bmatrix} x_1 \\ x_2 \\ x_3 \\ x_4 \\ x_5 \end{bmatrix} = \begin{bmatrix} 2s \\ s \\ -2s \\ 0 \\ 0 \end{bmatrix}.$$

13. Inconsistent if $\alpha \neq 4$. Otherwise,

$$x_3 = s, \; x_2 = \frac{2 - 2s}{3}, \; x_1 = \frac{1 - s}{3}.$$

(infinitely many solutions).

15. No solution for any α.

17.

$$x_1 = s, \; x_2 = \frac{2 + 3s}{6}, \; x_3 = -\frac{1 + 3s}{3}, \; x_4 = -\frac{1}{3}, \; x_5 = \frac{1}{3}.$$

19. (a) Infinitely many solutions.

  (b) Unique solution.

  (c) No solution.

  (d) Infinitely many solutions.

Exercises 1.4

  1. Yes: rank is 1.

  3. Yes: rank is 3.

  5. Yes: rank is 2.

  7. Consistent, infinitely many solutions.

  9. Consistent, infinitely many solutions.

  11. Consistent, unique solution.

  13. (a) All α.

  (b) $\alpha = 5$.

  15. (a) 3 columns, rank = 3, no free parameters.

  (b) 6 columns, rank = 3, 3 free parameters.

  (c) 6 columns, rank = 3, 3 free parameters.

  17. (Simple substitution)

  19. (Simple substitution)

  21. (Simple substitution)

  23. (a) At the left node, I_2 is *in*coming and I_1 and I_3 are *out*going currents. In the loop 1, the current I_3 passing through the resistance $R = 5\,(\Omega)$ drops $5I_3$ (V), followed by a voltage rise in the source of 5 (V). Thus

$$-(5)I_3 + 5 = 0 \quad \Leftrightarrow \quad 5I_3 = 5.$$

  And so on.

  (b) The echelon form is

$$\begin{bmatrix} 1 & -1 & 1 & 0 & 0 & 0 & \vdots & 0 \\ 0 & 1 & 0 & -1 & -1 & 1 & \vdots & 0 \\ 0 & 0 & 5 & 0 & 0 & 0 & \vdots & 5 \\ 0 & 0 & 0 & 3 & 0 & 0 & \vdots & 1 \\ 0 & 0 & 0 & 0 & 0 & 2 & \vdots & 6 \\ 0 & 0 & 0 & 0 & 0 & 0 & \vdots & 1 \end{bmatrix}$$

(c) The echelon form is changed to

$$
\left[
\begin{array}{cccccc:c}
1 & -1 & 1 & 0 & 0 & 0 & 0 \\
0 & 1 & 0 & -1 & -1 & 1 & 0 \\
0 & 0 & 5 & 0 & 0 & 0 & 6 \\
0 & 0 & 0 & 3 & 0 & 0 & 1 \\
0 & 0 & 0 & 0 & 0 & 2 & 6 \\
0 & 0 & 0 & 0 & 0 & 0 & 0
\end{array}
\right]
$$

(d) The parametrized solution shows that increasing I_5 by t units increases I_1 and I_2 by the same amount but does not affect the currents I_3, I_4, and I_6.

25. (a)

$$
\begin{array}{rrrcc}
x_1 & & -4x_3 & = & 0 \\
x_1 & -2x_2 & +x_3 & = & 0 \\
x_1 & +2x_2 & +4x_3 & = & 0.13,
\end{array}
$$

whose solution is

$$
x_3 = 0.01, \quad x_2 = 0.025, \quad x_1 = 0.04.
$$

(b) The equilibrium conditions for forces say that

$$
\begin{array}{rrrcl}
F_1 & & -F_3 & = & 0 \\
F_1 & +F_2 & +F_3 & = & 650
\end{array}
\quad\Leftrightarrow\quad
\left[
\begin{array}{ccc:c}
1 & 0 & -1 & 0 \\
1 & 1 & 1 & 650
\end{array}
\right]
$$

3 columns, rank = 2, so one free parameter (infinitely many solutions).

Exercises 1.5

1. Without pivoting, 4-digit computation yields

$$
x_2 = 9.962 \times 10^{-2}, \quad x_1 = 1.776 \times 10^4.
$$

With pivoting,

$$
x_2 = 1.000, \quad x_1 = 10.00.
$$

3. The system satisfying the given conditions is, in the matrix form,

$$\left[\begin{array}{cc:c} 0.004 & 2.8 & 252.8 \\ 1 & 1 & 290 \end{array}\right]$$

Without pivoting, 3-digit computation yields

$$x_2 = 89.9, \quad x_1 = 270.$$

With pivoting,

$$x_2 = 89.9, \quad x_1 = 200.$$

5. (b) The largest value is $2^{n-1} + s$, occurring in the $(n, n+3)$ entry.

   (c) For the first system, the solution form is typified by

$$x_5 = 1,$$
$$x_4 = 0,$$
$$x_3 = 0,$$
$$x_2 = 0,$$
$$x_1 = 0;$$

for the second system by

$$x_5 = s/16,$$
$$x_4 = -s/2,$$
$$x_3 = -s/4,$$
$$x_2 = -s/8,$$
$$x_1 = -s/16;$$

and for the third system, the value of x_4 will be the sum of values of this variable in the first two systems, namely, $x_4 = 0 + (-s/2) = -s/2.$

Exercises 2.1

1. (a)

$$\left[\begin{array}{cc} 4 & 4 \\ 2 & 5 \end{array}\right].$$

(b)

$$\begin{bmatrix} 4 & 0 \\ 10 & 15 \end{bmatrix}.$$

3. (a)

$$\mathbf{AB} = \begin{bmatrix} 18 & 14 & 4 \\ 16 & 20 & 4 \\ 15 & 23 & 4 \end{bmatrix}.$$

(b)

$$\mathbf{BA} = \begin{bmatrix} 1 & -1 & 9 \\ 14 & 23 & 9 \\ 15 & 22 & 18 \end{bmatrix}.$$

(c)

$$\mathbf{A}^2 = \begin{bmatrix} 8 & 12 & 12 \\ 9 & 8 & 12 \\ 11 & 12 & 12 \end{bmatrix}.$$

(d)

$$\mathbf{B}^2 = \begin{bmatrix} 16 & 3 & 3 \\ 9 & 24 & 3 \\ 25 & 27 & 6 \end{bmatrix}.$$

5. (a)

$$\mathbf{AB} = \begin{bmatrix} -20 & 3 & -15 \\ 23 & 3 & -3 \\ 10 & -6 & -6 \end{bmatrix}.$$

(b)

$$\mathbf{AC} = \begin{bmatrix} -7 & -11 & -15 \\ -7 & 1 & 25 \\ -8 & 8 & 20 \end{bmatrix}.$$

(c)

$$\begin{bmatrix} -27 & -8 & -30 \\ 16 & 4 & 22 \\ 2 & 2 & 14 \end{bmatrix}.$$

9. For example,

$$\mathbf{A} = \begin{bmatrix} 1 & -1 \\ 1 & -1 \end{bmatrix}.$$

11. For example, one can choose

$$\mathbf{A} = 2\mathbf{C}, \ \mathbf{B} = \mathbf{C} \qquad \text{or} \qquad \mathbf{A} = \mathbf{C} + \mathbf{I}, \ \mathbf{B} = \mathbf{I},$$

where $\mathbf{I}$ is the 2-by-2 multiplicative identity matrix.

13. **(a)** True **(b)** False **(c)** True **(d)** False

15. **(a)** We set

$$\mathbf{A} = \begin{bmatrix} 0 & 1 & 3 \\ -1 & 7 & -9 \\ 6 & -4 & 0 \end{bmatrix}, \quad \mathbf{x} = \begin{bmatrix} x_1 \\ x_2 \\ x_3 \end{bmatrix}, \quad \mathbf{b} = \begin{bmatrix} 0 \\ -9 \\ 2 \end{bmatrix}.$$

Then the given system is equivalent to $\mathbf{Ax} = \mathbf{b}$.

(b) Let

$$\mathbf{A} = \begin{bmatrix} -2 & 3 & 0 \\ -1 & 2 & 1 \\ -1 & -1 & 3 \end{bmatrix}, \quad \mathbf{t} = \begin{bmatrix} x \\ y \\ z \end{bmatrix}, \quad \mathbf{b} = \begin{bmatrix} -7 \\ 6 \\ -3 \end{bmatrix}.$$

With these notations, the system can be written in matrix form as $\mathbf{At} = \mathbf{b}$.

17. **(a)**

$$\mathbf{B} = \begin{bmatrix} 1 & 0 & 0 \\ 0 & 0 & 1 \\ 0 & 1 & 0 \end{bmatrix}.$$

(b)

$$\mathbf{B} = \begin{bmatrix} 1 & 0 & 0 \\ -1/2 & 1 & 0 \\ 0 & 0 & 1 \end{bmatrix}.$$

(c)

$$\mathbf{B} = \begin{bmatrix} 1 & 0 & 0 \\ 0 & 2 & 0 \\ 0 & 0 & 1 \end{bmatrix}.$$

19. (a)

$$\begin{bmatrix} 1 & 0 & 0 \\ -3 & 1 & 0 \\ 0 & 0 & 1 \end{bmatrix} \begin{bmatrix} 0 & 1 & 0 \\ 0 & 0 & 1 \\ 1 & 0 & 0 \end{bmatrix} \mathbf{A}.$$

(b)

$$\mathbf{B}^{(2)} = \begin{bmatrix} 1 & 0 & 0 \\ 0 & 1 & 0 \\ 0 & 3 & 1 \end{bmatrix} \begin{bmatrix} 1 & 0 & 0 \\ -1 & 1 & 0 \\ -3 & 0 & 1 \end{bmatrix} \mathbf{B}.$$

(c)

$$\mathbf{C}^{(2)} = \begin{bmatrix} 1 & 0 & 0 \\ 0 & 1 & 0 \\ 0 & 1 & 1 \end{bmatrix} \begin{bmatrix} 1 & 0 & 0 \\ -1 & 1 & 0 \\ 0 & 0 & 1 \end{bmatrix} \mathbf{C}.$$

21. (a) 0.2875.

(b)

$$\begin{bmatrix} 0.75 & 0.6 \\ 0.25 & 0.4 \end{bmatrix} \mathbf{x}_n.$$

(c) 1, 1.

(d) Yes.

23. (a) $\mathbf{A}(\mathbf{B}\mathbf{v})$ (b) $(\mathbf{A} + \mathbf{B})\mathbf{v}$

Exercises 2.2

1. (a)

$$\begin{bmatrix} \sqrt{3}/2 & -1/2 \\ 1/2 & \sqrt{3}/2 \end{bmatrix}.$$

(b)

$$\begin{bmatrix} -3/5 & -4/5 \\ -4/5 & 3/5 \end{bmatrix}.$$

(c)

$$\begin{bmatrix} 1/2 & 1/2 \\ 1/2 & 1/2 \end{bmatrix}.$$

3. The matrix **A** with $a_{1,1} = 1$ and all other zero entries shrinks the entire y-axis to the origin.

5. (a)

$$\frac{1}{10} \begin{bmatrix} 7 & -1 \\ -7 & 1 \end{bmatrix}.$$

(b)

$$\frac{1}{10} \begin{bmatrix} 3 + 4\sqrt{3} & -4 + 3\sqrt{3} \\ -4 + 3\sqrt{3} & -3 - 4\sqrt{3} \end{bmatrix}.$$

7.

$$\begin{bmatrix} 0 & -1/2 \\ 1/2 & 0 \end{bmatrix}.$$

9. Four counterclockwise rotations of a vector by $45°$ are equivalent to one rotation by $4 \times 45° = 180°$, whose rotation matrix is

$$\begin{bmatrix} -1 & 0 \\ 0 & -1 \end{bmatrix} = -\mathbf{I}.$$

11. (a)

$$\begin{bmatrix} 0 & 0 & 0 \\ 0 & 0 & 0 \\ 0 & 0 & 1 \end{bmatrix}.$$

(b)

$$\begin{bmatrix} 1/3 & 1/3 & 1/3 \\ 1/3 & 1/3 & 1/3 \\ 1/3 & 1/3 & 1/3 \end{bmatrix}.$$

Exercises 2.3

1.

$$\begin{bmatrix} 1 & 0 & 5 \\ 0 & 1 & 0 \\ 0 & 0 & 1 \end{bmatrix}.$$

3.

$$\begin{bmatrix} 1 & 0 & 0 \\ 0 & 1/5 & 0 \\ 0 & 0 & 1 \end{bmatrix}.$$

5.

$$\begin{bmatrix} 1 & 0 & 0 & 0 \\ 0 & 0 & 1 & 0 \\ 0 & 1 & 0 & 0 \\ 0 & 0 & 0 & 1 \end{bmatrix}.$$

7.

$$\begin{bmatrix} 4/9 & -1/9 \\ 1/9 & 2/9 \end{bmatrix}.$$

9. Not invertible.

11.

$$\begin{bmatrix} 0 & 1/2 & 0 & 0 \\ 0 & 0 & -1 & 1 \\ 0 & 0 & 1/2 & 0 \\ 1 & -3/2 & 0 & 0 \end{bmatrix}.$$

13.

$$\mathbf{A}^{-1} = \begin{bmatrix} 7 & -9/4 & -3/2 \\ 4 & -1 & -1 \\ -2 & 3/4 & 1/2 \end{bmatrix}.$$

(a)

$$\begin{bmatrix} 5/2 \\ 1 \\ -1/2 \end{bmatrix}.$$

(b)

$$\begin{bmatrix} -153/4 \\ -20 \\ 47/4 \end{bmatrix}.$$

(c)

$$\begin{bmatrix} 15/2 \\ 3 \\ -5/2 \end{bmatrix}.$$

15.

$$\begin{bmatrix} -3 & 3 & -5 \\ 9 & -8 & 15 \\ -5 & 5 & -7 \end{bmatrix}.$$

17. (a) False (b) True (c) True (d) True.

19. If $ad - bc = 0$ the rank of the matrix is less than 2:

    (i) if $d = 0$ then either the second column or the second row is zero;

    (ii) if $c = 0$ then either the first column or second row is zero;

    (iii) if $cd \neq 0$ then $a/c = b/d$ and one column is a multiple of the other.

21.

$$(\mathbf{ABC})^T = (\mathbf{A}(\mathbf{BC}))^T = (\mathbf{BC})^T \mathbf{A}^T = \mathbf{C}^T \mathbf{B}^T \mathbf{A}^T.$$

23.

$$\mathbf{M}_{\text{rot}}(\theta)^{-1} = \mathbf{M}_{\text{rot}}(-\theta) = \begin{bmatrix} \cos(-\theta) & -\sin(-\theta) \\ \sin(-\theta) & \cos(-\theta) \end{bmatrix}$$

$$= \begin{bmatrix} \cos\theta & \sin\theta \\ -\sin\theta & \cos\theta \end{bmatrix} = \mathbf{M}_{\text{rot}}(\theta)^T.$$

25.

$$\begin{bmatrix} 1 & 0 & 0 & 0 & 0 \\ 0 & 1 & 0 & 0 & 0 \\ 0 & 0 & 1 & 0 & 0 \\ 0 & 0 & 0 & 1 & 0 \\ 0.1 & 0.1 & 0.1 & 0.1 & 1.1 \end{bmatrix}^{-1} = \begin{bmatrix} 1 & 0 & 0 & 0 & 0 \\ 0 & 1 & 0 & 0 & 0 \\ 0 & 0 & 1 & 0 & 0 \\ 0 & 0 & 0 & 1 & 0 \\ -\dfrac{1}{11} & -\dfrac{1}{11} & -\dfrac{1}{11} & -\dfrac{1}{11} & \dfrac{10}{11} \end{bmatrix}.$$

Exercises 2.4

1. 0.

3. -6.

5. $66 + 6i$.

7. $abdg$.

9. $dx^3 - cx^2 + bx - a$.

11. $(x-1)^3(x+3)$.

13.

$$\det(\mathbf{A}) = ax + b,$$

a linear function of x. Examples demonstrating the possibilities are:

$$\det \begin{bmatrix} 1 & 0 & 0 \\ 0 & 0 & x \\ 0 & 0 & 0 \end{bmatrix} = 0 \quad \text{for any } x;$$

$$\det \begin{bmatrix} 1 & 0 & 0 \\ 0 & 1 & x \\ 0 & 0 & 1 \end{bmatrix} = 1 \neq 0 \quad \text{for any } x;$$

$$\det \begin{bmatrix} 1 & 0 & 0 \\ 0 & x & 0 \\ 0 & 0 & 1 \end{bmatrix} = x = 0 \quad \text{only for } x = 0.$$

15. 24

17. (a) $\det(\mathbf{A})$.

    (b) $3 \cdot \det(\mathbf{A})$.

    (c) $-\det(\mathbf{A})$.

    (d) $\det(\mathbf{A})$.

    (e) $81 \cdot \det(\mathbf{A})$.

    (f) $\det(\mathbf{A})$.

    (g) $3 \cdot \det(\mathbf{A})$.

    (h) $\det(\mathbf{A})$.

19. (a)

$$\det \begin{bmatrix} 1 & x & x^2 & \cdots & x^{n-1} \\ x & 1 & x & \cdots & x^{n-2} \\ & & \vdots & & \\ x^{n-2} & x^{n-2} & x^{n-3} & \cdots & x \\ x^{n-1} & x^{n-2} & x^{n-3} & \cdots & 1 \end{bmatrix} = \left(1 - x^2\right)^{n-1}.$$

for any n.

(b) Determinant $= 1$ for any n.

21. Let $P(x) = \alpha x^4 + \beta x^3 + \gamma x^2 + \delta x + \varepsilon$ denote the determinant. Then

$$\alpha = 1, \quad \beta = a + f + k + p,$$

$$\varepsilon = \det \begin{bmatrix} a & b & c & d \\ e & f & g & h \\ i & j & k & l \\ m & n & o & p \end{bmatrix}.$$

23. (b)

$$a_2 = \mathrm{cof}(1,3) = (-1)^{1+3}\det \begin{bmatrix} 1 & x_1 \\ 1 & x_2 \end{bmatrix} = x_2 - x_1.$$

(c) $z = x_1$ and $z = x_2$.

(d)

$$\det(\mathbf{V}) = (x_2 - x_1)(z - x_1)(z - x_2).$$

(e)

$$(x_1 - x_2)(x_1 - x_3)(x_2 - x_3)(z - x_1)(z - x_2)(z - x_3).$$

(The generalization is clear.)

Exercises 2.5

1. -6

3. ≈ -0.169748, accurate to four decimal places.

5.

$$\begin{bmatrix} 0 & 1 \\ 1 & 0 \end{bmatrix}.$$

7.

$$\begin{bmatrix} 1 & 0 & 0 \\ 0 & 0 & 1 \\ 0 & 1 & 0 \end{bmatrix}.$$

9.

$$\begin{bmatrix} 1 & 0 & 0 \\ 0 & 1/\beta & 0 \\ 0 & 0 & 1 \end{bmatrix}.$$

11.

$$\begin{bmatrix} 1 & 2 & 1 \\ 1 & 3 & 2 \\ 1 & 0 & 1 \end{bmatrix}.$$

13.

$$x_1 = 0, x_2 = 2.$$

15. $x_1 = 2, \quad x_2 = 0, \quad x_3 = -\frac{3}{2}.$

17.

$$\frac{dx_1}{d\alpha} = \frac{\text{cof}(2,1)}{|\mathbf{A}|}; \quad \frac{dx_2}{d\alpha} = \frac{\text{cof}(2,2)}{|\mathbf{A}|}; \quad \frac{dx_3}{d\alpha} = \frac{\text{cof}(2,3)}{|\mathbf{A}|}.$$

19. (a)

$$\frac{dx_1}{d\alpha} = \frac{8}{(7-2\alpha)^2},$$

(b)

$$\frac{dx_2}{d\alpha} = -\frac{4}{(7-2\alpha)^2}, (\alpha \neq 7/2).$$

21. The determinants are equal in the 4-by-4 and n-by-n cases.

23.

$$D_1 = \det[a] = a;$$
$$D_2 = a^2 - b^2;$$
$$D_3 = a^3 - 2ab^2.$$

25. (a)

$$F_n = aF_{n-1} + b^2 F_{n-2}.$$

(b)

$$F_1 = 1; \qquad F_2 = 2;$$
$$F_3 = 3; \qquad F_4 = 5;$$
$$F_5 = 8; \qquad F_6 = 13.$$

27. (a) True
   (b) False.
   (c) True.
   (d) True.

29. (a)

$$\begin{bmatrix} 1 & -1/2 & -1/12 \\ 0 & 1/4 & -5/24 \\ 0 & 0 & 1/6 \end{bmatrix}.$$

(b)

$$\begin{bmatrix} -13/4 & 5/2 & -1/4 \\ 5/2 & -2 & 1/2 \\ -1/4 & 1/2 & -1/4 \end{bmatrix}.$$

31.

$$\left| \mathbf{BAB}^{-1} \right| = |\mathbf{A}| \, ;$$
$$\left| \mathbf{BAB}^{T} \right| = |\mathbf{B}|^2 \, |\mathbf{A}| \, .$$

REVIEW PROBLEMS FOR PART I

1. The system is consistent if and only if $\gamma = 5$. In this case, the system has infinitely many solutions.

3. (a)

$$\begin{bmatrix} 10 \\ 34 \end{bmatrix}.$$

(b)

$$\begin{bmatrix} 0 & 4 & -4 \end{bmatrix}.$$

(c)

$$\begin{bmatrix} 4 & 8 & 12 \\ 6 & 12 & 18 \\ -2 & -4 & -6 \end{bmatrix}.$$

(d)

$$\begin{bmatrix} 2 & 3 & -1 \\ 4 & 2 & 3 \end{bmatrix}.$$

(e)

$$\begin{bmatrix} 0 & 1 & 0 & 0 \\ 1 & 0 & 0 & 0 \\ 0 & 0 & 0 & 1 \\ 0 & 0 & 1 & 0 \end{bmatrix}.$$

(f)

$$\begin{bmatrix} -1 & 0 & 25 \\ 0 & 1 & 0 \\ 0 & 0 & -1 \end{bmatrix}.$$

5. For example,

$$\mathbf{A} = \mathbf{B} = \begin{bmatrix} 0 & 1 \\ 0 & 0 \end{bmatrix}$$

7. (a) $\mathbf{E}$ results when the identical operation is performed on the rows of the identity.

   (b) $\mathbf{E}$ results when the inverse operation is performed on the rows of the identity.

   (c) The operations that $\mathbf{EA}$ performs on the rows of $\mathbf{A}$ are performed, instead, on the *columns* of $\mathbf{A}$.

   (d) Switch the corresponding columns of $\mathbf{A}^{-1}$.

   (e) Multiply the second column of $\mathbf{A}^{-1}$ by $1/7$.

   (f) Subtract the second column of $\mathbf{A}^{-1}$ from its first column.

9. Let $\mathbf{M}$ be an r-by-s matrix, $\mathbf{A} = [a_k]$ is the 1-by-r row vector with $a_i = 1$ and all other zero entries, and $\mathbf{B} = [b_k]$ is the s-by-1 column vector with $b_j = 1$ and all other zero entries.

11. (a) True. (b) True. (c) True. (d) False. (e) False. (f) False. For example,

$$\mathbf{A} = \begin{bmatrix} 1 & 0 \\ 1 & 0 \end{bmatrix}$$

13. (Multiply $(\mathbf{I} + \mathbf{K})\,\mathbf{x} = \mathbf{x} + \mathbf{K}\mathbf{x} = \mathbf{0}$ on the left by $\mathbf{x}^T$.)

15. (a)

$$\frac{dx_1}{d\gamma} = 1, \quad \frac{dx_2}{d\gamma} = \frac{6}{5}, \quad \frac{dx_3}{d\gamma} = \frac{3}{5}.$$

(b)

$$\frac{dx_1}{d\gamma} = -\frac{125}{(5\gamma - 5)^2}, \quad \frac{dx_2}{d\gamma} = -\frac{175}{(5\gamma - 5)^2}, \quad \frac{dx_3}{d\gamma} = -\frac{25}{(5\gamma - 5)^2}.$$

Exercises 3.1

1. (a) Yes   (b) No: $[1 \; -1 \; 0] + [1 \; 0 \; -1] = [2 \; -1 \; -1]$
   (c) Yes   (d) No: $[1 \; 1 \; 0] + [0 \; 0 \; 1] = [1 \; 1 \; 1]$

3. (a) Yes   (b) Yes
   (c) No: $f(x) \equiv 1$, multiplied by 2, does not belong.
   (d) Yes

5. (a) $p(x)$ in $\mathbf{P}_2$ implies $p(x) = p(1) + p'(1)(x - 1) + p''(1)(x - 1)^2/2$ (in $\mathbf{P}_2$).
   (b) Identify $a\mathbf{i} + b\mathbf{j} + c\mathbf{k}$ with $a + bx + cx^2$. Then linear combinations of vectors correspond to linear combinations of polynomials. But no 2 vectors span $\mathbf{R}^3$.

7. (a) Yes   (b) Yes   (c) Yes
   (d) No. The zero matrix does not belong.
   (e) Yes   (f) Yes

9. (a) True   (b) False (Take $\mathbf{v}_1 = \mathbf{0}$.)
   (c) True   (d) True $(\mathbf{0})$

11. (a) $[0 \; 0 \; 0 \; ...]$ (b) Yes
   (c) Yes   (d) Yes

13. (Verify the definition.)

15. (Verify the definition.)

17. A two-dimensional subspace must be the whole plane; a one-dimensional subspace must be a line through the origin; and the origin, alone, is a subspace. For $\mathbf{R}^3_{col}$, there is the whole space, planes through the origin, lines through the origin, and the origin.

19. No.

21. Test $x = c_1 e^x + c_2 \sin x$ at 0 and π.

Exercises 3.2

1. (a) No   (b) Yes   (c) Yes   (d) Yes

3. $\mathbf{b}_2$

5. Row reduce

$$
\begin{bmatrix}
1 & 0 & 2 & \vdots & 2 & 0 \\
0 & 1 & -3 & \vdots & -2 & -1 \\
0 & 1 & -3 & \vdots & -2 & -1 \\
1 & 3 & -7 & \vdots & -4 & -3
\end{bmatrix}
\quad \text{and} \quad
\begin{bmatrix}
2 & 0 & \vdots & 1 & 0 & 2 \\
-2 & -1 & \vdots & 0 & 1 & -3 \\
-2 & -1 & \vdots & 0 & 1 & -3 \\
-4 & -1 & \vdots & 1 & 3 & -7
\end{bmatrix}
$$

to show they are both consistent.

7. Independent

9. Independent

11. Dependent

13. Independent

15. Independent

17. $bc \neq 1$.

19. (a) True   (b) False   (c) False

(d) True   (e) True

21. If a nontrivial linear combination $a\mathbf{v}_1 + b\mathbf{v}_2 + c\mathbf{v}_3 + d\mathbf{v}_4$ were zero, then d_4 must be 0; so it would reduce to a nontrivial linear combination of $\mathbf{v}_1, \mathbf{v}_2, \mathbf{v}_3$.

23. A vanishing linear combination of $\mathbf{AD}, \mathbf{BD}, \mathbf{CD}$, postmultiplied by $\mathbf{D}^{-1}$, yields a vanishing linear combination of $\mathbf{A}, \mathbf{B}, \mathbf{C}$.

25. $\mathbf{Ax}$ is a linear combination of the columns of $\mathbf{A}$; $\mathbf{A}^2\mathbf{x}$ is a linear combination of the columns of $\mathbf{Ax}$; and so on. Now apply the linear combination composition principle.

27. (a) $a\mathbf{w}_1 + b\mathbf{w}_2 = \mathbf{0}$ implies $2a + 4b = 0$ and $-3a - 3b = 0$, which implies $a = b = 0$.

(b) $a\mathbf{w}_1 + b\mathbf{w}_2 + c\mathbf{w}_3 = \mathbf{0}$ implies $a + 4b = -3b + 2c = -5c = 0$, which implies $a = b = c = 0$.

29. The dot products are the entries of

$$
\begin{bmatrix}
\mathbf{v}_1^T \\
\mathbf{v}_2^T \\
\mathbf{v}_3^T
\end{bmatrix}
\begin{bmatrix}
\mathbf{n}
\end{bmatrix}.
$$

If these are zero for nonzero **n**, the vectors are linearly dependent, and conversely.

Exercises 3.3

1. (For example) $[1\ 0\ -1\ 0\ 0\ 0]$, $[0\ 1\ -1\ 0\ 0\ 0]$, $[0\ 0\ 0\ 1\ 0\ -1]$, and $[0\ 0\ 0\ 0\ 1\ -1]$.

3. (For example) $[1\ -1\ 0\ 0\ 0\ 0]$, $[0\ 0\ 1\ -1\ 0\ 0]$, and $[0\ 0\ 0\ 0\ 1\ -1]$.

5. (For example) rows 1–3 for the row space; columns 1–3 for the column space; the null space contains only **0**.

7. (For example) rows 1–4 for the row space; columns 1, 2, 4, and 5 for the column space; $[-1\ \ -0.5\ \ 1\ \ 0\ \ 0\ \ 0]^T$ for the null space.

9. (For example) rows 1–3 for the row space; columns 1–3 for the column space; the null space contains only **0**.

11. (For example) rows 1–4 for the row space; columns 1–4 for the column space; $[0\ \ -1\ \ 1\ \ -1\ \ 1]^T$ for the null space.

13. No.

15. Dimension = 15: all upper triangular matrices with a single "one" entry and the rest zeros.

17. Dimension = 10: all antisymmetric matrices with a single "one" above the diagonal, a corresponding "minus one" below the diagonal, and the rest zeros.

19. Dimension = 20: All matrices with a single "one" in the first four rows, a "minus one" in the corresponding column, and the rest zeros.

21. One.

23. Two. One. Two.

25. (For example) any two canonical basis vectors and [1 2 3].

27. Every column is a multiple of **v**; yes.

29.

$$\begin{bmatrix} 1 & 1 \\ 1 & 0 \\ 0 & 1 \end{bmatrix} \begin{bmatrix} 1 & 2 & 0 & -1 & 1 & 0 \\ 0 & 0 & 1 & 2 & -2 & 0 \end{bmatrix}$$

31. All $\alpha \neq 0$.

33. 4 and 13.

35. For example, Span {[1 1 0], [0 1 1] }.

37. **AB** comprises linear combinations of the columns of **A**; also it comprises linear combinations of the rows of **B**. Thus its rank cannot exceed the rank of either **A** or **B**.

39. (a) 1.   (b) 2.   (c) 2.

41. [1 −1 3].

43. [1/3, 2/3].

Exercises 4.1

1. $[5\ 10] = 11\,[3/5\ 4/5] + 2\,[-4/5\ 3/5]$.

3. $[3\ 6\ 9] = 11\,[1/3\ 2/3\ 2/3] + 2\,[-2/3\ -1/3\ 2/3] + (-1)\,[-2/3\ 2/3\ -1/3]$.

5. $\{[-4/5\ 3/5]\,, [3/5\ 4/5]\}$.

7. $\{[1/2\ -1/2\ 1/2\ -1/2]\,, [1/2\ 1/2\ 1/2\ 1/2]\}$.

9. $\{[1/2\ 1/2\ -1/2\ -1/2]\,, [1/2\ -1/2\ 1/2\ -1/2]\,,$
$[-1/2\ -1/2\ -1/2\ -1/2]\}$.

11. (a) (For example) $[1\ 0]$ and $[5\ 7]$. (b) (For example) $\{[1\ 2\ 3]\,, [1\ 0\ 0]\,, [0\ 1\ 0]\}$.

13. (For example) $[1\ -1\ -1\ 1]$.

15. (For example) $[1\ 0\ -1\ 0]$ and $[0\ 1\ 0\ -1]$.

17. (For example)

$$\left\{\left[\frac{\sqrt{5}}{5}\ \frac{2\sqrt{5}}{5}\ 0\ 0\right], \left[-\frac{2\sqrt{30}}{5}\ \frac{\sqrt{30}}{5}\ \frac{5\sqrt{30}}{5}\ 0\right],\right.$$
$$\left.\left[\frac{6\sqrt{370}}{370}\ -\frac{3\sqrt{370}}{370}\ -\frac{18\sqrt{370}}{370}\ \frac{\sqrt{370}}{370}\right]\right\}.$$

19. (For example)

$$\left\{\left[-\frac{2}{\sqrt{6}}\ \frac{1}{\sqrt{6}}\ \frac{1}{\sqrt{6}}\ 0\right]^T, \left[-\frac{4}{\sqrt{102}}\ -\frac{1}{\sqrt{102}}\ -\frac{7}{\sqrt{102}}\ \frac{6}{\sqrt{102}}\right]^T\right\}.$$

21.

$$\sin\phi_1\sin\phi_2\cos(\theta_1 - \theta_2) + \cos\phi_1\cos\phi_2 = 0.$$

23. Following the steps in Problem 22, it follows that the equality holds if and only if $f(x) = \|\mathbf{v} + x\mathbf{w}\|^2$ has a (real) root x_0. However, $\|\mathbf{v} + x_0\mathbf{w}\| = 0$ implies $\mathbf{v} + x_0\mathbf{w} = 0$.

25.

$$\|\mathbf{v} + \mathbf{w}\|^2 + \|\mathbf{v} - \mathbf{w}\|^2 = (\mathbf{v} + \mathbf{w})\cdot(\mathbf{v} + \mathbf{w}) + (\mathbf{v} - \mathbf{w})\cdot(\mathbf{v} - \mathbf{w})$$
$$= \mathbf{v}\cdot\mathbf{v} + 2(\mathbf{v}\cdot\mathbf{w}) + \mathbf{w}\cdot\mathbf{w} + \mathbf{v}\cdot\mathbf{v} - 2(\mathbf{v}\cdot\mathbf{w}) + \mathbf{w}\cdot\mathbf{w}$$
$$= 2\|\mathbf{v}\|^2 + 2\|\mathbf{w}\|^2.$$

Exercises 4.2

1. ± 1.

3. (a) $\begin{bmatrix} a & 0 \\ 0 & a \end{bmatrix}$ or $\begin{bmatrix} 0 & a \\ a & 0 \end{bmatrix}$, where $a = \pm 1$.

   (b) $\begin{bmatrix} a & -b \\ b & a \end{bmatrix}$, where $a^2 + b^2 = 1$.

5. $\begin{bmatrix} \cos\psi & 0 & \sin\psi \\ 0 & 1 & 0 \\ -\sin\psi & 0 & \cos\psi \end{bmatrix}$.

7. $\begin{bmatrix} 1 & 0 & 0 \\ 0 & 1 & 0 \\ 0 & 0 & 1 \end{bmatrix}, \begin{bmatrix} 1 & 0 & 0 \\ 0 & 0 & 1 \\ 0 & 1 & 0 \end{bmatrix}, \begin{bmatrix} 0 & 0 & 1 \\ 1 & 0 & 0 \\ 0 & 1 & 0 \end{bmatrix}, \begin{bmatrix} 0 & 1 & 0 \\ 1 & 0 & 0 \\ 0 & 0 & 1 \end{bmatrix}, \begin{bmatrix} 0 & 1 & 0 \\ 0 & 0 & 1 \\ 1 & 0 & 0 \end{bmatrix},$
   $\begin{bmatrix} 0 & 0 & 1 \\ 0 & 1 & 0 \\ 1 & 0 & 0 \end{bmatrix}$.

9. $\begin{bmatrix} 0.2720 & 0.6527 & 0.7071 \\ 0.6527 & 0.4148 & -0.6340 \\ 0.7071 & -0.6340 & 0.3132 \end{bmatrix}$, for example.

11. $\begin{bmatrix} 1/2 & 1/2 & 1/2 & 1/2 \\ 1/2 & 1/2 & -1/2 & -1/2 \\ 1/2 & -1/2 & 1/2 & -1/2 \\ 1/2 & -1/2 & -1/2 & 1/2 \end{bmatrix}$, for example.

13. $\begin{bmatrix} 0.5345 & 0.2673 & 0.8018 \\ 0.2673 & 0.8465 & -0.4604 \\ 0.8018 & -0.4604 & -0.3811 \end{bmatrix}$, for example. No.

15. $\begin{bmatrix} 1.0000 & 0 & 0 \\ 0 & 0.3162 & 0.9487 \\ 0 & 0.9487 & -0.3162 \end{bmatrix}$, for example. Yes.

17. No. (b) The point $[1\ 0]$, reflected about $y = x$, is sent to $[1\ 0]$, and then to itself when reflected about the y−axis. However, when reflected first about the y−axis, it is sent to $[-1\ 0]$, and then to $[0\ -1]$ when reflected about the line $y = x$.

19. (a) The hint, which is directly verified, shows that $(\mathbf{I} - \mathbf{A})$ has only the zero null vector.

Exercises 4.3

1. $\mathbf{x} = \begin{bmatrix} -1 \\ 1/2 \end{bmatrix}$; error$= \begin{bmatrix} 1/2 \\ -1/2 \\ -1/2 \\ 1/2 \end{bmatrix}$.

3. $\begin{bmatrix} 16 & 16 & 16 \\ 16 & 20 & 20 \\ 16 & 20 & 24 \end{bmatrix} \begin{bmatrix} x_1 \\ x_2 \\ x_3 \end{bmatrix} = \begin{bmatrix} -2 \\ -3 \\ -4 \end{bmatrix}.$ $\begin{bmatrix} x_1 \\ x_2 \\ x_3 \end{bmatrix} = \begin{bmatrix} 1/8 \\ 0 \\ -1/4 \end{bmatrix}.$

5. $x = [0.5\ 0.5\ 0.25]^T$,   error $= [0.25 - 0.25 - 0.25\ 0.25]^T$.

7. $m = \dfrac{S_{xy}}{S_{x^2}}$.

9. $m = \dfrac{S_{xy} - b_0 S_x}{S_{x^2}}$.

17.

$$b = \frac{1}{NS_{x^2} - S_x^2} \log \frac{y_1^{Nx_1} y_2^{Nx_2} \cdots y_N^{Nx_N}}{(y_1 y_2 \cdots y_N)^{S_x}}$$

$$\log c = \frac{1}{NS_{x^2} - S_x^2} \log \frac{(y_1 y_2 \cdots y_N)^{S_{x^2}}}{(y_1^{x_1} y_2^{x_2} \cdots y_N^{x_N})^{S_x}}.$$

19.

$$\begin{bmatrix} 1 & 1 & \cdots & 1 \\ e^{-x_1} & e^{-x_2} & \cdots & e^{-x_N} \\ \vdots & \vdots & \ddots & \vdots \\ e^{-kx_1} & e^{-kx_2} & \cdots & e^{-kx_N} \end{bmatrix} \begin{bmatrix} 1 & e^{-x_1} & \cdots & e^{-kx_1} \\ 1 & e^{-x_2} & \cdots & e^{-kx_2} \\ \vdots & \vdots & \ddots & \vdots \\ 1 & e^{-x_N} & \cdots & e^{-kx_N} \end{bmatrix} \begin{bmatrix} c_0 \\ c_1 \\ \vdots \\ c_k \end{bmatrix}$$

$$= \begin{bmatrix} 1 & 1 & \cdots & 1 \\ e^{-x_1} & e^{-x_2} & \cdots & e^{-x_N} \\ \vdots & \vdots & \ddots & \vdots \\ e^{-kx_1} & e^{-x_N} & \cdots & e^{-kx_N} \end{bmatrix} \begin{bmatrix} y_1 \\ y_2 \\ \vdots \\ y_N \end{bmatrix}.$$

Exercises 4.4

1. (a) No; (b) Yes; (c) No; (d) Yes; (e) Yes; (f) Yes.

3. $\left\{ 1, \sqrt{3}(2x - 1), \sqrt{5}(6x^2 - 6x + 1) \right\}$.

5. If $c_0 + c_1(x - 1) + \cdots + c_n(x - 1)^n = 0$, differentiate k times and set $x = 1$ to prove $c_k = 0$.

7. (a) $\frac{1}{2}(3x^2 - 1)$; (b) $x^3 - \frac{3}{5}x$.

9. The Vandermonde determinant is nonzero for distinct α_i. The second matrix is clearly nonsingular, so the condition implies that each c_i is zero.

11. (a) $c_1 + c_2 \ln x + c_3 e^x = 0 \implies c_3 = \lim_{x \to \infty} -\frac{c_1 + c_2 \ln x}{e^x} = 0 \implies c_2 = 0$
$\lim_{x \to \infty} -\frac{c_1}{\ln x} = 0 \implies c_1 = 0$.

(b) $c_1 e^x + c_2 x e^x + c_3 x^2 e^x = 0 \implies c_3 = \lim_{x \to \infty} -\frac{c_1 + c_2 x}{x^2} = 0 \implies c_2 = 0$
$\lim_{x \to \infty} -\frac{c_1}{x} = 0 \implies c_1 = 0$.

13. $(0) \sin x + \frac{3}{4} \cos x$.

15. (a) $\begin{bmatrix} \frac{x_2^3 - x_1^3}{3} & \frac{x_2^2 - x_1^2}{2} \\ \frac{x_2^2 - x_1^2}{2} & x_2 - x_1 \end{bmatrix} \begin{bmatrix} m \\ b \end{bmatrix} = \begin{bmatrix} \int_{x_1}^{x_2} x f(x)\, dx \\ \int_{x_1}^{x_2} f(x)\, dx \end{bmatrix}$

    (b) $x^2 \approx x - \frac{1}{6}$

    (c) $x^2 \approx x - \frac{1}{8}$

17. (Straightforward integration).

19. (a) (For example) add x^n to each element of the basis in Problem 4.

    (b) (For example) $\{x, x^2, \ldots, x^n\}$;

    (c) (For example) $\{(x-1), (x^2-1), \ldots, (x^n-1)\}$.

21. Since $\dim \mathbf{P}_n = n+1$, it suffices to show that the polynomials are linearly independent; apply the growth rate method.

REVIEW PROBLEMS FOR PART II

1. (a) (For example) $\begin{bmatrix} -1 & 1 & 0 & 0 & 0 \\ -1 & 0 & 1 & 0 & 0 \\ -1 & 0 & 0 & 1 & 0 \\ -1 & 0 & 0 & 0 & 1 \end{bmatrix} \mathbf{A} \begin{bmatrix} 1 \\ 1 \\ 1 \\ 1 \\ 1 \end{bmatrix} = \begin{bmatrix} 0 \\ 0 \\ 0 \\ 0 \end{bmatrix}$

    (b) (For example) $\begin{bmatrix} 1 & 1 & 1 & 1 & 1 \end{bmatrix} \mathbf{A} \begin{bmatrix} -1 & -1 & -1 & -1 \\ 1 & 0 & 0 & 0 \\ 0 & 1 & 0 & 0 \\ 0 & 0 & 1 & 0 \\ 0 & 0 & 0 & 1 \end{bmatrix} = \begin{bmatrix} 0 & 0 & 0 & 0 \end{bmatrix}$

    (c) The row sums, the column sums, the diagonal sum, and the reverse diagonal sum are all equal.

3. (For example) [22 191 213].

5. (a) No.   (b) Yes.   (c) No.   (d) No.   (e) No.

7. (For example) [1 0 0 0] and [0 1 0 0].

9. Let $\mathbf{v}_1$, $\mathbf{v}_2$, $\mathbf{v}_3$, and $\mathbf{w}_1$, $\mathbf{w}_2$, $\mathbf{w}_3$ be bases for the two subspaces. Because the dimension is 5, there must be a nontrivial relationship $c_1 \mathbf{v}_1 + c_2 \mathbf{v}_2 + c_3 \mathbf{v}_3 + c_4 \mathbf{w}_1 + c_5 \mathbf{w}_2 + c_6 \mathbf{w}_3 = \mathbf{0}$, which can be manipulated to show that the subspaces have a nonzero vector in common.

11. (For example) $\sin 2t$, $\cos 2t$, $\sin^2 t$.

13. By Problem 37, Exercises 3.3, the rank of $\mathbf{AB}$ can be no bigger than 3.

15. Yes.

17. First apply the equation with $\mathbf{u} = \begin{bmatrix} 1 & 0 & 0 & 0 \end{bmatrix}^T$ and $\begin{bmatrix} 0 & 1 & 0 & 0 \end{bmatrix}^T$ to conclude that $A_{11} = A_{22} = 1$. Then apply with $\mathbf{u} = \begin{bmatrix} 1 & 1 & 0 & 0 \end{bmatrix}^T$ to conclude that $A_{12} = A_{21} = 0$. Generalize.

19. -2.

Exercises 5.1

1. Any vector along the projection direction is an eigenvector with eigenvalue 1 and all vectors orthogonal to this vector are eigenvectors with eigenvalue 0.

3. Any vector with one 1, one -1, and five 0s is an eigenvector with eigenvalue -1. The vector with seven 1s is an eigenvector with eigenvalue 6.

5. The canonical basis vectors are eigenvectors; all have eigenvalue 1 except for the one with 1 in the row that gets multiplied.

7. One eigenvector is $\mathbf{x} = [1\,1\,0\,0]^T$. Another is $\mathbf{x} = [1\,0\,1\,0]^T$, if $\alpha = 0$. But for eigenvectors of the form $[y\,z\,1\,0]^T$, the second equation in the system $\mathbf{Ux} = 2\mathbf{x}$ is $2z + \alpha = 2z$, which necessitates $\alpha = 0$.

9. $\mathbf{Bu} = 2i\mathbf{u}$.

11. (a) $-1, 0$, and -2 respectively.

    (b) $[-i\,-1\,i\,1]^T$ is paired with eigenvalues $-i, -2i$, and 0, respectively. $[1\,1\,1\,1]^T$ is paired with eigenvalues $1, 0$, and 2, respectively.

13.

$$\mathbf{A} = \begin{bmatrix} 0 & 1 & 0 & 0 & 0 & 0 \\ 0 & 0 & 1 & 0 & 0 & 0 \\ 0 & 0 & 0 & 1 & 0 & 0 \\ 0 & 0 & 0 & 0 & 1 & 0 \\ 0 & 0 & 0 & 0 & 0 & 1 \\ 1 & 0 & 0 & 0 & 0 & 0 \end{bmatrix}, \quad \mathbf{B} = \begin{bmatrix} 0 & 1 & 0 & -1 & 0 & 1 \\ 1 & 0 & 1 & 0 & -1 & 0 \\ 0 & 1 & 0 & 1 & 0 & -1 \\ -1 & 0 & 1 & 0 & 1 & 0 \\ 0 & -1 & 0 & 1 & 0 & 1 \\ 1 & 0 & -1 & 0 & 1 & 0 \end{bmatrix}$$

$$\mathbf{C} = \begin{bmatrix} 0 & 1 & 0 & 1 & 0 & 1 \\ 1 & 0 & 1 & 0 & 1 & 0 \\ 0 & 1 & 0 & 1 & 0 & 1 \\ 1 & 0 & 1 & 0 & 1 & 0 \\ 0 & 1 & 0 & 1 & 0 & 1 \\ 1 & 0 & 1 & 0 & 1 & 0 \end{bmatrix}, \quad \mathbf{D} = \begin{bmatrix} d_1 & d_2 & d_3 & d_4 & d_5 & d_6 \\ d_6 & d_1 & d_2 & d_3 & d_4 & d_5 \\ d_5 & d_6 & d_1 & d_2 & d_3 & d_4 \\ d_4 & d_5 & d_6 & d_1 & d_2 & d_3 \\ d_3 & d_4 & d_5 & d_6 & d_1 & d_2 \\ d_2 & d_3 & d_4 & d_5 & d_6 & d_1 \end{bmatrix}$$

$[\omega\,\omega^2\,\omega^3\,\omega^4\,\omega^5\,\omega^6]^T$ is an eigenvector corresponding to the eigenvalues $\omega, 2\omega, 0$ and $(d_1 + d_2\omega + d_3\omega^2 + d_4\omega^3 + d_5\omega^4 + d_6\omega^5)$, respectively.

$[\omega^2\,\omega^4\,\omega^6\,\omega^8\,\omega^{10}\,\omega^{12}]^T$ is an eigenvector corresponding to the eigenvalues $\omega^2, -2\omega^2, 0$, and $(d_1 + d_2\omega^2 + d_3\omega^4 + d_4 + d_5\omega^2 + d_6\omega^4)$, respectively.

$[\omega^3\,\omega^6\,\omega^9\,\omega^{12}\,\omega^{15}\,\omega^{18}]^T$ is an eigenvector corresponding to the eigenvalues $-1, 1, 3$ and $(d_1 - d_2 + d_3 - d_4 + d_5 - d_6)$, respectively.

$[\omega^4\,\omega^8\,\omega^{12}\,\omega^{16}\,\omega^{20}\,\omega^{24}]^T$ is an eigenvector corresponding to the eigenvalues $\omega^4, 2\omega, 0$ and, $(d_1 - d_2\omega + d_3\omega^2 + d_4 + d_5\omega^4 + d_6\omega^2)$, respectively.

$\left[\omega^5\ \omega^{10}\ \omega^{15}\ \omega^{20}\ \omega^{25}\ \omega^{30}\right]^T$ is an eigenvector corresponding to the eigenvalues $-\omega^2, -2\omega^2, 0$, and $(d_1 + d_2\omega^5 + d_3\omega^4 + d_4\omega^3 + d_5\omega^2 + d_6\omega)$, respectively. $\left[\omega^6\ \omega^{12}\ \omega^{18}\ \omega^{24}\ \omega^{30}\ \omega^{36}\right]^T = [1\ 1\ 1\ 1\ 1\ 1]^T$ is an eigenvector corresponding to the eigenvalues 1, 1, 3, and $(d_1 + d_2 + d_3 + d_4 + d_5 + d_6)$, respectively.

15. If **u** is an eigenvector of **AB**, $(\mathbf{AB})(\mathbf{u}) = r(\mathbf{u}) \implies (\mathbf{BA})\mathbf{Bu} = (\mathbf{BAB})\mathbf{u} = \mathbf{B}(\mathbf{AB})\mathbf{u} = r\mathbf{Bu}$.

17. $\mathbf{Ax} = \mathbf{x}$ has only the trivial solution, implying $(\mathbf{I} - \mathbf{A})\mathbf{x} = \mathbf{0}$ has only the trivial solution.

19. (a)

$$\left(\mathbf{A} + \mathbf{u_1 b^T}\right)\mathbf{u_1} = \mathbf{Au_1} + \mathbf{u_1 b^T u_1} = r_1\mathbf{u_1} + \mathbf{u_1 b^T u_1} = \left(r_1 + \mathbf{b^T u_1}\right)\mathbf{u_1}.$$

(b)

$$\left(\mathbf{A} + \mathbf{u_1 b^T}\right)\left(\mathbf{u_2} + \frac{\mathbf{b^T u_2}}{r_2 - r_1'}\mathbf{u_1}\right) = \left(\mathbf{A} + \mathbf{u_1 b^T}\right)\mathbf{u_2} + \left(\mathbf{A} + \mathbf{u_1 b^T}\right)\left(\frac{\mathbf{b^T u_2}}{r_2 - r_1'}\mathbf{u_1}\right)$$
$$= \mathbf{Au_2} + \mathbf{u_1 b^T u_2} + \frac{\mathbf{b^T u_2}}{r_2 - r_1'}r_1'\mathbf{u_1}$$
$$= r_2\mathbf{u_2} + \left(\mathbf{b^T u_2} + \frac{r_1'\mathbf{b^T u_2}}{r_2 - r_1'}\right)\mathbf{u_1}$$
$$= r_2\left(\mathbf{u_2} + \frac{\mathbf{b^T u_2}}{r_2 - r_1'}\mathbf{u_1}\right).$$

Exercises 5.2

1. $[1\ 2]^T$ (3), and $[3\ -1]^T$ (10).
3. $[1\ 1]^T$ (−1), and $[1\ -1]^T$ (3).
5. $[1\ -1]^T$ (0), and $[1\ 1]^T$ (2).
7. $[1\ 1\ 1]^T$ (1), $[1\ 0\ 1]^T$ (2), and $[1\ 1\ 0]^T$ (3).
9. $[1\ -i]^T$ (i), and $[1\ i]^T$ (−i).
11. $[2\ 1]^T$ (1 + i), and $[i\ 1]^T$ (3).
13. The only eigenvalue is 6, corresponding to only one linearly independent eigenvector $[1\ 1]^T$.
15. The only eigenvalue is 5, corresponding to only one linearly independent eigenvector $[1\ 1]^T$.
17. The eigenvalues are 2, with multiplicity 1 and corresponding eigenvector $[1\ 1\ 1]^T$, and −1 with multiplicity 2, and the corresponding eigenvectors $[1\ -1\ 0]^T$ and $[1\ 0\ -1]^T$.

19. The eigenvalues are 1, with multiplicity 1 and corresponding eigenvector $[1 \ 3 \ -1]^T$, and 2 with multiplicity 2, and only one linearly independent corresponding eigenvector $[0 \ 1 \ 0]^T$.

21. (a) $\begin{bmatrix} 0 & 0 \\ 1 & 3 \end{bmatrix}$ $(0, 3)$.

    (b) $\begin{bmatrix} 1 & -1 & 2 \\ -1 & 1 & -2 \\ 0 & 0 & 0 \end{bmatrix}$, $(2, 0, 0)$.

23. (a) $\det(\mathbf{A}^T - r\mathbf{I}) = \det\left((\mathbf{A} - r\mathbf{I})^T\right) = \det(\mathbf{A} - r\mathbf{I})$.

    (b) $0 = \mathbf{v_1}^T(\mathbf{A}\mathbf{v_2}) - (\mathbf{A}^T\mathbf{v_1})\mathbf{v_2} = r_2\mathbf{v_1}^T\mathbf{v_2} - r_1\mathbf{v_1}^T\mathbf{v_2}$. Thus $r_1 \neq r_2 \implies \mathbf{v_1}^T\mathbf{v_2} = 0$.

    (c) (i) Eigenvalues are $r_1 = 1$, corresponding to $[1 \ -1]^T$ and $[1 \ 2]^T$, and $r_2 = 2$, corresponding to $[2 \ -1]^T$ and $[1 \ 1]^T$.

       (ii) Eigenvalues are $r_1 = -1$, corresponding to $[1 \ 1 \ 2]^T$ and $[1 \ 2 \ -2]^T$; $r_2 = -2$, corresponding to $[2 \ 0 \ 1]^T$ and $[1 \ 1 \ -1]^T$; and $r_3 = 1$, corresponding to $[0 \ 1 \ 1]^T$ and $[1 \ 3 \ -2]^T$.

25. $c_0 = \det(\mathbf{A} - 0\mathbf{I}) = \det(\mathbf{A})$.

27. -2 and 4.

Exercises 5.3

1. $\left\{ [1 \ 2]^T, [2 \ -1]^T \right\}$.

3. $\left\{ [0 \ 3 \ -4]^T, [5 \ -4 \ -3]^T, [5 \ 4 \ 3]^T \right\}$.

5. $\left\{ [1 \ -1 \ -1]^T, [1 \ 1 \ 0]^T, [1 \ -1 \ 2]^T \right\}$.

7. $\left\{ [i \ -1]^T, [i \ 1]^T \right\}$.

9. $\left\{ [1 \ 0 \ -(i+1)]^T, [0 \ 1 \ 0]^T, [(i-1) \ 0 \ -1]^T \right\}$.

11. $\left\{ [1 \ 1 \ 0]^T, \left[i \ -i \ \sqrt{2}\right]^T, \left[i \ -i \ -\sqrt{2}\right]^T \right\}$.

13. $\mathbf{A} = \begin{bmatrix} 2 & 1 \\ 1+i & 1+2i \end{bmatrix}$ provides a counterexample.

15. $\left\{ [i \ 0 \ 0], \frac{1}{\sqrt{2}}[0 \ 1 \ 1], \frac{1}{\sqrt{3}}\left[0 \ \frac{(3i-1)}{2} \ \frac{(i+1)}{2}\right] \right\}$.

17. (a) No. (b) No. (c) No. (d) Yes.

Exercises 6.1

1. $\mathbf{A} = \mathbf{P}^{-1}\mathbf{B}\mathbf{P} \implies \mathbf{A}^n = (\mathbf{P}^{-1}\mathbf{B}\mathbf{P})(\mathbf{P}^{-1}\mathbf{B}\mathbf{P})\cdots(\mathbf{P}^{-1}\mathbf{B}\mathbf{P}) = \mathbf{P}^{-1}\mathbf{B}^n\mathbf{P}$.

3. $\mathbf{A} = \mathbf{P}^{-1}\mathbf{DP}$, but $\mathbf{D} = a\mathbf{I}$. Thus $\mathbf{A} = \mathbf{P}^{-1}(a\mathbf{I})\mathbf{P} = a\mathbf{I}$.

5. Assume $\mathbf{A}_1$ is similar to $\mathbf{D}_1$ and $\mathbf{A}_2$ is similar to $\mathbf{D}_2$, with $\mathbf{D}_i$ as described. Note that the entries on a diagonal matrix can be rearranged by similarity transformations: switch rows by left-multiplying by the appropriate elementary row operator and switch columns by right-multiplying by the inverse of the same operator (which is identical to the operator itself). Thus $\mathbf{D}_1$ is similar to $\mathbf{D}_2$; therefore, $\mathbf{A}_1$ is similar to $\mathbf{A}_2$ (see paragraph after Definition 1).

7. $\mathbf{BA} = (\mathbf{A}^{-1}\mathbf{A})\mathbf{BA} = \mathbf{A}^{-1}(\mathbf{AB})\mathbf{A}$ *or* $\mathbf{AB} = (\mathbf{B}^{-1}\mathbf{B})\mathbf{AB} = \mathbf{B}^{-1}(\mathbf{BA})\mathbf{B}$.

Counterexample:

$$\mathbf{A} = \begin{bmatrix} 0 & 1 \\ 0 & 0 \end{bmatrix} \text{ and } \mathbf{B} = \begin{bmatrix} 1 & 0 \\ 0 & 0 \end{bmatrix}.$$

9. $\mathbf{P} = \begin{bmatrix} 1 & 3 \\ 2 & -1 \end{bmatrix}$, $\mathbf{D} = \begin{bmatrix} 3 & 0 \\ 0 & 10 \end{bmatrix}$;

$$\begin{bmatrix} 9 & -3 \\ -2 & 4 \end{bmatrix}^{-1} = \begin{bmatrix} 1 & 3 \\ 2 & -1 \end{bmatrix} \begin{bmatrix} 1/3 & 0 \\ 0 & 1/10 \end{bmatrix} \begin{bmatrix} 1 & 3 \\ 2 & -1 \end{bmatrix}^{-1}$$

$$e^{\begin{bmatrix} 9 & -3 \\ -2 & 4 \end{bmatrix}} = \begin{bmatrix} 1 & 3 \\ 2 & -1 \end{bmatrix} \begin{bmatrix} e^3 & 0 \\ 0 & e^{10} \end{bmatrix} \begin{bmatrix} 1 & 3 \\ 2 & -1 \end{bmatrix}^{-1}.$$

11. $\mathbf{P} = \begin{bmatrix} 1 & 1 \\ 1 & -1 \end{bmatrix}$, $\mathbf{D} = \begin{bmatrix} -1 & 0 \\ 0 & 3 \end{bmatrix}$;

$$\begin{bmatrix} 1 & -2 \\ -2 & 1 \end{bmatrix}^{-1} = \begin{bmatrix} 1 & 1 \\ 1 & -1 \end{bmatrix} \begin{bmatrix} -1 & 0 \\ 0 & 1/3 \end{bmatrix} \begin{bmatrix} 1 & 1 \\ 1 & -1 \end{bmatrix}^{-1}$$

$$e^{\begin{bmatrix} 1 & -2 \\ -2 & 1 \end{bmatrix}} = \begin{bmatrix} 1 & 1 \\ 1 & -1 \end{bmatrix} \begin{bmatrix} e^{-1} & 0 \\ 0 & e^3 \end{bmatrix} \begin{bmatrix} 1 & 1 \\ 1 & -1 \end{bmatrix}^{-1}.$$

13. $\mathbf{P} = \begin{bmatrix} 1 & 1 \\ -1 & 1 \end{bmatrix}$, $\mathbf{D} = \begin{bmatrix} 0 & 0 \\ 0 & 2 \end{bmatrix}$; The matrix is singular.

$$e^{\begin{bmatrix} 1 & 1 \\ 1 & 1 \end{bmatrix}} = \begin{bmatrix} 1 & 1 \\ -1 & 1 \end{bmatrix} \begin{bmatrix} 1 & 0 \\ 0 & e^2 \end{bmatrix} \begin{bmatrix} 1 & 1 \\ -1 & 1 \end{bmatrix}^{-1}.$$

15. $\mathbf{P} = \begin{bmatrix} 1 & 1 & 1 \\ 1 & 0 & 1 \\ 1 & 1 & 0 \end{bmatrix}$ $\mathbf{D} = \begin{bmatrix} 1 & 0 & 0 \\ 0 & 2 & 0 \\ 0 & 0 & 3 \end{bmatrix}$.

$$\begin{bmatrix} 4 & -1 & -2 \\ 2 & 1 & -2 \\ 1 & -1 & 1 \end{bmatrix}^{-1} = \begin{bmatrix} 1 & 1 & 1 \\ 1 & 0 & 1 \\ 1 & 1 & 0 \end{bmatrix} \begin{bmatrix} 1 & 0 & 0 \\ 0 & 1/2 & 0 \\ 0 & 0 & 1/3 \end{bmatrix} \begin{bmatrix} 1 & 1 & 1 \\ 1 & 0 & 1 \\ 1 & 1 & 0 \end{bmatrix}^{-1}.$$

$$e^{\begin{bmatrix} 4 & -1 & -2 \\ 2 & 1 & -2 \\ 1 & -1 & 1 \end{bmatrix}} = \begin{bmatrix} 1 & 1 & 1 \\ 1 & 0 & 1 \\ 1 & 1 & 0 \end{bmatrix} \begin{bmatrix} e & 0 & 0 \\ 0 & e^2 & 0 \\ 0 & 0 & e^3 \end{bmatrix} \begin{bmatrix} 1 & 1 & 1 \\ 1 & 0 & 1 \\ 1 & 1 & 0 \end{bmatrix}^{-1}.$$

17. $\mathbf{P} = \begin{bmatrix} 1 & 1 \\ -i & i \end{bmatrix}$, $\mathbf{D} = \begin{bmatrix} i & 0 \\ 0 & -i \end{bmatrix}$;

$$\begin{bmatrix} 0 & -1 \\ 1 & 0 \end{bmatrix}^{-1} = \begin{bmatrix} 1 & 1 \\ -i & i \end{bmatrix} \begin{bmatrix} -i & 0 \\ 0 & i \end{bmatrix} \begin{bmatrix} 1 & 1 \\ -i & i \end{bmatrix}^{-1}$$

$$e^{\begin{bmatrix} 0 & -1 \\ 1 & 0 \end{bmatrix}} = \begin{bmatrix} 1 & 1 \\ -i & i \end{bmatrix} \begin{bmatrix} e^i & 0 \\ 0 & e^{-i} \end{bmatrix} \begin{bmatrix} 1 & 1 \\ -i & i \end{bmatrix}^{-1}.$$

19. $\mathbf{P} = \begin{bmatrix} i & 2 \\ 1 & 1 \end{bmatrix}$, $\mathbf{D} = \begin{bmatrix} 3 & 0 \\ 0 & i+1 \end{bmatrix}$;

$$\begin{bmatrix} 1 & 2i \\ -1 & 3+i \end{bmatrix}^{-1} = \begin{bmatrix} i & 2 \\ 1 & 1 \end{bmatrix} \begin{bmatrix} 1/3 & 0 \\ 0 & 1/(i+1) \end{bmatrix} \begin{bmatrix} i & 2 \\ 1 & 1 \end{bmatrix}^{-1}$$

$$e^{\begin{bmatrix} 1 & 2i \\ -1 & 3+i \end{bmatrix}} = \begin{bmatrix} i & 2 \\ 1 & 1 \end{bmatrix} \begin{bmatrix} e^3 & 0 \\ 0 & e^{i+1} \end{bmatrix} \begin{bmatrix} i & 2 \\ 1 & 1 \end{bmatrix}^{-1}.$$

21. Let $\mathbf{A} = \begin{bmatrix} 0 & a \\ 0 & 0 \end{bmatrix}$. Clearly $\mathbf{A}^2 = \mathbf{0}$.

Suppose now $\mathbf{A} = \mathbf{B}^2$, $a \neq 0$, and $\mathbf{B} = \begin{bmatrix} x & y \\ z & t \end{bmatrix}$. Then

$$\begin{cases} x^2 + yz & = 0 \\ t^2 + yz & = 0 \\ z(x+t) & = 0 \\ y(x+t) & = 0. \end{cases}$$

From the first two equations, we see that $x = \pm t$. If $x = -t$, from the last equation, we get $0 = a$, a contradiction. If $x = +t$, the third equation implies

$zx = 0$. If $x = 0$, then $t = 0$, and the last equation yields again, $0 = a$. Finally, if $z = 0$, the first equation implies $x = 0$ and we again find a contradiction in the last equation.

23. The left- and right-hand members of (16) yield the same answer when multiplying an eigenvector $\mathbf{u}_i$, namely, $r_i\mathbf{u}_i$. Since the eigenvectors form a basis, they yield the same answer when multiplying any vectors in the space. For Hermitian matrices, the form is

$$\mathbf{A} = r_1\mathbf{u}_1\mathbf{u}_1^H + r_2\mathbf{u}_2\mathbf{u}_2^H + \cdots + r_n\mathbf{u}_n\mathbf{u}_n^H.$$

25.

$$\mathbf{C}_q\mathbf{v_i} = \begin{bmatrix} 0 & 1 & 0 & \cdots & 0 \\ 0 & 0 & 1 & \cdots & 0 \\ & & \vdots & & \\ 0 & 0 & 0 & \cdots & 1 \\ -a_0 & -a_1 & -a_2 & \cdots & -a_{n-1} \end{bmatrix} \begin{bmatrix} 1 \\ r_i \\ r_i^2 \\ \vdots \\ r_i^{n-1} \end{bmatrix}$$

$$= \begin{bmatrix} r_i \\ r_i^2 \\ r_i^3 \\ \vdots \\ -\sum_{i=0}^{n-1} a_i r_i^{n-1} \end{bmatrix} = \begin{bmatrix} r_i \\ r_i^2 \\ r_i^3 \\ \vdots \\ r_i^n \end{bmatrix} = r_i\mathbf{v_i}.$$

As in Section 6.1, this can be expressed $\mathbf{C}_q\mathbf{V}_q = \mathbf{V}_q\mathbf{D}$, or $\mathbf{V}_q^{-1}\mathbf{C}_q\mathbf{V}_q = \mathbf{D}$, where

$$\mathbf{D} = \begin{bmatrix} r_1 & 0 & \cdots & 0 \\ 0 & r_2 & & 0 \\ \vdots & & \ddots & \vdots \\ 0 & 0 & \cdots & r_n \end{bmatrix}.$$

Exercises 6.2

3. The change of coordinates generated by the matrix

$$\mathbf{Q} = \frac{1}{\sqrt{5}} \begin{bmatrix} -2 & 1 \\ 1 & 2 \end{bmatrix}$$

eliminates the cross terms. Maximum is 5, minimum is -10.

5. The change of coordinates generated by the matrix

$$\mathbf{Q} = \begin{bmatrix} 0 & \frac{\sqrt{2}}{2} & \frac{\sqrt{2}}{2} \\ \frac{3}{5} & -\frac{2\sqrt{2}}{5} & \frac{2\sqrt{2}}{5} \\ -\frac{4}{5} & -\frac{3\sqrt{2}}{10} & \frac{3\sqrt{2}}{10} \end{bmatrix}$$

eliminates the cross terms. Maximum is 6, minimum is -4.

7. The change of coordinates generated by the matrix

$$\mathbf{Q} = \begin{bmatrix} 1/\sqrt{3} & 1/\sqrt{2} & 1/\sqrt{6} \\ -1/\sqrt{3} & 1/\sqrt{2} & -1/\sqrt{6} \\ -1/\sqrt{3} & 0 & 2/\sqrt{6} \end{bmatrix}$$

eliminates the cross terms. Maximum is 3, minimum is 0.

9. (a) Let $\mathbf{Q}$ be an orthogonal matrix of eigenvectors corresponding to the ordering $r_1 \geq r_2 \geq \cdots \geq r_n$. The maximum of $\mathbf{v}^T\mathbf{Av}$ over all unit vectors $\mathbf{v}$ is the same as the maximum of $\mathbf{x}^T\mathbf{Q}^T\mathbf{AQx}$ over all unit $\mathbf{x}$. Then $\mathbf{x}^T\mathbf{Q}^T\mathbf{AQx} = r_1x_1^2 + r_2x_2^2 + \cdots + r_nx_n^2$ is a weighted average of the eigenvalues since $x_1^2 + x_2^2 + \cdots + x_n^2 = 1$. Its maximum (r_1) occurs when $\mathbf{x} = [1\,0\,0\ldots0]^T$, for which $\mathbf{v} = \mathbf{Qx}$ is the first column of $\mathbf{Q}$ or the first eigenvector.

(b) If $\mathbf{w}$ is a unit vector orthogonal to the first eigenvector $\mathbf{v}$, then $\mathbf{Q}^T\mathbf{w}$ is a unit vector orthogonal to $\mathbf{x} = [1\,0\,0\ldots0]^T$, that is, $\mathbf{Q}^T\mathbf{w}$ has the form $[0\,y_2\,y_3\ldots y_n]^T$ with $y_2^2 + y_3^2 + \cdots + y_n^2 = 1$. Therefore, $\mathbf{x}^T\mathbf{Q}^T\mathbf{AQx} = r_2y_2^2 + \cdots + r_ny_n^2$ is a weighted average of r_2 through r_n, whose maximum (r_2) is achieved when $\mathbf{y} = [0\,1\,0\ldots0]^T$. $\mathbf{Qy}$ is the second column of $\mathbf{Q}$.

(c) (Similarly for the remaining eigenvalues.)

(d) The Rayleigh quotient $\frac{\mathbf{v}^T\mathbf{Av}}{\mathbf{v}^T\mathbf{v}}$ has the same value as $\mathbf{v}_1^T\mathbf{Av}_1$, where $\mathbf{v}_1$ is the *unit* vector in the direction of $\mathbf{v}$, namely $\mathbf{v}_1 = \frac{\mathbf{v}}{|\mathbf{v}|}$, because $\mathbf{v}_1^T\mathbf{Av}_1 = \frac{\mathbf{v}^T}{|\mathbf{v}|}\mathbf{A}\frac{\mathbf{v}}{|\mathbf{v}|} = \frac{\mathbf{v}^T\mathbf{Av}}{|\mathbf{v}|^2} = \frac{\mathbf{v}^T\mathbf{Av}}{\mathbf{v}^T\mathbf{v}}$. Thus seeking the maximum Rayleigh quotient among all vectors (nonzero, of course) is the same as seeking the maximum $\mathbf{v}_1^T\mathbf{Av}_1$ among all unit vectors.

11. The matrix corresponding to the quadratic form is

$$\mathbf{A} = \frac{1}{2}\begin{bmatrix} 0 & 1 & \ldots & 1 \\ 1 & 0 & \ldots & 1 \\ \vdots & \vdots & \ddots & \vdots \\ 1 & 1 & \ldots & 0 \end{bmatrix}.$$

Using the theory of circulant matrices from Problem 14, Exercises 5.1, we find the eigenvalues of $\mathbf{A}$ to be $r_1 = (n-1)/2$, of algebraic multiplicity 1, and $r_2 = -1/2$, with algebraic multiplicity $(n-1)$.

Now using the reasoning of Problem 9 above, we conclude that the maximum of the given bilinear form, when $x_1^2 + x_2^2 + \cdots + x_n^2 = 1$, equals the largest eigenvalue, $(n-1)/2$.

13. (a) Since 1 is not an eigenvalue, $(\mathbf{I} - \mathbf{A})^{-1}$ exists, and the stationary equation $\mathbf{x} = \mathbf{A}\mathbf{x} + \mathbf{b}$ has the solution $(\mathbf{I} - \mathbf{A})^{-1}\mathbf{b}$.

   (b) $\mathbf{x}_1 = \mathbf{A}\mathbf{x}_0 + \mathbf{b}$, $\mathbf{x}_2 = \mathbf{A}\mathbf{x}_1 + \mathbf{b} = \mathbf{A}^2\mathbf{x}_0 + \mathbf{A}\mathbf{b} + \mathbf{b}$, $\mathbf{x}_3 = \mathbf{A}^3\mathbf{x}_0 + \mathbf{A}^2\mathbf{b} + \mathbf{A}\mathbf{b} + \mathbf{b}$, etc.

   (c) $\mathbf{x}_n \to \mathbf{0} + (\mathbf{I} - \mathbf{A})^{-1}\mathbf{b}$.

   (d), (e) straightforward.

   (f) (Eigenvalues of $\mathbf{B}$) = (Eigenvalues of $\mathbf{A}$) $\bigcup \{1\}$.

   (g) The hint reveals the answer.

   (h)

$$\mathbf{x}_n = (c_1 - d_1)\, r_1^n \mathbf{v}_1 + (c_2 - d_2)\, r_2^n \mathbf{v}_2 + \cdots + (c_m - d_m)\, r_m^n \mathbf{v}_m + (\mathbf{I} - \mathbf{A})^{-1}\mathbf{b}$$
$$= ((c_1 - d_1)r_1^n + d_1)\, \mathbf{v}_1 + ((c_2 - d_2)r_2^n + d_2)\, \mathbf{v}_2 + \ldots$$
$$+ ((c_m - d_m)r_m^n + d_m)\, \mathbf{v}_m$$

with

$$d_i = \frac{f_i}{r_i}, \ i = 1, \ldots, m,$$

where $\mathbf{b} = f_1\mathbf{v}_1 + f_2\mathbf{v}_2 + \cdots + f_m\mathbf{v}_m$.

Exercises 6.3

1. Possible answer:

$$\mathbf{Q} = \frac{1}{\sqrt{2}} \begin{bmatrix} 1 & 1 \\ -1 & 1 \end{bmatrix},$$
$$\mathbf{U} = \mathbf{Q}^H \mathbf{A} \mathbf{Q} = \frac{1}{2} \begin{bmatrix} 1 & 1 \\ -1 & 1 \end{bmatrix} \begin{bmatrix} 7 & 6 \\ -9 & -8 \end{bmatrix} \begin{bmatrix} 1 & -1 \\ 1 & 1 \end{bmatrix} = \begin{bmatrix} 1 & 15 \\ 0 & -2 \end{bmatrix}.$$

3. Possible answer:

$$\mathbf{Q} = \frac{1}{\sqrt{14}} \begin{bmatrix} 3i - 1 & 2 \\ 2 & 3i + 1 \end{bmatrix},$$

$$\mathbf{U} = \begin{bmatrix} -6i & 4 - 6i \\ 0 & 6i \end{bmatrix}.$$

5. Possible answer:

$$\mathbf{Q}_1 = \frac{1}{\sqrt{2}} \begin{bmatrix} 1 & 1 & 0 \\ 1 & -1 & 0 \\ 0 & 0 & \sqrt{2} \end{bmatrix},$$

$$\mathbf{Q}_2 = \begin{bmatrix} 1 & 0 & 0 \\ 0 & (i-1)/2 & \sqrt{2}/2 \\ 0 & \sqrt{2}/2 & (i+1)/2 \end{bmatrix},$$

$$\mathbf{U} = \mathbf{Q}_2^H (\mathbf{Q}_1^H \mathbf{A} \mathbf{Q}_1) \mathbf{Q}_2 = = \begin{bmatrix} -1 & 1 & (i+1)/\sqrt{2} \\ 0 & -i & -\sqrt{2}(i+1) \\ 0 & 0 & i \end{bmatrix}.$$

7. If the Schur decomposition is $\mathbf{U} = \mathbf{Q}^H \mathbf{A} \mathbf{Q}$, then $\mathrm{Trace}(\mathbf{U}) = \mathrm{Trace}(\mathbf{Q}^H \mathbf{A} \mathbf{Q}) = \mathrm{Trace}\left((\mathbf{Q}\mathbf{Q}^H)\mathbf{A}\right) = \mathrm{Trace}(\mathbf{I}\mathbf{A}) = \mathrm{Trace}(\mathbf{A})$ equals the sum of the eigenvalues of $\mathbf{A}$.

9. $\mathbf{A} = \begin{bmatrix} 1 & -1 \\ 1 & 1 \end{bmatrix}$ and $\mathbf{B} = \begin{bmatrix} 1 & 0 \\ 0 & -1 \end{bmatrix}$.

11.

$$\|\mathbf{A}\mathbf{v}\|^2 = \left(\overline{\mathbf{A}\mathbf{v}}\right)^T (\mathbf{A}\mathbf{v}) = \left(\bar{\mathbf{A}}\bar{\mathbf{v}}\right)^T (\mathbf{A}\mathbf{v}) = \bar{\mathbf{v}}^T \bar{\mathbf{A}}^T \mathbf{A}\mathbf{v} = \bar{\mathbf{v}}^T \mathbf{A}^H \mathbf{A}\mathbf{v}$$

$$= \bar{\mathbf{v}}^T \mathbf{A}\mathbf{A}^H \mathbf{v} = \left(\mathbf{A}^T \bar{\mathbf{v}}\right)^T (\mathbf{A}^H \mathbf{v}) = \left(\overline{\bar{\mathbf{A}}^T \mathbf{v}}\right)^T (\mathbf{A}^H \mathbf{v}) = \left(\overline{\mathbf{A}^H \mathbf{v}}\right)^T (\mathbf{A}^H \mathbf{v})$$

$$= \|\mathbf{A}^H \mathbf{v}\|^2$$

13. $\left\{ \left[\frac{1}{\sqrt{2}} \ \frac{1}{\sqrt{2}}\right]^T, \left[\frac{1}{\sqrt{2}} \ -\frac{1}{\sqrt{2}}\right]^T \right\}.$

15. $\left\{ \left[\frac{1}{3} \ \frac{2}{3} \ -\frac{2}{3}\right]^T, \left[\frac{2-6i}{3\sqrt{10}} \ \frac{4+3i}{3\sqrt{10}} \ 5\right]^T, \left[\frac{2+6i}{3\sqrt{10}} \ \frac{4-3i}{3\sqrt{10}} \ 5\right]^T \right\}$

17. Let

$$\mathbf{A} = [\mathbf{u}_1 \ \mathbf{u}_2 \ \dots \ \mathbf{u}_n] = \left[\mathbf{v}_1^T \ \mathbf{v}_2^T \ \dots \ \mathbf{v}_n^T\right]$$

be a normal $n \times n$ matrix, and let $\mathbf{e}_i$ denote the standard vector $[0 \dots 0 \ 1 \ 0 \dots 0]^T$, with the nonzero entry on the ith row. We have $\|\mathbf{A}\mathbf{e}_i\| = \|\mathbf{u}_i\|$, while $\|\mathbf{A}^H \mathbf{e}_i\| = \|\mathbf{v}_i\|$. Problem 11 showed that $\mathbf{A}$ normal implies $\|\mathbf{A}\mathbf{e}_i\| = \|\mathbf{A}^H \mathbf{e}_i\|$.

19. $\mathbf{Q} = \mathbf{Q}_1 \mathbf{Q}_2$, where

$$\mathbf{Q}_1 = \begin{bmatrix} 1/\sqrt{3} & 1/\sqrt{2} & -1/\sqrt{6} \\ 1/\sqrt{3} & -1/\sqrt{2} & -1/\sqrt{6} \\ 1/\sqrt{3} & 0 & 2/\sqrt{6} \end{bmatrix} \text{ and } \mathbf{Q}_2 = \begin{bmatrix} 1 & 0 & 0 \\ 0 & 1/\sqrt{2} & i/\sqrt{2} \\ 0 & i/\sqrt{2} & 1/\sqrt{2} \end{bmatrix}.$$

21.

$$\mathbf{A}^{-1} = \frac{1}{c_0} \left((-\mathbf{A})^{n-1} - c_{n-1}\mathbf{A}^{n-2} - \cdots - c_1\mathbf{I} \right).$$

Exercises 6.4

1. Norm = $\frac{1+\sqrt{5}}{2}$, cond = $\frac{3+\sqrt{5}}{2}$.
3. Norm = 16.848; cond = ∞.
5. Norm = 4; cond = 4.
7. $\mathbf{w}_1 = \mathbf{v}_1$; $\mathbf{w}_2 = \mathbf{u}_r$.
9. Minimize $(1 - 2t)^2 + (-t)^2 + (t)^2$.
11. $||\mathbf{AB}|| = \max_{||\mathbf{v}||=1}||(\mathbf{AB})\mathbf{v}|| \le ||\mathbf{A}||\,||\mathbf{Bv}|| \le ||\mathbf{A}||\,||\mathbf{B}||$ (1).
13. Equation (14) and Corollary 10.
15. The transpose of the SVD for $\mathbf{A}$ is an SVD for $\mathbf{A}^T$ and demonstrates that the nonzero singular values of $\mathbf{A}$ are the same as those of $\mathbf{A}^T$. Now apply Theorem 6 to $\mathbf{A}^T$.
17.

$$\begin{bmatrix} 0 & 1 \\ 0 & -1 \end{bmatrix}, \begin{bmatrix} 1 & 0 \\ 1 & 0 \end{bmatrix}, \text{ and } \begin{bmatrix} 1 & 1 \\ 0 & 0 \end{bmatrix}$$

are equally close rank 1 approximations.

19.

$$\begin{bmatrix} 1.0003 & 2.0003 & 3.0007 \\ 0.9997 & 1.9997 & 2.9993 \\ 1.0003 & 0.0003 & 1.0007 \\ 0.9997 & -0.0003 & 0.9993 \end{bmatrix}$$

is the closest rank 2 approximation. The Frobenius norms of the differences are 0.0012 and 0.0020.

21. (a) $r = 0.7395$, $s = 7.0263$, $t = 1.2105$.

   (b) $r = 3.1965$, $s = 8.0806$.

   (c) $r = 0$, $s = 6.7089$, $t = 1.5749$.

   (d) (a) 5.2632e-06 (b) 2.3226e-05 (c) 1.9846e-05. Part (b) constructed a rank 2 approximation by eliminating the third unknown (x_3); but part (c) utilized the *best* rank 2 approximation.

23. For square $\mathbf{A}$, all of the matrices have the same dimension; therefore, $\det\mathbf{A} = (\det\mathbf{U})(\det\mathbf{\Sigma})(\det\mathbf{U})$ and the unitary matrices have determinant ± 1. The simple 1-by-1 matrices [1] and [−1] demonstrate the two possibilities.

25. (a) $(\mathbf{A}\mathbf{v})^T(\mathbf{A}\mathbf{v}) = (\mathbf{v})^T(\mathbf{A}^T\mathbf{A}\mathbf{v})$.

(b) The directional derivative equals the dot product of the gradient with a unit vector in the direction in question. At a maximum on the sphere, all directional derivatives are zero in directions tangent to the sphere.

(c) $\nabla(\mathbf{x}^T\mathbf{A}^T\mathbf{A}\mathbf{x}) = 2\mathbf{A}^T\mathbf{A}\mathbf{x}$.

(d) $\mathbf{v}_2$ in tangent plane $\implies$ $\mathbf{v}_2$ orthogonal to gradient $\implies$ $\mathbf{A}\mathbf{v}_2$ orthogonal to $\mathbf{A}\mathbf{v}_1$.

27. (a) $\mathbf{W} = \mathbf{R}_{col}^n$.

(b) The subspace orthogonal to $\mathbf{u}_1$.

(c) The subspace orthogonal to $\mathbf{u}_1, \mathbf{u}_2, \ldots, \mathbf{u}_{k-1}$.

Exercises 6.5

1. Both iterations converge to the eigenvalue 4, eigenvector $[0\ 1\ 0]^T$.

3. The ratio of the inverse of the smallest-in-magnitude eigenvalue to that of the next smallest. Nearly-singular matrices have at least one very small eigenvalue (in magnitude).

5. Largest = 5; smallest = 1.

7. Largest = 2; smallest = -1.

9. Consider the differences of two consecutive iterates.

11. (a) 13 iterations: largest-in-magnitude eigenvalue ≈ 3.0068. (b) 10 iterations: smallest-in-magnitude eigenvalue ≈ 0.9987. (c) 4 iterations: eigenvalue ≈ 2.0000.

(d) The ratio of the largest-in-magnitude eigenvalue of $\mathbf{M}$ to its second largest is 1.5. The ratio of the largest-in-magnitude eigenvalue of $\mathbf{M}^{-1}$ to its second largest is 2. The ratio of the largest-in-magnitude eigenvalue of $(\mathbf{M} - 1.9\mathbf{I})^{-1}$ to its second largest is 9.

13. Two large eigenvalues of equal magnitude.

15. The normal equation is $\mathbf{v}^T\mathbf{v}r = \mathbf{v}^T\mathbf{A}\mathbf{v}$.

17. Apply $\mathbf{A}$ to the quadratic formula for the characteristic polynomial.

Exercises 7.1

3. (a)

$$\int \mathbf{A}(t)\, dt = \begin{bmatrix} t^2/2 + c_1 & e^t + c_2 \\ t + c_3 & e^t + c_4 \end{bmatrix}.$$

(b)

$$\int_0^1 \mathbf{B}(t)\, dt = \begin{bmatrix} \sin 1 & \cos 1 - 1 \\ 1 - \cos 1 & \sin 1 \end{bmatrix}.$$

(c)

$$\frac{d}{dt}[\mathbf{A}(t)\mathbf{B}(t)]$$
$$= \begin{bmatrix} (1 + e^t)\cos t + (e^t - t)\sin t & (e^t - t)\cos t - (e^t + 1)\sin t \\ e^t \cos t + (e^t - 1)\sin t & (e^t - 1)\cos t - e^t \sin t \end{bmatrix}.$$

13.

$$\begin{bmatrix} x \\ y \\ z \end{bmatrix}' = \begin{bmatrix} 1 & 1 & 1 \\ -1 & 0 & 2 \\ 0 & 4 & 0 \end{bmatrix} \begin{bmatrix} x \\ y \\ z \end{bmatrix}.$$

15.

$$\begin{bmatrix} x(t) \\ y(t) \end{bmatrix}' = \begin{bmatrix} 3 & -1 \\ -1 & 2 \end{bmatrix} \begin{bmatrix} x(t) \\ y(t) \end{bmatrix} + \begin{bmatrix} t^2 \\ e^t \end{bmatrix}.$$

21.

$$\begin{bmatrix} x_1'(t) \\ x_2'(t) \end{bmatrix} = \begin{bmatrix} 0 & 1 \\ 10 & 3 \end{bmatrix} \begin{bmatrix} x_1(t) \\ x_2(t) \end{bmatrix} + \begin{bmatrix} 0 \\ \sin t \end{bmatrix}.$$

23.

$$\begin{bmatrix} x_1'(t) \\ x_2'(t) \\ x_3'(t) \\ x_4'(t) \end{bmatrix} = \begin{bmatrix} 0 & 1 & 0 & 0 \\ 0 & 0 & 1 & 0 \\ 0 & 0 & 0 & 1 \\ -1 & 0 & 0 & 0 \end{bmatrix} \begin{bmatrix} x_1(t) \\ x_2(t) \\ x_3(t) \\ x_4(t) \end{bmatrix} + \begin{bmatrix} 0 \\ 0 \\ 0 \\ t^2 \end{bmatrix}.$$

33. $\begin{bmatrix} I_1 \\ I_3 \end{bmatrix}' = \begin{bmatrix} -45 & 45 \\ 90 & -110 \end{bmatrix} \begin{bmatrix} I_1 \\ I_3 \end{bmatrix} + \begin{bmatrix} 9/2 \\ 0 \end{bmatrix}.$

35. $\begin{bmatrix} I_2 \\ I_3 \end{bmatrix}' = \begin{bmatrix} -500 & -500 \\ -400 & -400 \end{bmatrix} \begin{bmatrix} I_2 \\ I_3 \end{bmatrix} + \begin{bmatrix} 500 \\ 400 \end{bmatrix}.$

37. $\begin{bmatrix} I_1 \\ I_2 \end{bmatrix}' = \begin{bmatrix} 0 & -8/5 \\ 10 & -10 \end{bmatrix} \begin{bmatrix} I_1 \\ I_2 \end{bmatrix} + \begin{bmatrix} 16/5 \\ 0 \end{bmatrix}.$

39. $\begin{bmatrix} I_1 \\ I_2 \end{bmatrix}' = \begin{bmatrix} 0 & -2 \\ 2 & -2 \end{bmatrix} \begin{bmatrix} I_1 \\ I_2 \end{bmatrix} + \begin{bmatrix} 0 \\ -3\sin 3t \end{bmatrix}.$

Exercises 7.2

1. $\begin{bmatrix} -3 & 1 \\ 1 & 2 \end{bmatrix} \begin{bmatrix} e^{10t} & 0 \\ 0 & e^{3t} \end{bmatrix} \begin{bmatrix} -3 & 1 \\ 1 & 2 \end{bmatrix}^{-1}$

3. $\begin{bmatrix} 1 & -1 \\ 1 & 1 \end{bmatrix} \begin{bmatrix} e^{-t} & 0 \\ 0 & e^{3t} \end{bmatrix} \begin{bmatrix} 1 & -1 \\ 1 & 1 \end{bmatrix}^{-1}$

5. $\begin{bmatrix} -1 & 1 \\ 1 & 1 \end{bmatrix} \begin{bmatrix} 1 & 0 \\ 0 & e^{2t} \end{bmatrix} \begin{bmatrix} -1 & 1 \\ 1 & 1 \end{bmatrix}^{-1}$

7. $\begin{bmatrix} 1 & 1 & 1 \\ 1 & 1 & 0 \\ 1 & 0 & 1 \end{bmatrix} \begin{bmatrix} e^{t} & 0 & 0 \\ 0 & e^{3t} & 0 \\ 0 & 0 & e^{2t} \end{bmatrix} \begin{bmatrix} 1 & 1 & 1 \\ 1 & 1 & 0 \\ 1 & 0 & 1 \end{bmatrix}^{-1}$

9. $\begin{bmatrix} i & -i \\ 1 & 1 \end{bmatrix} \begin{bmatrix} e^{it} & 0 \\ 0 & e^{-it} \end{bmatrix} \begin{bmatrix} i & -i \\ 1 & 1 \end{bmatrix}^{-1}$

11. $\begin{bmatrix} 2 & i \\ 1 & 1 \end{bmatrix} \begin{bmatrix} e^{(i+1)t} & 0 \\ 0 & e^{3t} \end{bmatrix} \begin{bmatrix} 2 & i \\ 1 & 1 \end{bmatrix}^{-1}$

13. $\begin{bmatrix} (1-t)e^{2t} & -te^{2t} \\ te^{2t} & (1+t)e^{2t} \end{bmatrix}$

15. $e^{-t} \begin{bmatrix} 1+3t-1.5t^2 & t & -t+0.5t^2 \\ -3t & 1 & t \\ 9t-4.5t^2 & 3t & 1-3t+1.5t^2 \end{bmatrix}$

17. $\begin{bmatrix} 1+t+0.5t^2 & t+t^2 & 0.5t^2 \\ -0.5t^2 & 1+t-t^2 & t-0.5t^2 \\ -t+0.5t^2 & -3t+t^2 & 1-2t+0.5t^2 \end{bmatrix} e^{-t}$

19. $\begin{bmatrix} t-2+2e^{-t}+4te^{-t} \\ t-1+4e^{-t} \\ 3t-6+6e^{-t}+12te^{-t} \end{bmatrix}$

21. $\begin{bmatrix} 2t+8-(8+7t+2t^2)e^{-t} \\ (t+1)(e^{-t}+2te^{-t}-1) \\ t-2+(2+t-2t^2)e^{-t} \end{bmatrix}$

23. $\begin{bmatrix} t^2-2t+2-2e^{-t} \\ -t^2+t/2-9/4+4e^{-t}/3+11e^{2t}/12 \end{bmatrix}$

25.

$$\begin{bmatrix} \sin(t) \\ \cos(t) - 1 \end{bmatrix}$$

27.

$$\begin{bmatrix} (4/13)e^{-3t} + 6te^{3t} + (2/13)(3\sin 2t - 2\cos 2t) \\ e^{3t} \\ (8/13)e^{-3t} + (3t - 1)e^{3t} + (4/13)(3\sin 2t - 2\cos 2t) \end{bmatrix}$$

29. (c) 56 terms (using MATLAB® R2012b)

(d) 202 terms (using MATLAB® R2012b)

33. $e^{\mathbf{A}t} \equiv \begin{bmatrix} e^t & 0 & 0 & 0 & 0 \\ 0 & e^{-t}(1+t) & te^{-t} & 0 & 0 \\ 0 & -te^{-t} & e^{-t}(1-t) & 0 & 0 \\ 0 & 0 & 0 & \cos(t) & \sin(t) \\ 0 & 0 & 0 & -\sin(t) & \cos(t) \end{bmatrix}$

35. $e^{\mathbf{A}t} \equiv \begin{bmatrix} e^{-t} & 0 & 0 & 0 & 0 \\ 0 & e^{-t}(1+t) & te^{-t} & 0 & 0 \\ 0 & -te^{-t} & e^{-t}(1-t) & 0 & 0 \\ 0 & 0 & 0 & e^{-2t}(1+2t) & te^{-2t} \\ 0 & 0 & 0 & -4te^{-2t} & e^{-2t}(1-2t) \end{bmatrix}$

Exercises 7.3

1. If the upper left corners of the Jordan blocks of the matrix occur in positions (i_p, i_p) with $1 \le p \le q$, then the column vectors with 1 in row i_p and zeros elsewhere are eigenvectors corresponding to eigenvalues r_p.

3. $\begin{bmatrix} r_1^{-1} & -r_1^{-2} & r_1^{-3} \\ 0 & r_1^{-1} & -r_1^{-2} \\ 0 & 0 & r_1^{-1} \end{bmatrix}$.

5.

$$\begin{bmatrix} r & 0 & 0 & 0 \\ 0 & r & 0 & 0 \\ 0 & 0 & r & 0 \\ 0 & 0 & 0 & r \end{bmatrix}, \begin{bmatrix} r & 0 & 0 & 0 \\ 0 & r & 0 & 0 \\ 0 & 0 & r & 1 \\ 0 & 0 & 0 & r \end{bmatrix}, \begin{bmatrix} r & 0 & 0 & 0 \\ 0 & r & 1 & 0 \\ 0 & 0 & r & 1 \\ 0 & 0 & 0 & r \end{bmatrix}, \begin{bmatrix} r & 1 & 0 & 0 \\ 0 & r & 0 & 0 \\ 0 & 0 & r & 1 \\ 0 & 0 & 0 & r \end{bmatrix},$$

$$\begin{bmatrix} r & 1 & 0 & 0 \\ 0 & r & 1 & 0 \\ 0 & 0 & r & 1 \\ 0 & 0 & 0 & r \end{bmatrix}.$$

7. $\begin{bmatrix} 4 & 1 \\ 0 & 4 \end{bmatrix}$ **9.** $\begin{bmatrix} 1 & 1 & 0 \\ 0 & 1 & 1 \\ 0 & 0 & 1 \end{bmatrix}$ **11.** $\begin{bmatrix} 3 & 0 & 0 \\ 0 & 5 & 1 \\ 0 & 0 & 5 \end{bmatrix}$ **13.** $\begin{bmatrix} 2 & 0 & 0 & 0 \\ 0 & 1 & 1 & 0 \\ 0 & 0 & 1 & 1 \\ 0 & 0 & 0 & 1 \end{bmatrix}$

15. $\begin{bmatrix} 0 & 1 & 0 & 0 \\ 0 & 0 & 0 & 0 \\ 0 & 0 & 0 & 1 \\ 0 & 0 & 0 & 0 \end{bmatrix}$ **17.** $\begin{bmatrix} 2 & 1 & 0 & 0 & 0 & 0 \\ 0 & 2 & 1 & 0 & 0 & 0 \\ 0 & 0 & 2 & 0 & 0 & 0 \\ 0 & 0 & 0 & 0 & 0 & 0 \\ 0 & 0 & 0 & 0 & 2 & 1 \\ 0 & 0 & 0 & 0 & 0 & 2 \end{bmatrix}$

19. $\begin{bmatrix} 3 & 1 \\ 0 & 3 \end{bmatrix}$; $e^{\mathbf{A}t} = \begin{bmatrix} -2 & 0 \\ 0 & 1 \end{bmatrix} \begin{bmatrix} e^{3t} & te^{3t} \\ 0 & e^{3t} \end{bmatrix} \begin{bmatrix} -0.5 & 0 \\ 0 & 1 \end{bmatrix}$.

21. $\begin{bmatrix} 3 & 0 & 0 \\ 0 & 1 & 1 \\ 0 & 0 & 1 \end{bmatrix}$; $e^{\mathbf{A}t} = \begin{bmatrix} 0 & 0 & 1 \\ 0.5 & 0 & -0.5 \\ 0.25 & -0.5 & -0.25 \end{bmatrix} \begin{bmatrix} e^{3t} & 0 & 0 \\ 0 & e^{t} & te^{t} \\ 0 & 0 & e^{t} \end{bmatrix} \begin{bmatrix} 1 & 2 & 0 \\ 0 & 1 & -2 \\ 1 & 0 & 0 \end{bmatrix}$.

23. $\begin{bmatrix} 2 & 0 & 0 & 0 \\ 0 & 1 & 1 & 0 \\ 0 & 0 & 1 & 1 \\ 0 & 0 & 0 & 1 \end{bmatrix}$; $e^{\mathbf{A}t} = \begin{bmatrix} 3 & 0 & -2 & -5 \\ 6 & -2 & -6 & -6 \\ 1 & 0 & 0 & 0 \\ 1 & 0 & 0 & -1 \end{bmatrix} \begin{bmatrix} e^{2t} & 0 & 0 & 0 \\ 0 & e^{t} & te^{t} & t^2 e^{t}/2 \\ 0 & 0 & e^{t} & te^{t} \\ 0 & 0 & 0 & e^{t} \end{bmatrix} \times$

$\begin{bmatrix} 0 & 0 & 1 & 0 \\ 1.5 & -0.5 & 3 & -4.5 \\ -0.5 & 0 & -1 & 2.5 \\ 0 & 0 & 1 & -1 \end{bmatrix}$

25.

$\begin{bmatrix} 2 & 0 & 0 & 0 & 0 \\ 0 & 3 & 1 & 0 & 0 \\ 0 & 0 & 3 & 1 & 0 \\ 0 & 0 & 0 & 3 & 1 \\ 0 & 0 & 0 & 0 & 3 \end{bmatrix}$; $e^{\mathbf{A}t} = \begin{bmatrix} 1 & 0 & 0 & 0 & 0 \\ -1 & 0 & 0 & 0 & 1 \\ 1 & 0 & 0 & 1 & -1 \\ -1 & 0 & 1 & -1 & 1 \\ 1 & 1 & -1 & 1 & -1 \end{bmatrix} \times$

$\begin{bmatrix} e^{2t} & 0 & 0 & 0 & 0 \\ 0 & e^{3t} & te^{3t} & t^2 e^{3t}/2 & t^3 e^{3t}/6 \\ 0 & 0 & e^{3t} & te^{3t} & t^3 e^{3t}/2 \\ 0 & 0 & 0 & e^{3t} & te^{3t} \\ 0 & 0 & 0 & 0 & e^{3t} \end{bmatrix} \begin{bmatrix} 1 & 0 & 0 & 0 & 0 \\ 0 & 0 & 0 & 1 & 1 \\ 0 & 0 & 1 & 1 & 0 \\ 0 & 1 & 1 & 0 & 0 \\ 1 & 1 & 0 & 0 & 0 \end{bmatrix}$.

31. The similarity transformation can best be described by example. If

$$\mathbf{A} = \begin{bmatrix} r & a & 0 & 0 & 0 & 0 & 0 \\ 0 & r & b & 0 & 0 & 0 & 0 \\ 0 & 0 & r & c & 0 & 0 & 0 \\ 0 & 0 & 0 & r & 0 & 0 & 0 \\ 0 & 0 & 0 & 0 & r & 0 & 0 \\ 0 & 0 & 0 & & 0 & r & e \\ 0 & 0 & 0 & 0 & 0 & 0 & r \end{bmatrix}, \text{ then take } \mathbf{V} = \begin{bmatrix} abc & 0 & 0 & 0 & 0 & 0 & 0 \\ 0 & bc & 0 & 0 & 0 & 0 & 0 \\ 0 & 0 & c & 0 & 0 & 0 & 0 \\ 0 & 0 & 0 & 1 & 0 & 0 & 0 \\ 0 & 0 & 0 & 0 & 1 & 0 & 0 \\ 0 & 0 & 0 & 0 & 0 & e & 0 \\ 0 & 0 & 0 & 0 & 0 & 0 & 1 \end{bmatrix}$$

and $\mathbf{V}^{-1}\mathbf{A}\mathbf{V}$ will have the proposed Jordan form.

33. $\begin{bmatrix} x_1 \\ x_2 \end{bmatrix} = \mathbf{A} \begin{bmatrix} x_1 \\ x_2 \end{bmatrix}$ where $\mathbf{A} = \begin{bmatrix} 0 & 1 \\ 0 & 0 \end{bmatrix}$, and $e^{\mathbf{A}t}\mathbf{C} = \begin{bmatrix} C_1 + C_2 t \\ C_2 \end{bmatrix}$

35. $e^{\mathbf{A}t} \begin{bmatrix} 1 \\ 0 \\ 1 \end{bmatrix} = \begin{bmatrix} 0 & 0 & 1 \\ 0.5 & 0 & -0.5 \\ 0.25 & -0.5 & -0.25 \end{bmatrix} \begin{bmatrix} e^{3t} & 0 & 0 \\ 0 & e^t & te^t \\ 0 & 0 & e^t \end{bmatrix} \begin{bmatrix} 1 & 2 & 0 \\ 0 & 1 & -2 \\ 1 & 0 & 0 \end{bmatrix} \begin{bmatrix} 1 \\ 0 \\ 1 \end{bmatrix}$.

Exercises 7.4

(See Exercises 7.3 for # 1–11 and 15–17.)

13. If r is an eigenvalue of $\mathbf{A}$, r^p is an eigenvalue of $\mathbf{A}^p$, but the latter are all zero. *Every* non zero vector is a generalized eigenvector of $\mathbf{A}$.

REVIEW PROBLEMS FOR PART III

1. Only $\mathbf{0}$ is similar to $\mathbf{0}$, and only $\mathbf{I}$ is similar to $\mathbf{I}$.

3. Eigenvectors for adding a nonzero multiple of the ith row to another row are all vectors with zero in the ith entry (eigenvalue is one); the matrix is defective.

   Eigenvectors for multiplication of the ith row by a nonzero scalar are all vectors with zero in the ith entry (eigenvalue equals one); and vectors whose only nonzero entry is its ith (eigenvalue equals the scalar).

   Eigenvectors for switching the ith and jth rows are all vectors with equal ith and jth entries (eigenvalue is one); and all vectors with opposite-signed ith and jth entries, all other entries zero (eigenvalue is -1).

5. $[2 \ -1]$ has eigenvalue -4; $[1 \ -1]$ has eigenvalue -5.

7. $[3 \ 1 \ -1]$ has eigenvalue 0; $[-1 \ 1 \ 1]$ has eigenvalue 2. (Defective matrix.)

9. $2 \pm i\sqrt{2}$

11. $[1 \ -1 \ 1 \ -1]$ has eigenvalue -2; $[1 \ 0 \ -1 \ 0]$ and $[0 \ 1 \ 0 \ -1]$ have eigenvalue 0; and $[1 \ 1 \ 1 \ 1]$ has eigenvalue 2.

13.

$$\mathbf{P} = \begin{bmatrix} 0.8944 & -0.7071 \\ -0.4472 & 0.7071 \end{bmatrix}.$$

15. Impossible

17.

$$\mathbf{P} = \begin{bmatrix} -0.2357 & 0.4364 & 0.4082 \\ 0.2357 & -0.2182 & -0.4082 \\ 0.9428 & -0.8729 & -0.8165 \end{bmatrix}.$$

19.

$$\mathbf{P} = \begin{bmatrix} i/\sqrt{2} & 0 & i/\sqrt{2} \\ -1/\sqrt{2} & 0 & 1/\sqrt{2} \\ 0 & 1 & i/\sqrt{2} \end{bmatrix}.$$

21.

$$\begin{bmatrix} x_1 \\ x_2 \end{bmatrix} = \frac{1}{\sqrt{2}} \begin{bmatrix} -1 & 1 \\ 1 & 1 \end{bmatrix} \begin{bmatrix} y_1 \\ y_2 \end{bmatrix}.$$

23.

$$\begin{bmatrix} x_1 \\ x_2 \\ x_3 \end{bmatrix} = \begin{bmatrix} 0.5774 & 0.2673 & 0.7715 \\ -0.5774 & 0.8018 & 0.15431 \\ -0.5774 & -0.5345 & 0.6172 \end{bmatrix} \begin{bmatrix} y_1 \\ y_2 \\ y_3 \end{bmatrix}.$$

25.

$$\begin{bmatrix} x_1 \\ x_2 \\ x_3 \end{bmatrix} = \begin{bmatrix} -0.7152 & 0.3938 & 0.5774 \\ 0.0166 & -0.8163 & 0.5774 \\ 0.6987 & 0.4225 & 0.5774 \end{bmatrix} \begin{bmatrix} y_1 \\ y_2 \\ y_3 \end{bmatrix}.$$

27. $\alpha = 1;$ $\begin{bmatrix} \dfrac{1}{6} - \dfrac{i}{2} \\ -\dfrac{1}{6} - \dfrac{i}{2} \\ \dfrac{2}{3} \end{bmatrix}$ $\begin{bmatrix} \dfrac{1}{6} + \dfrac{i}{2} \\ -\dfrac{1}{6} + \dfrac{i}{2} \\ \dfrac{2}{3} \end{bmatrix}$ $\begin{bmatrix} -\dfrac{2}{3} \\ \dfrac{2}{3} \\ \dfrac{1}{3} \end{bmatrix}.$

INDEX

adjoint, 103–4
affine transformations, 123–7
associative law, 61–3, 67
augmented coefficient matrix, 18

back substitution, 9
basis, 152, 155
best approximation, 182
block diagonal matrix, 317
block matrix, 98
boundary value problem, 2, 118–19

C(a,b), 135
canonical basis, 152
CAT (computerized axial tomography), 3
Cauchy–Schwarz inequality, 173
Cayley–Hamilton theorem, 262, 264
change of coordinate, 153–4
characteristic equation, polynomial, 217, 243
Cholesky factorization, 117, 343
circulant matrix, 215, 253, 345
coefficient matrix, 17
cofactor, 92–3
commutative law, 61
completing the square, 245
complex equations, 15

condition number, 54, 272
consistent systent, 29, 42
controllability, 199–200
Cramer's rule, 105
$\mathbf{C}_{\text{row}}^{m}$ ($\mathbf{C}_{\text{col}}^{n}$) ($\mathbf{C}_{m,n}$), 135

defective matrix, 224
deflated matrix, 216, 225
determinant, 90, 97, 101–2, 112, 148
diagonal matrix, 19
diagonalization of matrix, 238
diagonally dominant, 125
diet, antarctic, 1, 37
dimension, 156, 191, 202–4
discrete Fourier transform, 345
discrete time system, signal, 254
distributive law, 62–3
dot product, 59
Durer, A., 198

echelon form, 39
eigenvalue, eigenvector, 209, 223, 322
electrical circuit, 40, 304–6
elementary column operations, 25, 158
elementary row matrix operator, 63
elementary row operations, 20, 63, 158

Fundamentals of Matrix Analysis with Applications,
First Edition. Edward Barry Saff and Arthur David Snider.
© 2016 John Wiley & Sons, Inc. Published 2016 by John Wiley & Sons, Inc.

fast Fourier transform, 345
Fibonacci sequence, 163, 350
fixed point method, 123–7
Fredholm alternative, 189
free variable, 32, 39
Frobenius norm, 244
function space, 190
fundamental matrix, 312

Gauss elimination, 7–10, 27–34, 39
    back substitution, 9
    basic operations, 10
    operations count, 12–13, 16
Gauss, C. F., 7
Gauss–Jordan algorithm, 22–4
Gauss–Seidel, 126
generalized eigenvector, 224,
        322–3, 334
geometric series, 243, 255
global positioning systems, 122–3
golden ratio, 350
Goldstine, H., 47, 50
gradient, 246, 281
Gram matrix, 163, 196, 343
Gram–Schmidt algorithm, 169

Haar, 202–4
heat transfer, 302
Hermitian matrix, 228–9, 258
Hessenberg form, 344
Hilbert matrix, 54, 196
Hilbert space, 193
homogeneous system, 42
Hooke's law, 2, 45
Householder reflectors, 176,
        179, 201

identity matrix, 63
incidence matrix, 121
inconsistent system, 29, 45
inertia tensor, 255
inner product, 59
inner product, 193
integer equations, 16
integrating factor, 312
interconnected tanks, 296, 301
interpolation, 15
inverse, 76–80, 83, 85
invertible, 77

Jacobi, 125
Jordan block, 318

Jordan chain, 336
Jordan form, 318, 333
Jordan, W., 24

kernel, 136
Kirchhoff's laws, 44, 120–122

Lagrange interpolation problem, 15
least squares, 183, 185, 274, 348
    polynomial approximation, 26
least squares fit, straight line, 186
left inverse, 76–9
Legendre polynomial, 194
linear algebraic equations, 5
    graphs, 6, 7
linear combination, 137
    combination principle, 138
linear dependence (independence), 145, 147,
        191–3
linear systems, 5
lower trapezoidal matrix, 19
lower triangular matrix, 19
LU factorization, 116–18, 286

magic square, 198
mappings, 141
mass-spring system, 2, 45, 248–52, 297,
        302–3, 333
matrix
    addition, 62
    addresses, 25
    augmented coefficient, 18
    block, 98
    coefficient, 17
    column, 20
    diagonal, 19
    differential equation, 295, 307, 329, 336
    exponential, 241, 307, 308, 311, 314,
        323–5, 337
    incidence, 121
    inverse, 76
    multiplication, 58–61, 115–16
    nilpotent, 85
    representation, 17
    row, 20
    scalar multiplication, 62
    skew-symmetric, 110
    stochastic, 69
    symmetric, 226, 258
    trapezoidal, 19
    triangular, 19
matrix operator

elementary row, 63
   mirror reflection, 72–3, 76, 201
   orthogonal projection, 71–2, 75
   rotation, 70–71, 201
matrix square root, 240
minor, 92, 343
modal analysis, 252

nilpotent, 85
   matrix, 338
norm, 166, 174, 228, 244, 271
normal equations, 184
normal matrix, 260, 261, 264
normal modes, 251
null space, 136
nullity, 156, 158

operations count, 12–13, 16
orthogonal, 166, 228
   matrices, 174, 201
   subspace decomposition, 182

Padovan sequence, 351
parallelogram law, 173
parametrization of silutions, 32, 35–6
permutation, 97
pivot
   complete, 54
   element, 30, 39, 49, 52
   partial, 49, 52, 53
plastic constant, 351
P_n, 135
positive definite matrix, 342–3
power method, 283
primary decomposition theorem, 323
principal axes, 247
principal component analysis, 278
products of inertia, 255
pseudoinverse, 274

QR
   factorization, 201–2, 290
   method, 285
quadratic form, 244
quadratic polynomial, 245

range, 156, 158
rank, 40
   reduction, 278
rank one matrix, 162, 216
Rayleigh quotient, 253, 290
redundant system, 28

reflection, 201
right inverse, 76–9
rotation, 201, 221
roundoff, 46–50
row echelon form, 39–40
row space, 158
row-reduced echelon form, 39
$\mathbf{R}_{\text{row}}^m$ ($\mathbf{R}_{\text{col}}^m$) ($\mathbf{R}_{m,n}$), 135

scalar multiplication, 133
Schur decomposition, 258
self-adjoint, 228
shear matrix, 332
Sherman–Morrison–Woodbury, 85
shifted eigenvalue, 210
similarity transformation, 235
singular, 77
singular value decomposition, 266, 269, 279, 346
skew-symmetric, 110, 232
span, 138
spectral factorization, theorem, 239
speed, 16
stiffness matrix, 251, 255
stochastic matrix, 69, 290
subspace, 135
Sylvester, j. j., 17
Sylvester's criterion, 343
symmetric matrix, 226, 258
systems of linear equations, 5

time, computation, 16, 95
total least squares, 348
trace, 225
transpose, 69–70, 81, 83
trapezoidal matrix, 19
triangle inequality, 173
triangular matrix, 19, 213, 217

unitary matrix, 228, 258, 260
upper trapezoidal matrix, 19
upper triangular matrix, 19, 258

Vandermonde, 99, 195, 243
variation of parameters, 312
vector (row, column), 20
vector space, 133
von Neumann, 47, 50

wavelet, 202–4
weighted least squares, 188
Wilkinson, J., 47, 51